Practical Biostatistical Methods

Steve Selvin

University of California, Berkeley

An Alexander Kugushev Book

Duxbury Press
An Imprint of Wadsworth Publishing Company
I(T)P™   An International Thomson Publishing Company

Belmont • Albany • Bonn • Boston • Cincinnati • Detroit • London • Madrid • Melbourne
Mexico City • New York • Paris • San Francisco • Singapore • Tokyo • Toronto • Washington

Assistant Editor: Jennifer Burger
Editorial Assistant: Michelle O'Donnell
Production Services Coordinator: Gary Mcdonald
Production: Professional Book Center
Text and Cover Designer: Cloyce Wall

Print Buyer: Diana Spence
Copyediting: Professional Book Center
Illustrations: Professional Book Center
Electronic Composition: Professional Book Center
Printer: Arcata Graphics/Fairfield

*This book is printed on
acid-free recycled paper.*

For more information contact, contact Wadsworth Publishing Company:

Wadsworth Publishing Company
10 Davis Drive
Belmont, California 94002, USA

International Thomson Editores
Campos Eliseos 385, Piso 7
Col. Polanco
11560 México D. F. México

International Thomson Publishing Europe
Berkshire House 168-173
High Holborn
London, WC1V 7AA, England

International Thomson Publishing GmbH
Königswinterer Strasse 418
53227 Bonn, Germany

Thomas Nelson Australia
102 Dodds Street
South Melbourne 3205
Victoria, Australia

International Thomson Publishing Asia
221 Henderson Road
#05-10 Henderson Building
Singapore 0315

Nelson Canada
1120 Birchmount Road
Scarborough, Ontario
Canada M1K 5G4

International Thomson Publishing Japan
Hirakawacho Kyowa Building, 3F
2-2-1 Hirakawacho
Chiyoda-ku, Tokyo 102, Japan

Library of Congress Cataloging-in-Publication Data
Selvin, S.
        Practical biostatistical methods / Steve Selvin.
            p.   cm.
        "An Alexander Kugushev book."
        Includes bibliographical references and index.
        ISBN 0-534-23802-5 (alk. paper)
        1. Medicine—Research—Statistical methods.   2. Biometry.
    I. Title.
        [DNLM: 1. Biometry—methods.   WA 950 S469p 1995]
    R853.S7S45   1995
    610'.1'5195—dc20
    DNLM/DLC
    for Library of Congress   94-30582
                                                                CIP

Contents

3   Linear Regression with Two Independent Variables   67

4   Multivariable Regression   95

9   Log-Linear Models   293

10   Logistic Regression Analysis   365

11   Survival Data Analysis   **409**

12   Poisson Regression Analysis   **455**

Preface

Purpose

I developed this text in response to students who had taken a beginning one-year course in statistical methods and asked, "What do I take next?" Many users of modern statistical methods have neither the time nor the background to become involved in the mathematical development of intermediate or advanced statistical techniques. To fill the gap between a first year course and a rigorous mathematical presentation of multivariate statistics, *Practical Biostatistical Methods* presents material beyond the basic first-year topics. The material covered constitutes a second-year course in statistical methods with emphasis on approaches particularly useful for the analysis of biomedical data. Because modern computer software gives researchers easy access to almost all statistical methods regardless of their complexity, this text is a step toward dispelling a "black-box" approach to data analysis where answers are produced without knowledge of the statistical processes underlying the analytic procedures. With the goal of providing a basis for statistical thinking, nine analytic areas are explored:

Linear regression analysis

Analysis of covariance

Discriminant analysis

Principal component analysis

Contingency table analysis

Log-linear analysis

Logistic regression analysis

Survival analysis

Poissons regression analysis

Prerequisites

This text assumes knowledge of topics usually covered in a first-year introductory course, particularly a familiarity with the basic properties of the normal, *F*, chi-square, and binomial probability distributions. No knowledge of calculus (one small exception appears in Chapter 11) or matrix algebra is assumed or required. Although considerable statistical notation is introduced, no more than elementary algebra is used to show how and why an analytic approach works.

Approach

Practical Biostatistical Methods begins by reviewing a few selected univariate techniques such as the *t*-test, elementary chi-square analysis, and one-way analysis of variance. It then proceeds to describe a series of statistical methods at an intermediate level, which is the main topic of this book. These methods necessarily involve a number of multivariable techniques, which I discuss from an intuitive and practical viewpoint. I focus on the development and interpretation of the less complex two- and three-variable cases. The more mathematically complicated (higher dimensional) multivariable approaches are developed by analogy and further explored with "package" programs. I thus endeavor to bring some mathematical justification to the study of statistical methods, while keeping the mathematics at an elementary level. Each method is applied to a small numeric test case, as well as to a more complex and substantial data set. This process of exploring analytic techniques at a relatively simple mathematical level, followed by a test case, along with computer applications to more extensive and realistic data is intended to produce insights into a number of analytic approaches to data analysis. Every topic follows generally the same format: introduction, discussion and mathematical demonstrations, simple test case, research illustration, notation, and problem set.

Formal statistical tests are also treated in a uniform way. The principal analytic tool used throughout the text consists of comparing nested statistical models. Although the details differ, each test results from applying essentially the same statistical strategy.

Organization

The introduction to each chapter provides a general outline of the topic.

Mathematical Justification   In the remainder of the chapter the material within boxes adds mathematical justification to indicate the mechanism underlying a multivariable procedure. These mathematical demonstrations usually apply to the relatively simple two-dimensional case in keeping with the spirit of less emphasis on the higher order or general cases.

Test Case   Each chapter contains a simple test case in a separate section. The test case provides in concrete numeric terms an illustration of the analytic formulas and the statistical techniques discussed in the chapter. The test case also demonstrates the geometry and relationships within an analytic technique without the sometimes confusing statistical notation. In short, the test case provides a "numeric" review of the material in the chapter. This test case is also analyzed using a statistical computer "package" system, providing an introduction to the mechanics of implementation. It is critical to test computer analysis systems with a small and simple example to make sure the theory and methods described in statistical terms are coordinated with the computer input and output.

Realistic Example   Following the test case is a realistic example analyzed as part of each chapter to illustrate the issues of real-world interpretation. A single data set relating infant birth weight to parental characteristics is used in most chapters.

Notation   At the end of each chapter is a review of the notation introduced in that chapter.

Problems   Each chapter concludes with problems that consist of a few small data sets to be analyzed and some manipulations of the statistical expressions described in the chapter. These problems are, by and large, designed to be easily solved with elementary techniques or with a statistical software package.

References   The text throughout usually refers to readily available statistical textbooks rather than to the original articles. These secondary sources serve as a starting point for further pursuit of topics of interest, particularly for readers primarily interested in applied rather than theoretical issues.

The Computer Component   *Practical Biostatistical Methods* is compatible with a statistical analysis computer program called STATA (a statistics/graphics/data management system), because this system is easily used by a novice in either a PC or a workstation environment (UNIX or Macintosh). STATA is chosen as the system to implement the discussed statistical methods because it is a widely available, compact, fast-executing, and well-documented system that is simple to use. Additionally, STATA contains almost all statistical procedures found in the large and much more extensive software systems such as SAS or SPSS^X. Also an excellent supplementary text describing STATA for statistical applications exists [1]. To a lesser extent, the same approach and notation are also found in the SPSS^X system. Clearly, the principles and calculations presented do not depend on the computer software package chosen to analyze the data, and other systems are equally useful (such as SAS, BMDP, SYSTAT, or SPSS-PC).

Data   Included with this text is a computer disk containing all the data used in the illustrative analyses. The purpose of making these data available is to allow readers to duplicate easily the analyses for themselves as a way of further understanding the described statistical techniques. Perhaps more important the availability of the data encourages alternative approaches that can be contrasted to the ones proposed in the text.

Coverage

Chapter 1 briefly reviews a few fundamental concepts used repeatedly in the following chapters. These topics are the core of a first-year course and no attempt is made to develop fundamental material found in a large number of introductory texts (e.g., [2]–[12]). The general analytic strategy introduced in this first chapter is the basis of most statistical methods described in subsequent chapters.

Linear regression topics are addressed in the next four chapters. Chapter 2 describes simple linear regression analysis. Chapter 3 (bivariate regression) and Chapter 4 (multivariable linear regression) present more general linear regression models. Chapter 5 is a special application of regression techniques called analysis of covariance.

A presentation of the linear discriminant function constitutes Chapter 6, and a discussion of principal component analysis makes up Chapter 7. The next two chapters deal with the analysis of categorical data. Chapter 8 reviews some basic aspects of multivariate contingency table analysis, and Chapter 9 discusses the application of log-linear models to the analysis of contingency table data. The discussion of the log-linear approach consists of a detailed description of the analysis of the two- and three-dimensional tables, again in the spirit of concentrating on the less complicated lower dimensional cases, while simply indicating the generalization to the more complex higher dimensional tables.

Chapter 10 explores the logistic regression model and its applications, starting with the simplest case (a 2-by-2 table) and continuing to the general k-variate additive model. Chapter 11 covers a few topics from survival analysis with particular emphasis on estimating survival curves and mean survival times. A brief introduction to the proportional hazards model is also included. Finally, in Chapter 12 the topic of Poisson regression is discussed with particular emphasis on application to disease and mortality data.

References

1 Hamilton, L. C. *Statistics with STATA 3*. Belmont, Calif.: Duxbury Press, 1993.

2 Snedecor, G. W., and Cochran, W. G. *Statistical Methods*, 8th ed. Ames: The Iowa State University Press, 1989.

3 Sokal, R. R., and Rohlf F. J. *Biometry*. San Francisco, Calif.: W. H. Freeman, 1969.

4 Winer, B. J., Brown, D. R., and Michels, K. M. *Statistical Principles in Experimental Design*, 3rd ed. New York: McGraw-Hill, 1991.

5 Dixon, W. J., and Massey, F. J. *Introduction to Statistical Analysis*. New York: McGraw-Hill, 1960.

6 Freund, J. E. *Modern Elementary Statistics*, 7th ed. Englewood Cliffs, N. J.: Prentice Hall, 1988.

7 Woolson, R. F. *Statistical Methods for the Analysis of Biomedical Data.* New York: John Wiley, 1987.

8 Daniel, W. W. *Biostatistic: A Foundation for Analysis in the Health Sciences,* 5th ed. New York: John Wiley, 1991.

9 Anderson, T. W., and Sclove, S. L. *The Statistical Analysis of Data.* Palo Alto, Calif.: Scientific Press, 1986.

10 Brown, W. W., and Hollander, M. *A Statistics: A Biomedical Introduction.* New York: John Wiley, 1977.

11 Pagano, M., and Gauvreau, K. *Principles of Biostatistics.* Belmont, Calif.: Duxbury Press, 1993.

12 Rosner, B., *Fundamentals of Biostatistics,* 4th ed. Belmont, Calif.: Duxbury Press, 1995.

1

General Concepts

Background

At the onset, it is useful to review, several fundamental statistical concepts. This chapter briefly describes a few selected topics integral to many of the procedures discussed in subsequent chapters. This review is by no means complete. It is meant to supplement the knowledge of a "first-year" statistics course and is certainly not a replacement for the many beginning texts that cover the same topics (e.g., [1]–[5]).

The most basic element of a statistical analysis is a variable. A variable is simply the quantity counted or measured as part of the process of understanding the phenomenon under investigation. It can range from the number of genetic defects found in a series of cells to the distance between stars. A random variable is a variable that additionally has an associated probability. Random variables can be discrete (result from counting) or continuous (result from measuring). The simplest discrete random variable takes on the value 0 with probability p and the value 1 with probability $1 - p$, such as flipping a fair coin: heads = 0 with probability $p = 0.5$ and tails = 1 with probability $1 - p = 0.5$. Rolling a pair of dice provides another simple example. The discrete random variable is the sum of the faces that show on a pair of dice, and the probabilities associated with each possible value of the random variable are

s	2	3	4	5	6	7	8	9	10	11	12
p_s	1/36	2/36	3/36	4/36	5/36	6/36	5/36	4/36	3/36	2/36	1/36

where s is the sum of numbers on the faces of two dice and the probability of a specific sum S is $p_s = P(\text{sum} = s)$. The variable S is a random variable and the probabilities p_s constitute a discrete probability distribution.

An especially important continuous random variable comes from the normal probability distribution. Because a normally distributed random variable is continuous (takes on any value among a continuum of values), single observed values are not directly associated with a probability. Nevertheless, the probabilities associated with a continuous random variable, defined in terms of a cumulative probability, are found in tables or with computer programs. The probability distribution associated with a discrete or a continuous random variable is of primary interest in a statistical study. Usually the exact structure of the distribution is not known and it is necessary to estimate its parameters and infer its properties from a sample of data. In general terms, statistical analysis can be viewed as the study of these probability distributions using samples of data.

The discipline of statistics contains tools for analyzing a sample of data in terms of inferences about the probability distribution that produced the data. The objectives of statistical analysis can be classified into at least six categories:

1 **Data description:** condensing a sometimes large number of observed values into a few summary values to indicate important properties of the sampled population.

2 **Exploring relationships:** defining and describing the relationships among the sampled variables.

3 **Prediction:** using the collected data to predict values arising from sampled population under specific conditions.

4 **Classification:** the description and analysis of groups in terms of the common characteristics based on the sampled variables.

5 **Testing hypotheses:** assessing the likelihood that specific relationships exist among the variables (inferences about the sampled population).

6 **Creating hypotheses:** using steps 1–5 to understand better the variables being studied, potentially producing new ways of thinking about the population sampled.

These objectives are different ways of dealing with the same basic issue—variability. Statistical analyses involve observations that vary from sample to sample, bringing a degree of uncertainty to any specific analysis. A primary goal of almost all statistical methods is to identify and to understand systematic effects while accounting for the influence of this variability. To some degree each of these objectives is discussed in the following chapters; however, the greatest emphasis is on exploring relationships using statistical hypothesis testing tools.

Descriptive Statistics

A primary focus of a statistical investigation based on collected data is the description of specific properties of the population from which the sample was selected. The results from organizing and summarizing a sample of data are called descriptive statistics. The most fundamental descriptive statistic is the sample mean, defined as

$$\text{sample mean} = \bar{y} = \frac{y_1 + y_2 + y_3 + \cdots + y_n}{n} = \frac{\sum_{i=1}^{n} y_i}{n} \tag{1.1}$$

where n is the number of observations y_i in the sample. The estimate $\bar{y}$ is a description of the mean of the sampled population—an estimate of its "location."

Once a sample mean is calculated, a next question concerns its precision. Sample means vary from sample to sample because it is unlikely that any two samples, even from the same population, will be identical. The presence of sampling variation also implies that $\bar{y}$ will undoubtedly vary from the mean of the population. The only time the sample mean will always exactly equal the population mean is when the entire population is sampled. If the sampled values produce the same value for every sample, the sampling variability is zero; otherwise the magnitude of the sampling variation is a basic issue.

A measure of the population variability is also an important descriptive statistic and is typically estimated by the sample variance defined as

$$\text{sample variance} = S_Y^2 = \frac{\sum\limits_{i=1}^{n} (y_i - \bar{y})^2}{n - 1} . \tag{1.2}$$

The sample variance is an "average" of the squared deviations from the sample mean. There are theoretical reasons to divide by $n - 1$ rather than n, but they are not discussed at this point. The value S_Y^2 is an estimate of the variance of the sampled population (population variance). Another often used measure of sample variability is the standard deviation, which is simply the square root of the sample variance (i.e., $S_Y = \sqrt{S_Y^2}$).

The precision of $\bar{y}$ relates to both sample size and population variability. A sample mean constructed from a large number of observations more accurately reflects the location of a population than a mean calculated from few sample values. In addition, a sample mean calculated from a population with little variability more precisely estimates the population mean than a sample mean calculated from a highly variable population. The variance of a sample mean combines the influences of sample size and population variance σ_Y^2 to express the variance associated with $\bar{y}$ as

$$\textit{variance of a sample mean} = \frac{\sigma_Y^2}{n} \quad \text{and is estimated by} \quad S_{\bar{Y}}^2 = \frac{S_Y^2}{n}, \tag{1.3}$$

when the sampled values are independently selected (details in [1]–[5]). The population variance is commonly denoted as σ_Y^2. The meaning of statistical independence is discussed in a following section.

When a mean and variance are estimated with observations independently sampled from a normal population, two approaches are commonly used to account for the presence of sample variability—a statistical test or a confidence interval.

One version of Student's t-test (named after an early statistician W. S. Gosset [1878–1937], who used the pseudonym Student) is the statistical test given by

$$T = \frac{\bar{y} - \mu_0}{S_{\bar{y}}} = \frac{\bar{y} - \mu_0}{\sqrt{S_Y^2/n}}, \tag{1.4}$$

which allows an evaluation of the "distance" between the observed sample mean $\bar{y}$ and a postulated mean value μ_0. The test statistic T is either likely or unlikely under the conditions used to define μ_0 producing a valuable tool for considering $\bar{y}$ in light of sampling variation. The likelihood associated with T is judged using a t-distribution, which is derived from theoretical considerations (again see, [1]–[5]).

An α-level confidence interval for the population mean based on $\bar{y}$ is

$$\text{lower bound} = \bar{y} - t_{1-\alpha/2}S_{\bar{y}} \quad \text{and} \quad \text{upper bound} = \bar{y} + t_{1-\alpha/2}S_{\bar{y}} \tag{1.5}$$

where $t_{1-\alpha/2}$ is $(1 - \alpha/2)$-percentile of a t-distribution. The probability that the population mean is between these two bounds is $1 - \alpha$, producing a summary indicating the location of the population mean and at the same time reflecting the impact of sampling variation (i.e., the position and width of the interval).

Bivariate Sample

Often a sample of data consists of more than one measurement on each observation sampled. The next simplest case (after univariate sampling of single variable) is a sample in which two variables are collected jointly—measures x_i and y_i are part of the same observation. For example, a bivariate observation could consist of measuring the height and weight of pregnant women or measuring a worker's age and cumulative dose from exposure to radiation. Covariance is a statistical measure of the degree of linear relationship between two such variables. The estimated covariance between two numeric quantities represented by x and y is a number defined as

$$S_{XY} = \frac{\sum_{i=1}^{n} (x_i - \bar{x})(y_i - \bar{y})}{n - 1} \tag{1.6}$$

where n represents the number of pairs of observations (x_i, y_i) sampled and $\bar{x}$ and $\bar{y}$ represent the respective sample mean values. Technically, covariance is the first product-moment of two variables about their means. The estimated sample variance (equation 1.2) is a special case of covariance. When $X = Y$, then $S_{XY} = S_Y^2$.

The reason covariance quantifies the degree of linear association between two variables is explained by the following argument:

■ When $(x_i - \bar{x})$ is positive and $(y_i - \bar{y})$ is positive, then $(x_i - \bar{x})(y_i - \bar{y})$ is positive. Similarly, when $(x_i - \bar{x})$ is negative and $(y_i - \bar{y})$ is negative, then $(x_i - \bar{x})(y_i - \bar{y})$ is also positive. Both products add positive terms to S_{XY}. If these positive terms dominate the sum, S_{XY} will be a positive number, indicating that the pairs of values (x_i, y_i) tend to be simultaneously larger or simultaneously smaller than their respective mean values. This phenomenon is referred to as a positive association.

■ When $(x_i - \bar{x})$ is positive and $(y_i - \bar{y})$ is negative, then the product $(x_i - \bar{x})(y_i - \bar{y})$ adds a negative contribution to S_{XY}. Also, when $(x_i - \bar{x})$ is negative and $(y_i - \bar{y})$ is positive, the cross-product $(x_i - \bar{x})(y_i - \bar{y})$ adds a negative term to S_{XY}. Dominance of these negative terms yields a negative value of the covariance, indicating that pairs of large and small values measured relative to their respective means tend to occur frequently together. This phenomenon is called a negative association.

■ When the sum of the positive contributions to S_{XY} is about equal to the sum of the negative contributions, the covariance is close to zero. A covariance in the neighborhood of zero indicates that the variable X is not linearly related to the variable Y and vice versa.

The Pearson product-moment correlation coefficient (denoted r_{XY}) is a function of the expression for covariance (first described by Karl Pearson [1857–1936] who developed much of the foundations of modern statistics around the turn of the twentieth century). If S_{XY} is divided by the estimated standard deviation of X (S_X) and the estimated standard deviation of Y (S_Y), then

$$\text{product-moment correlation} = r_{XY} = \frac{S_{XY}}{S_X S_Y} \tag{1.7}$$

and

$$-1 \le \frac{S_{XY}}{S_X S_Y} \le 1 \quad \text{or} \quad -1 \le r_{XY} \le 1 .$$

Dividing by $S_X S_Y$ bounds the covariance between -1 and 1 and produces a unitless summary statistic. Values near -1 indicate the relationship between X and Y is accurately described by a straight line with a negative slope, values near 1 indicate a relationship accurately described by a straight line but with a positive slope, and values in the neighborhood of zero indicate that a straight line is of little use for describing the relationship between X and Y. Therefore, a correlation coefficient reflects the direction and degree of linear relationship between two variables that does not depend on the original measurement units. Both the covariance S_{XY} and the related quantity correlation r_{XY} play a central role in the description and analysis of multivariate data.

Expected Value (Population Mean)

The mean of the sampled population is so important that it gets a specific name: the expectation. Expectation is symbolized by $E(Y)$ and represents the mean of the distribution that generated the values symbolized by Y. It is the value estimated by the sample mean $\bar{y}$. Consider a single die where each face ($Y = 1, 2, 3, 4, 5,$ or 6) has the same probability of occurrence. Therefore, the probability of occurrence $= p_y = 1/6$, and

$$\text{expectation} = E(Y) = 1(1/6) + 2(1/6) + 3(1/6) + 4(1/6) + 5(1/6) + 6(1/6) = 3.5 .$$

That is, the mean of the population that generates the values found on the face of a die (values symbolized by Y) is 3.5 and $\bar{y}$, based on a sample of n tosses, is an estimate of this expected value.

A general definition of expectation for a discrete random variable is $E(Y) = \Sigma y p_{y_i}$ where the sum is over all possible values of the random variable Y and $p_y = P(Y = y)$ is the probability associated with each value of Y. A similar definition exists for a continuous random variable in terms of an integral. The concept and properties of expectation are completely developed in a variety of more theoretical texts (e.g., [6], [7]).

A basic property of the expectation, used throughout this text, is that the expectation of a sum of a series of random variables is the same as the sum of their expectations. In symbols,

$$E(Y_1 + Y_2 + \cdots + Y_k) = E(Y_1) + E(Y_2) + \cdots + E(Y_k) . \tag{1.8}$$

This addition property of population means shows that a sum of a series of observations sampled from different populations has a population mean equal to the sum of the mean values of each of the populations sampled. A concrete example comes again from rolling dice. The expectation of the population of possible outcomes associated with one toss of a die is $E(Y) = 3.5$ and the expectation associated with the sum of the faces on a toss of a pair of dice is $E(Y_1 + Y_2) = E(Y_1) + E(Y_2) = 3.5 + 3.5 = 7.0$. Therefore, a population from which a single roll of a pair of dice is a sample has a mean value of 7.0. Of course, the sum of values observed on a pair of dice is not usually 7.0, but in the long run the sample mean value of a large number of samples (rolls of the dice) will be close to 7.0. That is, the mean $\bar{y}$ becomes indistinguishable from $E(Y)$ when the sample size is large enough (i.e., $S_{\bar{Y}}^2 \approx 0$).

Multiplying each observation by a constant value or adding a constant value to each observation has the same affect on the expectation. In symbols,

$$E(aY + b) = aE(Y) + b , \tag{1.9}$$

where a and b are constants. In general, for the constants represented by $c_1, c_2, \cdots, c_k$, then

$$E(c_1 Y_1 + c_2 Y_2 + \cdots + c_k Y_k) = c_1 E(Y_1) + c_2 E(Y_2) + \cdots + c_k E(Y_k) . \tag{1.10}$$

The expectation of a random variable is an important parameter that is part of many statistical calculations, because it is often the population value that the sample is collected to estimate.

Multivariate Distributions

For the two-variable case, a bivariate observation (X, Y) has a distribution that depends on the properties of X, the properties of Y, and the joint properties of X and Y. The sample observations consist of pairs of values considered simultaneously. In most situations, knowledge about X and knowledge about Y are not sufficient to describe a bivariate distribution of X and Y. A simple example illustrates. Two simple univariate distributions with population means $E(X) = 1.5$ and $E(Y) = 0.5$ are

Distribution of X:

Distribution of Y:

$P(X = 0) = 0.25$ $P(Y = 0) = 0.5$

$P(X = 1) = 0.25$ $P(Y = 1) = 0.5$

$P(X = 2) = 0.25$

$P(X = 3) = 0.25$

but the bivariate distribution of X and Y cannot be determined without further information. The joint probabilities $P(X = i$ and $Y = j)$ remain free to vary between 0 and 1. A bivariate distribution, for example, might be

$P(X = 0$ and $Y = 0) = 0.05$ $P(X = 0$ and $Y = 1) = 0.20$

$P(X = 1$ and $Y = 0) = 0.10$ $P(X = 1$ and $Y = 1) = 0.15$

$P(X = 2$ and $Y = 0) = 0.15$ $P(X = 2$ and $Y = 1) = 0.10$

$P(X = 3$ and $Y = 0) = 0.20$ $P(X = 3$ and $Y = 1) = 0.05$.

This joint distribution of X and Y cannot be deduced from the two univariate distributions.

Note that this bivariate distribution has a negative association between X and Y because the pairs where both values exceed or both values do not exceed their expected values are less frequent (e.g., (3, 1) and (0, 0); frequency = 0.05) than the pairs where only one member of the pair exceeds its expected value (e.g., (3, 0) and (0, 1); frequency = 0.20). No hint of this negative association is given by the two univariate distributions.

A fundamental distribution occurs when the pair (X, Y) has a bivariate normal distribution. In this situation X has a normal distribution and Y has a normal distribution but a complete description of the bivariate distribution depends on the relationship between X and Y measured by the covariance or correlation. When the expectations (population means) and variances of the normal distributions associated with X and Y are known as well as the covariance, the bivariate normal distribution can be defined and, like all probability distributions, probabilities associated with specific pairs of X and Y, can be computed.

Figures 1.1 through 1.4 show two bivariate normal distributions from different perspectives. In both cases, the distribution of the x-values has expectation $E(X) = 2$ with variance of $X = 1$, and the distribution of y-values has expectation $E(Y) = 6$ with variance of $Y = 9$. However, the covariance is 0 in Figures 1.1 and 1.3, creating one bivariate normal distribution, and the covariance is 1.5 (correlation = 0.5) in Figures 1.2 and 1.4, creating another. The ellipses (Figures 1.1 and 1.2) represent the contours of equal heights of the bivariate distribution (similar to a topological map), which is one of many ways to display a two-dimensional bivariate distribution. Figures 1.3 and 1.4 show another view of the same joint distributions. As the correlation between X and Y increases, the probability contours become narrower. As the correlation between X and Y approaches 1 or −1, the ellipses approach a straight line.

FIGURE **1.1**

Bivariate normal distri-
bution: Covariance = 0.0
or correlation 0.0

Distribution of *X:*
$E(X) = 2$,
variance of $X = 1$
Distribution of *Y:*
$E(Y) = 6$,
variance of $Y = 9$

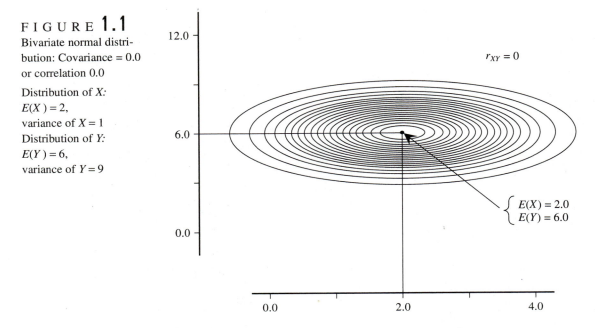

FIGURE **1.2**

Bivariate normal distri-
bution: Covariance = 1.5
or correlation 0.5

Distribution of *X:*
$E(X) = 2$,
variance of $X = 1$
Distribution of *Y:*
$E(Y) = 6$,
variance of $Y = 9$

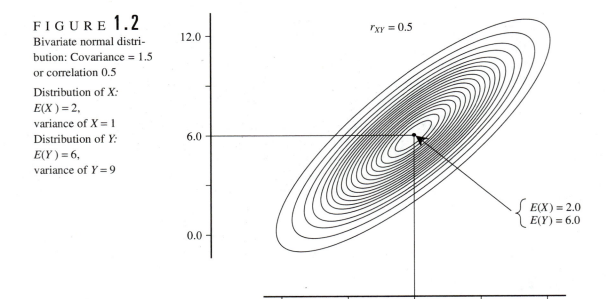

F I G U R E **1.3**

F I G U R E **1.3**
Bivariate normal distri-
bution: Covariance = 0.0
or correlation 0.0

Distribution of X:
$E(X) = 2$,
variance of $X = 1$
Distribution of Y:
$E(Y) = 6$,
variance of $Y = 9$

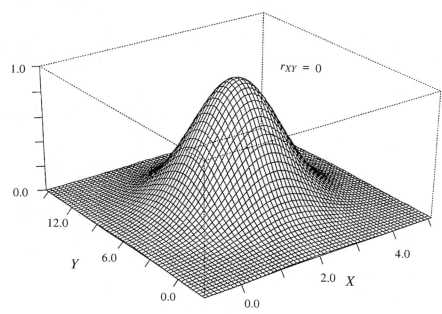

F I G U R E **1.4**
Bivariate normal distri-
bution: Covariance = 1.5
or correlation 0.5

Distribution of X:
$E(X) = 2$,
variance of $X = 1$
Distribution of Y:
$E(Y) = 6$,
variance of $Y = 9$

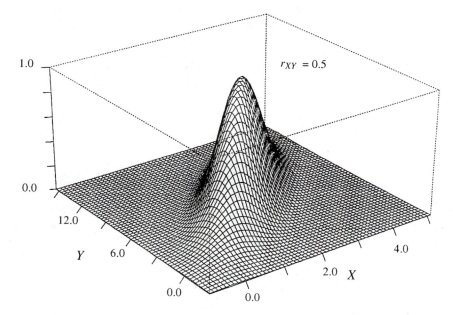

Another special case occurs when the variances of X and Y are equal and the covariance is 0; then the ellipses become circles describing a circular bivariate normal distribution centered at the point $(E(X), E(Y))$. Geometrically, the expectations $E(X)$ and $E(Y)$ determine the location of the bivariate distribution and the variances and covariance determine its shape.

When a number of measurements are made on a single observation, the resulting set of variables is called multivariate. A multivariate observation is a sample from a multivariate distribution. When the number of measured variables per observation is greater than two, simple geometric illustrations such as Figures 1.1 through 1.4 are not possible. Nevertheless, the issues and principles that apply to a bivariate observation generalize to a k-variate situation. For example, a series of observations made up of k variables from a multivariate normal distribution can be used to estimate population parameters, test hypotheses, and produce inferences about the sampled population. Like the bivariate case, the relationships among the k variables play an important role in characterizing the k-variate distribution—k means, k variances, and $k(k-1)/2$ covariances are necessary to define a multivariate normal distribution.

Why a Multivariable Approach?

The term *multivariable* refers to an approach based on multiple measurements on a single observational unit in which the analysis is directed at understanding the relationship of these usually correlated variables to a specific outcome variable. For example, measurements consisting of age, height, weight, and amount smoked might be taken on a pregnant woman to study the influence of these variables on a newborn child's birth weight (outcome).

The following 11 chapters are the long answer to the question, "Why a multivariable approach?" A short answer is a first step. Univariate comparisons of a series of variables among a number of groups could take one of two forms. At one extreme, the comparisons among groups could be based on the single most important variable (chosen somehow). At the other extreme, a series of separate comparisons could be conducted using each measured variable. Neither approach is satisfactory. Using a single variable makes no use of the other observed variables. Separate pairwise comparisons of a series of variables involve setting test-wise error rates and an overall error rate but, more importantly, do not lead to an understanding of how interrelated measurements jointly affect observed differences among the compared groups. An alternative is a multivariable approach.

Figure 1.5 illustrates why an approach that accounts for joint influences has distinct advantages when dealing with a series of correlated variables. The observations consist of two correlated variables, denoted X and Y, distributed into two groups. The separation of these two groups is clearly seen in the plot of the bivariate distribution (top left). The distribution of only X fails to indicate a presence of two groups (top right) and, similarly, the distribution of only Y (bottom left) also gives no indication of the nature of the bivariate structure. A multivariable approach might use the line labeled A to study the influences of both X and Y. The distribution of the variables X and Y projected on the line A is shown in the lower right of Figure 1.5. Here the "axis" A is used to describe the data rather than the variable X alone (the horizontal

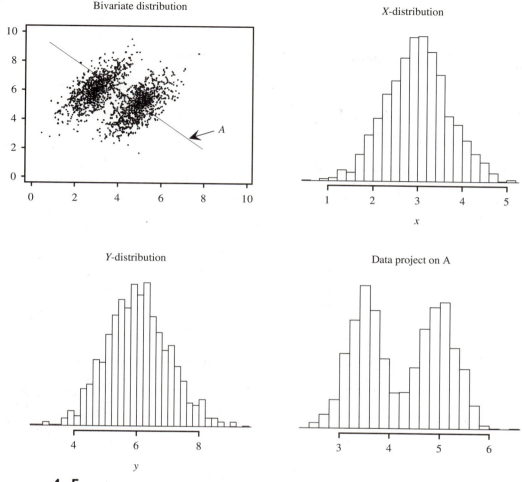

Bivariate distribution

X-distribution

Y-distribution

Data project on A

FIGURE **1.5**

Comparison of two groups viewed from three prospectives—bivariate, univariate, and multivariate

axis) or the variable *Y* alone (the vertical axis). Using the line *A*, the projected data clearly reflect two groups of observations (bottom right) using a single variable. This example is simple, perhaps simplistic, but starts to show the essence of a multivariable approach. Specifically, two variables are combined to form a single variable that communicates essential characteristics of a bivariate observation. When only two variables are involved, the need for multivariable techniques is minimal, but similar approaches and issues apply to any number of variables measured per observation compared among any number of groups. Employing multivariable approaches capitalizes on the joint information contained in correlated variables and at the same time avoids problems that arise when a series of related variables measured on a single observation are analyzed one at a time with univariate techniques.

Independence

Statistical independence is relevant to almost every statistical analysis. In general terms, independence between observations means the selection of any specific observation does not influence the selection of any other observation. For example, when individuals are chosen independently from a population, the height of any one person selected has no influence on the selection of any other individual. On the other hand, the height of an individual indicates, at least roughly, the weight of that person. Height and weight of humans are not independent. It is important to distinguish between independence of observations and independence of variables. The independence of individual observations is primarily determined by the way the data are collected. The independence of variables is a property of the sampled data and must be accounted for in the analysis.

Covariance, estimated by S_{XY} (expression 1.6), is related to the concept of independence. Two independent variables have a covariance of zero. The converse is not always true. Examples exist in which two variables are definitely related (not independent) but their covariance is zero. Covariance measures linear association, and nonlinear relationships are easily constructed to have a covariance of zero (see Box 1.1). A special situation arises when sampled data come from a multivariate normal distribution. In this case covariance is related to the degree of statistical independence between two variables. For example, two variables with zero covariance are independent when the pairs (x, y) are sampled from a bivariate normal distribution. Therefore, for methods that depend on the assumption or knowledge that the data come from a population with a multivariate normal or at least an approximate multivariate normal distribution, the covariance or correlation between variables reflects statistical independence.

B O X **1.1**   EXAMPLE: COVARIANCE = 0

Consider $y = x^2 - 5x + 7$:
then, $n = 6$ "observations"

$$x_1 = 0 \quad \text{gives} \quad y_1 = 7,$$
$$x_2 = 1 \quad \text{gives} \quad y_2 = 3,$$
$$x_3 = 2 \quad \text{gives} \quad y_3 = 1,$$
$$x_4 = 3 \quad \text{gives} \quad y_4 = 1,$$
$$x_5 = 4 \quad \text{gives} \quad y_5 = 3,$$
$$x_6 = 5 \quad \text{gives} \quad y_6 = 7,$$

where $\bar{x} = 2.50$, $\bar{y} = 3.67$, and $S_{XY} = 0.0$. Clearly, x and y are related.

To describe statistical independence in geometric terms, consider again the two variables X and Y with the bivariate normal distributions shown in Figures 1.1–1.4. The orientation of the contour ellipses is related to the degree of covariance between the two variables. If the major and minor axes of the ellipses formed by the probability contours are parallel to the coordinate x-axis and y-axis, the variables have covariance zero (Figures 1.1 and 1.3 show distributions with covariance = 0), which makes X and Y statistically independent for normally distributed data. Figures 1.2 and 1.4 (distributions with covariance $\neq 0$) show a bivariate normal distribution with axes not parallel to the x/y-coordinate axes, producing a positive covariance (covariance = 1.5 or correlation = $1.5/(\sqrt{1}\ \sqrt{9}) = 0.5$). The covariance between two variables is related to the degree to which the axes of the probability ellipses are not parallel to the coordinate axes.

Geometrically, area is synonymous with probability for a continuous univariate probability distribution. For a univariate distribution, the probability that an observation is greater than a specified point is equivalent to the area under the probability curve to the right of that point. For two jointly distributed variables, volume is the geometric equivalent of probability. The probability that simultaneously $X > c$ and $Y > c'$ is the volume of the joint probability distribution greater than c in the x-direction and, at the same time, greater than c' in the y-direction. When the ellipses of a bivariate normal distribution are oriented so that the major and minor axes are parallel to the coordinate axes (Figures 1.1 and 1.3; covariance = 0), the volume $P(X > c$ and $Y > c')$ equals $P(X > c)P(Y > c')$. For Figures 1.2 and 1.4 (covariance $\neq 0$), $P(X > c$ and $Y > c')$ is greater than $P(X > c)P(Y > c')$, implying a positive association between X and Y. Therefore, the probability of observing similar values of X and Y is greater than would be expected when two variables are independent. For a negative association, $P(X > c$ and $Y > c')$ is less than $P(X > c)P(Y > c')$, or the probability of observing similar values of X and Y is less than would be expected when the variables are independent.

Variance of a Sum

The variance of a sum of a series of variables depends on the variability of each component. Furthermore, the relationships among the variables that make up a sum, measured by the covariance, play a central role in determining the variance of a sum of random variables.

The simplest sum is the addition of two variables X and Y. The variance of the sum $X + Y$ is estimated by

$$S_{X+Y}^2 = S_X^2 + S_Y^2 + 2S_{XY} \qquad \text{(shown in Box 1.2)} . \tag{1.11}$$

The covariance term can be either large or small and either positive or negative or even zero, showing that the variance of the sum depends on the strength and direction of the relationship between X and Y as well as the variability of X and Y.

The role of the covariance in determining the variability of a sum of two variables is explained by the following argument:

- For two independent variables, the estimated covariance should be close to zero (not exactly zero because it is an estimate) and the variance of a sum of two independent variables is then estimated by the sum of the individual variances (i.e., $S_{X+Y}^2 = S_X^2 + S_Y^2$).

- A large and positive value of S_{XY} indicates that large values of X will be associated with large values of Y and small values of X will be similarly associated with small values of Y. The effect on the sum is to increase the frequency of extreme values, which is reflected by an increase in the variance of $X + Y$ relative to $S_X^2 + S_Y^2$ (i.e., $S_{X+Y}^2 = S_X^2 + S_Y^2 + 2S_{XY}$).

B O X   **1.2**   **THE ESTIMATED VARIANCE OF THE SUM $X+Y$**

$$S_{X+Y}^2 = \frac{1}{n-1} \Sigma \, [(x_i + y_i) - (\overline{x} + \overline{y})]^2$$

$$= \frac{1}{n-1} \Sigma \, [(x_i - \overline{x}) + (y_i - \overline{y})]^2$$

$$= \frac{1}{n-1} \Sigma (x_i - \overline{x})^2 + \frac{1}{n-1} \Sigma (y_i - \overline{y})^2 + \frac{2}{n-1} \Sigma (x_i - \overline{x})(y_i - \overline{y})$$

$$= S_X^2 + S_Y^2 + 2S_{XY}.$$

Similarly, where a and b represent constants, the variance of $aX + bY$ is estimated by

$$S_{aX+bY}^2 = a^2 S_X^2 + b^2 S_Y^2 + 2ab \, S_{XY}.$$

Properties:

1   If $b = 0$, then the estimated variance of a times X is

$$S_{aX}^2 = a^2 S_X^2.$$

2   If $a = 1$ and $b = -1$, then the estimated variance of the difference $X - Y$ is

$$S_{X-Y}^2 = S_X^2 + S_Y^2 - 2S_{XY}.$$

3   As noted, if $a = 1$ and $b = 1$, then the estimated variance of the sum $X + Y$ is

$$S_{X+Y}^2 = S_X^2 + S_Y^2 + 2S_{XY}.$$

In general, when $c_1, c_2, c_3, \cdots, c_k$ represent constants, then

$$S_{\Sigma c_i X_i}^2 = \Sigma c_i^2 S_{X_i}^2 + \underset{i \neq j}{\Sigma \Sigma} \, c_i c_j S_{X_i X_j}.$$

■ The opposite is true when X and Y are negatively associated. Large values of X are likely to be observed with small values of Y and vice versa. In this case, the sum of X and Y tends to be at least partially stabilized because the values of X and Y have an increased probability of balancing each other (i.e., reducing the likelihood of extreme values). The decreased likelihood of extreme values is reflected by a reduction in the variance of $X + Y$ relative to $S_X^2 + S_Y^2$ (i.e., $S_{X+Y}^2 = S_X^2 + S_Y^2 - 2S_{XY}$).

Figure 1.6 shows the distribution of 10,000 values $X + Y$ and $X - Y$ where X and Y are independent (top row; covariance = 0) and correlated (bottom row; correlation = 0.7). In both cases X and Y are sampled from a standard normal distribution, expectation = $\mu = 0$ with standard deviation = $\sigma = 1$ (the expectation or population mean of

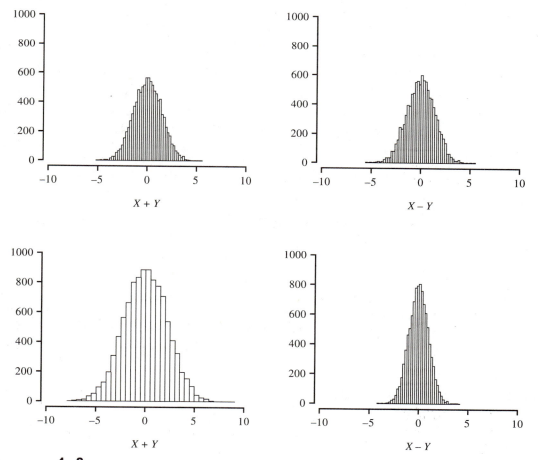

FIGURE **1.6**

The distribution of the sum $X + Y$ and difference $X - Y$ of two normally distributed variables

(top row: correlation = 0, bottom row: correlation = 0.7)

a normal distribution is conventionally represented by the special symbol μ and the variance by σ^2).

Variance of More Than Two Variables

Covariances play similar roles in determining the variance of a sum of more than two variables but in a more complex fashion. Each of a series of possible covariance terms contributes to the overall variability. There are $k(k-1)/2$ different covariance terms associated with a sum of k variables. In symbols,

$$S^2_{X_1+X_2+\cdots+X_k} = S^2_{X_1} + S^2_{X_2} + \cdots + S^2_{X_k}$$
$$+ 2\,S_{X_1X_2} + 2\,S_{X_1X_3} + \cdots + 2\,S_{X_{k-1}X_k}$$

or

$$S^2_{\Sigma X_i} = \Sigma S^2_{X_i} + \underset{i\neq j}{\Sigma\Sigma}\, S_{X_iX_j} \, .$$

(1.12)

For example, if $S^2_{X_1} = 4$, $S^2_{X_2} = 6$, $S^2_{X_3} = 10$ and X_1, X_2, X_3 are independent, then

$$S^2_{X_1+X_2+X_3} = S^2_{X_1} + S^2_{X_2} + S^2_{X_3} = 4 + 6 + 10 = 20 \, .$$

However, if these variables are not independent where $S_{X_1X_2} = 1$, $S_{X_1X_3} = 4$, and $S_{X_2X_3} = 5$, then

$$S^2_{X_1+X_2+X_3} = S^2_{X_1} + S^2_{X_2} + S^2_{X_3} + 2S_{X_1X_2} + 2S_{X_1X_3} + 2S_{X_2X_3}$$
$$= 4 + 6 + 10 + 2 + 8 + 10 = 40 \, .$$

Similarly, if the variables X_1, X_2, and X_3 are not independent but have negative covariances of $S_{X_1X_2} = -1$, $S_{X_1X_3} = -2$, and $S_{X_2X_3} = -2$, then

$$S^2_{X_1+X_2+X_3} = 4 + 6 + 10 - 2 - 4 - 4 = 10 \, .$$

Sometimes the variances and covariances are displayed in an array in which the rows and columns represent the specific variables and the array contains the variances (diagonal elements) and the covariances (nondiagonal elements). For example, the previous illustration gives the following three variance/covariance arrays:

(1)	x_1	x_2	x_3
x_1	4	0	0
x_2	0	6	0
x_3	0	0	10

(2)	x_1	x_2	x_3
x_1	4	1	4
x_2	1	6	5
x_3	4	5	10

(3)	x_1	x_2	x_3
x_1	4	-1	-2
x_2	-1	6	-2
x_3	-2	-2	10

When the variances and covariances of k variables are displayed in a variance/covariance array, the variance of the sum of the k variables is the sum of all values contained in the array. Such arrays are a compact and convenient way to display the variances and covariances among the k variables and are found in the output of statistical computer systems.

Distribution of Sums of Squares

First, the chi-square distribution is defined by the following theorem:

If $Z_1, Z_2, \cdots, Z_m$ are m independent random variables with standard normal distributions (mean $= \mu = 0$ and variance $= \sigma^2 = 1.0$), then the variable $SS = Z_1^2 + Z_2^2 + \cdots + Z_m^2$ has a chi-square distribution with m degrees of freedom.

The symbol SS stands for sum of squares and, under the conditions of the theorem, probabilities associated with any value of SS ($0 \leq SS \leq \infty$) can be calculated with a computer program or looked up in a chi-square table for a given number of degrees of freedom (m). Developing a sum of squares with a chi-square distribution from normally distributed values is a typical analytic pattern and a basic tool in assessing questions concerning sampled populations.

Fundamental to a number of statistical techniques is the assumption or knowledge that the observations are independently sampled from a normal distribution. When a sample consists of independent values selected from a normal parent population, the sum of squared deviations of the observed values from the population mean divided by the variance associated with the sampled normal distribution has a chi-square distribution. Specifically, consider a series of n independent values $y_1, y_2, \cdots, y_n$:

when y_i has a normal distribution (mean $= \mu$ and variance $= \sigma^2$),

then $z_i = (y_i - \mu)/\sigma$ also has a normal distribution (mean $= 0$ and variance $= 1.0$),

giving $z_i^2 = (y_i - \mu)^2/\sigma^2$ a chi-square distribution with one degree of freedom

(special case of the chi-square theorem; when $m = 1$) and, summarizing the n variables,

$SS = \Sigma z_i^2 = \Sigma(y_i - \mu)^2/\sigma^2$ has a chi-square distribution with n degrees of freedom.

Usually the mean μ is unknown and is estimated. Using information from the sampled data to estimate μ affects the distribution of the sum of squares, SS. This influence is accounted for by reducing the degrees of freedom by 1 for each independent estimate used to specify the mean value μ. Therefore,

$$SS = \frac{1}{\sigma^2} \sum_{i=1}^{n} (y_i - [\text{estimated mean}])^2$$

(1.13)

still has a chi-square distribution but with $n - s$ degrees of freedom where s represents the number of independent estimates used to specify the mean. Again, σ^2 represents the variance of the sampled normally distributed population. For example, the simplest case involves using the sample mean $\bar{y}$ as an estimate of the population mean μ; then, if y_i represents one of a series of n independent and normally distributed values with associated variance σ^2, the quantity

$$SS = \frac{1}{\sigma^2} \sum_{i=1}^{n} (y_i - \bar{y})^2$$

(1.14)

has a chi-square distribution with $n - 1$ degrees of freedom ($s = 1$).

A general method to evaluate hypotheses concerning the underlying structure of the data involves comparing the consequences of estimating the mean value of the sampled population under different conditions. The mean is estimated under one set of conditions and then restrictions are placed on these conditions and the mean value is again estimated. The first set of conditions produces a sum of squares measuring the fit of the mean to the data, or

$$SS_1 = \sum_{i=1}^{n} (y_i - [\text{estimated mean for a specific set of conditions}])^2 ,$$

(1.15)

called the residual sum of squares generated under one set of conditions. The impact of a second, more restrictive set of conditions is also measured by the fit of the mean to the data and produces a second residual sum of squares generated under a set of more restrictive conditions or

$$SS_0 = \sum_{i=1}^{n} (y_i - [\text{estimated mean under more restricted conditions}])^2 .$$

(1.16)

The difference between these two residual sums of squares ($SS_0 - SS_1$) reflects the influence of the added restrictions. A formal evaluation of this difference depends on the property that each sum of squares has a chi-square distribution when divided by the variance σ^2 and is the topic of the next section. A systematic comparison of such sums of squares produces an analysis of variance strategy that applies in a large variety of situations.

F-Statistic for Nested Hypotheses

A formal derivation of the F-distribution is beyond the scope of this text. Instead, the following theorem serves as an introduction:

If a sum of squares denoted as SS_1 has a chi-square distribution with df_1 degrees of freedom under a set of conditions called hypothesis 1 (symbolized as H_1), and if a sum of squares denoted as SS_0 has a chi-square distribution with df_0 degrees of freedom under a set of conditions called hypothesis 0 (symbolized as H_0), and, furthermore, hypothesis 0 (H_0) is created by a restriction of hypothesis 1 (H_1), then

$$F = \frac{(SS_0 - SS_1)/(df_0 - df_1)}{SS_1/df_1}$$

(1.17)

has an F-distribution with $(df_0 - df_1)$ and df_1 degrees of freedom when SS_1 and SS_0 differ only because of random variation associated with the variable under investigation.

This theorem allows the difference observed between two sums of squares to be statistically assessed, which means the difference is related to a probability using an F-distribution. The F-distribution is named for R. A. Fisher (1890–1962), who provided much of the mathematical basis for early statistical theory and developed many of the most fundamental statistical procedures.

The phrase "H_0 is created by a restriction of hypothesis 1 (H_1)" means that H_0 is a special case of H_1—sometimes called a nested hypothesis. Often, the nested hypothesis H_0 is created by setting some of the parameters in the model generated by H_1 to zero. For example, H_0 is a restriction of H_1 if

$$H_1: \text{ the data are described by the model } a + b_1 x_1 + b_2 x_2$$

and

$$H_0: \text{ the data are described by the model } a + b_1 x_1 .$$

The second hypothesis is a special case of the first (i.e., the parameter $b_2 = 0$). The simpler hypothesis H_0 is nested within H_1 and is necessarily described using fewer parameters.

Any time data come from a normal or approximately normal distribution, the sums of squares SS_0 and SS_1 have at least an approximate chi-square distribution under specific conditions. When these two sums of squares measure the fit of two nested hypotheses, a formal statistical test of the observed difference is possible using an F-distribution.

The property that H_0 is a restriction of H_1 guarantees that SS_0 is always larger than SS_1 and df_0 is always larger than df_1. The difference $SS_0 - SS_1$ is always positive because a simpler model always fits the data less well than a more complex model, causing an increase in the sum of squares. The restrictive hypothesis H_0 always involves a reduction in the number of parameters. The fewer parameters needed to describe the data, the larger the degrees of freedom associated with the residual sum of squares (remember, $df = n - s$). Therefore, $df_0 - df_1$ is also always a positive number. The value $df_0 - df_1$ can be alternatively viewed as the number of additional parameters necessary to differentiate hypothesis H_1 from H_0. Many times, $df_0 - df_1$ is simply the number of parameters set to zero to generate H_0.

The question arises: Is the increase in the sum of squares likely caused by the ability of the more complex model to capitalize on random variation and fit the data more closely, or is the more complex model necessary to reflect systematic properties of the data?

Relating $SS_0 - SS_1$ to an F-distribution gives an assessment of the probability that an observed increase arose strictly from capitalizing on chance variation. When the difference $SS_0 - SS_1$ is small (greater values have a high probability of occurring by chance), the inference is made that the more complex model is not worthwhile. When $SS_0 - SS_1$ is large (greater values have a small probability of occurring by chance), then the inference is made that the parameters removed from the model to generate H_0 are a useful part of the description of the data.

To be specific, consider the previous example (H_1: $a + b_1x_1 + b_2x_2$ versus H_0: $a + b_1x_1$). If the sum of squares measuring the fit of both models is similar (F has a high probability of occurring by chance), then it is inferred that the additional variable x_2 adds little to describing the data ($b_2 = 0$). Conversely, if the difference $SS_0 - SS_1$ is judged unlikely to be a result of random variation (F has a small probability of occurring by chance), then the inference is made that x_2 is an important part of the model describing the variables under investigation ($b_2 \neq 0$). This approach to assessing the impact of x_2 is called the "F-to-remove" test of the variable x_2. That is, the F-statistic becomes a tool in deciding whether the variable x_2 should be retained or removed from a model describing the sampled data.

Example of the *F*-to-Remove Approach

To illustrate the F-to-remove analysis of variance strategy in a simple case, two hypotheses are considered: the mean of a sampled normal distribution is not equal to zero ($\mu \neq 0$) versus the mean of the same sampled distribution is zero ($\mu = 0$). Data (y_i) collected to discriminate between these two alternatives are given in Table 1.1.

The analysis of variance strategy, as mentioned, is a general approach designed to contrast the sums of squares generated under each of two hypotheses. Hypothesis 1 postulates that the sample comes from a normal population with mean not equal to zero ($\mu \neq 0$), and hypothesis 0 postulates that the same population is sampled but the mean is zero ($\mu = 0$). A value of $\mu = 0$ is a special case of all possible values of μ, making the hypotheses nested. Specifically,

H_1: the data are sampled from a normal distribution, mean $= \mu$

and

H_0: the data are sampled from the same normal distribution, mean $= 0$.

The sum of squares based on $n = 25$ observations (Table 1.1) generated under the hypothesis $\mu \neq 0$ (H_1) is

$$SS_1 = \Sigma \, (y_i - \overline{y}\,)^2 = 22.845$$

with $df_1 = 24$ where the mean μ is estimated by $\overline{y} = 0.175$. The more restrictive hypothesis $\mu = 0$ (H_0: often called the null hypothesis) generates

$$SS_0 = \Sigma(y_i - 0)^2 = \Sigma y_i^2 = 23.614$$

with $df_0 = 25$. The F-statistic contrasting these two sums of squares is then

$$F = \frac{(SS_0 - SS_1)/(df_0 - df_1)}{SS_1/df_1} = \frac{(23.614 - 22.845)/(25 - 24)}{22.845/24} = 0.808 \; .$$

T A B L E **1.1**

Twenty-five observations from a normal distribution to test the conjecture that $\mu = 0$

0.813	1.708	0.051	−0.869	−0.789	1.531	−0.138	−0.266	0.871	1.307
1.406	−0.090	−1.455	0.021	−0.203	−0.605	0.193	0.295	1.085	0.871
0.482	1.387	−1.435	−1.722	−0.065					

The value F has an F-distribution when the mean of the sampled population is $\mu = 0$ and the data have a normal distribution. For the illustrative data, $F = 0.808$ with $df_0 - df_1 = 25 - 24 = 1$ and $df_1 = n - 1 = 24$ degrees of freedom and yields a p-value of 0.378, showing no persuasive evidence that the mean of the population sampled differs from zero. Incidentally, in this simple case the F-to-remove statistic is identical to (gives the same p-value as) the t-test of the null hypothesis $\mu = 0$, because

$$F = \frac{(SS_0 - SS_1)/1}{SS_1/(n-1)} = \left[\frac{\bar{y} - 0}{S_{\bar{y}}}\right]^2 = T^2$$

where T^2 is the squared value resulting from a t-test of the same data (expression 1.4).

Throughout, the term *p-value* is usually used instead of the more formal term *significance probability*. A p-value is defined as the probability of observing a more extreme result than the one observed under the hypothesis that the observed difference is due to chance. There is some controversy [8] over the exact interpretation of a p-value—the purist should consider a significance probability as a numerical guide to declaring an observed difference as due to nonrandom influences.

The term *null hypothesis* is used, perhaps overused, in the analysis of data. The null hypothesis takes a variety of forms but all have the same basic underlying meaning. The null hypothesis postulates that any differences or influences observed in values calculated from the data are strictly due to the random nature of the variable or variables being analyzed. When the null hypothesis is false, it means the data show an influence consistent with a systematic effect. In the context of an F-to-remove test, the null hypothesis states that the difference in the sum of squares $SS_0 - SS_1$ arises only because random variation allows the more complicated model to better fit the data.

One-Way Analysis of Variance

A common pattern of data collection involves sampling a series of populations and measuring a single characteristic on each observation. Each observation is denoted y_{ij} where j represents the jth observation sampled from the ith population (e.g., y_{23} is the third value sampled from the second population). Using these data, mean values are estimated from each of k populations ($\bar{y}_i$) and become the focus of an analysis of variance. Mean values will certainly differ because of the intrinsic variability in a statistical sample. However, the basic issue is whether systematic differences exist among the populations sampled, reflected by the estimated mean values. Data classified into a series of k categories is called a one-way classification and the associated analysis of these estimated means is called a one-way analysis of variance.

When the series of sources sampled have the same population mean (μ), the sample means should be similar. Conversely, if one or more of the mean values of the sampled populations differ, it is likely that the sample means reflect this difference. A useful measure to compare differences among the k sampled means is $\bar{y}_i - \bar{y}$ where $\bar{y}$ represents the overall mean based on the entire sample. An effective summary of these deviations, called the between sum of squares, is

$$\text{between sum of squares} = SS_{between} = \sum_{i=1}^{k} n_i (\bar{y}_i - \bar{y})^2 \tag{1.18}$$

where n_i values are selected from each of k populations. Note that

$$\bar{y}_i = \frac{1}{n_i} \sum_{j=1}^{n_i} y_{ij} \quad \text{and} \quad \bar{y} = \frac{1}{N} \sum_{i=1}^{k} n_i \bar{y}_i = \frac{1}{N} \sum_{i=1}^{k} \sum_{j=1}^{n_i} y_{ij} \quad \text{where} \quad N = \sum_{i=1}^{k} n_i .$$

To evaluate the relative magnitude of the between sum of squares, it is compared to a measure of the "background" variation formed by combining the usual estimates of variability (expression 1.2), one from each of the k sources of data. This means that a pooled estimate of variability formed by combining variance estimates from each of the k populations is

$$\text{within sum of squares} = SS_{within} = \sum_{i=1}^{k} (n_i - 1) S_i^2 \quad \text{where} \quad S_i^2 = \frac{1}{n_i - 1} \sum_{j=1}^{n_i} (y_{ij} - \bar{y}_i)^2 .$$

The same within sum of squares is often expressed as

$$SS_{within} = \sum_{i=1}^{k} \sum_{j=1}^{n_i} (y_{ij} - \bar{y}_i)^2 . \tag{1.19}$$

More formally, a null hypothesis postulates that the population means do not differ (H_0: $\mu_1 = \mu_2 = \cdots = \mu_k = \mu$) versus the alternative hypothesis H_1 that one or more population means differ among the k normally distributed populations under investigation. Furthermore, it is assumed or known that the variances associated with each sampled population are the same (denoted σ^2). A test statistic that allows an assessment of the null hypothesis is the ratio of the between sum of squares to the within sum of squares each divided by its degrees of freedom, or

$$F = \frac{SS_{between}/(k-1)}{SS_{within}/(N-k)} . \tag{1.20}$$

The value F has an F-distribution with $k - 1$ and $N - k$ degrees of freedom when the variables analyzed (y_{ij}) have the same mean value (null hypothesis true), the same variance, and are normally distributed in all k populations sampled ([9] or [10] present a more complete and general approach to analysis of variance techniques).

One-Way Analysis of Variance Using Nested Hypotheses

The identical one-way analysis of variance to evaluate differences arising among a series of sample means (expression 1.20) can be described in terms of contrasting nested hypotheses. If the first hypothesis (H_1) postulates that at least some of the sampled populations have different means, then a sum of squares measure of fit is

$$SS_1 = \sum_{i=1}^{k} \sum_{j=1}^{n_i} (y_{ij} - \bar{y}_i)^2 .$$

(1.21)

The sum of squares SS_1/σ^2 has a chi-square distribution with $df_1 = N - k$ degrees of freedom when the values y_{ij} are selected from k populations with normal distributions and the same variance (σ^2). As mentioned, the degrees of freedom are the total number of observations minus the number of independent estimates (k mean values are estimated), giving $df_1 = N - s = N - k$ degrees of freedom (N is the total number of observations).

A nested hypothesis (H_0) postulates that a series of identical normal populations are sampled. The common mean is then estimated by $\bar{y}$. The sum of squares measuring the fit of this more restrictive hypothesis is

$$SS_0 = \sum_{i=1}^{k} \sum_{j=1}^{n_i} (y_{ij} - \bar{y})^2 .$$

(1.22)

The value SS_0/σ^2 has a chi-square distribution with $df_0 = N - 1$ degrees of freedom when the data are sampled from k identical normally distributed populations (same mean μ and same variance σ^2). Again, the degrees of freedom are the number of observations minus the number of independent estimates, giving $df_0 = N - s = N - 1$ degrees of freedom (one mean value is estimated). The difference $SS_0 - SS_1$, therefore, measures the impact of postulating that the sampled data come from k identical populations.

Using the general F-statistic (expression 1.17) applied to the two nested hypotheses is no different from comparing the between and within sum of squares (expression 1.20), or

$$F = \frac{SS_{between}/(k - 1)}{SS_{within}/(N - k)} = \frac{(SS_0 - SS_1)/(df_0 - df_1)}{SS_1/df_1}$$

because

$$SS_0 - SS_1 = SS_{between} = \sum_{i=1}^{k} n_i(\bar{y}_i - \bar{y})^2 .$$

T A B L E   **1.2**

One-way classification of cholesterol levels by age groups

Age Categories	n_i	$\bar{y}_i$	s_i^2
< 25	6	225.3	163.47
25–34	8	220.5	66.86
35–45	10	225.4	121.38
> 45	6	232.7	147.07
Total	30	225.5	—

A small data set illustrates. A total of $N = 30$ individuals measured for cholesterol levels and classified into $k = 4$ groups by age have the following values:

age < 25: 225, 237, 219, 211, 216, and 244

age 25–34: 216, 222, 218, 226, 224, 205, 220, and 233

age 35–45: 224, 225, 237, 204, 217, 221, 223, 243, 235, and 225

age > 45: 210, 235, 230, 241, 236, and 244.

A summary of the cholesterol data is given in Table 1.2. The question addressed: Is there evidence of systematic differences among mean cholesterol levels for the four age groups? Typically the results from an analysis of variance are summarized in a table, such as Table 1.3.

The sum of squares $SS_0 = 3621.467$ and $SS_1 = 3113.067$ with $N - 1 = 29$ and $N - k = 26$ degrees of freedom, respectively, give

$$F = \frac{(3621.467 - 3113.067)/3}{3113.067/26} = 1.415 ,$$

yielding a p-value $= 0.261$. Because a difference of $SS_0 - SS_1 = 508.400$ is likely to arise by chance (p-value $= 0.261$), the data present no strong evidence of a difference among the four age categories in mean level of cholesterol.

T A B L E   **1.3**

Analysis of variance table for the cholesterol data on 30 individuals classified by four age groups

Source	Sum of Squares	Degrees of Freedom
Between sum of squares	$SS_0 - SS_1 = \Sigma n_i(\bar{y}_i - \bar{y})^2 = 508.400$	$k - 1 = 3$
Within sum of squares	$SS_1 = \Sigma\Sigma(y_{ij} - \bar{y}_i)^2 = 3113.067$	$N - k = 26$
Total sum of squares	$SS_0 = \Sigma\Sigma(y_{ij} - \bar{y})^2 = 3621.467$	$N - 1 = 29$

General Concepts

A small test case using the cholesterol data (Table 1.2) demonstrates the computer implementation of the one-way analysis of variance technique. The following STATA code uses the command "infile group chol using example.dat" to read the cholesterol data into the STATA system. All computer-based statistical analysis systems require a series of keywords to invoke a specific process. For example, "infile" reads the data, "oneway" does the analysis of variance, and "exit, clear" ends the execution for the STATA system. Each system has a manual in which the keywords are defined and the commands explained producing the program "language." In this test case and subsequent test cases, a small data set is used so the reader can clearly see the computer syntax and check the results with hand calculations. More importantly, the test cases are annotated so the issues discussed in the text are easily identified in the computer output (not always a simple task). Once a small test case is understood, it is generally not difficult to "scale up" to a larger and more realistic application. Details of the STATA system are left to the reference manual and it is assumed that the reader can explore these issues independently.

```
example.anova

. infile g y using anov.dat
(30 observations read)

. list
            g        y
    1.      1       225
    2.      1       237
    3.      1       219
    4.      1       211
    5.      1       216
    6.      1       244
    7.      2       216
    8.      2       222
    9.      2       218
   10.      2       226
   11.      2       224
   12.      2       205
   13.      2       220
   14.      2       233
   15.      3       224
   16.      3       225
   17.      3       237
   18.      3       204
   19.      3       217
   20.      3       221
   21.      3       223
   22.      3       243
   23.      3       235
   24.      3       225
   25.      4       210
   26.      4       235
   27.      4       230
   28.      4       241
   29.      4       236
   30.      4       244
```

The annotations on the listing indicate y_{ij}, with braces grouping y_{1j} (observations 1–6), y_{2j} (observations 7–14), y_{3j} (observations 15–24), and y_{4j} (observations 25–30).

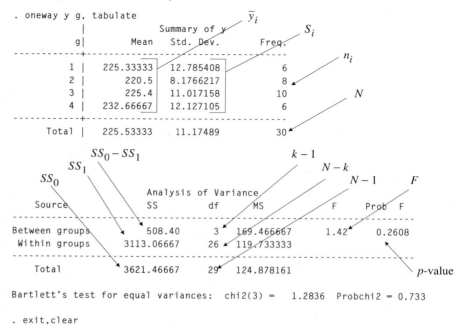

```
. oneway y g, tabulate
                    |         Summary of y
                  g |      Mean    Std. Dev.          Freq.
      --------------+------------------------------------------
                  1 |  225.33333    12.785408              6
                  2 |     220.5     8.1766217              8
                  3 |     225.4     11.017158             10
                  4 |  232.66667    12.127105              6
      --------------+------------------------------------------
              Total |  225.53333    11.17489              30


                            Analysis of Variance
          Source              SS          df       MS            F      Prob F
      ------------------------------------------------------------------------
      Between groups        508.40         3    169.466667      1.42     0.2608
      Within groups      3113.06667        26   119.733333
      ------------------------------------------------------------------------
              Total      3621.46667        29   124.878161


Bartlett's test for equal variances:   chi2(3) =   1.2836   Probchi2 = 0.733

. exit,clear
```

Labels on the output: $\bar{y}_i$, S_i, n_i, N, SS_0, SS_1, $SS_0 - SS_1$, $k-1$, $N-k$, $N-1$, F, *p*-value

Notation

Summary values:

The sample size $= n$, n_i, or N

Expectation:

$$E(Y) = \Sigma y p_y \quad \text{and} \quad E(\Sigma\, c_i Y_i) = \Sigma c_i E(Y_i)$$

The estimated variance based on a sample size n:

$$S_Y^2 = \frac{\sum_{i=1}^{n} (y_i - \bar{y})^2}{n-1}$$

The estimated standard deviation:

$$S_Y = \sqrt{S_Y^2} = \sqrt{\frac{\Sigma(y_i - \bar{y})^2}{n-1}}$$

The estimated covariance between two variables X and Y:

$$S_{XY} = \frac{\sum_{i=1}^{n} (x_i - \bar{x})(y_i - \bar{y})}{n-1}$$

Independence:

$$P(X > c \text{ and } Y > c') = P(X > c)P(Y > c')$$

Positive association:

$$P(X > c \text{ and } Y > c') > P(X > c)P(Y > c')$$

Negative association:

$$P(X > c \text{ and } Y > c') < P(X > c)P(Y > c')$$

F-statistic:

$$F = \frac{(SS_0 - SS_1)/(df_0 - df_1)}{SS_1/df_1}$$

and F has an F-distribution with $(df_0 - df_1)$ and df_1 degrees of freedom.

Problems

T A B L E **1.4**
Data from women over 40 years of age used for exercises in Chapters 1–5

Obs.	Gestation	Sex	Birth Weight	Cigs./Day	Height	$Weight_0$	$Weight_1$
1	36	0	3300	0	160.0	67.3	82.7
2	38	0	3300	60	167.6	52.7	76.0
3	38	0	4100	20	167.6	64.2	79.6
4	38	1	2900	10	163.9	72.7	95.8
5	39	0	2820	0	161.3	50.0	63.3
6	39	0	3040	0	158.8	49.1	61.5
7	39	0	4120	0	160.0	57.7	73.5
8	39	0	4200	0	174.0	68.0	86.8
9	39	1	3100	0	171.5	67.3	85.6
10	39	1	3330	0	160.0	74.0	90.5
11	39	1	3410	0	165.1	55.9	70.7
12	39	1	3420	0	162.6	52.3	66.0
13	40	0	2450	20	167.6	61.4	72.5
14	40	0	2885	0	167.7	60.0	78.6
15	40	0	3235	0	170.2	50.0	65.5
16	40	0	3320	0	165.1	63.6	80.2
17	40	0	3600	0	165.1	53.2	68.7
18	40	0	3720	0	165.0	57.7	74.4
19	40	0	3720	0	172.7	61.4	80.0
20	40	0	3820	0	175.3	60.8	78.1

T A B L E   **1.4**
(continued)

Obs.	Gestation	Sex	Birth Weight	Cigs./Day	Height	$Weight_0$	$Weight_1$
21	40	0	3840	0	167.0	60.5	83.9
22	40	0	3880	0	156.2	57.3	73.7
23	40	0	3960	0	157.5	52.7	68.2
24	40	0	4465	0	157.5	51.4	66.4
25	40	1	2980	0	160.0	47.7	55.2
26	40	1	3040	0	162.0	49.0	60.3
27	40	1	3060	20	157.5	61.0	75.0
28	40	1	3100	0	170.2	55.5	64.6
29	40	1	3120	0	160.3	56.8	75.4
30	40	1	3205	0	172.7	58.2	75.5
31	40	1	3220	0	170.0	64.6	86.0
32	40	1	4100	40	167.0	67.0	85.0
33	41	0	3100	0	168.9	61.4	69.2
34	41	0	3720	0	170.2	57.7	67.7
35	41	0	3720	20	170.2	57.7	80.5
36	41	0	3900	0	167.0	68.0	85.4
37	41	0	3990	0	165.1	52.3	71.2
38	41	0	4050	0	167.6	61.0	78.5
39	41	0	4080	0	162.6	59.1	83.1
40	41	0	4100	0	165.1	60.5	86.5
41	41	0	4460	20	165.1	56.8	88.0
42	41	0	5220	0	157.5	56.8	68.2
43	41	1	3300	40	162.6	74.1	89.7
44	41	1	3400	0	172.7	71.4	87.8
45	41	1	4000	0	165.1	90.0	100.8
46	41	1	4030	0	166.0	63.0	95.3
47	43	1	3220	0	166.4	60.9	72.0
48	43	1	4270	0	162.6	54.5	70.3

Note: The data in Table 1.4 consist of seven measured variables on women over 40 years old who gave birth to their first infant. The infant's gestation (weeks), sex (0 = male and 1 = female), and birth weight (grams) as well as maternal smoking exposure (cigarettes per day) and height (cm) are included. Of particular interest is the amount of weight gained during prepregnancy—$weight_0$ is the woman's pre-pregnancy weight (kilograms) and $weight_1$ is her weight at the last prenatal visit.

1   Using the data in Table 1.4 conduct a *t*-test to investigate the difference in mean birth weights (grams) between male and female births:

male: $n_1 = 29, \bar{y}_1 = 3728.1$

female: $n_2 = 19, \bar{y}_2 = 3379.2$

Recall that $T = \dfrac{\bar{y}_2 - \bar{y}_1}{\sqrt{S_p^2\left(\dfrac{1}{n_1} + \dfrac{1}{n_2}\right)}}$ where $S_p^2 = \dfrac{(n_1 - 1)S_1^2 + (n_2 - 1)S_2^2}{n_1 + n_2 - 2}$

with degrees of freedom $= n_1 + n_2 - 2$.

2 Compute the within sum of squares for the weights of males and females in Table 1.4—*within sum of squares* $= W = \Sigma\Sigma(y_{ij} - \bar{y}_i)^2$.

Compute the between sum of squares for the weights of males and females in Table 1.4—*between sum of squares* $= B = \Sigma n_i(\bar{y}_i - \bar{y})^2$.

Show that the F-test is identical to the squared t-test, or

$$F = \frac{B}{W/(n_1 + n_2 - 2)} = T^2$$

where T is the t-statistic calculated in problem 1.

3 Show from the definition of a sample variance that $S_{aX}^2 = a^2 S_X^2$.

4 When is $S_{X+Y}^2 < S_{X-Y}^2$?

5 For the dice example, if Y_1 and Y_2 are two independent tosses of a die and $S = Y_1 + Y_2$, then

S	2	3	4	5	6	7	8	9	10	11	12
p_s	1/36	2/36	3/36	4/36	5/36	6/36	5/36	4/36	3/36	2/36	1/36

Verify that $E(Y_1 + Y_2) = E(Y_1) + E(Y_2) = E(S) = 7.0$.

6 Construct a variance/covariance array for the 10 following values:

	x_1	x_2	x_3
1	2	12	5.00
2	11	30	6.09
3	14	40	4.00
4	16	12	5.33
5	18	2	4.25
6	18	5	3.75
7	18	20	5.75
8	22	30	4.66
9	23	75	3.83
10	26	40	3.75

7 Compute the estimated variance S_{X+Y}^2 two ways from the following data: first, directly from the values $X + Y$ and, second, from the expression $S_X^2 + S_Y^2 + 2S_{XY}$.

x	1	2	3	4	5	6	7	8	9	10
y	10	9	8	7	6	5	4	3	2	1

Repeat the calculations for the values

x	2	5	3	1	5	7	7	2	9	1
y	3	1	5	7	8	4	2	6	9	5

8 Consider 30 pregnant women classified by race where the weight gain (kilograms) during pregnancy is recorded:

White	12.4	13.5	16.4	18.1	12.7	11.3	19.1	20.0	
Black	13.5	14.1	12.6	23.0	25.6	15.5	12.7	18.8	19.3
Hispanic	12.3	13.2	15.1	19.0	17.3	12.5			
Asian	12.5	13.1	17.1	14.2	15.1	16.2	13.9		

Use a one-way analysis of variance to assess the differences in weight gain related to race.

References

1 Snedecor, G. W., and Cochran, W. G. *Statistical Methods*, 8th ed. Ames: The Iowa State University Press, 1989.

2 Sokal, R. R., and Rohlf, F. J. *Biometry*. San Francisco, Calif.: W. H. Freeman, 1969.

3 Winer, B. J., Brown, D. R., and Michels, K. M. *Statistical Principles in Experimental Design*, 3rd ed. New York: McGraw-Hill, 1991.

4 Dixon, W. J., and Massey, F. J. *Introduction to Statistical Analysis*. New York: McGraw-Hill, 1960.

5 Freund, J. E. *Modern Elementary Statistics*, 7th ed. Englewood Cliffs, N. J.: Prentice Hall, 1988.

6 Mosteller, F., Rouke, R. E., and Thomas, G. B. *Probability with Statistical Applications*. Menlo Park, Calif.: Addison-Wesley, 1973.

7 Larsen, R. J., and Marx, M. L. *An Introduction to Mathematical Statistics and Its Applications*, 2nd ed., Englewood Cliffs, N. J.: Prentice Hall, 1986.

8 Goodman, S. N. *P*-Values, Hypothesis Tests and Likelihood: Implications for Epidemiology of a Neglected Historical Debate. *American Journal of Epidemiology* 137 (1993): 485–500.

9 Rosner, B. *Fundamental of Biostatistics*, 3rd ed. Boston: Duxbury Press, 1990.

10 Dunn, O. J., and Clark, V. A. *Applied Statistics: Analysis of Variance and Regression*. New York: John Wiley, 1974.

2

Simple Linear Regression

Background

A single straight line often effectively describes a two-dimensional scatter of data collected to investigate the relationship between two variables. A straight-line summary serves two purposes: It simply describes the relationship between two variables and, in addition, provides a basis for statistically evaluating this description. For example, the amount of carbon monoxide (CO) found in a person's lungs and the number of cigarettes reported smoked each day can be summarized by a straight line. Variation exists among the CO levels, even for individuals reporting the same amounts smoked, potentially obscuring what might be a simple relationship (a case of "trees obscuring the forest"). It is usually profitable to condense data in a parsimonious fashion. Many statistical models exist to represent succinctly and assess the relationship between two variables such as CO and the number of cigarettes smoked. The simplest is a straight line. The effectiveness of a line depends on how well the linear model reflects the data. When a line accurately describes the relationship between CO and reported smoking levels, inferences drawn from analyzing the properties of the line lead to a better understanding of smoking exposure. The strategy fails when the model does not adequately represent the data. The correspondence between a statistical model and the observed data is always a critical element of an analysis. The choice of a mathematical model to represent the relationships under investigation is always equivocal, but a good starting point is the simple linear model, discussed in the following sections.

Method: Criterion for Estimation

A simple linear model relates a variable x to a random variable Y using two parameters (a, b) and the mathematical form

$$E(Y \mid x_i) = a + bx_i .$$

(2.1)

31

The symbol x denotes a nonrandom quantity called the independent variable. The independent variable is given other names, including predictor variable, explanatory variable, regressor variable, or sometimes simply x-variable. The term *independent* in the linear model context is potentially confused with the concept of independence in the probability sense—they are not related. The random variable represented by Y is called the dependent variable. A sample of dependent variables with the same value of the independent variable x_i produces a distribution of Y-values: y_1, y_2, y_3, $\cdots$. The symbol $E(Y \mid x_i)$ represents the mean of the population that produced the Y-values. Furthermore, the population mean $E(Y \mid x_i)$ lies on a straight line (expression 2.1).

The parameter represented by b is the key element of a simple linear regression analysis and represents the slope of the line that characterizes the linear model. The slope b measures the expected change in the dependent variable Y from a one-unit increase in the independent variable x, or

$$E(Y \mid x_i) = a + bx_i \text{ and } E(Y \mid x_i + 1) = a + b(x_i + 1)$$

giving

$$slope = E(Y \mid x_i + 1) - E(Y \mid x_i) = b \; .$$

The parameter b is also referred to as a regression coefficient. The parameter represented by a is the y-intercept, which is the value of $E(Y \mid x_i)$ when $x_i = 0$. Linear models are referred to as regression models, and the accompanying analysis is called regression analysis. The simple linear model (expression 2.1) is a special case of regression analysis.

Estimation: The Method of Least Squares

A criterion is necessary to select a "best" line to represent the linear relationship between x and Y based on the sampled data. Selecting a particular line means choosing a specific pair of values a and b. The application of one such criterion, called the method of least squares, gives estimates of the parameters a and b producing a "best" line.

The least squares criterion requires that the parameters a and b be chosen so the sum of the squared distances (d_i's shown in Figure 2.1) between the estimated line and the n data points is minimum, or

$$Q = \Sigma d_i^2 = \Sigma (y_i - [a + bx_i])^2 \tag{2.2}$$

is minimum. The value d_i is the distance between a line and a point (x_i, y_i) measured parallel to the y-axis (again, Figure 2.1).

The variables x_i and y_i are measured values making up the n paired observations (x_i, y_i) and constitute the bivariate data set to be summarized by a straight line. The choices of a and b that give the smallest possible value of Q are called the least squares estimates, denoted by $\hat{a}$ and $\hat{b}$, and the least squares estimated line is then

$$\hat{y}_i = \hat{a} + \hat{b}x_i \; . \tag{2.3}$$

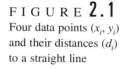

FIGURE **2.1**

Four data points (x_i, y_i) and their distances (d_i) to a straight line

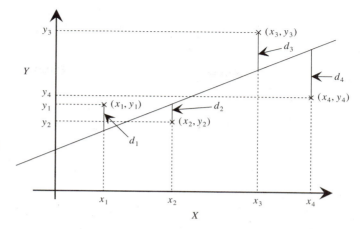

The "hat" or circumflex placed over a symbol denotes an estimator of a population parameter—$\hat{a}$ (reads "a hat") and $\hat{b}$ estimate the population parameters represented by a and b. All choices for the intercept a and/or the slope b other than $\hat{a}$ and $\hat{b}$ give a larger value for the criterion Q.

By algebraic or calculus arguments the values $\hat{a}$ and $\hat{b}$ that minimize Q come from solving the following equations based on n data pairs, called normal equations,

$$\hat{a}n + \hat{b}\Sigma x_i = \Sigma y_i \quad \text{or} \quad \Sigma(y_i - \hat{y}_i) = 0 \qquad \text{first normal equation}$$

$$\hat{a}\Sigma x_i + \hat{b}\Sigma x_i^2 = \Sigma x_i y_i \quad \text{or} \quad \Sigma(y_i - \hat{y}_i)x_i = 0 \qquad \text{second normal equation} \qquad (2.4)$$

where again $\hat{y}_i = \hat{a} + \hat{b}x_i$. The solutions to these two normal equations produce estimates of the slope and intercept as

$$\text{estimated slope} = \hat{b} = \frac{\Sigma(x_i - \bar{x})(y_i - \bar{y})}{\Sigma(x_i - \bar{x})^2} = \frac{S_{XY}}{S_X^2}$$

and

$$\text{estimated intercept} = \hat{a} = \bar{y} - \hat{b}\bar{x}. \qquad (2.5)$$

The least squares process guarantees that the estimates $\hat{a}$ and $\hat{b}$ produce the minimum value of Q, or

$$Q_{\min} = \Sigma(y_i - [\hat{a} + \hat{b}x_i])^2 = \Sigma(y_i - \hat{y}_i)^2. \qquad (2.6)$$

In this sense, the estimated straight line $\hat{y}_i = \hat{a} + \hat{b}x_i$ is the "best" summary of the observed scatter of the sampled observations.

An intuitive justification of the least squares estimate of the slope b is:

1 Each data pair (x_i, y_i) yields an estimate of the slope b, namely

$$\hat{b}_i = (y_i - \bar{y})/(x_i - \bar{x}).$$

2   A reasonable estimate of b is a weighted average of these n individual estimated slopes, namely $\hat{b} = \Sigma w_i \hat{b}_i / \Sigma w_i$.

3   One sensible choice for the weights is $w_i = (x_i - \bar{x})^2$, giving the greatest weight to values of x most distant from $\bar{x}$. Then

$$\hat{b} = \frac{\Sigma w_i \hat{b}_i}{\Sigma w_i} = \frac{\Sigma(x_i - \bar{x})(y_i - \bar{y})}{\Sigma(x_i - \bar{x})^2} = \frac{S_{XY}}{S_X^2},$$

which is the least squares estimate of the slope b.

Underlying Requirements

Five requirements are necessary to allow statistical inference regarding the least squares estimated line. It should first be noted that the computation of the "best" line (minimum Q) does not depend on any assumptions about the structure of the data. It is the statistical assessment of the properties of the estimated line that requires the assumption (or knowledge) that the data are sampled from populations with specific structures. The requirements are:

1   Each observation y_i is independently sampled.

2   The dependent variable Y has a normal distribution at each point x_i.

3   Each value Y comes from a population with the same variance regardless of the value x_i, denoted by $\sigma_{Y|x}^2$.

4   The expected values of the dependent variable Y lie on a straight line: in symbols, $E(Y \mid x_i) = a + bx_i$.

5   The x-values are fixed quantities (known exactly), which means x is not a statistical quantity but rather an "input" variable with values often selected by the researcher.

These requirements constitute in a broad sense the simple linear regression model. Schematically, the regression requirements are shown in Figure 2.2. In short, the dependent variable Y is an independently sampled, normally distributed observation from one of a series of populations with mean $E(Y \mid x_i) = a + bx_i$ and variance $\sigma_{Y|x}^2$.

The five requirements above are the mathematically exact conditions used to develop the theory and properties of linear regression analysis. Obviously sample data never perfectly conform to these requirements. Practical experience and some statistical theory show that strict adherence to these requirements is not necessary. For example, data with approximate normal distributions do not cause important distortions in the evaluation of the estimated parameters, particularly when the sample size exceeds 30 or so. The requirements of equal variance and linearity have the same property—moderate deviance from the ideal does not cause major problems in applications to collected data. One requirement, however, is critical: The observations must be independently sampled [1]. If the data are not independent, inferences

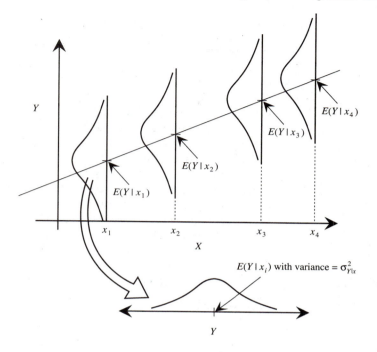

F I G U R E 2.2
The display of four populations and the underlying simple linear regression model

drawn from a regression analysis are not to be trusted. Although techniques are available to investigate the validity of the underlying statistical model (Chapter 4), determining the degree of correspondence between the theoretical model (requirements 1–5) and the population that generated the data almost always remains a difficult task.

Measuring Variation: Regression Case

A fundamental quantity in any statistical analysis is a measure of the variation associated with the variable being studied. For simple linear regression, an estimate of this "background" sampling variation associated with the dependent variable Y is

$$S_{Y|x}^2 = \frac{\Sigma(y_i - \hat{y}_i)^2}{n - 2}.$$

(2.7)

The sum of squared deviations of the observed values from each predicted value (based on the linear model) divided by $n - 2$ gives an unbiased estimate of the variance of the normally distributed parent populations from which the sample was selected (i.e., $\sigma_{Y|x}^2$; see requirement 3 and Figure 2.2). The estimated variance (expression 2.7) refers to the variability of Y associated with a distribution defined by a specific value of x, as indicated by the notation $Y|x$ (reads "Y given x"). For example, the variation among levels of CO (Y) for individuals who smoke a specific number of cigarettes per day (x) is $\sigma^2_{Y|x}$ and is estimated by $S^2_{Y|x}$. The estimate of the variability taking x into account ($S^2_{Y|x}$; expression 2.7) is distinctly different from the previous measure of variability (S^2_Y; expression 1.2), where any influence from x is ignored.

An essential element in a statistical analysis is the estimation of parameters along with an assessment of their sampling variation. An estimated parameter, such as $\hat{b}$, varies from sample to sample because the composition of each sample differs, causing different values of the estimated parameter. It follows that this sampling variability is a direct function of the variability within the sampled population and is estimated by $S^2_{Y|x}$. Such "background" variation is generally unavoidable and any conclusions drawn from a regression analysis must be tempered by the knowledge that variation influences the estimated quantities. Even for large samples, the estimate $\hat{b}$ will not exactly equal the parameter b, and measuring the precision of the estimated slope is at the heart of a simple linear regression analysis.

Variance of the Estimated Slope

First note that,

$$\text{if } w_i = \frac{x_i - \bar{x}}{\Sigma(x_i - \bar{x})^2},$$

then the estimated regression coefficient can be represented as

$$\hat{b} = \frac{\Sigma(x_i - \bar{x})(y_i - \bar{y})}{\Sigma(x_i - \bar{x})^2} = \Sigma\left[\frac{x_i - \bar{x}}{\Sigma(x_i - \bar{x})^2}\right](y_i - \bar{y}) = \Sigma w_i(y_i - \bar{y}) = \Sigma w_i y_i \quad (\bar{y}\,\Sigma w_i = 0),$$

showing that the estimated slope of a regression line is a weighted sum of the normally distributed values y_i (requirement 2). Therefore, the value $\hat{b}$ has a normal distribution because sums of normally distributed variables are themselves normally distributed [2]. The fact that the estimated slope has a normal distribution makes it possible to conduct t-tests and construct confidence intervals using an estimate of the variability of $\hat{b}$.

An estimated variance of the estimated slope $\hat{b}$ is

$$S^2_{\hat{b}} = \frac{S^2_{Y|x}}{\Sigma(x_i - \bar{x})^2} = \frac{S^2_{Y|x}}{(n-1)\,S^2_X}.$$

$$(2.8)$$

The expression for the estimated variance of $\hat{b}$ (derived in Box 2.1) shows that the variation of Y directly influences the variance of the estimated slope. Increases in the variability of the dependent variable expectedly reduce the precision of the estimate $\hat{b}$ (increase $S^2_{\hat{b}}$). However, increases in the spread of the independent variable (increases in $\Sigma(x_i - \bar{x})^2$) increase the precision of the estimated slope (decrease $S^2_{\hat{b}}$). The increased precision comes from the natural property that data points farther apart more accurately estimate a straight line than do points that lie close together.

That is, variation in data points that are close together has a greater impact on the estimated slope of a line than similar variation in points that are farther apart. In addition, increases in the sample size n certainly decrease the sampling variability of $\hat{b}$, improving the precision of the estimated slope. The relationship of these three elements ($S_{Y|x}^2$, S_X^2, and n) to the precision of the estimate of $\hat{b}$ is clearly seen in expression 2.8. The estimated standard deviation of $\hat{b}$ (the square root of the estimated variance) is called the standard error of $\hat{b}$, denoted $S_{\hat{b}}$, and is an important part of a *t*-test or confidence interval based on $\hat{b}$.

The estimate of the variance of $\hat{b}$ and the fact that $\hat{b}$ is normally distributed makes it possible to statistically evaluate the estimated slope $\hat{b}$ against specified alternatives. A typical hypothesis is that the estimated slope represents a random fluctuation from the value of zero (H_0: $b = 0$). The test statistic

$$T = \frac{\hat{b} - 0}{S_{\hat{b}}}$$

(2.9)

has a *t*-distribution with $n - 2$ degrees of freedom when x is linearly unrelated to Y ($b = 0$). Also, a $(1 - \alpha)\%$-confidence interval for b based on the estimate $\hat{b}$ is

$$\text{lower bound} = \hat{b} - t_{1-\alpha/2}\, S_{\hat{b}} \quad \text{and} \quad \text{upper bound} = \hat{b} + t_{1-\alpha/2}\, S_{\hat{b}}$$

(2.10)

where $t_{1-\alpha/2}$ is the $(1 - \alpha/2)$-percentile of a *t*-distribution with $n - 2$ degrees of freedom. Like all inferences drawn from an estimated simple linear regression equation, the validity of a *t*-test and the accuracy of a confidence interval depend on the requirements placed on the sampled population (independence, normality, equal variance, and linearity) being at least approximately correct.

B O X **2.1**   VARIANCE OF THE ESTIMATED SLOPE

$$\hat{b} = \frac{S_{XY}}{S_X^2} = \frac{1}{\Sigma(x_i - \bar{x})^2}\,[\Sigma(x_i - \bar{x})(y_i - \bar{y})] = \frac{1}{\Sigma(x_i - \bar{x})^2}\,\Sigma(x_i - \bar{x})y_i$$

because $\bar{y}\,\Sigma(x_i - \bar{x}) = 0$, then

$$S_{\hat{b}}^2 = \left\{ \frac{1}{\Sigma(x_i - \bar{x})^2} \right\}^2 \Sigma(x_i - \bar{x})^2 S_{Y|x}^2$$

where the variance of Y is estimated by $S_{Y|x}^2$. Furthermore,

$$S_{\hat{b}}^2 = \frac{S_{Y|x}^2}{\Sigma(x_i - \bar{x})^2} = \frac{S_{Y|x}^2}{(n-1)\,S_X^2}.$$

Note: Four requirements for a linear regression (1, 3, 4, and 5) are necessary for $S_{\hat{b}}^2$ to accurately estimate the variance of $\hat{b}$.

Partitioning the Total Sum of Squares

The total sum of squares, defined as $\Sigma(y_i - \bar{y})^2$, partitions into two meaningful pieces:

total sum of squares = residual sum of squares + regression sum of squares

or, specifically,

$$\Sigma(y_i - \bar{y})^2 = \Sigma(y_i - \hat{y}_i)^2 + \Sigma(\hat{y}_i - \bar{y})^2 \quad \text{(details in Box 2.2)}. \tag{2.11}$$

Each sum of squares reflects important and different aspects of using a linear model. Functionally, the quantity $y_i - \hat{y}_i$ is the distance from an observation to the estimated line and the quantity $\hat{y}_i - \bar{y}$ is the distance from the estimated line to the horizontal line $\bar{y}$ (Figure 2.3), producing measures of different aspects of a simple linear regression analysis.

Residual Sum of Squares

The residual sum of squares, $\Sigma(y_i - \hat{y}_i)^2$, measures the agreement between the data points y_i and the corresponding points $\hat{y}_i$ estimated on the basis of a regression equation, giving a direct measure of the lack of fit of the linear model to the data. The residual sum of squares is strictly a function of the sampling variability associated with the dependent variable when the requirements 1–4 hold. If the relationship

B O X   **2.2**   PARTITIONING OF $\Sigma(y_i - \bar{y})^2$

$$(y_i - \bar{y}) = (y_i - \hat{y}_i) + (\hat{y}_i - \bar{y})$$

$$(y_i - \bar{y})^2 = (y_i - \hat{y}_i)^2 + (\hat{y}_i - \bar{y})^2 + 2(y_i - \hat{y}_i)(\hat{y}_i - \bar{y})$$

$$\Sigma(y_i - \bar{y})^2 = \Sigma(y_i - \hat{y}_i)^2 + \Sigma(\hat{y}_i - \bar{y})^2 + 2\Sigma(y_i - \hat{y}_i)(\hat{y}_i - \bar{y})$$

$$= \Sigma(y_i - \hat{y}_i)^2 + \Sigma(\hat{y}_i - \bar{y})^2 \quad \text{because} \quad \Sigma(y_i - \hat{y}_i)(\hat{y}_i - \bar{y}) = 0 \quad \text{(see below)}.$$

To show that $\Sigma(y_i - \hat{y}_i)(\hat{y}_i - \bar{y}) = 0$:

$$\Sigma(y_i - \hat{y}_i)(\hat{y}_i - \bar{y}) = \Sigma(y_i - \hat{y}_i)\hat{y}_i - \bar{y}\,\Sigma(y_i - \hat{y}_i)$$

$$= \Sigma(y_i - \hat{y}_i)(\hat{a} + \hat{b}x_i) - \bar{y}\,\Sigma(y_i - \hat{y}_i)$$

$$= \hat{a}\Sigma(y_i - \hat{y}_i) + \hat{b}\Sigma(y_i - \hat{y}_i)x_i - \bar{y}\Sigma(y_i - \hat{y}_i).$$

Both $\hat{a}\Sigma(y_i - \hat{y}_i) = 0$ and $\bar{y}\Sigma(y_i - \hat{y}_i) = 0$ from first normal equations (expression 2.4). Also, $\hat{b}\Sigma(y_i - \hat{y}_i)x_i = 0$, from second normal equations (expression 2.4). Therefore,

$$\Sigma(y_i - \hat{y}_i)(\hat{y}_i - \bar{y}) = 0.$$

T A B L E **2.1**

Analysis of variance summaries for linear regression

Source	Sum of Squares	Degrees of Freedom	Mean Square
Regression	$SS_0 - SS_1 = \Sigma(\hat{y}_i - \bar{y})^2$	1	$\Sigma(\hat{y}_i - \bar{y})^2/1$
Residual	$SS_1 = \Sigma(y_i - \hat{y}_i)^2$	$n-2$	$\Sigma(y_i - \hat{y}_i)^2/(n-2) = S_{Y|x}^2$
Total	$SS_0 = \Sigma(y_i - \bar{y})^2$	$n-1$	$\Sigma(y_i - \bar{y})^2/(n-1) = S_Y^2$

under investigation is nonlinear, the residual sum of squares measures simultaneously random fluctuation and the nonlinear influence. The contribution to the residual sum of squares that comes from applying a linear model to nonlinear data is called "wrong-model" bias. That is, $\Sigma(y_i - \hat{y}_2)^2$ measures both random variables plus "wrong-model" bias. For data where $E(Y|x_i)$ is a linear function of x (no wrong-model bias) then, as already noted, the residual sum of squares divided by $n-2$ ($S_{Y|x}^2$; expression 2.7) estimates the background variation of the variable Y ($\sigma_{Y|x}^2$).

Regression Sum of Squares

The regression sum of squares, $\Sigma(\hat{y}_i - \bar{y})^2$, is directly related to the estimated slope $\hat{b}$ because

$$\Sigma(\hat{y}_i - \bar{y})^2 = \Sigma([\hat{a} + \hat{b}x_i] - \bar{y})^2 = \Sigma([\bar{y} - \hat{b}\bar{x} + \hat{b}x_i] - \bar{y})^2 = \hat{b}^2 \Sigma(x_i - \bar{x})^2 . \quad (2.12)$$

The regression sum of squares is zero only when the estimated regression line is exactly horizontal ($\hat{b} = 0$); otherwise the regression sum of squares is directly proportional to the magnitude of the squared value of the estimated slope of the least squares estimated line. Therefore, the regression sum of squares indicates the magnitude of the linear association between the independent variable x and the dependent variable Y. In the simplest terms, the residual sum of squares measures the lack of fit and the regression sum of squares measures the fit of the estimated values to the data.

Partitioning the total sum of squares is a specific application of an analysis of variance. The analysis of variance associated with a simple linear regression model generates the summary statistics given in Table 2.1. The mean square column contains the estimated variances $S_{Y|x}^2$ and S_Y^2, showing the change in variability from including the variable x as part of the description of Y; thus the name analysis of variance. The components of the analysis of variance are illustrated in Figure 2.3.

F-Test

The analysis of variance sums of squares make it possible to compare the gains from using a straight line with a nonzero slope compared to a horizontal line ($b = 0$) as a

FIGURE **2.3**

The partitioning of $y_i - \bar{y}$ into pieces $y_i - \hat{y}_i$ and $\hat{y}_i - \bar{y}$

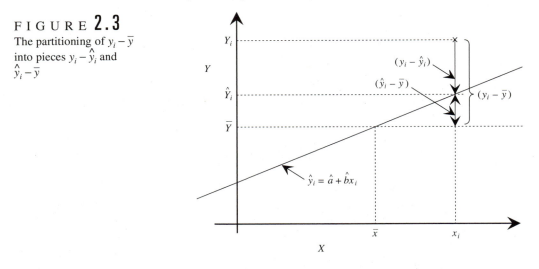

summary of the n data points. Stated more formally, the hypothesis that x is linearly related to Y is tested against the hypothesis that no linear relationship exists between x and Y. The analysis of variance table (Table 2.1) contains the components necessary to assess statistically the worth of a regression equation as a tool for describing the data set.

The residual sum of squares (denoted SS_1) divided by $\sigma^2_{Y|x}$ has a chi-square distribution (see Box 2.3) with $n - 2$ degrees of freedom and measures the fit of the data to a linear model with a nonzero slope, or

$$H_1: \text{ if } E(Y | x_i) = a + bx_i, \text{ then } \frac{SS_1}{\sigma^2_{Y|x}} = \frac{1}{\sigma^2_{Y|x}} \Sigma(y_i - \hat{y}_i)^2$$

has a chi-square distribution with $n - 2$ degrees of freedom for any value of b. The least squares estimate of the mean value at each x_i corresponding to each value of y_i is $\hat{y}_i = \hat{a} + \hat{b}x_i$.

When $b = 0$, the total sum of squares (denoted SS_0) divided by $\sigma^2_{Y|x}$ also has a chi-square distribution and measures the fit of the data to a horizontal line, or

$$H_0: \text{ if } E(Y | x_i) = a, \text{ then } \frac{SS_0}{\sigma^2_{Y|x}} = \frac{1}{\sigma^2_{Y|x}} \Sigma(y_i - \bar{y})^2$$

has a chi-square distribution with $n - 1$ degrees of freedom. The least squares estimate of the mean value corresponding to each x_i is $\hat{y}_i = \hat{a} = \bar{y}$, a horizontal line.

The degrees of freedom generated by H_1 and H_0 are, respectively, $n - 2$ and $n - 1$. In the first case, the mean values of Y are assumed to be located on the regression line. The estimate of the mean of Y for any value x_i under conditions set by H_1 is

$\hat{y}_i = \hat{a} + \hat{b}x_i$, which requires two independent estimated quantities, namely $\hat{a}$ and $\hat{b}$. The degrees of freedom associated with SS_1 are, therefore, $n - 2$. Under the conditions set by H_0, the mean values are assumed to lie on a horizontal line estimated by a single value, $\hat{a} = \bar{y}$. The degrees of freedom associated with SS_0 are correspondingly $n - 1$.

Because H_0: $E(Y \mid x_i) = a$ $(b = 0)$ is a special case of H_1: $E(Y \mid x_i) = a + bx_i$, then an F-statistic formally measures the differences between the two nested hypotheses H_0 and H_1, or

---

BOX **2.3** CHI-SQUARE DISTRIBUTION OF THE TOTAL AND RESIDUAL SUMS OF SQUARES

When requirements 1–4 for a linear regression hold, the following is true:

Total Sum of Squares (SS_0)

y_i	has a normal distribution
$[y_i - E(Y \mid x_i)]/\sigma_{Y \mid x}$	has a normal distribution; mean = 0 and variance = 1.0
$[y_i - E(Y \mid x_i)]^2/\sigma^2_{Y \mid x}$	has a chi-square distribution with 1 degree of freedom
$\dfrac{1}{\sigma^2_{Y \mid x}} \Sigma[y_i - E(Y \mid x_i)]^2$	has a chi-square distribution with n degrees of freedom

when $y_1, y_2, \ldots, y_n$, are independent observations .

Then ,

$$\frac{SS_0}{\sigma^2_{Y \mid x}} = \frac{1}{\sigma^2_{Y \mid x}} \Sigma(y_i - \bar{y})^2$$

has a chi-square distribution with $n - 1$ degrees of freedom when $b = 0$.

Note: The mean value $E(Y \mid x_i)$ is estimated by $\bar{y}$ (a loss of 1 degree of freedom) when the mean values $E(Y \mid x_i)$ lie on the horizontal line estimated by $\bar{y}$.

Residual Sum of Squares (SS_1)
Similar to the total sum of squares, because

$$\frac{1}{\sigma^2_{Y \mid x}} \Sigma[y_i - E(Y \mid x_i)]^2$$

has a chi-square distribution with n degrees of freedom, then

$$\frac{SS_1}{\sigma^2_{Y \mid x}} = \frac{1}{\sigma^2_{Y \mid x}} \Sigma(y_i - \hat{y}_i)^2$$

has a chi-square distribution with $n - 2$ degrees of freedom when $b \neq 0$.

Note: The mean value $E(Y \mid x_i)$ is estimated by $\hat{y}_i$ (a loss of 2 degrees of freedom) when the mean values $E(Y \mid x_i)$ lie on a straight line with a nonzero slope.

$$F = \frac{(SS_0 - SS_1)/[(n-1)-(n-2)]}{SS_1/(n-2)}$$

$$= \frac{[\Sigma(y_i - \bar{y})^2 - \Sigma(y_i - \hat{y}_i)^2]/[(n-1)-(n-2)]}{\Sigma(y_i - \hat{y}_i)^2/(n-2)}$$

$$= \frac{\Sigma(\hat{y}_i - \bar{y})^2}{S_{Y|x}^2} = \frac{\text{mean square regression}}{\text{mean square residual}} \tag{2.13}$$

where F has an F-distribution with 1 and $n-2$ degrees of freedom when SS_0 and SS_1 differ only because of random variation (i.e., $b = 0$). Notice that the divisor $\sigma_{Y|x}^2$ of both the total sum of squares and the residual sum of squares cancels out of the numerator and denominator of the F-statistic.

The F-statistic is the ratio of the regression mean square to the residual mean square given in the analysis of variance table (Table 2.1) and is directly calculated from the observed values y_i. When the value of F is no larger than would be expected by chance variation, the inference is made that no evidence exists that a linear model based on x is worthwhile (i.e., $b = 0$). Conversely, if the value of F is large (unlikely to have occurred by chance variation), then a linear model based on x is potentially a useful description of the sampled observations (i.e., $b \neq 0$). As before, the analysis of variance F-test depends on the validity of the four basic requirements underlying linear regression (independence, normality, equal variance, and linearity).

It is important to note that the squared value of the t-statistic (expression 2.9) is identical to the F-statistic (expression 2.13). It is always true that the square of a two-tail t-statistic with k degrees of freedom is equal to an F-statistic with 1 and k degrees of freedom. This property can be verified by comparing T^2 for a level of $1 - \alpha/2$ from a t-table to the F-value from an F-table at the probability level $(1 - \alpha)$. For example, for $1 - \alpha/2 = 0.975$, $T = 2.228$ with 10 degrees of freedom; for $1 - \alpha = 0.95$, $F = 4.964$ with 1 and 10 degrees of freedom, but $T^2 = (2.228)^2 = 4.964 = F$.

Multiple Correlation Coefficient

A single indicator of the effectiveness of a linear model is the squared multiple correlation coefficient. One definition of the squared multiple correlation coefficient is

$$R^2 = \frac{\text{regression sum of squares}}{\text{total sum of squares}} = \frac{\Sigma(\hat{y}_i - \bar{y})^2}{\Sigma(y_i - \bar{y})^2} = \frac{SS_0 - SS_1}{SS_0}. \tag{2.14}$$

The value R^2 is the proportion of the total sum of squares accounted for by the regression sum of squares. The value $1 - R^2$, therefore, is the proportion of sum of squares not explained by the linear model. Like all proportions, R^2 is bounded between 0 and 1 (i.e., $0 \leq R^2 \leq 1$). For simple linear regression, R^2 is identical to the squared product-moment correlation coefficient r_{XY}^2 (expression 1.7, see Box 2.4). Therefore, the squared product-moment correlation can be interpreted as an expression of the proportion of the total variation of Y explainable by a simple linear model based on x.

Another useful expression for R^2 linking the total sum of squares to the residual sum of squares is

$$(1 - R^2)\Sigma(y_i - \bar{y})^2 = \Sigma(y_i - \hat{y}_i)^2 \quad \text{or} \quad (1 - R^2) SS_0 = SS_1 \tag{2.15}$$

(also see Box 2.5). This expression shows that the magnitude of R^2 is determined by the difference between using $\hat{y}_i$ rather than $\bar{y}$ to estimate the mean values of the sampled populations. In other words, the summary R^2 reflects the gain in fit (decrease in total sum of squares) achieved by using $\hat{y}_i$ instead of $\bar{y}$ as an estimate of the population mean values $E(Y \mid x_i)$.

BOX **2.4** RELATIONSHIP BETWEEN R^2 AND r^2

r_{XY} = Pearson product-moment correlation coefficient (expression 1.7)

$$= \frac{\Sigma(x_i - \bar{x})(y_i - \bar{y})}{\sqrt{\Sigma(x_i - \bar{x})^2 \Sigma(y_i - \bar{y})^2}}$$

and

$$R^2 = \frac{\Sigma(\hat{y}_i - \bar{y})^2}{\Sigma(y_i - \bar{y})^2} = \frac{\hat{b}^2 \Sigma(x_i - \bar{x})^2}{\Sigma(y_i - \bar{y})^2} = \left[\frac{\Sigma(x_i - \bar{x})(y_i - \bar{y})}{\Sigma(x_i - \bar{x})^2} \right]^2 \frac{\Sigma(x_i - \bar{x})^2}{\Sigma(y_i - \bar{y})^2}$$

$$= \frac{[\Sigma(x_i - \bar{x})(y_i - \bar{y})]^2}{\Sigma(x_i - \bar{x})^2 \Sigma(y_i - \bar{y})^2} = r_{XY}^2 .$$

BOX **2.5** R^2 REPRESENTS THE REDUCTION IN TOTAL VARIATION

The proportional reduction in the total sum of squares from using $\hat{y}_i$ instead of $\bar{y}$ is

$$\frac{\Sigma(y_i - \bar{y})^2 - \Sigma(y_i - \hat{y}_i)^2}{\Sigma(y_i - \bar{y})^2} = \frac{\Sigma(\hat{y}_i - \bar{y})^2}{\Sigma(y_i - \bar{y})^2} = R^2$$

Alternatively,

$$\Sigma(y_i - \bar{y})^2 = \Sigma(y_i - \hat{y}_i)^2 + \Sigma(\hat{y}_i - \bar{y})^2$$

$$1 = \frac{\Sigma(y_i - \hat{y}_i)^2}{\Sigma(y_i - \bar{y})^2} + R^2$$

$$1 - R^2 = \frac{\Sigma(y_i - \hat{y}_i)^2}{\Sigma(y_i - \bar{y})^2}$$

and

$$(1 - R^2)\Sigma(y_i - \bar{y})^2 = \Sigma(y_i - \hat{y}_i)^2 \quad \text{or} \quad (1 - R^2) SS_0 = SS_1 .$$

Prediction from a Regression Equation

Perhaps the most useful function of the estimated parameters $\hat{a}$ and $\hat{b}$ is the estimation of the dependent variable, $\hat{y}$. For example, estimated values are clearly instrumental to calculate the residual sum of squares. Values of the estimated dependent variables also are used in a variety of other contexts. An important application is the prediction of a dependent variable from a specific independent variable or

$$\text{given } x_0, \text{ then } \hat{y}_0 = \tilde{y}_0 = \hat{a} + \hat{b}x_0. \tag{2.16}$$

The estimated dependent variable at x_0 has two interpretations. For a particular x_0, $\hat{y}_0$ is an estimate of the mean of the population of values associated with x_0, or $\tilde{y}_0$ is the estimate of a specific observation selected from the population defined by x_0. The tilde ($\sim$) is a reminder that a specific observation is being predicted but functionally both estimates are identical (expression 2.16). For example, the estimate $\hat{y}_0$ could be an estimate of the mean level of concentration of CO for a group of smokers who smoke $x_0 = 20$ cigarettes per day, or $\tilde{y}_0$ could be an estimate of the level of CO in the lungs of a specific person who smokes $x_0 = 20$ cigarettes per day. For either interpretation, a statistical evaluation requires an estimate of the variance.

For estimating a mean value from a regression line, an estimate of variance of $\hat{y}_0$ is

$$\text{variance}(\hat{y}_0) = S_{Y_0}^2 = S_{Y|x}^2 \left[\frac{1}{n} + \frac{(x_0 - \bar{x})^2}{\Sigma(x_i - \bar{x})^2} \right]. \tag{2.17}$$

The quantity $S_{Y|x}^2$ is the estimated variance of the dependent variable discussed earlier (expression 2.7). Confusion can arise when it is noted that the magnitude of the variance of $\hat{y}_0$ depends on the value x_0. This property of $\hat{y}_0$, at first glance, seems to contradict the requirement that the variance of Y be constant for all values of x (requirement 3). The "equal-variance" requirement, however, applies to the dependent variable Y and not to a predicted mean value estimated by $\hat{y}_0$. It is natural that these two quantities have different variances because one is a single sample value and the other is an estimated value based on the entire data set ($S_{\hat{y}_0}^2 \leq S_{Y|x}^2$).

A $(1 - \alpha)\%$-confidence interval for a predicted mean value based on n observations and the estimate $\hat{y}_0$ is constructed in the usual way using the estimated variance (expression 2.17) and is

$$\text{lower bound} = \hat{y}_0 - t_{1-\alpha/2} \, S_{\hat{Y}_0} \quad \text{and} \quad \text{upper bound} = \hat{y}_0 + t_{1-\alpha/2} \, S_{\hat{Y}_0} \tag{2.18}$$

where $t_{1-\alpha/2}$ again represents the $(1 - \alpha/2)$-percentile of a T-distribution with $n - 2$ degrees of freedom.

Predicting a Single Value

When predicting a single observation (denoted $\tilde{y}_0$) rather than a mean (denoted $\hat{y}_0$), the estimated value is the same as $\hat{y}_0$ (i.e., $\hat{y}_0 = \tilde{y}_0$) but, not surprisingly, the variance is larger. The estimated variance of $\tilde{y}_0$ when a simple linear regression equation is used to predict a single value Y (expression 2.16) is

$$\text{variance}(\tilde{y}_0) = S_{\tilde{Y}_0}^2 = S_{Y|x}^2 \left[1 + \frac{1}{n} + \frac{(x_0 - \bar{x})^2}{\Sigma(x_i - \bar{x})^2} \right].$$

(2.19)

The confidence interval associated with the estimate $\tilde{y}_0$ has the same form as the confidence interval for the predicted mean but uses $S_{\tilde{Y}_0}^2$, or

$$\text{lower bound} = \tilde{y}_0 - t_{1-\alpha/2} S_{\tilde{Y}_0} \quad \text{and} \quad \text{upper bound} = \tilde{y}_0 + t_{1-\alpha/2} S_{\tilde{Y}_0}.$$

(2.20)

A $(1 - \alpha)$-confidence band, a band that has a probability of $1 - \alpha$ of containing the true regression line, is not constructed from a series of confidence intervals based on $\hat{y}_0$ (expression 2.18). The width of the geometric region at the point x_0 that has a probability of $1 - \alpha$ of containing the true regression line is given by

$$\text{lower bound} = \hat{y}_0 - \sqrt{2 F_{1-\alpha}} S_{\hat{Y}_0} \quad \text{and} \quad \text{upper bound} = \hat{y}_0 + \sqrt{2 F_{1-\alpha}} S_{\hat{Y}_0}$$

(2.21)

where $F_{1-\alpha}$ is the $(1 - \alpha)$-percentile of an F-distribution with 2 and $n - 2$ degrees of freedom. Using the estimated variance of the predicted mean value $\hat{y}_0$ $(S_{\hat{Y}_0}^2)$, a series of such bounds forms a $1 - \alpha$ confidence band. As before, $\hat{y}_0$ refers to a predicted mean value based on x_0 and $S_{\hat{Y}_0}^2$ is the estimated variance of $\hat{y}_0$ at that point. The distinction between a series of confidence intervals and a confidence band is lost in some elementary texts. The geometry of a confidence band is discussed more extensively elsewhere ([3] or [4]), and a simple illustration of these three types of confidence intervals is given in Figure 2.4.

The issue of the precision of the estimate $\hat{y}_0$ is relatively straightforward. More difficult issues concern the accuracy of the estimate. The accuracy of the estimated value depends on the accuracy of the linear model as a representation of the data. An accurate regression equation (R^2 high) produces potentially useful values, particularly within the range of the x-values. Outside the range of the independent variable neither the data nor the analysis address the question of linearity. Almost always, a linear model describes the data only over a limited range. Estimates from extrapolating outside the range of the data are probably not useful (likely subject to wrong-model bias), and without evidence of linearity such estimates, at best, are suspect. Interpolation of values within the range of the measured x-values are much less likely to be misleading and evidence from the analysis indicates the worth of the estimates $\hat{y}_0$ or $\tilde{y}_0$ (e.g., R^2, $S_{\hat{Y}_0}$, and confidence intervals).

Test for Linearity

The assumption of linearity (requirement 4) can be formally evaluated when several y-values (dependent variables) are observed for at least some of the x-values (independent variables). This pattern of data collection is illustrated in Figure 2.5. Note, as before:

$$\bar{y}_i = \Sigma_j y_{ij}/n_i, \; \bar{y} = \Sigma\Sigma y_{ij}/N, \text{ and } N = \Sigma n_i \; (i = 1, 2, \cdots, k, \text{ and } j = 1, 2, \cdots, n_i)$$

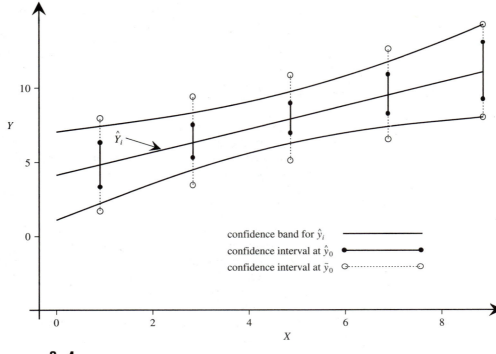

FIGURE **2.4**

A series of confidence intervals for values $\hat{y}_0$ and $\tilde{y}_0$ and a confidence band for the estimated line $\hat{y}_i$

where n_i represents the number of observations at each x_i and k represents the number of different values of x_i.

If data are sampled from a population with an underlying linear relationship, then $\hat{y}_i$ estimates the population mean of the dependent variable at each value x_i. The sample mean $\bar{y}_i$ estimates the population mean of the dependent variable at each value x_i regardless of the underlying relationship. Evidence of linearity, therefore, results from comparing $\hat{y}_i$ and $\bar{y}_i$. If all values $\hat{y}_i$ and $\bar{y}_i$ are similar, it supports the inference that the mean values fall on a straight line and, conversely, if one or more of the values $\hat{y}_i$ differ dramatically from $\bar{y}_i$, the linearity of the data becomes questionable. An example of a nonlinear relationship between x and Y (dashed line) is displayed in Figure 2.6, where $\hat{y}_i$ is likely to differ from $\bar{y}_i$ for at least some of the values x_i (x_4 particularly).

Contrasting Sums of Squares

Differences generated by the two ways of estimating the mean values are evaluated by contrasting two sums of squares. The within group sum of squares $\Sigma\Sigma(y_{ij} - \bar{y}_i)^2$ measures fit regardless of the relationship between x and Y because the mean values are estimated by $\bar{y}_i$. The residual sum of squares $\Sigma\Sigma(y_{ij} - \hat{y}_i)^2$ measures fit when the

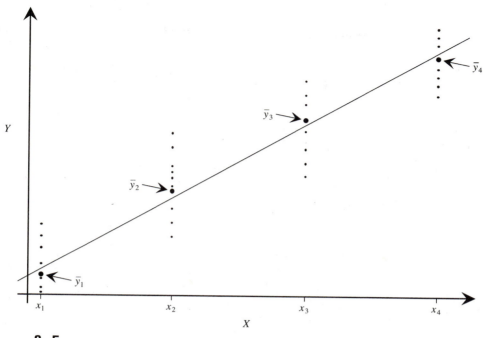

FIGURE 2.5

Several dependent variables Y for each independent variable x_i

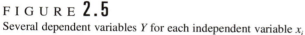

underlying relationship between x and Y is postulated to be linear because the mean values are estimated by $\hat{y}_i$. An F-test is typically used to formally assess differences between these two sums of squares, which should be similar if the data are accurately described by a straight line. In other words, little observed difference exists between $\bar{y}_i$ and $\hat{y}_i$.

The linear regression requirements of normality (requirement 2) and equal variance (requirement 3) give

$$z_{ij} = \frac{y_{ij} - E(Y \mid x_i)}{\sigma_{Y \mid x}} .$$

A standard normal distribution; then,

$$z_{ij}^2 = \frac{[y_{ij} - E(Y \mid x_i)]^2}{\sigma_{Y \mid x}^2} ,$$

has a chi-square distribution with 1 degree of freedom. Because all observations y_{ij} are independent (requirement 1), then

$$\Sigma\Sigma\, z_{ij}^2 = \frac{\Sigma\Sigma[y_{ij} - E(Y \mid x_i)]^2}{\sigma_{Y \mid x}^2}$$

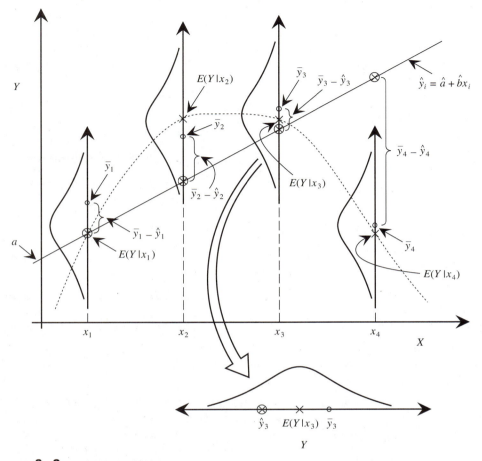

F I G U R E **2.6**

An example of a nonlinear relationship between x and Y (dotted line)

has a chi-square distribution with N degrees of freedom where $N = \Sigma n_i$ is the total number of observations.

When no assumption is made about the relationship between x and Y, the k unknown population mean values are estimated by $\bar{y}_1, \bar{y}_2, \cdots, \bar{y}_k$ at a loss of k degrees of freedom, giving

$$\frac{SS_1}{\sigma_{Y|x}^2} = \frac{1}{\sigma_{Y|x}^2} \Sigma\Sigma (y_{ij} - \bar{y}_i)^2 \tag{2.22}$$

a chi-square distribution with $N - k$ degrees of freedom. That is, $\bar{y}_i$ estimates $E(Y \mid x_i)$.

When x is linearly related to Y (i.e., $E(Y \mid x_i) = a + bx_i$), the k unknown population means are estimated by $\hat{y}_i = \hat{a} + \hat{b}x_i$ at a loss of 2 degrees of freedom; then

$$\frac{SS_0}{\sigma_{Y|x}^2} = \frac{1}{\sigma_{Y|x}^2} \Sigma\Sigma (y_{ij} - \hat{y}_i)^2 \tag{2.23}$$

has a chi-square distribution with $N - 2$ degrees of freedom. That is, $\hat{y}_i$ estimates $E(Y \mid x_i)$.

An F-test helps evaluate the difference between these two measures of fit where one hypothesis allows any specific relationship between x and Y (SS_1) and the other nested hypothesis postulates a linear relationship between x and Y (SS_0). When $E(Y \mid x_i) = a + bx_i$, then

$$F = \frac{(SS_0 - SS_1)/(k - 2)}{SS_1/(N - k)} \tag{2.24}$$

has an F-distribution with $k - 2$ and $N - k$ degrees of freedom. A large F-statistic leads to the inference that the data are not accurately described by a straight line.

If the hypothesis of linearity is accepted, then typically the analysis proceeds as though the relationship between x and Y is linear. Remember, a small value of F does not "prove" x is linearly related to Y; it simply implies that no strong evidence exists to the contrary. There is no difference in principle between estimating a least squares regression equation with several y-values per x-value and a linear regression with one y-value per x-value. The data are simply treated as N pairs of values (x_i, y_{ij}) ignoring the fact that more than one y-value is measured for some of the x-values and, as before, the ratio of the mean square regression to the mean square residual (expression 2.13) has an F-distribution with 1 and $N - 2$ degrees of freedom when $b = 0$. Accurate results, of course, depend on the requirements for linear regression being at least approximately fulfilled by the sampled data.

Example: Exercise and Weight Loss

Twenty athletes of about equal height and weight were divided randomly into four groups of five individuals. Each group participated in a special exercise program for different lengths of time (x_i = 15, 30, 45, and 60 minutes). The loss in body weight (Y = ounces) is given in Table 2.2.

From the data, $k = 4$, $n_i = 5$, and $N = 20$; therefore $\Sigma\Sigma y_{ij} = 443$ with $\bar{y} = 22.15$ and

$$\Sigma\Sigma (y_{ij} - \bar{y})^2 = 2586.55, \ \Sigma\Sigma (x_{ij} - \bar{x})(y_{ij} - \bar{y}) = 3712.5, \text{ and } \Sigma\Sigma (x_{ij} - \bar{x})^2 = 5625.0,$$

yielding estimates $\hat{a} = -2.60$ and $\hat{b} = 0.660$. (Note: $\hat{a} = -2.60$ is not physically possible, indicating that a regression line is only meaningful over a limited range.) The estimated mean values for each group are

$$H_1: \text{weight loss is not necessarily linear}$$
$$\bar{y}_1 = 6.6, \quad \bar{y}_2 = 18.0, \quad \bar{y}_3 = 27.6, \quad \bar{y}_4 = 36.4,$$

or

$$H_0: \text{weight loss is linear}$$
$$\hat{y}_1 = 7.3, \quad \hat{y}_2 = 17.2, \quad \hat{y}_3 = 27.1, \quad \hat{y}_4 = 37.0.$$

T A B L E   **2.2**

Weight loss for $k = 4$ groups each with $n_i = 5$ observations from an exercise experiment

	Group 1 (15 minutes)		Group 2 (30 minutes)		Group 3 (45 minutes)		Group 4 (60 minutes)	
	x	y	x	y	x	y	x	y
	15	4	30	14	45	31	60	38
	15	8	30	21	45	27	60	32
	15	6	30	20	45	26	60	40
	15	10	30	16	45	24	60	35
	15	5	30	19	45	30	60	37
x_i	15		30		45		60	
$\bar{y}_i$	6.6		18.0		27.6		36.4	
$\sum_j (y_{ij} - \bar{y}_i)^2$	23.2		34.0		33.2		37.2	

The sums of squares that allow evaluation of the difference between these two sets of estimated means (test of linearity) are

$$SS_1 = \Sigma\Sigma \, (y_{ij} - \bar{y}_i)^2 = 127.600, \text{ and } SS_0 = \Sigma\Sigma \, (y_{ij} - \hat{y}_i)^2 = 136.300 \, .$$

The test for linearity (i.e., the mean values of weight loss fall on a straight line) is

H_0: Loss of body weight is linearly related to length of exercise time—
$$E(Y \mid x_i) = a + bx_i \, .$$

The test for a linear loss of weight associated with exercise time gives

$$F = \frac{(136.300 - 127.600)/2}{127.600/16} = \frac{4.350}{7.975} = 0.545$$

yielding a p-value of 0.590 (degrees of freedom are $k - 2 = 2$ and $N - k = 16$). The test of a linear association between weight loss and exercise time is

H_0^*: Loss of body weight is not related to time of exercise—$b = 0$.

The test statistic

$$F = \frac{(2586.550 - 136.300)/1}{136.300/18} = \frac{2450.250}{7.572} = 323.584$$

yields a p-value < 0.001 (degrees of freedom are 1 and $N - 2 = 18$). Therefore, the data show no evidence of nonlinearity and a strong positive linear relationship between loss of weight and exercise time (shown in Figure 2.7).

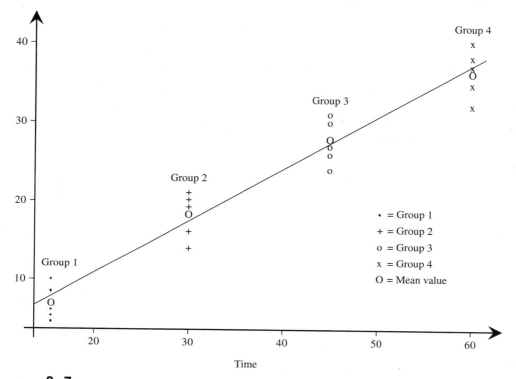

Data on weight loss from the exercise experiment ($N = 20$ and $k = 4$)

Estimated Regression Line with Intercept Zero

Situations arise in which the y-intercept is logically zero. An appropriate linear model is then $E(Y \mid x_i) = bx_i$. For example, in an anthropological study of the relationship between island size and population, an island of area zero necessarily has a population of zero. It is sometimes desirable in situations where the intercept must be zero to estimate a straight line with an intercept exactly zero. Such a line is estimated by least squares techniques and differs little in principle from the two-parameter model (expression 2.1). The intercept is, of course, $a = 0$ with the slope estimated by

$$\hat{b} = \frac{\Sigma x_i y_i}{\Sigma x_i^2},$$

(2.25)

giving the estimated line $\hat{y}_i = \hat{b} x_i$. The residual sum of squares is again $\Sigma(y_i - \hat{y}_i)^2$ where now $\hat{y}_i = \hat{b} x_i$.

An estimate of the variance of the estimate $\hat{b}$ is

$$S_{\hat{b}}^2 = \frac{\Sigma(y_i - \hat{y}_i)^2 / (n-1)}{\Sigma x_i^2}$$

(2.26)

allowing *t*-tests and confidence intervals to be constructed following the same pattern as the two-parameter regression analysis (expressions 2.9 and 2.10).

The estimate $\hat{b}$ is in terms of deviation from zero rather than from the mean values $\bar{x}$ and $\bar{y}$. This property is not surprising because the estimated line is forced to pass through zero and is, therefore, expressed in terms of deviations from the origin. Perhaps the only important difference between the two-parameter regression line and a regression line through zero is that the estimated line $\hat{y}_i = \hat{b}x_i$, in general, does not pass through the point $(\bar{x}, \bar{y})$. For simple linear regression, the least squares estimated line always passes through the point $(\bar{x}, \bar{y})$. When $x_i = \bar{x}$, then, $\hat{y}_i = \hat{a} + \hat{b}\bar{x} = \bar{y} - \hat{b}\bar{x} + \hat{b}\bar{x} = \bar{y}$.

An Alternative Estimate of the Regression Line

The presence of extreme values can disproportionately influence the least squares estimate of a slope $\hat{b}$. Even a single outlier observation can substantially change the estimated slope $\hat{b}$. It is worth briefly discussing an approach to estimating a linear regression equation designed to be relatively unaffected by extreme observations, called a resistant estimate.

To achieve resistant estimates of the intercept and slope, the data are divided into three groups based on the ordered values of the independent variable *x*. Each group contains approximately one-third of the data (e.g., if the total number of observations is $n = 3k$, then the leftmost group has *k* members, the middle group has *k* members, and the rightmost group has *k* members; if the number of observations is not divisible by three, then the observations are allocated as closely as possible to the ideal of $n/3$). Using these "thirds" a point is constructed based on the median of the *x*-values and the median of the *y*-values calculated separately from within each group. The pairs of median values

$$(x_L, y_L), \quad (x_M, y_M), \quad \text{and} \quad (x_R, y_R) \tag{2.27}$$

become representative values of the left, middle, and right groups, respectively.

Estimates of the slope and intercept are then

$$\hat{b}^* = \frac{y_R - y_L}{x_R - x_L} \quad \text{and} \quad \hat{a}^* = \frac{1}{3}(y_L + y_M + y_R) - \hat{b}^*\frac{1}{3}(x_L + x_M + x_R). \tag{2.28}$$

These two quantities produce an estimated regression equation of $\hat{y}_i = \hat{a}^* + \hat{b}^*x_i$. Because the estimate of the slope depends only on the median values from the left and right groups, it is relatively unaffected by extreme values. The estimated intercept $\hat{a}^*$ is similarly unaffected by outlier observations. The values $\hat{a}^*$ and $\hat{b}^*$ produce a resistant estimate of the regression equation. An alternative estimate of the regression equation is achieved by using $\hat{b}^*$ in conjunction with another estimate of the intercept, namely $\hat{a}' = \bar{y} - \hat{b}^*\bar{x}$ where $\bar{y}$ and $\bar{x}$ are the usual mean values. The estimate $\hat{a}'$ guarantees that the estimated regression line goes through the point $(\bar{x}, \bar{y})$ but is slightly less resistant to the influences of extreme values.

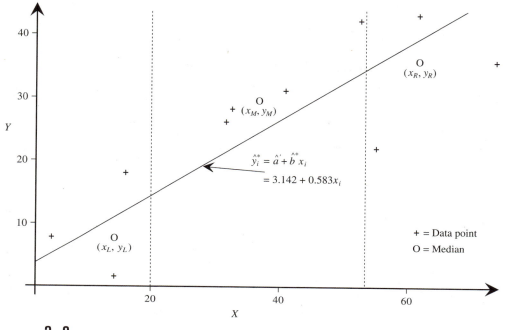

FIGURE **2.8**

Ten data points divided into three groups, with median of x and median of Y calculated for each group

Resistant estimates protect against disruptive influence of outlier values but do not make full use of the data. Least squares estimates use information from each data value but can be distorted by extreme values. The tradeoff between using a sensitive and efficient estimate versus an estimate that is less efficient but not highly influenced by extreme data is a typical tradeoff in many statistical situations. Other resistant estimates and refinements are described elsewhere (e.g., [5] and [6]).

A small data set illustrates:

Data	1	2	3	4	5	6	7	8	9	10
x	4	14	16	32	33	41	53	55	62	75
Y	8	2	18	26	28	31	42	22	43	36

where the data are ordered by the value of x (smallest to largest) and are displayed in Figure 2.8. The resistant estimate of the slope is $\hat{b}^* = (36 - 8)/(62 - 14) = 0.583$ because $x_R = 62$, $x_L = 14$, $y_R = 36$, and $y_L = 8$. The intercept is estimated by $\hat{a}' = \bar{y} - \hat{b}^* = 25.6 - 0.583(38.5) = 3.142$, giving a resistant estimated line $\hat{y}_i^* = 3.142 + 0.583x_i$. These resistant estimates are somewhat different from the least squares estimates of $\hat{b} = 0.489$ and $\hat{a} = 6.758$. It is not difficult, however, to create examples showing a considerable difference between these two approaches by including a few extremely nonlinear points.

Test Case

Computer Implementation

Simple Linear Regression

Example: Dependent variable Y = carbon monoxide levels measured in an individual's lungs and the independent variable x_i = the reported number of cigarettes smoked per day for $n = 10$ individuals. These data are a small subset of a large data set analyzed as part of the Multiple Risk Factor Intervention Trial [7]. These 10 observations are a small part of a set of 94 individuals who failed to quit smoking after an extensive smoking cessation program.

Data	1	2	3	4	5	6	7	8	9	10
x = cigs./day	12	30	40	12	2	5	20	30	75	40
Y = CO	2	11	14	16	18	18	18	22	23	26

	Mean	Standard Deviation	Covariance-Correlation
x	$\bar{x} = 26.6$	$S_X = 21.773$	$S_{XY} = 60.911$
Y	$\bar{y} = 16.8$	$S_Y = 6.795$	$r_{XY} = 0.412$

The least squares estimates are

$$\hat{b} = 60.911/474.044 = 0.128 \quad \text{and} \quad \hat{a} = 16.8 - (0.128)\,(26.6) = 13.382 \,,$$

giving an estimated regression equation of $\hat{y}_i = 13.382 + 0.128x_i$.

Analysis of Variance Table

Source	Sum of Squares	Degrees of Freedom	Mean Square
Regression	$\Sigma(\hat{y}_i - \bar{y})^2 = 70.440$	1	70.440
Residual	$\Sigma(y_i - \hat{y}_i)^2 = 345.160$	$n - 2 = 8$	43.145
Total	$\Sigma(y_i - \bar{y})^2 = 415.600$	$n - 1 = 9$	46.178

$$\text{Regression sum of squares} = \hat{b}^2\Sigma(x_i - \bar{x})^2 = (0.1285)^2(4266.4) = 70.440$$

$$\text{Total sum of squares} = 415.600 \text{ and residual} = \text{total} - \text{regression} = 345.160$$

Squared Multiple Correlation Coefficient

$$R^2 = 70.440/415.600 = 0.170$$

Also,

$$S_Y^2 = 46.178 \text{ and } S_{Y|x}^2 = 43.145$$

Ninety-five Percent Confidence Intervals for the Slope *b*

$$\text{lower bound} = 0.128 - 2.307(0.101) = -0.103 \text{ and}$$
$$\text{upper bound} = 0.128 + 2.307(0.101) = 0.360$$

where $t_{0.975} = 2.307$ with degrees of freedom = 8.

Test of Regression (*b* = 0)

$$F = \frac{(415.600 - 345.160)/(9 - 8)}{345.160/8} = \frac{70.440}{43.145} = 1.633$$

has an *F*-distribution with 1 and 8 degrees of freedom and yields a *p*-value = 0.237. To repeat, estimated regression coefficient $\hat{b} = 0.128$, and the variability of this estimate is measured by

$$S_{\hat{b}}^2 = \frac{43.145}{4266.4} = 0.0101$$

with $S_{\hat{b}} = 0.101$.

Note:

$$t\text{-test for } H_0: b = 0$$

$$t = \frac{0.1285 - 0}{\sqrt{0.0101}} = 1.278 \text{ and } (1.278)^2 = 1.633 = F.$$

Therefore, again the *p*-value = 0.237.

Simple Linear Regression

The computer implementation using the STATA system is illustrated for the small data set (*n* = 10) where the independent variable is reported levels of smoking (*x* = cigarettes per day) and the dependent variable is level of carbon monoxide found in a person's lungs (*y*; measured on parts per million, or ppm). The annotated STATA output follows.

```
example.regression

. infile co cigs using col.dat
(10 observations read)

. list
```

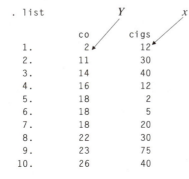

	co	cigs
1.	2	12
2.	11	30
3.	14	40
4.	16	12
5.	18	2
6.	18	5
7.	18	20
8.	22	30
9.	23	75
10.	26	40

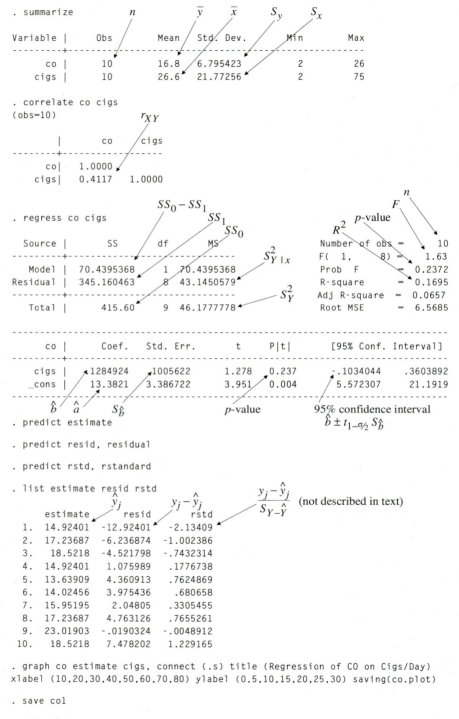

```
. summarize
                              n        ȳ       x̄       S_y      S_x

Variable |      Obs        Mean    Std. Dev.        Min        Max
---------+-----------------------------------------------------
      co |       10        16.8     6.795423          2         26
    cigs |       10        26.6     21.77256          2         75

. correlate co cigs
(obs=10)
                              r_XY

         |       co       cigs
---------+------------------
      co |   1.0000
    cigs |   0.4117     1.0000


                          SS_0 - SS_1
. regress co cigs                 SS_1
                                    SS_0                              n
                                                                F    p-value
  Source |       SS       df       MS         R²        Number of obs =      10
---------+------------------------------           S²_Y|x  F( 1,     8) =    1.63
   Model |  70.4395368     1   70.4395368                Prob > F      =  0.2372
Residual |  345.160463     8   43.1450579                R-square      =  0.1695
---------+------------------------------           S²_Y  Adj R-square  =  0.0657
   Total |     415.60      9   46.1777778                Root MSE      =  6.5685

----------------------------------------------------------------------
      co |     Coef.   Std. Err.       t      P|t|     [95% Conf. Interval]
---------+------------------------------------------------------------
    cigs |   .1284924   .1005622     1.278    0.237    -.1034044     .3603892
   _cons |   13.3821    3.386722     3.951    0.004     5.572307     21.1919
----------------------------------------------------------------------
        b̂     â      S_b̂              p-value        95% confidence interval
. predict estimate                                   b̂ ± t_{1-α/2} S_b̂

. predict resid, residual

. predict rstd, rstandard

. list estimate resid rstd
                      ŷ_j          y_j - ŷ_j        y_j - ŷ_j     (not described in text)
          estimate       resid        rstd          S_{Y-Ŷ}
    1.    14.92401   -12.92401    -2.13409
    2.    17.23687    -6.236874   -1.002386
    3.     18.5218    -4.521798    -.7432314
    4.    14.92401     1.075989    .1776738
    5.    13.63909     4.360913    .7624869
    6.    14.02456     3.975436    .680658
    7.    15.95195     2.04805     .3305455
    8.    17.23687     4.763126    .7655261
    9.    23.01903     -.0190324   -.0048912
   10.     18.5218     7.478202    1.229165

. graph co estimate cigs, connect (.s) title (Regression of CO on Cigs/Day)
xlabel (10,20,30,40,50,60,70,80) ylabel (0,5,10,15,20,25,30) saving(co.plot)

. save co1

. exit,clear
```

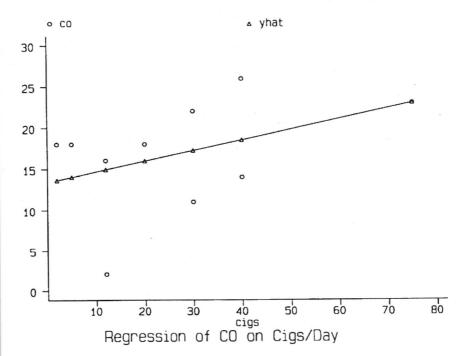

Regression of CO on Cigs/Day

Applied Example

Simple Linear Regression

Introduction

Data extracted from the Child Health and Development Study (CHDS) [8] are analyzed and interpreted at the end of each chapter with the goal of better understanding the application of each statistical technique.

The Child Health and Development Study was conducted on members of the Kaiser Foundation Health Plan—a prepaid medical care program—who reside in the San Francisco–East Bay area. All women who reported for prenatal care to three clinics of the Kaiser Hospitals in northern California constitute the study population. In all, there are 20,500 pregnancies and 18,600 live-born children in the study population. Participation in the study was nearly 100%.

The study families represent a broad range of economic, social, and educational characteristics. About two-thirds are white, one-fifth are black, 3% to 4% are Asian, and the remaining are members of other ethnic groups or interracial marriages. Some 30% of the husbands are in professional occupations. A large number are members of various unions. Nearly 10% are employed at the University of California, Berkeley, in academic and administrative positions, and about 20% are in government service. The educational and income levels of this group are somewhat higher than those of California as a whole. Thus, the study population is broadly based and is typical of an employed population. It is deficient in the poor and the affluent segments of the population because members of these groups are not frequent in this prepaid medical care program.

As part of the study, a lengthy prenatal interview was conducted with each female participant. Extensive information was gathered on a large number of pregnancy-related variables. During the interview, which generally took place early in pregnancy, great effort was made to determine as reliably as possible the date of the first day of her last normal menstrual period. From this date and the date of termination of pregnancy, an estimate of the length of gestation was derived.

The following analyses employ a subset of the Child Health and Development data. Live-born, white, male infants make up a study sample of 680 observations. Employing a subset of data homogeneous with respect to sex and race increases the sensitivity of the analyses by removing sources of variation. The practice of examining data separately for sex-race categories is a typical analytic strategy. Table 2.3 shows the 12 variables selected from each infant-mother-father set.

The parental observations in the following chapters serve as independent variables, and their relationship to the characteristics of the infants (principally birth weight) are assessed employing various statistical techniques (regression, discriminant, principal component, covariance, log-linear, and logistic analysis).

In the following STATA output summarizing the CHDS sample, a column is labeled "Std. Dev." The abbreviation "Std. Dev." refers to the estimate of the standard deviation of the population sampled (S_y) and should not be confused with the

T A B L E **2.3**

	Variable Code	Range	Mean	Standard Deviation
Infant Measurements				
Head circumference (inches)	HEADCIR	11–15	13.22	0.62
Birth weight (pounds)	BWT	3.3–11.4	7.52	1.09
Gestation (weeks)	GESTWKS	29–48	39.77	1.88
Maternal Measurements				
Age (years)	MAGE	15–42	25.86	5.46
Cigarettes (number smoked/day)	MNOCIG	0–50	7.43	11.27
Height (inches)	MHEIGHT	57–71	64.43	2.48
Pre-pregnancy weight (pounds)	MPPWT	85–246	126.90	17.88
Paternal Measurements				
Age (years)	FAGE	18–52	28.80	6.13
Father's education (years)	FEDYRS	6–16	13.38	2.20
Cigarettes (number smoked/day)	FNOCIG	0–50	14.44	14.17
Height (inches)	FHEIGHT	62–79	70.62	2.64

standard error of the mean, which measures the variability of the sample mean (i.e., $S_{\bar{y}} = \sqrt{S_Y^2/n}$).

Description of CHDS Data

The following STATA code is used to create a STATA file by means of a "save" option. Once a STATA data file is established, the "use" option retrieves it for analysis. For example, the following output shows a simple description ("summarize") of the 12 CHDS variables followed by the command "save chds." The CHDS data will now be maintained in an STATA data save-file named chds.dta. Most statistical analysis systems have a similar feature because reading and processing the original data file for each computer run is often costly and time-consuming.

chds.description

```
. infile id headcir length bwt gestwks mage mnocig mheight mppwt fage fedyrs
fn ocig fheight using  chds.dat
(680 observations read)

. summarize headcir length bwt gestwks mage mnocig mheight mppwt fage fedyrs
fn ocig fheight

Variable |     Obs        Mean   Std. Dev.        Min         Max
---------+-----------------------------------------------------------
 headcir |     680    13.21912    .6263148         11          15
  length |     680    20.27941    .9821018         17          23
     bwt |     680    7.516471    1.092346        3.3        11.4
 gestwks |     680    39.77059    1.875433         29          48
    mage |     680    25.85735    5.463382         15          42
  mnocig |     680    7.430882    11.27202          0          50
 mheight |     680    64.43382    2.483235         57          71
   mppwt |     680    126.8956    17.87766         85         246
    fage |     680        28.8    6.133133         18          52
  fedyrs |     680    13.37941    2.202593          6          16
  fnocig |     680    14.43824     14.1703          0          50
 fheight |     680    70.61912    2.638324         62          79

. rm chds.dta

. save chds
file chds.dta saved

. exit,clear
```

Simple Linear Regression

Two simple linear regression analyses are presented using the Child Health and Development data (computer output follows). The dependent variable in each case is birth weight (pounds; BWT) and the independent variables are length of gestation period (weeks; GESTWKS) and amount of maternal smoking (reported number of cigarettes smoked per day; MNOCIG).

T A B L E   **2.4**

Analysis of variance for birth weight and gestation

Source	Sum of Squares	Degrees of Freedom	Mean Square	F
Regression	146.931	1	146.931	150.195
Residual	663.265	678	0.978	
Total	810.196	679	1.193	

It is assumed that birth weight is normally distributed with equal variance at each value of the independent variable. Each pregnancy represents a separate mother-father-infant set, which assures statistical independence among observations. Genetic relationships among a few of the study participants are a minor source of non-independence that is ignored because any effects are inconsequential.

The percentage of variation in birth weight explainable by gestation employing a straight line is 18.1% (R^2 = {regression sum of squares}/{total sum of squares} = 146.9/810.2 = 0.181). The association between gestational age and infant birth weight is positive and significant (GESTWKS; $\hat{b} = 0.248$ with p-value < 0.001). Another quantitative expression of the same phenomenon is that gestational age is positively correlated with the infant's birth weight ($r_{XY} = 0.426$). The estimated slope $\hat{b} = 0.248$ implies that for an increase in one week of gestation, the estimated increase in infant's birth weight is 0.248 pounds, or about 4 ounces assuming that the relationship is adequately described by a straight line (a not very realistic assumption for the earlier weeks of pregnancy). Employing the estimated regression equation $\hat{y}_i = -2.348 + 0.248x_i$ yields the summary values given in the analysis of variance Table 2.4.

When the relationship between maternal smoking and birth weight is represented by a linear regression model, the percentage of total variation explained by postulating a simple linear relationship between mother's smoking exposure (MNOCIG) and her infant's birth weight is 3.2% ($R^2 = 0.032$). Summarizing the scatter of observations with a single line is, at best, moderately successful. Maternal smoking is negatively and significantly associated with birth weight (MNOCIG; $\hat{b} = -0.017$ and p-value < 0.001). Employing the estimated regression equation $\hat{y}_i = 7.646 - 0.017x_i$ yields the summary values given in the analysis of variance Table 2.5.

Because gestational time and maternal smoking are associated (women who smoke tend to have infants of shorter gestation), a more comprehensive analysis is desirable. The interrelationship between these two independent variables makes interpretation of two separate simple linear regression analyses unsatisfactory because neither analysis accounts for the possible influence of the other variable. Multiple linear regression analysis is one way to take into account and assess simultaneous influences of a number of interrelated independent variables on a dependent

T A B L E **2.5**

Analysis of variance for birth weight and smoking

Source	Sum of squares	Degrees of Freedom	Mean Square	F
Regression	26.078	1	26.078	22.548
Residual	784.118	678	1.157	
Total	810.196	679	1.193	

variable such as birth weight. The next chapter introduces a linear regression model with two independent variables as a way to examine both the separate and the joint influences of variables such as gestation and maternal smoking on birth weight.

Simple Linear Regression

The following STATA code and output use the previously saved CHDS data ("use chds") to illustrate two simple linear regression analyses: $Y =$ birth weight in pounds and $x =$ number of weeks of gestation or $x =$ the number of reported cigarettes smoked each day.

```
chds.regression

. use chds

. correlate bwt gestwks mnocig
(obs=680)

        |     bwt   gestwks   mnocig
--------+-----------------------------
    bwt|   1.0000
gestwks|   0.4259   1.0000
 mnocig|  -0.1794  -0.0708   1.0000

. regress bwt gestwks

   Source |      SS       df       MS              Number of obs =     680
----------+------------------------------           F(  1,   678) =  150.19
    Model |  146.930653     1   146.930653          Prob  F       =  0.0000
 Residual |  663.264893   678   .978266805          R-square      =  0.1814
----------+------------------------------           Adj R-square  =  0.1801
    Total |  810.195546   679   1.19321877          Root MSE      =  .98907

------------------------------------------------------------------------------
    bwt  |     Coef.   Std. Err.      t     P|t|      [95% Conf. Interval]
---------+--------------------------------------------------------------------
 gestwks |   .2480389   .0202391    12.255   0.000      .2083     .2877778
   _cons |  -2.348182   .805816     -2.914   0.004    -3.930377  -.7659874
------------------------------------------------------------------------------
```

```
. regress bwt mnocig

  Source |       SS       df       MS              Number of obs =     680
---------+------------------------------           F(  1,   678) =   22.55
   Model | 26.0775015      1  26.0775015           Prob > F      =  0.0000
Residual | 784.118045    678  1.15651629           R-square      =  0.0322
---------+------------------------------           Adj R-square  =  0.0308
   Total | 810.195546    679  1.19321877           Root MSE      =  1.0754

------------------------------------------------------------------------------
     bwt |      Coef.   Std. Err.       t     P|t|     [95% Conf. Interval]
---------+--------------------------------------------------------------------
  mnocig |  -.0173859   .0036613     -4.749   0.000    -.0245748   -.0101969
   _cons |   7.645663   .0494062    154.751   0.000     7.548655    7.742671
------------------------------------------------------------------------------

. graph bwt gestwks, saving(gest)

. graph bwt mnocig, saving(mnocig)

. exit
```

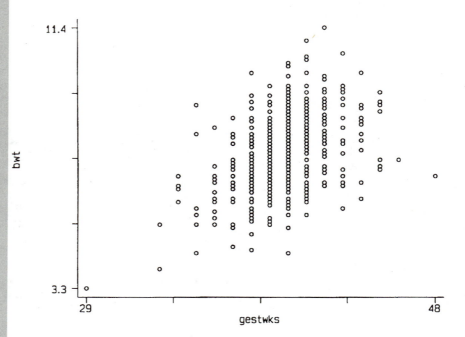

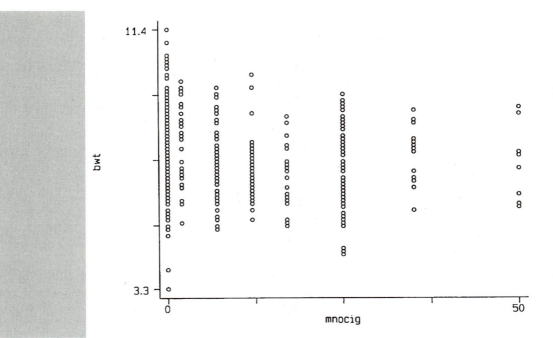

Notation

Linear model:

$$E(Y \mid x_i) = a + bx_i$$

Estimated linear model:

$$\hat{y}_i = \hat{a} + \hat{b}x_i$$

Estimated slope (regression coefficient):

$$\hat{b} = \frac{\Sigma(x_i - \bar{x})(y_i - \bar{y})}{\Sigma(x_i - \bar{x})^2}$$

Estimated intercept:

$$\hat{a} = \bar{y} - \hat{b}\bar{x}$$

Estimated variance of Y:

$$S_Y^2 = \frac{\Sigma(y_i - \bar{y})^2}{n - 1}$$

Estimated variance of Y accounting for x:

$$S_{Y\mid x}^2 = \frac{\Sigma(y_i - \hat{y}_i)^2}{n - 2}$$

Estimated variance of $\hat{b}$:

$$S_{\hat{b}}^2 = \frac{S_{Y|x}^2}{\Sigma(x_i - \bar{x})^2}$$

Analysis of variance summaries:

$$\text{Regression sum of squares} = SS_0 - SS_1 = \Sigma(\hat{y}_i - \bar{y})^2$$
$$\text{Residual sum of squares} = SS_1 = \Sigma(y_i - \hat{y}_i)^2$$
$$\text{Total sum of squares} = SS_0 = \Sigma(y_i - \bar{y})^2$$

Squared multiple correlation coefficient:

$$R^2 = \frac{\text{regression sum of squares}}{\text{total sum of squares}} = \frac{SS_0 - SS_1}{SS_0}$$

Problems

1 Using data from Table 1.4, let the dependent variable be infant birth weight (column 4) and the independent variable be the mother's weight gained during pregnancy ($gain = weight_1 - weight_0$). Apply a simple linear regression analysis to these data. Explain why the resulting squared multiple correlation coefficient is upward biased and the slope $\hat{b}$ is also biased.

2 Produce a simple linear regression analysis relating birth weight to weight gain using the data from Table 1.4 that gives an unbiased value of R^2 and slope b.

3 Using the data in Table 1.4, compute resistant estimates of the slope $\hat{b}^*$ and intercept $\hat{a}^*$ (or $\hat{a}'$) where again the independent variable Y = infant birth weight and x = maternal weight gain. Compare these results to the least squares estimates in problem 2.

4 Consider the following values: x: 2, 4, 6, and 8 and y: 26, 74, 154, and 266. Using a simple linear regression analysis, show that the squared multiple correlation coefficient is $R^2 = 0.969$. Note that $y = 10 + 4x^2$ fits the data perfectly.

5 Show that if the estimated slope $\hat{b} = 0$, then the correlation coefficient must also be $r_{XY} = 0$.

6 Show that $\Sigma(y_i - \hat{y}_i)^2 \le \Sigma(y_i - \bar{y})^2$ but $S_{Y|X}^2$ is not always less than S_Y^2.

7 Verify that $\hat{y} = \hat{a} + \hat{b}x_i = \bar{y} + \hat{b}(x_i - \bar{x})$. Because $S_{c_1X_1 + c_2X_2}^2 = c_1^2 S_{X_1}^2 + c_2^2 S_{X_2}^2$ when X_1 and X_2 are independent variables, show that

$$S_Y^2 = S_{Y|X}^2 \left[\frac{1}{n} + \frac{(x_i - \bar{x})^2}{\Sigma(x_i - \bar{x})^2} \right].$$

Note: $\bar{y}$ is independent of $\hat{b}$.

8 Show that $\Sigma(x_i - \bar{x}) = 0$ and $\Sigma(y_i - \hat{y}_i) = 0$.

9 Demonstrate that the residual sum of squares can be calculated without estimating the values $\hat{y}_i$ or, specifically,

$$\text{residual sum of squares} = \Sigma(y_i - \hat{y}_i)^2 = \Sigma(y_i - \bar{y})^2 - \hat{b}\Sigma(x_i - \bar{x})(y_i - \bar{y}).$$

Hint: Use expression 2.11.

10 Using the estimates in problem 2, predict the mean birth weight of infants of mothers who gain 12 kilograms of weight and predict the birth weight of a specific infant of a mother who gains 12 kilograms of weight. Construct a 95% confidence interval based on these two estimates. Construct and plot a 95% confidence band based on the estimated line in problem 2.

11 Estimate the slope and intercept for the following data using both least squares estimation and resistant estimates:

x	1	2	3	4	5	6	7	8	9	100
Y	12	25	33	38	55	64	71	85	101	1000

Make the same estimates from the following data:

x	1	2	3	4	5	6	7	8	9	50
Y	9	19	29	48	65	64	88	91	101	50

Both data sets have extreme observations (the last observation), but results are only highly affected in the second case. Explain why.

References

1 Lehmann, E. *Testing Statistical Hypotheses*. New York: John Wiley, 1966.

2 Snedecor, G. W. and Cochran, W. G. *Statistical Methods*, 8th ed. Ames: The Iowa State University Press, 1989.

3 Miller, R. G. *Simultaneous Statistical Inference*. New York: McGraw-Hill, 1966.

4 Draper, R. N., and Smith, H. *Applied Regression Analysis*, 2nd ed. New York: John Wiley, 1981.

5 Hoaglin, D. C., Mosteller, F., and Tukey, J. W. (editors). *Exploring Data Tables, Trends, and Shapes*. New York: John Wiley, 1985.

6 Hoaglin, D. C., Mosteller, F., and Tukey, J. W. (editors). *Understanding Robust and Exploratory Data Analysis*. New York: John Wiley, 1985.

7 Multiple Risk Factor Intervention Trial Study Group. The Multiple Risk Factor Intervention Trial (MRFIT): A National Study of Primary Prevention of Coronary Disease. *JAMA* 235 (1976): 815–827.

8 Yerushalmy, J. The California Child Health and Development Studies—Study Design, and Some Illustrative Findings on Congenital Heart Disease. In *Congenital Malformations, Proceedings of the Third International Conference* (pp. 299–306). International Congress Series No. 204. New York: Excerpta Medica, 1970.

3

Linear Regression with Two Independent Variables

Background

The complexities of most research issues require multivariable statistical models to begin to reflect the biological or physical situation. The next step in model complexity, after a simple linear model, is a linear model with two independent variables. In the study of the relationship between carbon monoxide (CO) and smoking, for example, an additional factor might be the time since the measured individual last smoked a cigarette. Recent smoking elevates levels of CO regardless of the number of cigarettes reported smoked per day. To effectively analyze the relationship between the dependent variable CO and the amount reported smoked, the time since a person last smoked must be taken into account (additional independent variable). The time since last smoked and the amount a person smokes are clearly related (correlation = −0.286; see the analysis of the CHDS data set described in Chapter 2). An effective analysis of the CO measurements, therefore, takes into account the joint and individual influences of these two variables. A linear model with two independent variables provides an opportunity to separate and study the individual effects on the dependent variable from each independent variable. For example, a measure of the influence of the number of cigarettes smoked per day on CO levels is described while accounting for recent smoking that would otherwise bias a direct assessment.

Although many techniques exist to describe relationships among a series of variables, an important analytic structure to study two independent variables and one dependent variable is again a linear model. As before, the success or failure of the approach depends on how well the necessarily simple mathematical form (a weighted sum) reflects the data. Most of the principles discussed in the next sections in the context of a bivariate linear model generalize to the multiple linear regression model, which is the topic of Chapter 4.

Method: Criterion for Estimation

A linear regression model with two independent variables is an extension of the simple linear regression model (one independent variable). The bivariate linear regression equation is

$$E(Y \mid x_1, x_2) = a + b_1 x_1 + b_2 x_2, \tag{3.1}$$

which geometrically defines a plane (Figure 3.1). Specifically, the expected values of the dependent variable lie on a plane with intercept represented by a and the slopes associated with the incline (or decline) of the two edges of the plane represented by b_1 (x_1-direction) and b_2 (x_2-direction). Analogous to the simple linear regression equation, the slopes b_1 and b_2 are called partial regression coefficients or sometimes just regression coefficients.

The regression coefficient b_i is the change in the dependent variable Y associated with a one-unit change in the independent variable x_i, whereas the other independent variable is held constant. In this sense a partial regression coefficient measures the expected influence on Y from each independent variable separately. The two coefficients b_1 and b_2, therefore, are assessments of the individual contributions of each independent variable to the observed response in Y with the overlapping influences of x_1 and x_2 statistically disentangled. For example, the influence on levels of CO from the number of cigarettes smoked (x_1), measured by a regression coefficient, is "free" from the obscuring effects of the time a person last smoked (x_2). Algebraically,

$$E(Y \mid x_1 + 1, x_2) - E(Y \mid x_1, x_2) = [a + b_1(x_1 + 1) + b_2 x_2] - [a + b_1 x_1 + b_2 x_2] = b_1,$$

which demonstrates that b_1 measures the influence of x_1 on Y for any constant value of x_2. The partial regression coefficient b_1 reflects the change in the dependent variable Y as though all individuals sampled have identical values of the dependent variable x_2. The coefficient b_2 has a similar interpretation—b_2 reflects the impact on Y as though all individuals compared have the same values of x_1. The ability to separate the influences of each x-variable depends on the properties of a linear model and, clearly, is useful only when such a model adequately represents the relationships within the sampled data.

Method of Least Squares

The model parameters a, b_1, and b_2 are usually estimated by the method of least squares. The criterion in the bivariate case

$$Q = \Sigma \, d_i^2 = \Sigma(y_j - [a + b_1 x_{1j} + b_2 x_{2j}])^2 \tag{3.2}$$

is minimized to yield estimates $\hat{a}$, $\hat{b}_1$, and $\hat{b}_2$ where d_i is the distance from a data point to the estimated plane (Figure 3.1).

To estimate these three parameters, the observations are collected in sets of three measurements (y_j, x_{1j}, x_{2j}) and the values a, b_1, and b_2 that minimize Q are the solution to the following normal equations (derived with calculus techniques [1]):

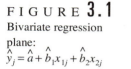

F I G U R E **3.1**
Bivariate regression plane:
$\hat{y}_j = \hat{a} + \hat{b}_1 x_{1j} + \hat{b}_2 x_{2j}$

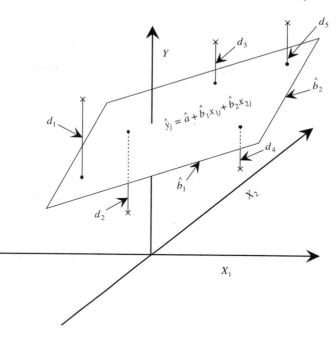

$$\hat{a}n + \hat{b}_1 \Sigma x_{1j} + \hat{b}_2 \Sigma x_{2j} = \Sigma y_j \qquad \text{or} \qquad \Sigma(y_j - \hat{y}_j) = 0$$

$$\hat{a}\Sigma x_{1j} + \hat{b}_1 \Sigma x_{1j}^2 + \hat{b}_2 \Sigma x_{1j}x_{2j} = \Sigma x_{1j}y_j \qquad \text{or} \qquad \Sigma x_{1j}(y_j - \hat{y}_j) = 0$$

$$\hat{a}\Sigma x_{2j} + \hat{b}_1 \Sigma x_{1j}x_{2j} + \hat{b}_2 \Sigma x_{2j}^2 = \Sigma x_{2j}y_j \qquad \text{or} \qquad \Sigma x_{2j}(y_j - \hat{y}_j) = 0 \qquad (3.3)$$

where $\hat{y}_j = \hat{a} + \hat{b}_1 x_{1j} + \hat{b}_2 x_{2j}$. Dividing both sides of the first normal equation by the sample size n and solving for the estimated intercept $\hat{a}$ yields $\hat{a} = \bar{y} - \hat{b}_1\bar{x}_1 - \hat{b}_2\bar{x}_2$. Substituting $(\bar{y} - \hat{b}_1\bar{x}_1 - \hat{b}_2\bar{x}_2)$ for $\hat{a}$ into the second and third normal equations and performing some manipulations yields two reduced normal equations, which are

$$\hat{b}_1 S_1^2 + \hat{b}_2 S_{12} = S_{Y1}$$

$$\hat{b}_1 S_{12} + \hat{b}_2 S_2^2 = S_{Y2}. \qquad (3.4)$$

The complete notation for the components of these expressions is given at the end of the chapter.

The solution to the reduced normal equations produces the least squares estimates of the regression coefficients as

$$\hat{b}_1 = \frac{S_2^2 S_{Y1} - S_{Y2}S_{12}}{S_1^2 S_2^2 - S_{12}^2} \quad \text{and} \quad \hat{b}_2 = \frac{S_1^2 S_{Y2} - S_{Y1}S_{12}}{S_1^2 S_2^2 - S_{12}^2} \qquad (3.5)$$

with $\hat{a} = \bar{y} - \hat{b}_1\bar{x}_1 - \hat{b}_2\bar{x}_2$. The three estimates define a plane minimum distance from n observed data points (y_j, x_{1j}, x_{2j}) where distance is measured parallel to the y-axis (Figure 3.1). That is,

$$Q_{\min} = \Sigma(y_j - [\hat{a} + \hat{b}_1 x_{1j} + \hat{b}_2 x_{2j}])^2 = \Sigma(y_j - \hat{y}_j)^2$$

is minimum. Using the estimates $\hat{a}$, $\hat{b}_1$, and $\hat{b}_2$ gives the "best" plane or the "best" estimated regression equation of

$$\hat{y}_j = \hat{a} + \hat{b}_1 x_{1j} + \hat{b}_2 x_{2j}\ . \tag{3.6}$$

As before, the word *best* simply means the sum of squares Q is minimum and will be larger for all other choices of a, b_1, or b_2. It does not mean the equation is automatically a useful description of the relationship of x_1 and x_2 to Y. The estimated expression is the best of its kind (linear) but its kind may be of little use. The question of usefulness of the estimated linear model is part of the regression analysis to be discussed.

Partitioning the Total Sum of Squares

The partitioned total sum of squares for a bivariate linear regression analysis has the same form as a simple linear regression analysis, which is

$$\Sigma(y_j - \bar{y})^2 = \Sigma(y_j - \hat{y}_j)^2 + \Sigma(\hat{y}_j - \bar{y})^2\ . \tag{3.7}$$

The residual and regression sums of squares have the same interpretation as the simple linear regression case. The basic difference is that the estimated dependent variable $\hat{y}_j = \hat{a} + \hat{b}_1 x_{1j} + \hat{b}_2 x_{2j}$ is made up of three estimates $(\hat{a}, \hat{b}_1,$ and $\hat{b}_2)$ rather than two $(\hat{a}, \hat{b})$. Similarly, the analysis of variance table differs only in the degrees of freedom column, as Table 3.1 illustrates.

As always, the variance of the dependent variable Y refers to a population of values. Although this population is rarely well described, it is envisioned that each observation y_j is a sample from a distribution defined by a specific pair of values x_1

T A B L E **3.1**

Analysis of variance for a bivariate regression

Source	Sum of Squares	Degrees of Freedom	Mean Squares
Regression	$SS_0 - SS_1 = \Sigma(\hat{y}_j - \bar{y})^2$	2	$\Sigma(\hat{y}_j - \bar{y})^2/2$
Residual	$SS_1 = \Sigma(y_j - \hat{y}_j)^2$	$n - 3$	$S^2_{Y|x_1, x_2} = \Sigma(y_j - \hat{y}_j)^2/(n - 3)$
Total	$SS_0 = \Sigma(y_j - \bar{y})^2$	$n - 1$	$S^2_Y = \Sigma(y_i - \bar{y})^2/(n - 1)$

and x_2 and that the variance of this distribution is denoted by $\sigma^2_{Y|x_1,x_2}$. An estimate of the variance of the dependent variable Y (i.e., an estimate of $\sigma^2_{Y|x_1,x_2}$) is the residual sum of squares divided by its degrees of freedom (residual mean square), or

$$S^2_{Y|x_1,x_2} = \frac{\Sigma(y_j - \hat{y}_j)^2}{n-3}.$$

(3.8)

The accuracy of the estimated variance, not surprisingly, depends on requirements underlying a linear regression analysis—particularly the requirements of linearity and equal variance. A linear relationship between x_1, x_2, and Y is necessary so that all deviations $(y_j - \hat{y}_j)$ strictly reflect the random nature of the dependent variable, sometimes called pure error. If the relationship under investigation is in reality nonlinear (not a plane), then the residual sum of squares is likely inflated, making $S^2_{Y|x_1,x_2}$ an overstatement of the actual variance that results from wrong-model bias. Furthermore, the variance of Y must be the same for all values of the dependent variables because $S^2_{Y|x_1,x_2}$ is an "average" of the observed squared deviations, which then produces an estimate of the common variance, namely $\sigma^2_{Y|x_1,x_2}$.

The squared multiple correlation coefficient, introduced in the context of simple linear regression (expression 2.14), has the same form and interpretation in the bivariate case: $R^2 = \{regression\ sum\ of\ squares\}/\{total\ sum\ of\ squares\}$. Once again, R^2 summarizes the fit achieved by including the independent variables x_1 and x_2 in the bivariate linear regression equation. The relationship between the multiple correlation coefficient and the pairwise product-moment correlation coefficients is more complex than the simple linear regression case and is discussed in a following section.

Statistical Tests of Estimated Regression Coefficients

Several statistical tests of the estimated regression coefficients arise in a bivariate linear regression analysis. Interest can be focused on the analytic worth of the entire model as a description of the variation of Y or the independent variables can be assessed individually ($\hat{b}_1$ as an assessment of the influence of x_1 or $\hat{b}_2$ as an assessment of the influence of x_2). The requirements continue to be necessary that the dependent variable Y is sampled independently, from one of a series of linearly related normal distributions with the same variance.

F-Test: Comparing Nested Models

To test whether the bivariate regression model (employing both x_1 and x_2) describes the data to an important extent, two nested models are compared. The first model postulates that the data vary randomly from the bivariate linear model, or

$$H_1: \ E(Y|x_1,x_2) = a + b_1x_1 + b_2x_2,$$

and a measure of fit of the estimated dependent variable based on x_1 and x_2 is the residual sum of squares given by

$$SS_1 = \Sigma(y_j - \hat{y}_j)^2 . \tag{3.9}$$

The quantity $SS_1/\sigma^2_{Y|x_1, x_2}$ has a chi-square distribution under H_1 with $n - 3$ degrees of freedom (loss of one degree of freedom for each least squares estimate, $\hat{a}$, $\hat{b}_1$, and $\hat{b}_2$).

A second model postulates that the data are no more than random fluctuations from a constant value, implying that x_1 and x_2 are not relevant to the description of the observations, or

$$H_0:\ E(Y \mid x_1 , x_2) = a \quad (b_1 = b_2 = 0) .$$

The constant value a is estimated by the mean of all n observations ($\hat{a} = \overline{y}$). The measure of fit of the dependent variable ignoring x_1 and x_2 is the sum of squares calculated under this more restrictive condition ($H_0: b_1 = b_2 = 0$), giving

$$SS_0 = \Sigma(y_j - \overline{y})^2 . \tag{3.10}$$

The quantity $SS_0/\sigma^2_{Y|x_1, x_2}$ has a chi-square distribution under H_0 with $n - 1$ degrees of freedom (loss of one degree of freedom for the least squares estimate of a by $\hat{a} = \overline{y}$). A formal comparison of these two sums of squares is

$$F = \frac{(SS_0 - SS_1)/[(n-1) - (n-3)]}{SS_1/(n-3)} = \frac{(SS_0 - SS_1)/2}{S^2_{Y|x_1, x_2}}$$

$$= \frac{\text{regression mean square}}{\text{residual mean square}} \tag{3.11}$$

where F has an F-distribution with 2 and $n - 3$ degrees of freedom when the difference $SS_0 - SS_1$ results only from the random nature of the data.

It is important to keep in mind that the increased flexibility of the more complicated model always takes advantage of chance relationships and produces a smaller residual sum of squares ($SS_0 > SS_1$), even when H_0 is true. Also note once again that the F-statistic is a ratio so that the value $\sigma^2_{Y|x_1, x_2}$ cancels out of the numerator and denominator, making it unnecessary to deal with the variance parameter. When F is small (greater values are judged probable), then the inference is made that no important gains are achieved by considering information from x_1 and x_2. When F is large (greater values are judged unlikely), then the inference is made that x_1 and/or x_2 are relevant to the description of Y. This F-test (expression 3.11) does not indicate the relative importance of x_1 and x_2. The test of hypothesis H_0 against H_1 is equivalent to testing the hypothesis that b_1 and b_2 are simultaneously equal to zero against the alternative that at least one coefficient is not zero.

Influences of Independent Variables

A similar approach is used to investigate the individual influence of an independent variable. The hypothesis H_1 remains unchanged but a different nested hypothesis is generated. The nested hypothesis postulates that the dependent variable is influenced by only one independent variable. Two possibilities exist:

$$H_0': \; E(Y \mid x_1, x_2) = a + b_1 x_1 \quad (b_2 = 0)$$

and

$$H_0'': \; E(Y \mid x_1, x_2) = a + b_2 x_2 \quad (b_1 = 0) \; .$$

Assume interest is focused on x_2 (the procedure is the same for investigating x_1). The hypothesis H_0' postulates that no influence on the dependent variable exists associated with the variable represented by x_2. A comparison of hypothesis H_0' to hypothesis H_1 measures exclusively the amount of response observed in Y attributable to the variable x_2, as x_1 is in both models. The comparison of H_0' to H_1 is equivalent to testing the hypothesis that $b_2 = 0$ while accounting for the influence of x_1. The estimated dependent variable disregarding x_2 produces the estimated linear model

$$\hat{y}_j' = \hat{a}' + \hat{b}_1' x_{1j} \tag{3.12}$$

where x_{1j} represents the jth value of the independent variable x_1 corresponding to the observed dependent variable y_j. The estimates $\hat{a}'$ and $\hat{b}_1'$ are generally different from the estimates $\hat{a}$ and $\hat{b}_1$ made from the bivariate model. The fact that estimated values depend on the model being considered has important consequences that are discussed (confounder bias—following section).

Fit of the Dependent Variable

The measure of the fit of the estimated dependent variable based on only x_1 (H_0') is the residual sum of squares

$$SS_0' = \Sigma(y_j - \hat{y}_j')^2 \tag{3.13}$$

where the value $SS_0'/\sigma^2_{Y \mid x_1 x_2}$ has a chi-square distribution with $n-2$ degrees of freedom when x_2 is unrelated to Y. The sum of squares SS_0' is exactly the residual sum of squares from a simple linear regression analysis, making the degrees of freedom again $n-2$. Because H_0' is a special case of H_1, a formal comparison of H_1 with H_0' is

$$F = \frac{(SS_0' - SS_1)/[(n-2)-(n-3)]}{SS_1/(n-3)} = \frac{(SS_0' - SS_1)/1}{S^2_{Y \mid x_1, x_2}} \tag{3.14}$$

where F has an F-distribution with 1 and $n-3$ degrees of freedom when SS_0' and SS_1 differ only because of random variation. As mentioned, the test, sometimes referred to as the F-to-remove test, is used to assess the possibility of removing the variable x_2 from the regression equation.

It is worth reemphasizing the general nature of an F-test:

The F-statistic is a ratio of two measures of variability, one generated under a "true" model (denominator) and one generated under a restriction of this "true" model (numerator). The denominator is a measure of the unavoidable intrinsic variation in the observed data. The numerator indicates the increase in variability caused by more stringent conditions. The ratio reflects the impact of the restricted conditions and the F-statistic provides a measure easily related to a probability (from tables or computer programs). Comparing changes in variability (analysis of variance) is a basic part of much of statistical analysis.

Variance of the Estimated Regression Coefficient

The estimated variance of the estimated regression coefficient $\hat{b}_i$ is given by

$$S_{\hat{b}_i}^2 = \frac{S_{Y|x_1,x_2}^2}{\sum_j (x_{ij} - \bar{x}_i)^2 (1 - r_{12}^2)} = \frac{S_{Y|x_1,x_2}^2}{(n-1)S_i^2(1 - r_{12}^2)} \text{ (justified elsewhere [1]) .}$$

(3.15)

Much like the simple linear regression case, the estimated variance $S_{Y|x_1,x_2}^2$ directly influences the estimated variance of $\hat{b}_i$. The denominator of the expression for the variance contains components that reflect the sample size $(n-1)$, the range of the x_i-values (S_i^2), and the degree of association between the two independent variables (r_{12}). Increases in either the sample size or the range again lead to increases in the precision of the estimate $\hat{b}_i$ ($S_{\hat{b}_i}^2$ decreases). The factor $(1 - r_{12}^2)$ shows that the association between the independent variables x_1 and x_2, measured by the squared correlation coefficient r_{12}^2, also influences the variance of the estimated regression coefficients. Increases in correlation (r_{12}) decrease the precision of the estimate ($S_{\hat{b}_i}^2$ increases). If x_1 and x_2 are highly correlated (r_{12}^2 near 1.0), then the estimated variance of the regression coefficients is substantially increased (multiplied by a factor of $1/(1 - r_{12}^2)$). Such highly correlated variables are said to be nearly collinear (perfectly collinear variables have a correlation of exactly 1.0). Conversely, the variability of $\hat{b}_i$ is relatively unaffected by uncorrelated independent variables ($r_{12}^2 = 0$) or weakly correlated independent variables (r_{12}^2 near 0).

Geometric Explanation

The reason for the increased variability in the estimated regression coefficient caused by correlated independent variables can be described geometrically. First, it should be pointed out that a regression coefficient b_i reflects the response in Y to an increase in the independent variable measured parallel to the x_i-axis. The coefficient $\hat{b}_1$, for example, is the amount of change in Y for a one-unit increase in x_1 measured parallel to the x_1-axis. A crucial element in the precision of the estimated regression coefficients $\hat{b}_1$ and $\hat{b}_2$ is the length and breadth (spread in x_1 and x_2 directions measured parallel to the coordinate axes) of the area covered by the sampled points (x_1, x_2). When x_1 and x_2 are not strongly correlated, the data tend to be spread in both x_1 and x_2 directions, allowing relatively precise estimates of the regression coefficients (Fig-

FIGURE **3.2**
Area: X_1 and X_2
independent

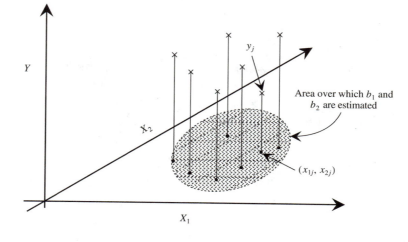

At least the relationship between x_1 and x_2 has minimal effects on the estimates of the regression coefficients.

For two correlated variables, the length and breadth of the area covered by the independent variables are reduced. As correlation increases, this area becomes smaller and begins to resemble a straight line with no spread. That is, given a specific value of one variable, the distance spanned by the other correlated variable is reduced. For example, the ages of married couples are strongly correlated ($r \approx 0.8$). Both variables have distributions that range over 60 years but when a wife's age is, say, 35, the age of the husband is not likely to differ much from age 37. It is the fact that slopes of the regression plane are estimated over this restricted span that causes

FIGURE **3.3**
Area: X_1 and X_2 correlated (not independent)

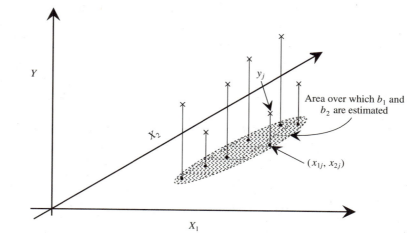

the decrease in the precision of the estimated regression coefficients (Figure 3.3). For a given wife's age (x_1), the restricted range available to estimate the impact of the husband's age (x_2) reduces the precision of the estimate of the slope $(\hat{b}_2)$. This decrease is algebraically related to $1/(1 - r_{12}^2)$, which translates into an increase in the estimated variance $(S_{\hat{b}_2}^2)$. Similarly, the precision of the estimate of b_1 is related to the span of the x_1 values measured parallel to the x_1-axis at each value of x_2 (again, reflected by r_{12}). The most extreme case occurs when x_1 and x_2 are perfectly correlated. Then, the values x_1 and x_2 occur on a single line (i.e., no spread), making it impossible to estimate the regression coefficients in either the x_1-direction or the x_2-direction because of perfect collinearity.

The following intuitive argument describes why a loss of precision is incurred from correlated x-variables:

■ A basic purpose of a bivariate linear regression model is to evaluate the separate influences of two variables on the dependent variable. The regression coefficients b_1 and b_2 measure the individual impact of variables x_1 and x_2, as noted. When x_1 and x_2 are highly correlated, the bivariate regression equation becomes nearly a "univariate" regression equation. That is,

$$E(Y \mid x_1, x_2) = a + b_1 x_1 + b_2 x_2$$

becomes approximately

$$E(Y \mid x_1, x_2) = a + (b_1 + b_2)x_1$$

because $r_{12} \approx 1$, meaning that x_1 accurately predicts x_2. Highly correlated variables produce a regression equation that contains two coefficients but information on essentially one variable. The lack of precision (high variability of $\hat{b}_1$ and $\hat{b}_2$) stems from the inability of the mathematical mechanism of least squares estimation to separate the relative influences of the two variables by estimating two regression coefficients. This inability to assign accurate values to b_1 and b_2 translates into an increase in variability. Ultimately, when two independent variables become perfectly collinear, it becomes impossible to assign meaningful estimates at all.

Regression Coefficient: Test and Confidence Interval

The estimated variance of a regression coefficient $(S_{\hat{b}}^2)$ is a key element in the evaluation of the estimate $\hat{b}_i$. Two typical approaches to assessing $\hat{b}_1$ or $\hat{b}_2$ while accounting for the impact of sampling variation are once again a t-test and a confidence interval. The form and interpretation of both approaches are the same as described for simple linear regression (Chapter 2). Estimation of the coefficient $\hat{b}_i$ and the variance of $\hat{b}_i$ $(S_{\hat{b}_i}^2)$ are a bit more complicated; nevertheless,

$$t\text{-}statistic = T = \frac{\hat{b}_i - b_0}{S_{\hat{b}_i}}$$

(3.16)

and T has a T-distribution with $n - 3$ degrees of freedom when $b_i = b_0$. Squaring the value T makes this t-test identical to the previous F-to-remove test of variable x_i, because $T^2 = F$ (expression 3.14). Additionally, a $(1 - \alpha)\%$-confidence interval is

$$\text{lower bound} = \hat{b}_i - t_{1-\alpha/2} S_{\hat{b}_i} \quad \text{and} \quad \text{upper bound} = \hat{b}_i + t_{1-\alpha/2} S_{\hat{b}_i} \qquad \text{(3.17)}$$

where the $t_{1-\alpha/2}$-percentile comes from a T-distribution with $n - 3$ degrees of freedom. The confidence intervals for b_1 and b_2 are usually related. The joint confidence region of the two parameters b_1 and b_2 involves an elliptical region about the point $(\hat{b}_1, \hat{b}_2)$ at the center. In general, an α-level joint confidence region cannot be constructed from the two individual confidence intervals (see [2] for a detailed discussion).

Partial Correlation Coefficient

The product-moment correlation coefficient quantifies the degree of linear association between two variables. A partial correlation coefficient arises from applying this correlation coefficient but after adjusting for the influence of a third variable. If x_1 and x_2 represent two variables related to a variable Y, then it is natural to measure the correlation between x_1 and Y while accounting for the influences of x_2. For example, a partial correlation measures the association between reported cigarettes smoked per day and CO levels adjusted for the time since an individual last smoked a cigarette. Similar to a partial regression coefficient, the partial correlation coefficient is designed to reflect the association between smoking and CO levels as though the data were collected from individuals with identical times since they last smoked.

The adjustment of x_1 and Y for the influence of x_2 is accomplished by simple linear regression techniques. An estimated value $\hat{x}_1$ is calculated using x_1 as a dependent variable and x_2 as an independent variable. Similarly, x_2 is used to predict Y, again with a simple linear regression equation. In symbols, $\hat{x}_1 = \hat{A} + \hat{B}x_2$ and $\hat{y} = \hat{a} + \hat{b}x_2$. Two "new" variables are created, namely the residual values $x_1' = x_1 - \hat{x}_1$ and $y' = y - \hat{y}$. These two residual variables are adjusted so the linear influence of x_2 is removed. Evidence that the residual values are free from the influence of x_2 is found in the fact that the correlation of x_2 with both x_1' and y' is zero. The partial correlation coefficient is the product-moment correlation applied to the constructed variables x_1' and y', or

$$\text{partial correlation} = \text{correlation}(x_1 - \hat{x}_1, y - \hat{y})$$

$$= \text{correlation}(x_1', y') = \frac{S_{x_1'y'}}{S_{x_1'} S_{y'}}. \qquad \text{(3.18)}$$

The correlation between residual values x_1' and y' quantifies the linear association between x_1 and Y with the influence of x_2 statistically removed.

A test case of 10 observations (Table 3.2) illustrates the calculation of a partial correlation coefficient. The variable Y represents the level of carbon monoxide found in the lungs of a smoker who reports smoking x_1 cigarettes per day and also reports x_2 hours since last smoked.

T A B L E   **3.2**

Illustration of a partial correlation coefficient (Y = CO levels, x_1 = cigarettes/day and x_2 = time since last smoked in hours)

	y	x_1	x_2	$\hat{x}_1$	$\hat{y}'$	$x_1' = x_1 - \hat{x}$	$y' = y - \hat{y}'$
1	2	12	5.00	24.015	15.426	−12.015	−13.426
2	11	30	6.09	16.167	11.254	13.833	−0.254
3	14	40	4.00	31.215	19.253	8.785	−5.253
4	16	12	5.33	21.639	14.163	−9.639	1.837
5	18	2	4.25	29.415	18.297	−27.415	−0.297
6	18	5	3.75	33.015	20.210	−28.015	−2.210
7	18	20	5.75	18.615	12.555	1.385	5.445
8	22	30	4.66	26.463	16.727	3.537	5.273
9	23	75	3.83	32.439	19.904	42.561	3.096
10	26	40	3.75	33.015	20.210	6.985	5.790

The following correlation coefficients are calculated in the usual manner (expression 1.7) from the values in Table 3.2:

$$r_{12} = -0.286, \quad r_{Y1} = 0.412, \quad r_{Y2} = -0.488 \; ;$$

$$r_{x_1' x_2} = 0, \quad r_{Y' x_2} = 0, \text{ and}$$

$$Y' r_{Y' x_1'} = 0.325$$

where $r_{Y' x_1'} = 0.325$ is the partial correlation between y and x_1 adjusted for the influence of x_2 (the correlation between x_1' and y'; columns 6 and 7 in Table 3.2). The directly measured correlation between x_1 and y (r_{Y1}) is 0.412, including the indirect the influence of x_2. The partial correlation coefficient ($r_{Y' x_1'}$) is 0.325. Not surprisingly, the correlation between CO and levels of reported smoking is reduced when the time since a person last smoked is taken into account. Values of x_2 and y could be analogously adjusted for the influence of x_1 and the partial correlation coefficient calculated (e.g., $r_{Y2} = -0.488$ and $r_{Y' x_2'} = -0.423$, from the data in Table 3.2). More generally, the same process produces a partial correlation coefficient between any two variables adjusted for the influence of a third variable.

Multiple Correlation Coefficient

Another view of the partial correlation comes from noting that the squared partial correlation coefficient ($r^2_{Y' x_1'}$) is related to that part of the squared multiple correlation coefficient (R^2) strictly attributable to x_1 after accounting for x_2. The squared multiple correlation coefficient can be represented as the sum of the squared correlation coefficient of y and x_2 (r^2_{Y2}) and the squared partial correlation coefficient of y and x_1 ($r^2_{Y' x_1'}$) multiplied by $1 - r^2_{Y2}$. Specifically,

$$R^2 = r_{Y2}^2 + (1 - r_{Y2}^2)\, r_{y'x_1'}^2 .$$

(3.19)

From the example data (Table 3.2), the squared multiple correlation coefficient is $R^2 = 0.318$, calculated directly from a bivariate regression equation relating independent variables x_1 and x_2 to Y. Then, the partition of $R^2 = 0.318 = (-0.488)^2 + (1 - (-0.488)^2)\,(0.325)^2 = 0.238 + 0.080$ shows the partial correlation viewed as a way of evaluating the relative roles of x_1 and x_2 in the makeup of the multiple correlation coefficient. The quantity 0.080 is an important measure of the impact of an independent variable called a part correlation, to be discussed in the next chapter. The squared multiple correlation coefficient is not simply related to a product-moment correlation, like the simple linear regression case; nevertheless, R^2 is a function of two components, each based on specific product-moment correlation coefficients (expression 3.19).

Often a partial correlation coefficient is expressed by the "shortcut" formula

$$r_{Yi.j} = \frac{r_{Yi} - r_{ij} r_{Yj}}{\sqrt{(1 - r_{Yj}^2)\,(1 - r_{ij}^2)}}$$

(3.20)

where $r_{Yi.j}$ represents the partial correlation coefficient measuring the association between variable x_i and Y after adjusting for the influence of variable x_j (i.e., $r_{Yi.j} = r_{Y'x'}$). The concept of partial correlation can be extended to estimate the correlation between two variables while adjusting for several variables but is not explored here [3].

Coefficients from a Bivariate Linear Regression Equation

Partial correlation coefficients are related to the coefficients from a linear regression equation. The exact relationship derived from the estimate $\hat{b}_i$ (expression 3.5) and the partial correlation coefficient (expression 3.20) is

$$\hat{b}_i = r_{Yi.j} \sqrt{\frac{S_Y^2\,(1 - r_{Yj}^2)}{S_i^2\,(1 - r_{ij}^2)}} .$$

(3.21)

Expression 3.21 shows that $\hat{b}_i = 0$ means $r_{Yi.j} = 0$ and vice versa. More important, expression 3.21 implies that the residual variables x_i' and y' relate to a bivariate regression coefficient. If a simple linear regression equation relating x_i' (independent variable) to Y' (dependent variable) is estimated, the estimated slope is identical to the estimated partial regression coefficient $\hat{b}_i$ calculated from a bivariate regression equation (expression 3.5). In other words, removing the linear influence of x_2 from x_1 and removing the linear influence of x_2 from Y and then estimating the slope b from a simple linear regression using the two residual values is the same as estimating the coefficient b_1 directly from a bivariate linear regression equation. To be specific, the estimated coefficient for the illustrative data (Table 3.2) $\hat{b}_1 = 0.093$ regardless of whether it is directly estimated from the bivariate regression equation (expression 3.5) or estimated using a simple linear regression equation employing the residual variables x_1' and Y' (columns 6 and 7, Table 3.2). The word *partial* describes both adjusted correlations and regression coefficients because both arise from essentially the same statistical process.

As noted, the regression coefficient $\hat{b}_1$ estimates the expected response in Y for a one-unit increase in x_1 while accounting for the influence of x_2. Viewing the coefficient $\hat{b}_1$ as resulting from a linear regression based on the residual values x_1' and Y' shows exactly what is meant by "accounting for the influence of x_2."

Confounder Bias

The absence of a specific variable from the regression model can bias the estimated values of the coefficients included in the model. This bias is referred to as confounder or confounding bias, and the bivariate case illustrates.

When the bivariate regression equation fails to account for x_2, the confounder bias is estimated by

$$\text{confounder bias} = \hat{b}_2 r_{12} S_2 / S_1 . \tag{3.22}$$

More specifically, if the regression coefficient b_1 is estimated from n pairs of data points (y, x_1) and b_1 is also estimated from the same n pairs but with the variable x_2 added (y, x_1, x_2), then the difference between the first and second estimates of b_1 is $\hat{b}_2 r_{12} S_2 / S_1$ (justified in Box 3.1).

B O X **3.1**   CONFOUNDER BIAS (ILLUSTRATION FROM THE BIVARIATE CASE)

The bivariate regression model is

$$E(Y \mid x_1, x_2) = a + b_1 x_1 + b_2 x_2 .$$

If x_2 is ignored, the regression equation is

$$E(Y \mid x_1) = a + b_1 x_1$$

and the estimate of b_1 based on Y and x_1 is

$$\hat{b}_1' = \frac{S_{Y1}}{S_1^2} .$$

When x_2 is included, the estimate of b_1 based on Y, x_1, and x_2 is

$$\hat{b}_1 = \frac{S_{Y1} S_2^2 - S_{Y2} S_{12}}{S_1^2 S_2^2 - S_{12}^2} .$$

Then, the difference is

$$\text{confounder bias} = \hat{b}_1' - \hat{b}_1 = \frac{\hat{b}_2 r_{12} S_2}{S_1} .$$

Expression 3.22 identifies two situations in which no confounder bias arises from omitting x_2 from the analysis. If the excluded variable x_2 is unrelated to the dependent variable Y (i.e., $\hat{b}_2 = 0$), no bias results. That is, the estimation of b_1 produces the same value whether or not x_2 is included in the analysis. The term *unrelated* means x_2 is unrelated to the dependent variable after adjustment for any influence from x_1 (i.e., $\hat{b}_2 = r_{y2.1} = 0$). A correlation of zero between Y and x_2 ($r_{y2} = 0$) will not ensure the absence of confounder bias because $\hat{b}_2$ ($r_{y2.1}$) is not necessarily zero when x_2 and Y are uncorrelated (see expression 3.20). Also, when the excluded variable x_2 is unrelated to the retained variable x_1 (i.e., $r_{12} = 0$), no confounder bias results. Otherwise, the failure to include variable x_2 in the regression analysis will bias the results inferred from the regression coefficient associated with x_1. The bias can be large or small and positive or negative depending on $\hat{b}_2$ and r_{12}. Although the expression for the confounder bias refers to the bivariate case, the same phenomenon is an issue in regression equations with any number of variables.

The time since an individual last smoked (x_2) is certainly related to the CO levels found in that person's lungs (i.e., $b_2 \neq 0$) and is also related to the number of cigarettes smoked per day ($r_{12} \neq 0$), so the analysis of the number of reported cigarettes smoked per day (x_1) should account for this influence. Time since a person last smoked is a confounding variable; therefore, the time variable is an important component of the model. Without a measure of the time since a person last smoked in the regression model, the impact on CO levels from number of cigarettes smoked is difficult to interpret because the regression coefficient associated with reported smoking exposure (b_1) reflects a mixture of influences from two variables (confounder bias). Specifically, the impact of the number of cigarettes smoked per day (x_1) is estimated by $\hat{b}_1' = 0.128$ when the variable time since last smoked (x_2) is not included in the regression analysis (test case in Chapter 2). The estimate 0.128 overstates the relationship between reported number of cigarettes smoked per day and CO because it also indirectly reflects an association with recent smoking. The same regression coefficient is substantially reduced to $\hat{b}_1 = 0.093$ when the time since last smoked is included in the analysis (test case, Chapter 3), adjusting $\hat{b}_1$ for the influence of x_2. The confounder bias is $\hat{b}_1' - \hat{b}_1 = 0.128 - 0.093 = 0.036$ or $bias = \hat{b}_2 r_{12} S_2 / S_1 = (-3.161)(-0.286)(0.866)/21.773 = 0.036$. In general, confounder bias is defined and measured as the difference in two estimates, where the two estimates differ because they are estimated from models containing differing sets of independent variables. Confounder bias is sometimes called incomplete-model bias.

Precision of Regression Coefficients

The precision of an estimated regression coefficient depends on the relationships to the other independent variables—those included as well as those excluded from the analysis—much like the previously described confounder bias. Even when no confounder bias exists, the decision as to whether or not to include an independent variable influences the estimated variance of the regression coefficients. Consider the estimated simple linear regression equation $\hat{y}_j = \hat{a}' + \hat{b}_1' x_{1j}$ where the estimated variance of $\hat{b}_1'$ (expression 2.13) is

$$\text{variance}(\hat{b}_1') = S_{\hat{b}_1'}^2 = \frac{S_{Y|x_1}^2}{(n-1)S_1^2} .$$

The estimated bivariate model with the variable x_2 added is $\hat{y}_j = a + \hat{b}_1 x_{1j} + \hat{b}_2 x_{2j}$ and the estimated variance of $\hat{b}_1$ (expression 3.15) becomes

$$\text{variance}(\hat{b}_1) = S_{\hat{b}_1}^2 = \frac{S_{Y|x_1,x_2}^2}{(n-1)S_1^2(1-r_{12}^2)} .$$

A bit of manipulation shows that the ratio of these two variances is approximately (sample size n large)

$$\text{efficiency} = v = \frac{\text{variance}(\hat{b}_1')}{\text{variance}(\hat{b}_1)} = \frac{S_{\hat{b}_1'}^2}{S_{\hat{b}_1}^2} \approx \frac{1-r_{12}^2}{1-r_{Y2.1}^2} . \tag{3.23}$$

The impact on precision of the estimate $\hat{b}_1$ from including or excluding x_2 from the regression analysis can be inferred from this "efficiency" ratio. Four relevant cases are:

1. $r_{12} = 0$ and $\hat{b}_2 = 0$ ($r_{Y2.1} = 0$)
2. $r_{12} = 0$ and $\hat{b}_2 \neq 0$ ($r_{Y2.1} \neq 0$)
3. $r_{12} \neq 0$ and $\hat{b}_2 = 0$ ($r_{Y2.1} = 0$)
4. $r_{12} \neq 0$ and $\hat{b}_2 \neq 0$ ($r_{Y2.1} \neq 0$).

When x_1 is not related to x_2 and x_2 is not related to Y (case 1), then x_2 can be excluded from the analysis and the estimate $\hat{b}_1$ and the variance of $\hat{b}_1$ are not affected. If the independent variables are unrelated and x_2 is related to Y after adjustment for x_1 (case 2), then the ratio v is greater than 1 making the *variance*$(\hat{b}_1')$ greater than *variance*$(\hat{b}_1)$ implying a gain in efficiency in the estimation of b_1 is achieved by including x_2 in the regression equation. Note that this gain is achieved despite the fact that the estimate of b_1 is free from confounder bias. If x_1 and x_2 are related but x_2 is unrelated to Y (case 3), then a loss of efficiency is incurred when x_2 is included in the regression analysis because v is then less than 1. That is, v less than 1 makes variance$(\hat{b}_1')$ less than variance$(\hat{b}_1)$. In this case, the estimate $\hat{b}_1$ is also free of confounder bias but it is advantageous to exclude x_2. When x_2 is related to x_1 and also related to the dependent variable Y (case 4), x_2 is usually retained in the model to avoid confounder bias regardless of the impact on the precision of the estimate of b_1.

Generally, if a strong direct association exists between Y and x_2, then the precision of b_1 is usually increased by including the variable x_2 in the analysis (decreases $S_{\hat{b}_1}$), whereas a strong correlation between x_2 and x_1 can have a detrimental influence on the precision of the estimate of $\hat{b}_1$ (increases $S_{\hat{b}_1}$). Therefore, like confounder bias, the variance of $\hat{b}_1$ depends on the properties of x_2 and, again, the question of whether x_2 has a substantial effect on estimation precision of b_1 depends on r_{12} and $\hat{b}_2$.

The last two sections describe confounder bias and precision issues in terms of the variable x_1. Parallel arguments and expressions can be developed where the influences on b_2 are considered from excluding or including x_1 in the regression equation. Furthermore, these same issues also apply to regression analyses with more than two independent variables (next chapter).

Test Case

Computer Implementation

Bivariate Regression Analysis

Data continued from Chapter 2 test case: carbon monoxide levels (Y) analyzed by cigarettes per day (x_1) and time since last cigarette smoked (x_2) for 10 individuals [4].

Data Number	1	2	3	4	5	6	7	8	9	10	Mean
Carbon monoxide = Y	2	11	14	16	18	18	18	22	23	26	16.8
Cigarettes/day = x_1	12	30	40	12	2	5	20	30	75	40	26.6
Time since last smoked = x_2 (in hrs. & fraction)	5.00	6.09	4.00	5.33	4.25	3.75	5.75	4.66	3.83	3.75	4.64

$$S_1^2 = \frac{\Sigma(x_{1j} - \bar{x}_1)^2}{n-1} = 474.044 \qquad S_1 = 21.773$$

$$S_2^2 = \frac{\Sigma(x_{2j} - \bar{x}_2)^2}{n-1} = 0.749 \qquad S_2 = 0.866$$

$$S_{12} = \frac{\Sigma(x_{1j} - \bar{x}_1)(x_{2j} - \bar{x}_2)}{n-1} = -5.394 \qquad r_{12} = -0.286$$

$$S_Y^2 = \frac{\Sigma(y_j - \bar{y})^2}{n-1} = 46.178 \qquad S_Y = 6.795$$

$$S_{Y1} = \frac{\Sigma(y_j - \bar{y})(x_{1j} - \bar{x}_1)}{n-1} = 60.911 \qquad r_{Y1} = 0.412$$

$$S_{Y2} = \frac{\Sigma(y_j - \bar{y})(x_{2j} - \bar{x}_2)}{n-1} = -2.868 \qquad r_{Y2} = -0.488$$

Normal Equations

$$\hat{b}_1 S_1^2 + \hat{b}_2 S_{12} = S_{Y1} \quad \text{or} \quad \hat{b}_1(474.044) + \hat{b}_2(-5.394) = 60.911$$

$$\hat{b}_1 S_{12} + \hat{b}_2 S_2^2 = S_{Y2} \quad \text{or} \quad \hat{b}_1(-5.394) + \hat{b}_2(0.749) = -2.868$$

Estimates of Regression Coefficients

$$\hat{b}_1 = \frac{S_{Y1}S_2^2 - S_{Y2}S_{12}}{S_1^2 S_2^2 - S_{12}^2} = 0.093$$

$$\hat{b}_2 = \frac{S_{Y2}S_1^2 - S_{Y1}S_{12}}{S_1^2 S_2^2 - S_{12}^2} = -3.161$$

$$\hat{a} = \bar{y} - \hat{b}_1\bar{x}_1 - \hat{b}_2\bar{x}_2 = 16.8 - (0.093)(26.6) - (-3.161)(4.64) = 29.011$$

Regression Equation

$$\hat{y}_j = \hat{a} + \hat{b}_1 x_{1j} + \hat{b}_2 x_{2j} = 29.011 + 0.093x_{1j} - 3.161x_{2j}$$

Multiple Correlation Coefficient

$$R^2 = 132.309/415.600 = 0.318$$
$$R = \sqrt{0.318} = 0.564$$

Predicted Values

	1	2	3	4	5	6	7	8	9	10	Mean
Predicted = $\hat{y}_j$	14.31	12.53	20.07	13.27	15.76	17.62	12.68	17.05	23.84	20.86	16.8
Observed = y_j	2	11	14	16	18	18	18	22	23	26	16.8
Residual = $y_j - \hat{y}_j$	−12.31	−1.53	−6.07	2.73	2.24	0.38	5.32	4.95	−0.84	5.14	0

$$S_{Y|x_1,x_2}^2 = \frac{\Sigma(y_j - \hat{y}_j)^2}{n-3} = 40.470$$

Analysis of Variance Table

	Sum of Squares	Degrees of Freedom	Mean Squares
Regression	$\Sigma(\hat{y}_j - \bar{y})^2 = 132.309$	$k = 2$	66.155
Residual	$\Sigma(y_j - \hat{y}_j)^2 = 283.291$	$n - k - 1 = 7$	40.470
Total	$\Sigma(y_j - \bar{y})^2 = 415.600$	$n - 1 = 9$	46.178

$$H_0: \ b_1 \text{ and } b_2 \text{ simultaneously} = 0$$

and

$$F = 66.155/40.470 = 1.635 \qquad F_{2,7}(5\%) = 4.74 \qquad p\text{-value} = 0.262$$

Tests of Individual Regression Coefficients

$$\text{Set } b_1 = 0$$

Analysis of variance table for $E(Y \mid x_2) = a_2 + b_2 x_2$:

	Sum of Squares	Degrees of Freedom	Mean Squares
Regression	98.781	1	98.781
Residual	316.819	8	39.602
Total	415.600	9	46.178

$H_0': b_1 = 0$

$$F_1 = \frac{(316.819 - 283.291)/(8-7)}{283.291/7} = \frac{33.528}{40.470} = 0.826, \quad p\text{–value} = 0.393$$

$$\text{Set } b_2 = 0$$

Analysis of variance table for $E(Y \mid x_1) = a_1 + b_1 x_1$:

	Sum of Squares	Degrees of Freedom	Mean Squares
Regression	70.440	1	70.440
Residual	345.160	8	43.145
Total	415.600	9	46.178

$H_0'': b_2 = 0$

$$F_2 = \frac{(345.160 - 283.291)/(8-7)}{283.291/7} = \frac{61.869}{40.470} = 1.530, \quad p\text{–value} = 0.256$$

Note: The two analysis of variance tables have the same form as the table produced in Chapter 2 for simple linear regression.

Variance and *t*-Test of Regression Coefficients

$$S_{\hat{b}_1}^2 = \frac{S_{Y|x_1,x_2}^2}{(n-1)S_1^2(1-r_{12}^2)} = \frac{40.470}{9(474.044)\,(1-[-0.286]^2)} = 0.0103$$

$$T = \frac{\hat{b}_1 - 0}{S_{\hat{b}_1}} = \frac{0.093}{\sqrt{0.0103}} = 0.910 \qquad \text{Note: } T^2 = (0.910)^2 = 0.826 = F_1$$

$$S_{\hat{b}_2}^2 = \frac{S_{Y|x_1,x_2}^2}{(n-1)S_2^2(1-r_{12}^2)} = \frac{40.470}{9(0.749)\,(1-[-0.286]^2)} = 6.538$$

$$T = \frac{\hat{b}_2 - 0}{S_{\hat{b}_2}} = \frac{-3.161}{\sqrt{6.538}} = -1.236 \qquad \text{Note: } T^2 = (-1.236)^2 = 1.530 = F_2\,.$$

Linear Regression: Two Independent Variables

The following annotated STATA code produces a bivariate regression analysis using the test data. Again, Y = carbon monoxide levels, x_1 = cigarettes smoked per day, and x_2 = time since a person last smoked. A STATA "two-way" plot of the variables Y, x_1, and x_2 is also included.

example.bivariate

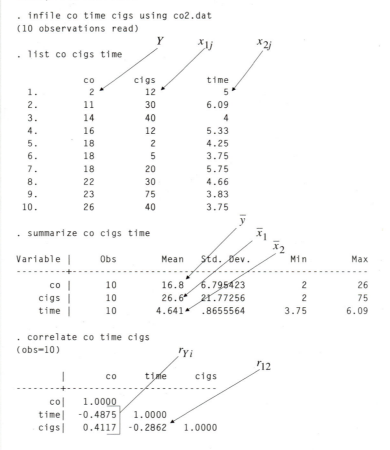

```
. infile co time cigs using co2.dat
(10 observations read)

. list co cigs time

            co      cigs      time
  1.         2        12         5
  2.        11        30      6.09
  3.        14        40         4
  4.        16        12      5.33
  5.        18         2      4.25
  6.        18         5      3.75
  7.        18        20      5.75
  8.        22        30      4.66
  9.        23        75      3.83
 10.        26        40      3.75

. summarize co cigs time

Variable |     Obs        Mean    Std. Dev.       Min        Max
---------+-----------------------------------------------------
      co |      10        16.8    6.795423          2         26
    cigs |      10        26.6    21.77256          2         75
    time |      10       4.641    .8655564       3.75       6.09

. correlate co time cigs
(obs=10)

         |      co       time      cigs
---------+---------------------------
      co|   1.0000
    time|  -0.4875     1.0000
    cigs|   0.4117    -0.2862     1.0000
```

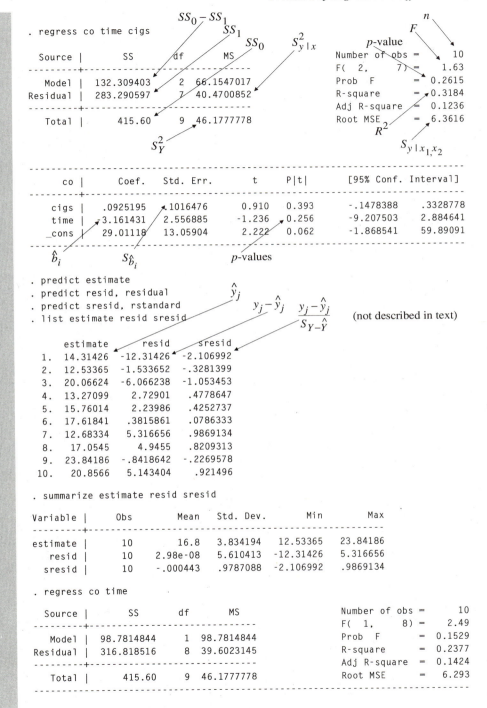

. regress co time cigs

Annotations: $SS_0 - SS_1$, SS_1, SS_0, $S^2_{y|x}$, p-value, n, F

Source	SS	df	MS			
Model	132.309403	2	66.1547017	Number of obs =	10
Residual	283.290597	7	40.4700852	F(2, 7) =	1.63
				Prob F =	0.2615
				R-square =	0.3184
				Adj R-square =	0.1236
Total	415.60	9	46.1777778	Root MSE =	6.3616

Annotations: S^2_Y, R^2, $S_{y|x_1,x_2}$

co	Coef.	Std. Err.	t	P|t|	[95% Conf. Interval]
cigs	.0925195	.1016476	0.910	0.393	-.1478388 .3328778
time	-3.161431	2.556885	-1.236	0.256	-9.207503 2.884641
_cons	29.01118	13.05904	2.222	0.062	-1.868541 59.89091

Annotations: $\hat{b}_i$, $S_{\hat{b}_i}$, p-values

. predict estimate
. predict resid, residual
. predict sresid, rstandard
. list estimate resid sresid

Annotations: $\hat{y}_j$, $y_j - \hat{y}_j$, $\dfrac{y_j - \hat{y}_j}{S_{Y-\hat{Y}}}$ (not described in text)

	estimate	resid	sresid
1.	14.31426	-12.31426	-2.106992
2.	12.53365	-1.533652	-.3281399
3.	20.06624	-6.066238	-1.053453
4.	13.27099	2.72901	.4778647
5.	15.76014	2.23986	.4252737
6.	17.61841	.3815861	.0786333
7.	12.68334	5.316656	.9869134
8.	17.0545	4.9455	.8209313
9.	23.84186	-.8418642	-.2269578
10.	20.8566	5.143404	.921496

. summarize estimate resid sresid

Variable	Obs	Mean	Std. Dev.	Min	Max
estimate	10	16.8	3.834194	12.53365	23.84186
resid	10	2.98e-08	5.610413	-12.31426	5.316656
sresid	10	-.000443	.9787088	-2.106992	.9869134

. regress co time

Source	SS	df	MS			
Model	98.7814844	1	98.7814844	Number of obs =	10
Residual	316.818516	8	39.6023145	F(1, 8) =	2.49
				Prob F =	0.1529
				R-square =	0.2377
				Adj R-square =	0.1424
Total	415.60	9	46.1777778	Root MSE =	6.293

```
      co |      Coef.   Std. Err.        t    P|t|      [95% Conf. Interval]
---------+----------------------------------------------------------------
    time |  -3.827553    2.423504    -1.579   0.153     -9.416162    1.761057
   _cons |   34.56367    11.42217     3.026   0.016      8.224092    60.90325
---------+----------------------------------------------------------------

. regress co cigs

  Source |       SS       df       MS                  Number of obs =     10
---------+----------------------------               F(  1,     8) =    1.63
   Model |  70.4395368    1   70.4395368              Prob  F       = 0.2372
Residual | 345.160463     8   43.1450579              R-square      = 0.1695
---------+----------------------------               Adj R-square  = 0.0657
   Total |     415.60     9   46.1777778              Root MSE      = 6.5685

      co |      Coef.   Std. Err.        t    P|t|      [95% Conf. Interval]
---------+----------------------------------------------------------------
    cigs |   .1284924    .1005622     1.278   0.237     -.1034044    .3603892
   _cons |    13.3821    3.386722     3.951   0.004      5.572307     21.1919
---------+----------------------------------------------------------------

. graph co time cigs, matrix label saving(example)

. save co2
file co2.dta saved

. exit,clear
```

Regression -- Gestation and smoking

Applied Example

Linear Regression with Two Independent Variables

Using a linear regression model with gestational age (GESTWKS) and maternal smoking (MNOCIG) simultaneously as independent variables shows 20.4% ($R^2 = 0.204$) of the total variation in birth weight (BWT) explained by a linear bivariate model. That is, the total variation (810.196) separates into two pieces with 20.4% associated with the regression sum of squares and the remaining 79.6% associated with the residual sum of squares. The regression sum of squares indicates the strength of the joint influence of the two independent variables ($F = 86.611$ with p-value < 0.001) on an infant's birth weight.

The separate effects of the independent variables are inferred from the regression coefficients, GESTWKS: $\hat{b}_1 = 0.242$ with p-value < 0.001 and MNOCIG: $\hat{b}_2 = -0.0145$ with p-value < 0.001. Each additional week of gestational age is expected to increase an infant's birth weight an estimated 0.242 pounds = 3.8 ounces. This increase represents the influence on birth weight from gestational time adjusted for the effects of maternal smoking. The unadjusted value is slightly larger (see previous test case in Chapter 2). The influence of gestation is not adjusted for other associated variables. Not suprisingly, the effects of the independent variables are now adjusted by a regression model for only those variables included in the linear equation.

Maternal smoking (MNOCIG) shows a negative influence on infant birth weight, which is again inferred from the regression coefficient. For a constant gestational age, smoking a pack of cigarettes per day (20 cigarettes) reduces an infant's birth weight, on the average, an estimated 20(0.0145) = 0.300 lbs. = 4.7 ounces. The estimated changes in birth weight from increases in weeks of gestation or numbers of cigarettes smoked are a direct application of the linear model and only meaningful to the extent that the model reflects the relationships within the sampled data.

The t-statistic serves as a useful way to compare the magnitude of the relative contributions to the variability in birth weight from the two independent variables free from the fact that they are measured in different units (weeks and cigarettes/day). A t-statistic is unitless, yielding a standardized form of the regression coefficients (i.e., $\hat{b}_i / S_{\hat{b}_i}$). Comparison shows that maternal smoking exposure has considerably less influence on birth weight than gestational age (GESTWKS: $t = 12.1$ and MNOCIG: $t = -4.4$). The analysis of variance summaries of the regression analysis accounting for the joint influences of gestational age and maternal smoking exposure on birth weight are shown in Table 3.3. The values displayed in Table 3.3 are based on the estimated regression equation $\hat{y}_j = -1.994 + 0.242x_{1j} - 0.0145x_{2j}$.

The influence of the confounder bias associated with these two independent variables is seen by making comparisons with the results from the corresponding simple linear regression analyses. For gestational age, $\hat{b}_1 = 0.242$ when maternal smoking is included in the analysis and $\hat{b}_1' = 0.248$ when maternal smoking is not included (Chapter 2). Similarly, $\hat{b}_2 = -0.0145$ when gestational age is included and $\hat{b}_2' = -0.0174$ when gestation is not included in the model (again, Chapter 2). Neither variable causes much confounder bias.

Bivariate regression analysis illustrates two important issues that apply to general k-variate regression analyses. The squared multiple correlation coefficient increases with each variable added to the regression equation ($R^2 = 0.181$,

T A B L E   **3.3**

Analysis of variance table—birth weight by gestation and maternal smoking

Source	Sum of Squares	Degrees of Freedom	Mean Squares	F
Regression	165.067	2	82.533	86.611
Residual	645.129	677	0.953	
Total	810.196	679	1.193	

GESTWKS only; $R^2 = 0.204$, GESTWKS + MNOCIG), and the meaning of the regression coefficients becomes better defined (less confounder bias) as other related variables are added to the analysis at, perhaps, a cost of decreased estimation precision. The influence of eight variables, recorded for both the mother and the father, on an infant's birth weight is explored using a k-variate regression equation in the next chapter.

Linear Regression: Two Independent Variables

Using the CHDS data ("use chds") and the STAT regression command ("regress") produces a bivariate regression analysis with dependent variable Y = birth weight (BWT) and two independent variables x_1 = gestational age (GESTWKS) and x_2 = number of reported cigarettes smoked per day (MNOCIG).

```
chds.bivariate

. use chds

. correlate bwt gestwks mnocig
(obs=680)

        |    bwt   gestwks   mnocig
--------+---------------------------
    bwt |  1.0000
gestwks |  0.4259   1.0000
 mnocig | -0.1794  -0.0708   1.0000


. regress bwt gestwks mnocig

   Source |      SS       df       MS                Number of obs =     680
----------+------------------------------           F(  2,   677) =   86.61
    Model | 165.066938      2  82.5334689           Prob  F       =  0.0000
 Residual | 645.128609    677  .952922612           R-square      =  0.2037
----------+------------------------------           Adj R-square  =  0.2014
    Total | 810.195546    679  1.19321877           Root MSE      = .97618
```

```
-----------------------------------------------------------------------------
    bwt |     Coef.   Std. Err.       t    P|t|     [95% Conf. Interval]
--------+--------------------------------------------------------------------
 gestwks |   .2418504   .0200256    12.077   0.000     .2025307    .2811701
  mnocig |  -.0145355   .0033318    -4.363   0.000    -.0210775   -.0079935
   _cons |  -1.994051   .7994411    -2.494   0.013    -3.563733   -.4243686
-----------------------------------------------------------------------------

. graph bwt gestwk mnocig, matrix label saving(bwt)

. exit
```

Notation

Summary values:

Total number of observations $= n$.

Variances and covariances:

$$S_1^2 = \frac{\Sigma(x_{1j} - \overline{x}_1)^2}{n-1} \qquad S_2^2 = \frac{\Sigma(x_{2j} - \overline{x}_2)^2}{n-1}$$

$$S_{12} = \frac{\Sigma(x_{1j} - \overline{x}_1)\,(x_{2j} - \overline{x}_2)}{n-1}$$

$$S_{Y1} = \frac{\Sigma(y_j - \overline{y})\,(x_{1j} - \overline{x}_1)}{n-1} \quad \text{and} \quad S_{Y2} = \frac{\Sigma(y_j - \overline{y})\,(x_{2j} - \overline{x}_2)}{n-1}$$

Bivariate regression equation:

$$E(Y \mid x_1, x_2) = a + b_1 x_1 + b_2 x_2$$

Estimated bivariate regression equation:

$$\hat{y}_j = \hat{a} + \hat{b}_1 x_{1j} + \hat{b}_2 x_{2j}$$

Estimated variance of dependent variable Y:

$$S_{Y \mid x_1, x_2}^2 = \frac{\Sigma(y_j - \hat{y}_j)^2}{n-3}$$

Partial correlation coefficient:

$$r_{Yi.j} = \frac{r_{Yi} - r_{Yj} r_{ij}}{\sqrt{(1 - r_{Yj}^2)\,(1 - r_{ij}^2)}}$$

F-test for contribution from b_i ("to remove b_i"):

$SS_1 = $ residual sum of squares for the bivariate model; degrees of freedom $= n - 3$

$SS_0' = $ residual sum of squares for the bivariate model with x_i removed;
    degrees of freedom $= n - 2$

$$F = \frac{(SS_0' - SS_1)/[(n-2) - (n-3)]}{SS_1/(n-3)} = \frac{SS_0' - SS_1}{S_{Y|x_1,x_2}^2}$$

and F has F-distribution with degrees of freedom = 1 and $n - 3$.

Problems

1   Conduct a bivariate regression analysis using the data in Table 1.4 where Y = birth weight, $x_1 = weight_0$ and $x_2 = weight_1 -$ birth weight. Compare these results with a simple linear regression analysis using the variable netgain (problem 2, Chapter 2).

2   The amount a mother gains during pregnancy is to some extent related to her size (big women tend to gain more weight than small women). Conduct a bivariate regression analysis where Y = birth weight, x_1 = netgain, and x_2 = height. Compute the confounder bias associated with the variable height.

3   Using the expression for $\hat{b}_1'$ (expression 2.5, simple linear regression) and the expression for $\hat{b}_1$ (expression 3.5, bivariate linear regression) show that

$$\hat{b}_1' - \hat{b}_1 = \frac{\hat{b}_2 r_{12} S_2}{S_1}.$$

4   Demonstrate that the correlation between the variables x_2 and $\hat{x}_1$ is exactly 1.0 ($r_{x_2\hat{x}_1} = 1$), where $\hat{x}_1 = \hat{a} + \hat{b}x_2$.

5   Compute the residual values x_1' and y' from the following data:

Y = score on a word recognition examination, x_1 = weight of a child on his fifth birthday (pounds), and x_2 = the child's height (inches) on his fifth birthday.

Obs.	Y	x_1	x_2
1	42	35	40
2	50	40	20
3	110	26	44
4	48	39	48
5	63	44	44
6	100	50	26
7	83	45	43
8	91	28	46
9	52	42	43
10	46	37	44
11	67	40	47
12	23	61	29

Compute the partial correlation coefficient between the score variable and the child's weight adjusted for the height of the child from the residuals y' and x_1'. Compute the partial correlation coefficient with the shortcut formula and compare the results to correlation(x_1', y'). Compute the coefficients from the bivariate model $a + b_1 x_1 +$

$b_1 x_2$. Compare $\hat{b}_1$ from this analysis to the coefficient computed from a regression analysis of the residuals ($y' =$ the dependent variable and $x_1' =$ the independent variable).

6 Conduct a bivariate regression analysis using the data in Table 1.4 where $Y =$ birth weight, $x_1 = weight_0$ and $x_2 = weight_1 -$ birth weight. Also conduct a regression analysis of Y without x_2; that is, $Y =$ birth weight, $x_1 = weight_0$. Compare and discuss the estimate of regression coefficient b_1 and the variance of this estimate.

7 Show that the squared multiple correlation coefficient is a simple function of the product-moment correlation coefficients r_{y1} and r_{y2} when x_1 and x_2 are uncorrelated; that is, show that $R^2 = r_{Y1}^2 + r_{Y2}^2$ when $r_{12} = 0$.

References

1 Draper, N. R., and Smith, H. *Applied Regression Analysis*, 2nd ed. New York: John Wiley, 1981.
2 Acton, S. F. *Analysis of Straight Line Data*. New York: Dover, 1966.
3 Neter, J., Wasserman, W., and Kutner, M. H. *Applied Linear Statistical Models: Regression, Analysis of Variance and Experimental Designs*. Homewood, Ill.: Irwin, 1990.
4 Multiple Risk Factor Intervention Trial Study Group. The Multiple Risk Factor Intervention Trial (MRFIT): A National Study of Primary Prevention of Coronary Disease. *JAMA* 235 (1976): 815–827.

4

Multivariable Regression

Background

Multivariable regression analysis parallels the bivariate case (Chapter 3) because the two-variable regression equation is simply extended to include any number of independent variables. For example, a multiple regression approach is one way to study the relationship between levels of carbon monoxide found in a person's lungs and a series of responses to questions concerning smoking exposure [1]. Many interrelated and, therefore, correlated items appear on a typical smoking exposure questionnaire, such as: How many cigarettes do you smoke per day? When was the last time you smoked? How deeply do you inhale? How many times have you quit? What percent tar is present in your brand of cigarettes? At what age did you start smoking? A multiple linear regression analysis explores such issues as which questions are good indicators of CO levels or how well do particular combinations of questions predict CO levels. Using several independent variables complicates the estimation of the regression coefficients and increases the number of possible comparisons. The fundamental purpose, however, does not change. As in a bivariate regression analysis, the goal of a k-variate linear model is to measure the separate influences of a series of related independent variables and to evaluate their individual as well as their joint impact on the variability observed in the dependent variable.

As part of a linear regression analysis, a number of questions are addressed. Some examples are:

How well do the values calculated from the estimated linear model reflect the observed data?

What expected response in the dependent variable is associated with each individual independent variable while accounting for the other measured variables?

What is the relative influence of each independent variable on the variability in Y?

When a regression equation is used to predict specific values, what improvement is expected from the addition of a new variable to the regression equation?

Is there reason to doubt the assumptions underlying a multiple regression analysis?

Furthermore, measures are developed in the course of the analysis that give some idea of the worth of the answers to these questions in the face of random variation and potential biasing effects.

Method: Criterion

A multivariable linear regression equation with an arbitrary number of independent variables (represented by k) is

$$E(Y \mid x_1, x_2, \cdots, x_k) = a + b_1 x_1 + b_2 x_2 + \cdots + b_k x_k .$$

$$(4.1)$$

The $k + 1$ parameters $(a, b_1, b_2, \cdots, b_k)$ are usually estimated by applying the method of least squares. To perform a multiple regression analysis with k independent variables, each of n observations must consist of $k + 1$ measurements $(y_j, x_{1j}, x_{2j}, \cdots, x_{kj})$. Minimizing the distance (parallel to the y-axis) between the points defined by the regression equation and the observations y_j requires calculus and yields a set of normal equations (given in Box 4.1). The solution to these normal equations yields k estimated regression parameters and a constant term. Employing these estimates, the estimated regression equation is

$$\hat{y}_j = \hat{a} + \hat{b}_1 x_{1j} + \hat{b}_2 x_{2j} + \cdots + \hat{b}_k x_{kj} \quad j = 1, 2, \cdots, n ,$$

$$(4.2)$$

called a k-dimensional hyperplane in mathematical contexts.

The interpretation of the regression coefficient b_i is not different from the two-variable model; a one-unit increase in the independent variable x_i produces a b_i-unit change in the dependent variable Y, while the other $k - 1$ variables are held constant.

Partitioning of the Total Sum of Squares

The partitioning of the total sum of squares into a residual sum of squares and a regression sum of squares remains functionally the same as the previous two regression models (i.e., $\Sigma(y_j - \overline{y})^2 = \Sigma(y_j - \hat{y}_j)^2 + \Sigma(\hat{y}_j - \overline{y})^2$). The basic difference is that $\hat{y}_j$ represents an estimated dependent variable based on an estimated k-variate regression equation. The requirements that the dependent variable be normally distributed, be independently sampled, and have equal variance remain necessary to draw valid inferences from the partitioned sum of squares. The summary statistics from an analysis of variance generated by a k-variable regression analysis are given in Table 4.1.

The residual sum of squares divided by its degrees of freedom yields a residual mean square value $= \Sigma(y_j - \hat{y}_j)^2 / (n - (k + 1))$, denoted $S^2_{Y \mid x_1, x_2, \cdots, x_k}$. As before, the residual mean square is an estimate of the "background" variation in Y when the data at least approximately meet the requirements of the linear regression model.

F-Test

An F-statistic again serves to test the overall analytic worth of the k-variable regression equation (expression 4.2). Like the bivariate regression analysis, the fit of the

T A B L E **4.1**

Analysis of variance table for a k-variate regression analysis

Source	Sum of Squares	Degrees of Freedom	Mean Square
Regression	$SS_0 - SS_1 = \Sigma(\hat{y}_j - \bar{y})^2$	k	Regression SS/k
Residual	$SS_1 = \Sigma(y_j - \hat{y}_j)^2$	$n - (k+1)$	Residual SS/$(n - (k+1))$
Total	$SS_0 = \Sigma(y_j - \bar{y})^2$	$n - 1$	

B O X **4.1**   **NORMAL EQUATIONS FOR k INDEPENDENT VARIABLES**

For the linear model

$$E(Y \mid x_1, x_2, \cdots, x_k) = a + b_1 x_1 + b_2 x_2 + \cdots + b_k x_k,$$

the reduced normal equations are:

$$\hat{b}_1 S_1^2 + \hat{b}_2 S_{21} + \hat{b}_3 S_{31} + \cdots + \hat{b}_k S_{k1} = S_{Y1}$$
$$\hat{b}_1 S_{12} + \hat{b}_2 S_2^2 + \hat{b}_3 S_{32} + \cdots + \hat{b}_k S_{k2} = S_{Y2}$$
$$\hat{b}_1 S_{13} + \hat{b}_2 S_{23} + \hat{b}_3 S_3^2 + \cdots + \hat{b}_k S_{k3} = S_{Y3}$$
$$\vdots$$
$$\hat{b}_1 S_{1k} + \hat{b}_2 S_{2k} + \hat{b}_3 S_{3k} + \cdots + \hat{b}_k S_k^2 = S_{Yk}.$$

In general, the mth normal equation (notation completely defined at the end of the chapter) is

$$\sum_i \hat{b}_i S_{im} = S_{Ym} \quad \text{where } S_{mm} = S_m^2.$$

The solution to these equations yields the k least squares estimated regression coefficients

$$\hat{b}_1, \hat{b}_2, \hat{b}_3, \cdots, \hat{b}_k$$

and a constant term

$$\hat{a} = \bar{y} - \hat{b}_1 \bar{x}_1 - \hat{b}_2 \bar{x}_2 - \hat{b}_3 \bar{x}_3 - \cdots - \hat{b}_k \bar{x}_k.$$

estimated dependent variable from the k-variate model is measured by the residual sum of squares, or

$$SS_1 = \Sigma(y_j - \hat{y}_j)^2 . \tag{4.3}$$

The value $\hat{y}_j$ is specified by estimating $k + 1$ parameters $(\hat{a}, \hat{b}_1, \hat{b}_2, \cdots, \hat{b}_k)$, leaving $n - (k + 1)$ degrees of freedom associated with the residual sum of squares.

When all k independent variables have no systematic impact on Y, making the data random fluctuations from a constant value (i.e., the model $E(Y \mid x_1, x_2, \cdots, x_k) = a$ or an equivalently $b_1 = b_2 = \cdots = b_k = 0$), then the fit of this constrained model is measured by

$$SS_0 = \Sigma(y_j - \overline{y})^2 . \tag{4.4}$$

The mean of the y-values estimates the constant a $(\hat{a} = \overline{y})$, which requires a single estimate leaving $n - 1$ degrees of freedom. Because the model generating SS_0 is a special case of the model generating SS_1 (nested models), the F-statistic to assess H_0 is

$$F = \frac{(SS_0 - SS_1)/[(n - 1) - (n - (k + 1))]}{SS_1/(n - (k + 1))}$$

$$= \frac{(SS_0 - SS_1)/k}{SS_1/(n - (k + 1))} = \frac{\text{regression mean square}}{\text{residual mean square}} . \tag{4.5}$$

The test statistic F has an F-distribution with k and $n - (k + 1)$ degrees of freedom when SS_0 and SS_1 differ only because of random variation; all regression coefficients are zero. The significance of using all k variables to predict the dependent variable is rarely in doubt. More important questions concern the relative roles of the k independent variables.

The Influence of a Selected Subset of Variables

An F-test is also used to assess the influence of a selected subset of the k independent variables, called the F-test "to remove q variables." The symbol q represents the number of variables to be evaluated. If interest is focused on any subset of variables $x_1, x_2, \cdots, x_q$ where $q \leq k$, then two regression models are estimated—one including all k variables (H_1: all k variables are relevant) and one with q specific variables removed (H_0': only $k - q$ variables are relevant). Although the q-subscripts are denoted in order (i.e., 1, 2, 3, $\cdots$, q), it is meant that any q variables can be removed from the regression equation to assess their joint impact. The hypothesis H_0' postulates that q removed independent variables have no influence on the dependent variable Y (i.e., the corresponding regression coefficients are zero) and summary sums of squares reflect the tractability of this hypothesis.

The k-variable regression model (H_1) also yields a measure of fit but based on all k independent variables (SS_1; expression 4.3). The $(k - q)$-variable model (H_0') yields a measure of fit based on a subset of the independent variables where

$$SS_0' = \Sigma(y_j - \hat{y}_j')^2 . \tag{4.6}$$

For the model generated under H_0', the estimated regression equation is

$$\hat{y}_j' = \hat{a}' + \hat{b}'_{q+1} x_{q+1,j} + \hat{b}'_{q+2} x_{q+2,j} + \cdots + \hat{b}_k' x_{kj} \tag{4.7}$$

where $k - q + 1$ estimates are made to specify $\hat{y}_j'$, leaving $n - (k - q + 1)$ degrees of freedom associated with the residual sum of squares SS_0'.

The quantity SS_1 is generated under the hypothesis that all k variables are necessary to describe the relationship between the independent variables and the dependent variable and, similar to the other regression models, has a chi-square distribution when divided by the variance of Y. The quantity SS_0' is the residual sum of squares generated under the nested hypothesis H_0' that only $k - q$ variables are necessary to describe the dependent variable and also, when divided by the variance of Y, has a chi-square distribution when H_0' is true. The two hypotheses are nested (H_0' is a special case of H_1), which means an F-statistic serves to test formally the observed differences between the fit of these two models, or

$$F = \frac{(SS_0' - SS_1)/[(n - (k - q + 1)) - (n - (k + 1))]}{SS_1/(n - (k + 1))} = \frac{(SS_0' - SS_1)/q}{SS_1/(n - (k + 1))} \tag{4.8}$$

where F has an F-distribution with q and $n - (k + 1)$ degrees of freedom when H_0 and H_0' do not systematically differ. The accuracy of this F-test, like the previous F-tests, depends on the extent to which the sampled population meets the requirements of the linear regression model.

A careful look at the test to remove q variables shows that previous F-tests are no more than special cases of H_0'. For example, the overall test of a regression equation consists of removing all k variables so $q = k$, producing the previous F-statistic with k and $n - (k + 1)$ degrees of freedom (expression 4.5).

Further Interpretation of the Squared Multiple Correlation Coefficient, R^2

The definition of the squared multiple correlation coefficient is unchanged in the multivariable regression analysis context—the value R^2 equals the regression sum of squares divided by the total sum of squares. Yet another interpretation of R^2 is possible in terms of the usual product-moment correlation coefficient (see Box 4.2). The product-moment correlation between the observed values y and the estimated values $\hat{y}$ is the multiple correlation coefficient R, or, in symbols,

$$\text{correlation}\,(y, \hat{y}) = r_{Y,\hat{Y}} = R = \sqrt{\frac{\text{regression sum of squares}}{\text{total sum of squares}}} = \sqrt{\frac{\Sigma(\hat{y}_j - \bar{y})^2}{\Sigma(y_j - \bar{y})^2}} . \tag{4.9}$$

The multiple correlation coefficient directly reflects the extent to which the model predicts the observed values. A value of $R^2 = 1$ occurs when every estimated dependent variable equals the corresponding observed values ($\hat{y}_j = y_j$), producing a residual sum of squares equal to zero ($SS_1 = 0$). Geometrically, $R^2 = 1$ means that the n points (y_j, $\hat{y}_j$) fall exactly on a straight "line." At the other extreme, $R^2 = 0$ means the independent variables used in the linear regression equation are useless in a description of the observed data ($SS_1 = SS_0$).

The correlation between the observation y and the residual quantity ($y - \hat{y}$) is $\sqrt{1 - R^2}$, or

$$\text{correlation}(y, [y - \hat{y}]) = r_{Y, (Y-\hat{Y})} = \sqrt{1 - R^2}$$

$$= \sqrt{\frac{\text{residual sum of squares}}{\text{total sum of squares}}} = \sqrt{\frac{\Sigma(y_j - \hat{y}_j)^2}{\Sigma(y_j - \bar{y})^2}}. \quad \text{(4.10)}$$

Additionally, the correlation between the estimate $\hat{y}$ and residual $y - \hat{y}$ is zero, or

$$\text{correlation}(\hat{y}, [y - \hat{y}]) = r_{\hat{Y}, (Y-\hat{Y})} = 0. \quad \text{(4.11)}$$

It is noteworthy that the product-moment correlation between the estimated values and the deviations of the estimated values from the observed values is zero ($r_{\hat{Y}, (Y-\hat{Y})} = 0$). The estimated regression equation "explains" the data in terms of a linear relationship ($\hat{y}$) and the "unexplained" portion ($y - \hat{y}$) contains only information not "explainable" by a linear relationship, leaving the residual values uncorrelated with the estimated values.

The multiple correlation coefficient R produces a natural summary of the effectiveness of a linear model to represent the relationships under investigation. It simply answers the question: How well does the equation work as a description of the collected data? Specifically, R^2 is the proportion of the variability explained by the estimated regression equation. A value of $R > 0.7$ usually means the regression equation is potentially an effective statistical tool. Clearly, for R in the neighborhood of zero, the regression equation is not useful as a description of the dependent variable. The worth of a regression equation for intermediate values of R is not easily assessed and depends to a large extent on subject matter considerations.

Addition of a Variable

The addition of any variable to the regression equation always increases the squared multiple correlation coefficient. No variable can be added that reduces the value of R^2. The addition of a particular independent variable may increase R^2 only slightly but nevertheless R^2 never decreases. The magnitude of this increase is a function of (1) the variables already in the regression equation and (2) the relationship between the added variable and the dependent variable Y. The relative importance of these two components must be separated by further analysis (discussed in the following section on part correlation). A variable such as father's age, when added to a regression equation containing the highly correlated variable mother's age, adds only information relevant to Y not already reflected by the variable maternal age or any other correlated variables in the regression equation. The increase in R^2 measures the additional worth of the added variable but depends on the variables already in the equation.

Also note: The F-to-remove test can also be viewed as a test of the change induced in the squared multiple correlation coefficient by the deletion of q variables. If R_1^2 represents the squared multiple correlation coefficient from the k-variate model and R_0^2 represents the squared multiple correlation for the model with q variables removed, then

$$F = \frac{(R_1^2 - R_0^2)/q}{R_1^2/(n - (k + 1))}$$

(4.12)

is identical (same F-value) to the previously described F-statistic (expression 4.8).

Shrinkage: R^2 (Adjusted)

The method of least squares produces the minimum possible residual sum of squares, maximizing the regression sum of squares; therefore, the value of R^2 is the largest possible for the analyzed data set. The observed value of the squared multiple correlation coefficient is inflated because the "best" linear equation is estimated from the data. When an estimate of the population value of R^2 (the expected value of R^2) is desired, the estimate R^2 from a regression analysis is likely too large. In other words, an estimate of the actual proportion of variation explainable by the estimated linear model is likely overstated by R^2 computed from the sampled data. In statistical terms, R^2 is a biased estimate of the expected amount of variation explainable by a linear model. An alternative way of viewing the upward bias of R^2 is to consider a regression equation computed from one data set applied to another data set sampled from the same population. The maximizing property of the least squares method produces an estimated value of R^2 likely larger than an R^2 calculated from a second sample using the regression equation estimated from the first data set. Other summaries calculated from multivariate data have the same property—values calculated from a sample perform less well when applied to newly collected data.

A quantity "R^2 adjusted" is defined as

$$R_{adj}^2 = R^2 - \frac{k}{n - k - 1}(1 - R^2) \quad [2].$$

(4.13)

This reduced squared multiple correlation coefficient is approximately corrected for the upward bias of the sample estimate R^2. The estimate R^2 is reduced to provide a better estimate of the expected value; thus the term *shrinkage* is used. An obvious property of R_{adj}^2 is that $R_{adj}^2 \approx R^2$ when the number of observations n is much larger than the number of variables k and, when n and k are similar, R_{adj}^2 will differ from R^2 and the correction becomes important.

B O X   **4.2**   THE MULTIPLE CORRELATION $= R = r_{Y\hat{Y}}$

Here $\hat{y}_j = \hat{a} + \hat{b}x_j$ (simple linear regression) is used to demonstrate that $R = r_{Y\hat{Y}}$. The more general multiple regression case, not presented, follows the same pattern but involves more extensive notation.

1   Correlation $\{y, \hat{y}\} = r_{Y\hat{Y}}$

$$r_{Y\hat{Y}} = \frac{\Sigma(y_j - \bar{y})(\hat{y}_j - \bar{\hat{y}})}{\sqrt{\Sigma(y_j - \bar{y})^2}\sqrt{\Sigma(\hat{y}_j - \bar{\hat{y}})^2}}$$

Note 1: $\bar{\hat{y}} = \bar{y}$;   the mean of the estimated values = the mean of the observations

Note 2: Regression sum of squares $= \Sigma(\hat{y}_j - \bar{\hat{y}})^2 = \Sigma(\hat{y}_j - \bar{y})^2$

Note 3: $\Sigma(y_j - \bar{y})(\hat{y}_j - \bar{\hat{y}}) = \Sigma(y_j - \bar{y})(\hat{y}_j - \bar{y})$

$\qquad = \Sigma(y_j - \bar{y})(\hat{a} + \hat{b}x_j - \bar{y}) = \Sigma(y_j - \bar{y})(\hat{b}[x_j - \bar{x}])$ because $\hat{a} = \bar{y} - \hat{b}\bar{x}$

$\qquad = \hat{b}\,\Sigma(y_j - \bar{y})(x_j - \bar{x}) = \{\hat{b}\}^2\,\Sigma(x_j - \bar{x})^2 = \Sigma(\hat{y}_j - \bar{y})^2$

$\qquad = $ regression sum of squares

$$r_{Y\hat{Y}} = \frac{\text{regression sum of squares}}{\sqrt{\text{total sum of squares}}\;\sqrt{\text{regression sum of squares}}}$$

$$r_{Y\hat{Y}} = \sqrt{\frac{\text{regression sum of squares}}{\text{total sum of squares}}} = \sqrt{R^2} = R$$

2   Correlation $\{y, y - \hat{y}\} = r_{Y, Y-\hat{Y}}$

$$r_{Y, Y-\hat{Y}} = \frac{\Sigma(y_j - \bar{y})(y_j - \hat{y}_j)}{\sqrt{\Sigma(y_j - \bar{y})^2}\;\sqrt{\Sigma(y_j - \hat{y}_j)^2}}$$

Note 4: $\Sigma(y_j - \bar{y})(y_j - \hat{y}_j)$

$\qquad = \Sigma(y_j - \bar{y})([y_j - \bar{y}] - [\hat{y}_j - \bar{y}])$

$\qquad = \Sigma(y_j - \bar{y})^2 - \Sigma(\hat{y}_j - \bar{y})^2$

$\qquad = $ total sum of squares $-$ regression sum of squares

$\qquad = $ residual sum of squares

$$r_{Y, Y-\hat{Y}} = \frac{\text{residual sum of squares}}{\sqrt{\text{total sum of squares}}\;\sqrt{\text{residual sum of squares}}}$$

$$= \sqrt{\frac{\text{residual sum of squares}}{\text{total sum of squares}}} = \sqrt{1 - R^2}$$

A simple product-moment correlation coefficient is a biased estimate of the population correlation because it is a special case of the multiple correlation coefficient. Just as in the case of R^2 in general, the sample value r_{XY}^2 (expression 1.7) is likely to overstate the true population value and should be corrected, especially for small samples. This correction is rarely done in practice.

Consider, for example, the data on smoking given for the test case in Chapter 3, where the value $R^2 = 0.318$ was estimated using a bivariate regression equation. Then

$$R_{adj}^2 = 0.318 - \frac{2}{7}(1 - 0.318) = 0.124 \ .$$

The substantially reduced value 0.124 is a better estimate of the actual proportion of the variation explainable by the estimated regression equation.

Occasionally a data set contains a large number of independent variables and it is advantageous to choose an efficient subset of these variables rather than working with the full regression equation. Because R^2 increases with the addition of each variable to the regression equation, it provides little help in selecting an "optimum" subset of independent variables. However, R_{adj}^2 increases to a maximum and then decreases as different combinations of independent variables are added to a regression equation. This property provides one basis for selecting a subset of variables that is optimum in a specific sense. It is argued that the "optimum" regression equation employs the subset of independent variables that maximizes R_{adj}^2. Other methods and measures are available to select an optimum subset of independent variables. This rather sophisticated topic of variable selection is not discussed in detail [see 3], but a bit more discussion is included in Chapter 11.

Path Coefficients

An important purpose of a regression analysis is the comparison of the estimated regression coefficients. Rarely are all the independent variables measured in same units. For example, one x-value might be weight measured in pounds and another might be age recorded in years. To directly compare the effects reflected by each regression coefficient associated with variables measured in different units, a standardization is necessary. Without a standardization, the relative magnitude of compared coefficients depends on the units—a comparison of coefficients based on pounds versus years differs from the same comparison based on ounces versus months. Path coefficients are a standardization that allows direct comparisons of the influences on the dependent Y from independent variables measured in different units.

Consider, for example, the response in Y from a one standard deviation increase in the independent variable x_1. Then,

$$E(Y \mid x_1, x_2, \cdots, x_k) = a + b_1 x_1 + b_2 x_2 + \cdots + b_k x_k$$

and

$$E(Y \mid x_1 + \sigma_1, x_2, \cdots, x_k) = a + b_1(x_1 + \sigma_1) + b_2 x_2 + \cdots + b_k x_k$$

yielding

$$E(Y \mid x_1 + \sigma_1, x_2, \cdots, x_k) - E(Y \mid x_1, x_2, \cdots, x_k) = b_1 \sigma_1.$$

The path coefficient (denoted p_i) is the expected change produced in Y from an increase of one standard deviation of x_i divided by the standard deviation of the dependent variable (i.e., $p_i = b_i \sigma_i / \sigma_Y$). In symbols, the estimated path coefficient is $\hat{p}_i = \hat{b}_i S_i / S_Y$. A path coefficient is unitless and serves to summarize the influence of x_i on Y in such a way that the impact of each independent variable can be compared directly, which is said to be commensurate. Expressing the influence of x_i per standard deviation of Y (divided by S_Y) leads to several interesting interpretations.

Commensurate Measure of Response

Path coefficients are commensurate in the sense that the estimated response in Y comes from "equal" increases in the x-values. Each x-value is incremented by one standard deviation, which represents approximately equal quantities in terms of probabilities when the independent variables have at least approximately normal distributions. One standard deviation in pounds is equivalent to one standard deviation in years of age. It is this equivalence that makes the path coefficient a commensurate measure of the relative contributions to the response in Y, regardless of the original measurement units.

An alternative view of the commensurate nature of the path coefficient comes from applying linear regression techniques to standardized data. Each independent variable and the dependent variable are standardized so all mean values are zero and all sample variances are 1.0. The new variables

$$y_j^* = (y_j - \bar{y})/S_Y$$
$$z_{1j} = (x_{1j} - \bar{x}_1)/S_1$$
$$z_{2j} = (x_{2j} - \bar{x}_2)/S_2$$
$$\cdot$$
$$\cdot$$
$$z_{kj} = (x_{kj} - \bar{x}_k)/S_K$$

are commensurate in the sense that each standardized variable is unitless with mean $= 0.0$ and variance $= 1.0$. When a set of normal equations (Box 4.3) derived from standardized data is solved to yield k estimated coefficients, the estimated regression equation is

$$\hat{y}_j^* = \hat{p}_1 z_{1j} + \hat{p}_2 z_{2j} + \cdots + \hat{p}_k z_{kj} \quad \text{with } j = 1, 2, \cdots, n. \tag{4.14}$$

More important, the estimated regression coefficient $\hat{p}_i = \hat{b}_i S_i / S_Y$ is the estimated path coefficient associated with the independent variable, x_i.

The usual analysis of variance table is given in Table 4.2 (top). Due to the standardization, several simplifications occur (see Box 4.4), yielding the expressions for the sums of squares given in the analysis of variance Table 4.2 (bottom).

The squared multiple correlation coefficient can be interpreted as a sum of correlation coefficients between each x_i and Y weighted by the corresponding path coefficient, or

$$R^2 = \Sigma \hat{p}_i r_{Yi} . \tag{4.15}$$

B O X **4.3**   NORMAL EQUATIONS FOR k INDEPENDENT STANDARDIZED VARIABLES *

$$E(Y^* \mid z_1, z_2, z_3, \cdots, z_k) = p_1 z_1 + p_2 z_2 + p_3 z_3 + \cdots + p_k z_k$$

and the normal equations are:

$$\hat{p}_1 + \hat{p}_2 r_{21} + \hat{p}_3 r_{31} + \cdots + \hat{p}_k r_{k1} = r_{Y1}$$

$$\hat{p}_1 r_{12} + \hat{p}_2 + \hat{p}_3 r_{32} + \cdots + \hat{p}_k r_{k2} = r_{Y2}$$

$$\hat{p}_1 r_{13} + \hat{p}_2 r_{23} + \hat{p}_3 + \cdots + \hat{p}_k r_{k3} = r_{Y3}$$

$$\vdots$$

$$\hat{p}_1 r_{1k} + \hat{p}_2 r_{2k} + \hat{p}_3 r_{3k} + \cdots + \hat{p}_k = r_{Yk} .$$

In general, the mth normal equation is

$$\Sigma_i \hat{p}_i r_{im} = r_{Ym} \quad \text{where } r_{mm} = 1.0 .$$

The values r_{ij} and r_{yj} are the usual Pearson correlation coefficients (expression 1.7). The solution to the k normal equations yields the k estimated path coefficients

$$\hat{p}_1, \hat{p}_2, \hat{p}_3, \cdots, \hat{p}_k$$

and the constant term

$$\hat{a} = \bar{y}^* - \hat{p}_1 \bar{z}_1 - \hat{p}_2 \bar{z}_2 - \hat{p}_3 \bar{z}_3 - \cdots - \hat{p}_k \bar{z}_k = 0$$

$\bar{y}^* = 0$ and $\bar{z}_1 = \bar{z}_2 = \cdots = \bar{z}_k = 0$ because the data are standardized.

*(for standardized data; mean = 0 and variance = 1)

T A B L E   **4.2**

Analysis of variance table from standardized data

Source	Sum of Squares	Degrees of Freedom
Regression	$\Sigma(\hat{y}_j^* - \bar{y}^*)^2 = \Sigma(\hat{y}_j^*)^2$	k
Residual	$\Sigma(y_j^* - \hat{y}_j^*)^2$	$n - (k + 1)$
Total	$\Sigma(y_j^* - \bar{y}^*)^2 = \Sigma(y_j^*)^2$	$n - 1$

An alternative form of the analysis variance table for standardized data

Source	Sum of Squares	Degrees of Freedom
Regression	$(n - 1)\Sigma\hat{p}_i r_{Yi}$	k
Residual	$(n - 1)[1 - \Sigma\hat{p}_i r_{Yi}]$	$n - (k + 1)$
Total	$n - 1$	$n - 1$

The value R^2 is the identical squared multiple correlation coefficient that would have been calculated from a regression analysis using the original (unstandardized) data. Standardizing the data neither increases nor decreases the fit of the estimated values to the data. When R^2 is viewed as a weighted sum of values r_{Yi}, the path coefficients are an assessment of the contribution of the correlation between each independent variable and the dependent variable to the overall correlation between predicted values from the linear model and the observed values (i.e., $R = r_{\hat{Y}Y}$).

Path coefficients play a variety of roles in statistical analysis. For example, path coefficients are an effective way to describe genetic data. Geneticist Sewell Wright invented path coefficients to study the degree of genetic relationship among a series of related individuals. His pedigree analysis and its relationship to a regression equation are discussed elsewhere [3]. Path analysis also has been applied to problems arising in the social sciences [4], where path coefficients are traditionally referred to as beta coefficients.

Two Measures of the Contribution from an Independent Variable

A number of measures lead to a detailed understanding of the role of a specific independent variable in the regression equation. Two, among many, measures are tolerance and part correlation.

Tolerance

Formally, tolerance is defined as $1 - R_{(i)}^2$ where $R_{(i)}$ is a specialized multiple correlation coefficient reflecting the correlation between the ith independent variable and the

$k - 1$ other independent variables. Specifically, the quantity $R_{(i)}^2$ represents the squared multiple correlation coefficient that results from a regression analysis where x_i plays the role of a dependent variable and the remaining $k - 1$ independent variables act as independent variables. The value $R_{(i)}^2$ indicates the amount of information contained in a linear combination of the other $k - 1$ independent variables that is also contained in the variable x_i. The value $R_{(i)}^2$ does not involve the dependent variable Y. For the two independent variables in a bivariate regression equation, $R_{(i)}$ is the squared simple correlation between x_1 and x_2 ($R_{(1)}^2 = r_{12}^2 = R_{(2)}^2$).

B O X **4.4**   RELATIONSHIP BETWEEN THE PATH COEFFICIENT
AND THE MULTIPLE CORRELATION COEFFICIENT

For the bivariate regression ($k = 2$), then

$$\Sigma(\hat{y}_j - \bar{y})^2 = \Sigma(\hat{a} + \hat{b}_1 x_{1j} + \hat{b}_2 x_{2j} - \bar{y})^2$$
$$= \Sigma(\bar{y} - \hat{b}_1 \bar{x}_1 - \hat{b}_2 \bar{x}_2 + \hat{b}_1 x_{1j} + \hat{b}_2 x_{2j} - \bar{y})^2$$
$$= \Sigma(\hat{b}_1 [x_{1j} - \bar{x}_1] + \hat{b}_2 [x_{2j} - \bar{x}_2])^2$$
$$= \hat{b}_1^2 (n-1) S_1^2 + \hat{b}_2^2 (n-1) S_2^2 + 2\hat{b}_1 \hat{b}_2 (n-1) S_{12}$$

$$R^2 = \frac{\Sigma(\hat{y}_j - \bar{y})^2}{\Sigma(y_j - \bar{y})^2}$$
$$= \frac{\hat{b}_1^2 S_1^2}{S_Y^2} + \frac{\hat{b}_2^2 S_2^2}{S_Y^2} + \frac{2\hat{b}_1 \hat{b}_2 S_{12}}{S_Y^2}$$
$$= \{\hat{p}_1\}^2 + \{\hat{p}_2\}^2 + \frac{2\hat{b}_1 S_{12} \hat{b}_2 S_2 S_1}{S_1 S_2 S_Y^2}$$
$$= \{\hat{p}_1\}^2 + \{\hat{p}_2\}^2 + 2\hat{p}_1 \hat{p}_2 r_{12} = \hat{p}_1 (\hat{p}_1 + \hat{p}_2 r_{12}) + \hat{p}_2 (\hat{p}_2 + \hat{p}_1 r_{12}).$$

Note: $\hat{p}_1 + \hat{p}_2 r_{12} = r_{Y1}$ (first normal equation)
$\hat{p}_1 r_{12} + \hat{p}_2 = r_{Y2}$ (second normal equation)

Therefore,

$$R^2 = \hat{p}_1 (\hat{p}_1 + \hat{p}_2 r_{12}) + \hat{p}_2 (\hat{p}_2 + \hat{p}_1 r_{12})$$
$$= \hat{p}_1 r_{Y1} + \hat{p}_2 r_{Y2}.$$

In general, the relationship between the path coefficient and the squared multiple correlation coefficient is

$$R^2 = \Sigma \hat{p}_i r_{Yi}.$$

If the tolerance is close to 1.0 ($R^2_{(i)}$ is close to zero), then the independent variable x_i potentially measures a dimension of Y not measured by the other independent variables. More precisely, the other independent variables in the regression equation are not linearly related to x_i, making it possibly a useful addition to the analysis of Y. Low values of tolerance ($R^2_{(i)}$ close to 1) occur when the ith variable contains little information not already accounted for by the other variables in the regression equation. That is, a high correlation exists between x_i and a linear combination of the other independent variables. In this case, x_i has little potential for adding new information useful in explaining the variation in Y.

One method for deciding whether a variable should or should not be included in a regression equation is to set a limit on the tolerance $1 - R^2_{(i)}$ so that no variable is included in the regression equation unless its tolerance exceeds a given limit (e.g., tolerance = 0.1). If the tolerance of a variable is 0.0 ($R^2_{(i)} = 1.0$), then the variable is perfectly collinear with respect to some combination of the other independent variables and adds no new information to a regression analysis. Therefore, tolerance calculated for each variable is an indicator of the presence of collinearity. Setting a tolerance limit at a specific level gives protection against including redundant information from independent variables in the regression equation. Including highly correlated variables in a regression equation, as noted, reduces the reliability associated with the estimation of the regression coefficients (increases $S^2_{b_j}$). Restricting the analysis to just those variables with a tolerance above a set level is another way to produce an optimum subset of independent variables.

The reciprocal of the tolerance value is called the variation inflation factor. This term comes from the property that the estimated variance of $\hat{b}_i$ contains the factor $1/(1 - R^2_{(i)})$, as noted in the next section (expression 4.18). A plot of $1 - R^2_{(i)}$ (Figure 4.1) shows that the tolerance is small only for values of $R_{(i)}$ close to 1. For example, $1 - R^2_{(i)} < 0.1$ only when $R_{(i)} > 0.95$.

Part Correlation

The squared part correlation (denoted $C^2_{(i)}$) is equal to the increase in the squared multiple correlation coefficient achieved by adding a specific variable (x_i) to the regression equation. This quantity is also called a semipartial correlation coefficient. The part correlation is a function of two quantities—the relationship of the dependent variable to the added variable (measured by the path coefficient, $\hat{p}_i$) and the relationship of the other variables in the regression equation to the added variable (measured by the tolerance, $1 - R^2_{(i)}$). Symbolically, the increase in R^2 from the addition of the ith variable is

$$\text{squared part correlation} = C^2_{(i)} = R^2_{x_i-\text{included}} - R^2_{x_i-\text{excluded}} = \hat{p}^2_i(1 - R^2_{(i)}) . \qquad (4.16)$$

Expression 4.16 shows that the part correlation associated with x_i (increases in R^2) is the squared path coefficient for the variable x_i decreased by the amount equal to the tolerance $(1 - R^2_{(i)})$.

FIGURE **4.1**
A plot of $R_{(i)}$ against
$1 - R^2_{(i)}$

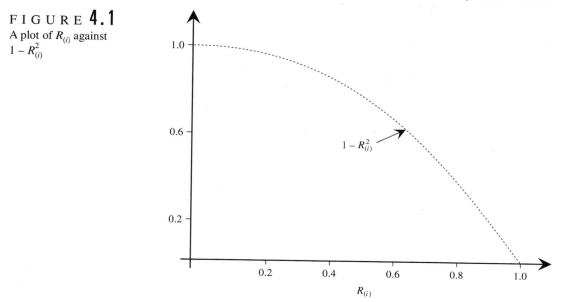

The part correlation indicates the influence not already accounted for by other variables included in the regression equation of a specific independent variable on the dependent variable. For example, if the squared multiple correlation coefficient using only x_1 to estimate Y is $R^2 = 0.169$ (see previous test case, Chapter 2 and Chapter 3) and the squared part correlation $C^2_{(2)}$ associated with a second independent variable x_2 is 0.149, then the squared multiple correlation coefficient for the regression equation using both x_1 and x_2 is $R^2 + C^2_{(2)} = 0.169 + 0.149 = 0.318$. The value 0.149 indicates the influence of x_2 on the dependent variable Y after including x_1 in the regression equation. Because $\hat{p}^2_2 = 0.162$ and $1 - R^2_{(2)} = 0.918$, the part correlation calculated directly is $C^2_{(2)} = (0.162)(0.918) = 0.149$.

The squared part correlation is directly related to the t-statistic associated with each regression coefficient. That is,

$$t\text{-statistic} = t_i = \frac{\hat{b}_i}{S_{\hat{b}_i}} \quad \text{and}$$

$$\text{the squared part correlation} = C^2_{(i)} = \frac{t^2_i(1 - R^2)}{(n - (k + 1))} . \tag{4.17}$$

Therefore, t^2_i is directly related to the increase in R^2 achieved by adding x_i to a regression equation containing $k - 1$ other independent variables. All computer regression

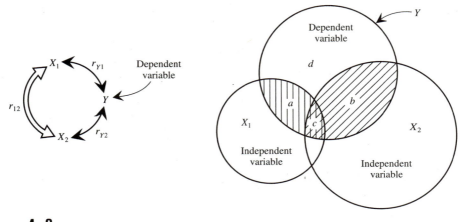

F I G U R E **4.2**

A partitioning of the relationship between two independent variables in a regression equation

programs produce values of t_i but only a few calculate the part correlation. Clearly, the relationship between the squared t-statistic (t_i^2) and the squared part correlation ($C_{(i)}^2$) makes the part correlation easily calculated for a given variable from computer output.

The part correlation is the product of two components: the path coefficient and the tolerance. Therefore, small increases in R^2 (small values of the squared part correlation) could result because the added variable is unrelated to Y ($\hat{p}_i$ is small) or because the added variable brings little new information to the analysis ($1 - R_{(i)}^2$ is small). In short, the part correlation measures the increased "unique" contribution to R^2 conditional on the other variables already in the equation. The path coefficient and the tolerance reflect the magnitude of the two separate components that make up the total contribution.

The roles of the part correlation, the partial correlation, and the multiple correlation in a bivariate regression analysis are illustrated geometrically (Figure 4.2) by three overlapping circles, where one circle represents the dependent variable Y and the other two circles represent the independent variables x_1 and x_2. The areas a, b, and c in Figure 4.2 have values:

$$a = \hat{p}_1^2(1 - R_{(1)}^2) = \hat{p}_1^2(1 - r_{12}^2) = R^2 - r_{Y2}^2 = C_{(1)}^2$$
$$b = \hat{p}_2^2(1 - R_{(2)}^2) = \hat{p}_2^2(1 - r_{12}^2) = R^2 - r_{Y1}^2 = C_{(2)}^2$$
$$c = 2\hat{p}_1\hat{p}_2 r_{12} + r_{12}^2(\hat{p}_1^2 + \hat{p}_2^2)$$

and $a + c = r_{Y1}^2$, $b + c = r_{Y2}^2$. Also, the squared partial correlations are depicted by

$$r^2_{Y1.2} = a/(a + d) \text{ and } r^2_{Y2.1} = b/(b + d).$$

The total variation "explained" is

$$a + b + c = \hat{p}_1^2 + \hat{p}_2^2 + 2\hat{p}_1\hat{p}_2 r_{12} = \hat{p}_1 r_{Y1} + \hat{p}_2 r_{Y2} = R^2,$$

and $d = 1 - R^2$ is the proportion of the variability not related to x_1 and x_2. Such a diagram is helpful in sorting out the various ways x_1 and x_2 contribute to a regression analysis but should not be taken as a literal representation of a bivariate regression analysis. It is a valid representation when all correlations are positive. The quantity c, however, can be negative and the geometric representation of a negative area is not possible in a simple figure made up of three overlapping circles. Figure 4.3 displays four ways $R^2_{(i)}$ and $C^2_{(i)}$ play a role in a k-variate regression analysis, again using overlapping circles to represent the regression variables.

A question remains: Why is the part correlation called a correlation coefficient? The part correlation is the correlation between the dependent variable Y and an adjusted independent variable, $x_i' = x_i - \hat{x}_i$. The value $\hat{x}_i$ is the estimated value of x_i based on a regression equation consisting of the other $k - 1$ independent variables, making x_i' the ith residual value. The part correlation is the *correlation*$(y, x_i - \hat{x}_i)$ or *correlation*(y, x_i'). A partial correlation coefficient is *correlation*$(y - \hat{y}, x_i - \hat{x}_i)$, showing why a part correlation is sometimes called a semipartial correlation because x_i is adjusted but the dependent variable Y is not. Viewing the part correlation as a product-moment correlation between Y and that part of x_i not reflected by the variables already in the regression equation produces yet another perspective on the increase in R^2 associated with a specific independent variable.

The part correlation is illustrated with data used from the previous test case in Chapter 3 and repeated in Table 4.3. The part correlation is the *correlation*$(y, x_1') = r_{Yx_1'} = 0.284$, which is the correlation between the dependent variable Y and the independent variable x_1 adjusted for the linear influence of x_2 (columns 1 and 6 in Table 4.3). The part correlation can also be directly calculated because $R^2_{x_1-\text{excluded}} = 0.238$ when the bivariate regression equation contains only x_2 and $R^2_{x_1-\text{included}} = 0.318$ when both x_1 and x_2 are included in the regression equation. The part correlation is then

$$C_{(1)} = \sqrt{0.318 - 0.238} = 0.284 = r_{Yx_1'}.$$

The two methods of estimating the part correlation are algebraically identical. Note that this particular part correlation was also discussed briefly in Chapter 3, where $C^2_{(1)} = (0.284)^2 = 0.080$.

To summarize, Table 4.4 (called a correlation array) reviews the previously mentioned correlations for the data given in Table 4.3—the variables form the columns and rows whereas the table contains all possible correlations (e.g., $r_{yx_1'} = 0.284$; row 1 and column 6).

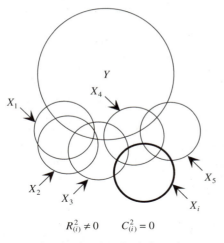

$$R_{(i)}^2 \neq 0 \qquad C_{(i)}^2 = 0$$

X_i is related to the other independent
variables and not related to Y.

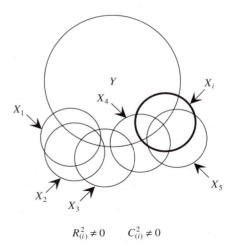

$$R_{(i)}^2 \neq 0 \qquad C_{(i)}^2 \neq 0$$

X_i is related to the other independent
variables and related to Y.

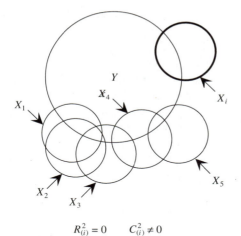

$$R_{(i)}^2 = 0 \qquad C_{(i)}^2 \neq 0$$

X_i is not related to the other independent
variables and related to Y.

$$R_{(i)}^2 = 0 \qquad C_{(i)}^2 = 0$$

X_i is not related to the other independent
variables and not related to Y.

FIGURE **4.3**

Four ways $C_{(i)}$ and $R_{(i)}$ could behave in a k-variate regression analysis

T A B L E   **4.3**

Illustration of a part correlation (data repeated)

	y	x_1	x_2	$\hat{x}_1$	$\hat{y}$	$x_1' = x_1 - \hat{x}_1$	$y' = y - \hat{y}'$
1	2	12	5.00	24.015	15.426	−12.015	−13.426
2	11	30	6.09	16.167	11.254	13.833	−0.254
3	14	40	4.00	31.215	19.253	8.785	−5.253
4	16	12	5.33	21.639	14.163	−9.639	1.837
5	18	2	4.25	29.415	18.297	−27.415	−0.297
6	18	5	3.75	33.015	20.210	−28.015	−2.210
7	18	20	5.75	18.615	12.555	1.385	5.445
8	22	30	4.66	26.463	16.727	3.537	5.273
9	23	75	3.83	32.439	19.904	42.561	3.096
10	26	40	3.75	33.015	20.210	6.985	5.790

T A B L E   **4.4**

Review of a series of correlations referring to the data given in Table 4.3

	y	x_1	x_2	$\hat{x}_1$	$\hat{y}'$	x_1'	y'
y	1.000	0.412	−0.488	0.488	0.488	0.284	0.873
x_1	—	1.000	−0.286	0.286	0.286	0.958	0.312
x_2	—	—	1.000	1.000	1.000	0.000	0.000
$\hat{x}_1$	—	—	—	1.000	1.000	0.000	0.000
$\hat{y}'$	—	—	—	—	1.000	0.000	0.000
x_1'	—	—	—	—	—	1.000	0.325
y'	—	—	—	—	—	—	1.000

Variance of the Estimated Slope

An estimated variance of the estimated slope $\hat{b}_i$ based on n observations is given by

$$S_{b_i}^2 = \frac{S_{Y|x_1, x_2, \cdots, x_k}^2}{\sum_j (x_{ij} - \overline{x}_i)^2 (1 - R_{(i)}^2)} = \frac{S_{Y|x_1, x_2, \cdots, x_k}^2}{(n-1)S_i^2(1 - R_{(i)}^2)} \qquad \text{justified in [5].}$$

$$(4.18)$$

This expression is a generalization of the estimated variance for an estimated regression coefficient given earlier (expression 3.15) in the bivariate case and is made up of analogous components. The principal difference is that r_{12} is replaced by $R_{(i)}$. Parallel to the previous regression models, the mean square residual $S^2_{Y|x_1, x_2, \cdots, x_k}$ is an estimate of the background variation associated with the dependent variable Y under the linear regression model. The quantity S^2_i reflects the range of the x_i-variable associated with the regression coefficient b_i. As noted, the multiple correlation $R_{(i)}$ measures the correlation between variable x_i and the other $k - 1$ independent variables, making the quantity $1 - R^2_{(i)}$ the tolerance for variable x_i defined in the last section.

If x_i is strongly related to the other variables in the regression equation (tolerance near 0.0), then the estimated variance of the regression coefficient is substantially increased, multiplied by the variation inflation factor of $1/(1 - R^2_{(i)})$. The reason for the increased variability is similar to the two-dimensional case. In simple terms, the span over which b_i is estimated is small because of the high correlation of x_i with the other variables in the regression equation. If the variable x_i brings no new information to the analysis (linearly related to the other $k - 1$ variables), a precise estimate of its separate impact on Y is not possible. By "no new information," it is meant that the geometric region covered by the $k - 1$ independent variables is not expanded by the addition of x_i.

Two sometimes useful solutions exist to the problem of dealing with nearly collinear variables (*tolerance* ≈ 0). Transformations can be used to make related variables statistically independent, eliminating collinearity. Such transformations are briefly discussed in the chapter on principal components (Chapter 7). A more pragmatic approach is to exclude the collinear variable from the analysis on the grounds that keeping a collinear variable in the analysis adds no useful new information to the regression equation and only increases the variability of the estimates of the coefficients of the correlated variables. Most computer regression systems eliminate collinear variables automatically. Other than noting the problem of collinear variables and identifying two simple solutions, more detailed exploration of this issue is left to other texts ([5] or [6], for example).

Residual Value Analysis

By way of review, the four basic properties of the dependent variable Y underlying a typical regression analysis are:

1   Each observation $(y_j, x_{1j}, x_{2j}, \cdots, x_{kj})$ is sampled independently.

2   The y-values have normal distributions.

3   The sampled y-values have the same variance regardless of the values of the independent variables.

4   A linear model describes the data, or $E(Y | x_1, x_2, \cdots, x_k) = a + b_1 x_1 + b_2 x_2 + \cdots + b_k x_k$.

A regression model relies on a relatively simple mathematical structure that at best approximates the real-world complexities. As emphasized, the technique of summarizing a set of data with a linear model succeeds or fails to the extent that the theoretical structure reflects the relationships under study. Specifically, requirements 1–4 must be at least approximately fulfilled. The inspection of the residual values $(y_j - \hat{y}_j)$ helps assess the validity of the basic requirements for a regression analysis.

Normality

The quantity

$$r_j = \frac{y_j - \hat{y}_j}{S_{Y|x_1, x_2, \cdots, x_k}}$$

(4.19)

is one way to calculate the standardized residual value generated by each observation. If the dependent variable fulfills requirements 1–4, then the n standardized residual values $r_1, r_2, \cdots, r_n$ have an approximate normal distribution with mean = 0 and variance = 1.0. The r-values are not independent because $\Sigma r_j = 0$, but the effects of the nonindependence are minor and can usually be ignored, particularly if the sample size n is much larger than the number of variables k. Other versions of standardized residuals can be considered. For example, the difference between the observed and estimated values can be divided by the standard error of the difference between observed and estimated values ($S_{Y-\hat{Y}}$), generally producing a value similar to r_j.

Large residual values indicate possible outlier observations that are candidates for exclusion from the analysis. Typographical errors, machine malfunctions, recording errors, and the like produce observations that disproportionally affect the analytic results. Statistical tests for outlier values exist ([7] or [8]) but usually a formal test to identify large residual values is not necessary. Extreme values are easily located, but whether or not they indicate values that should be excluded from the analysis is a more difficult question. Extreme observations should only be rejected out of hand if they can be traced to a cause; otherwise the rejection of an extreme observation should be carefully considered. Extreme observations arise in at least two ways: An observation is an outlier observation resulting from some unknown error or represents a source of unique information that is likely important in the overall analysis. The problem, of course, is that it is usually difficult to determine which of these two possibilities produced the extreme value. Common sense is a powerful approach to recognizing and dealing with outlier observations.

Because about 95% of standard normal variates lie between –2 and 2, it is expected that about 5% of the standardized residual values would be more extreme when properties 1–4 apply. If substantially more or less than 5% of the r-values are found beyond these limits, the requirements of normality become suspect.

Other more sensitive tests of normality exist. Residual values can be evaluated with parametric or nonparametric procedures [9]. A simple plot of the r_j-values them-

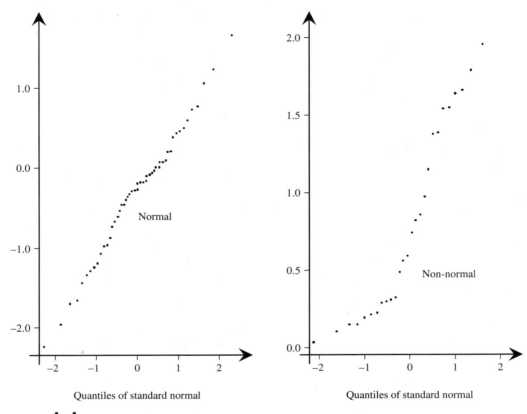

FIGURE **4.4**

An illustrative quantile plot of normal (left) and non-normal (right) data

selves (e.g., histogram) is often revealing. Specialized plots are also useful to assess normality. When a sample from a normal distribution is plotted with special techniques (called a quantile plot) [10], the data fall on a straight line, illustrated in Figure 4.4. Two quantile plots, one from normally distributed data (left) and one from data that are not normally distributed (right), are illustrated in Figure 4.4.

However, visually determining whether the nonlinearity observed in a normal plot is more than would be expected solely from random variation is sometimes difficult. Normal data produce measures of skewness (a measure of symmetry) and kurtosis (a measure of the peak of a unimodal distribution) close to zero, providing another way to examine the requirement of normality. These two statistical measures reflecting the shape of the sampled distribution are routinely produced by computer package programs but are only effective when large numbers of observations are available ($n > 100$).

Investigation of residual values only shows whether the assumed structure underlying the dependent variable seems to be inappropriate for the observed data. The

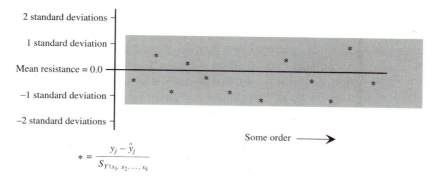

$$* = \frac{y_j - \hat{y}_j}{S_{Y|x_1, x_2, \ldots, x_k}}$$

FIGURE **4.5**

A plot of a series of random standardized residual values

question of whether the assumptions are correct is not answered. When the hypothesis that the residual values are normally distributed is accepted it means, as usual, that insufficient evidence exists to disprove the hypothesis. Lack of evidence is not synonymous with the hypothesis being correct. This spirit of searching for violations of assumptions equally applies to the following considerations of residual values.

Residual Plots

Perhaps the most effective way to investigate possible violations of the assumptions of independence, equal variance, and linearity is to plot the r_j-values for each observation. The order in which the residual values are plotted is chosen by the investigator. For example, the observations might have occurred in a time sequence; then it makes sense to plot the standardized residual values in the order of first occurred to last occurred. Occasionally little attention is paid to the sequence and the residual values are plotted in the order the data were listed or coded. Computer programs usually plot residual values, for example, in the order the data are input, which is not necessarily useful. A typical residual plot is shown in Figure 4.5.

Residual values reflect only random variation when the requirements of a linear regression analysis are met. A plot of residual values when the requirements of a linear regression are fulfilled, therefore, should show no systematic pattern. For example, standardized residual values are likely to fluctuate randomly above and below a horizontal line at zero (Figure 4.5).

Residual values can be plotted against a number of possibilities (values on the horizontal axes) to explore the adequacy of a regression model to represent the data. A few examples are:

1 Independent variables ordered by time (as mentioned) or other kinds of natural sequences.

2 The estimated dependent value $(\hat{y})$.

3 Specific independent variables (x_i) or even variables not included in the regression equation.

Mostly the choice depends on the situation and, of course, a number of different plots can be inspected. Some plots are not very helpful. For example, plotting residual values against the original observations can be misleading. The correlation between residual values and dependent variables is proportional to $1 - R^2$ (expression 4.10) and, therefore, a scatter of points ($y_j - \hat{y}_j$, y_j) will approach a straight line when the regression model is a poor representation of the data ($R \approx 0$).

A wedge-shaped pattern of residual values is usually taken to indicate that the variances are not equal over the range of the plotted variable. A note of caution about this interpretation is worthwhile. If the number of plotted points increases or decreases over the range of the plot, the variability of a set of random residual values can appear to change for reasons unrelated to requirements 1–4. One cause of an apparent nonrandom pattern is the greater likelihood of observing extreme values as more observations are plotted. The observed range of a variable tends to increase with increasing sample size. Another source of apparently nonrandom residuals is the difference in variability related to the distance from the center of the data values (expression 2.17 for simple linear regression—the further x_j is from $\bar{x}$, the greater the variability of $\hat{y}_j$). Both these sources of increased variability usually have small effects, particularly for regression analyses based on moderate or large sample sizes. Figure 4.6 shows a series of residual values in which an association exists between the plotted variable and the variance of Y; the variance increases as the value of the dependent variable increases.

When a plot shows a nonrandom pattern of residual values potentially indicating unequal variances, certain transformations of the independent variable tend to stabilize the variance, producing "data" that better conform to the requirement of equal variance. The application of such transformations does not follow rigorous rules, because such rules do not exist. The usual approach is to perform the analysis with a transformed dependent variable and inspect the residual values to determine the

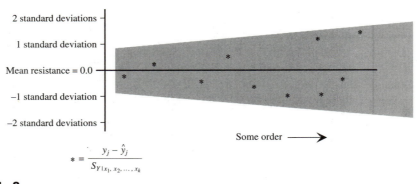

$$* = \frac{y_j - \hat{y}_j}{S_{Y|x_1, x_2, \dots, x_k}}$$

F I G U R E 4.6

A plot of a series of nonrandom standardized residual values degree of improvement. Two common transformations are the logarithm of the dependent variable and the square root of the dependent variable. Both transformations tend to make skewed distributions more symmetric by decreasing values found in the right-hand tail to a greater extent than values from the rest of the distribution. Computer systems exist that try a large number of different transformations on a data set and allow the selection of an "optimal" transformation (e.g., STATA). The use of transformations is an art form discussed in detail in other texts (e.g., [11]).

Testing Residual Values

Patterns of residual values are occasionally observed that indicate the lack of specific terms in the regression model. The lack of higher-order terms or more complicated components (e.g., $x_i x_j$ or x_i^2) can cause nonrandom patterns of residual values. A U-shaped pattern of residual values could indicate, for example, that the model lacks a quadratic term. Some rules exist for selecting a more appropriate model but, typically, the modification of a model to better fit the linear regression requirements is basically a nonstructured problem and relies, like transformations, to a large extent on intuition as well as trial and error.

To evaluate a pattern of residual values a statistical test is often a useful tool. One such formal test may detect patterns of nonrandomness by comparing the number of residual values occurring above and below the line $r = 0$. When the data fulfill the four regression requirements, the number of residual values above zero (positive) is expected to equal the number of values below zero (negative). The binomial test (sign test) or the normal approximation to the sign test ($n > 20$) is one way to assess formally this property of residual values [9]. If $P =$ the number of positive residual values occurring among the n observations, then the test statistic

$$z = \frac{P - n/2}{\sqrt{n/4}}$$

(4.20)

has an approximate standard normal distribution when the estimated residual values are only random deviations from the regression equation. That is, $n/2$ positive values are expected and the observed value P has a variance of $n/4$, giving z an approximate standard normal distribution when the residual values are randomly distributed about the line $r = 0$. Extreme values of z lead to the inference that the residual values are nonrandomly distributed.

A more sensitive approach to examining residual values for a trend is achieved by applying the Wald-Wolfowitz run test. This test, described in detail in most texts on nonparametric statistics [9], can be applied to detect patterns among the residual values. A run is defined as a sequence of identical signs. For example, the sequence

$$\underset{1}{\underline{+}} \quad \underset{2}{\underline{-\,-}} \quad \underset{3}{\underline{+\;+\;+}} \quad \underset{4}{\underline{-}} \quad \underset{5}{\underline{+\;+}} \quad \underset{6}{\underline{-\,-\,-}} \quad \underset{7}{\underline{+}}$$

contains a total of seven runs. Too few runs implies that the sequence of residual values shows a pattern and the data cannot be safely assumed to fit the structure required for a standard linear regression analysis. For a specific number of positive values (n_1) and negative values (n_2), tables exist giving the probability of observing a particular number of runs under the condition that the sequence of positive and negative signs is random [9]. Like most nonparametric procedures, a normal approximation exists for large samples ($n_1 > 10$ and $n_2 > 10$), making specialized tables unnecessary. Let T be the number of runs observed in a sequence of n_1 positive values and n_2 negative values; then

$$E(T) = \frac{2n_1 n_2}{n_1 + n_2} + 1$$

(4.21)

and the variance of T is

$$\sigma_T^2 = \frac{2n_1 n_2 (2n_1 n_2 - n_1 - n_2)}{(n_1 + n_2)^2 (n_1 + n_2 - 1)} .$$

(4.22)

The test statistic

$$z = \frac{T - E(T)}{\sigma_T}$$

(4.23)

has an approximate standard normal distribution when the signs of the residual values are distributed in a random sequence. Extreme values of z (unlikely under the hypothesis of randomness) lead to rejecting the notion that the residual values are randomly distributed, implying that the regression model needs to be modified to better reflect the structure underlying the data. For example, if the pattern of residual values is

$$\begin{array}{ccccc} + + + & - & + & - - - - - - - - - & + + + + + + + \\ 1 & 2 & 3 & 4 & 5 \end{array}$$

where $n_1 = 11$, $n_2 = 9$, and $T = 5$, then

$$E(T) = \frac{2(11)\,(9)}{20} + 1 = 10.9 \quad \text{and}$$

$$\sigma_T^2 = \frac{198(178)}{400(19)} = 4.637 ,$$

and

$$z = \frac{5 - 10.9}{\sqrt{4.637}} = -2.740 \quad \text{(one-sided, } p\text{-value} = 0.003) .$$

The value z leads to rejecting the hypothesis that the pattern of signs is consistent with a random sequence of residual values. These two statistical assessments of residual values are representative of a large number of approaches to evaluating randomness.

Linearity Transformations

Some models might be called intrinsically linear. These models become linear after applying a simple transformation to the dependent variable (requirement 4). Certain functional forms are easily transformed to be linear, making it possible to apply linear regression techniques. For example, if

$$E(Y \mid z_j) = A\, z_j^b, \text{ then } \log[E(Y \mid z_j)] = \log(A) + b \log(z_j) = a + b\, x_j \qquad (4.24)$$

is linear where $a = \log(A)$ and $x_j = \log(z_j)$. When a linear regression model results from the transformed dependent variable (e.g., $\log[y_j]$), the least squares estimates (e.g., $\hat{a}$ and $\hat{b}$) can be used to estimate the components of the original model. For example, for expression 4.24, $\hat{A} = e^{\hat{a}}$, making $\hat{y}_j = \hat{A} z_j^{\hat{b}}$ where $\hat{a}$ and $\hat{b}$ are least squares estimates from the transformed data. Other examples of intrinsically linear models are

$$E(Y \mid z_1, z_2, \cdots, z_k) = a z_1^{b_1} z_2^{b_2} \cdots z_k^{b_k},$$

$$E(Y \mid z_1, z_2, \cdots, z_k) = e^{a + b_1 z_1 + b_2 z_2 + \cdots + b_k z_k},$$

or

$$E(Y \mid z_1, z_2, \cdots, z_k) = \frac{1}{1 + e^{a + b_1 z_1 + b_2 z_2 + \cdots + b_k z_k}}.$$

The analysis of a transformed dependent variable is not different from the usual linear regression analysis (e.g., the same computer programs apply). However, it should be noted that the basic linear regression requirements 1–4 apply to the transformed variable and not the variable itself. For example, it is required that the $\log(y_j)$ (expression 4.24) have at least an approximate normal distribution with equal variance for all x_j-values, which means the original observed values y_j must have an asymmetric distribution skewed to the right at each value of x_j.

Polynomial Regression

The parameters of a polynomial regression model can be estimated by the same least squares process used to estimate coefficients for the k-variate regression model. The relatively simple polynomial model of second degree (a quadratic model) serves to illustrate the principles and the concepts, which easily generalize to a polynomial model of any degree.

A polynomial regression model of degree k is

$$E(Y \mid x) = a + b_1 x + b_2 x^2 + b_3 x^3 + \cdots + b_k x^k \qquad (4.25)$$

and reduces to a quadratic model when $k = 2$ or $E(Y \mid x) = a + b_1 x + b_2 x^2$. The estimation of the coefficients a, b_1, and b_2 is identical to the least squares estimation procedure discussed earlier. The data consist of a dependent variable and an independent

T A B L E   **4.5**

Polynomial regression applied to 10 pairs of observations (x, y)

Obs.	y	x	x^2	$\hat{y}^{(1)}$	$\hat{y}^{(2)}$	residuals$^{(1)}$	residuals$^{(2)}$
1	70	1	1	11.600	65.645	58.400	4.355
2	82	2	4	40.133	58.148	41.867	23.852
3	35	3	9	68.667	59.659	−33.667	−24.659
4	52	4	16	97.200	70.177	−45.200	−18.177
5	75	5	25	125.733	89.703	−50.733	−14.703
6	122	6	36	154.267	118.236	−32.267	3.764
7	180	7	49	182.800	155.777	−2.800	24.223
8	220	8	64	211.333	202.326	8.667	17.674
9	260	9	81	239.867	257.882	20.133	2.118
10	304	10	100	268.400	322.445	35.600	−18.445

x-value formed into sets of three values (i.e., y_j, x_j, and x_j^2). The independent variables x and x^2 play the same role as x_1 and x_2 in the bivariate regression. However, it is important to note that x and x^2 are likely correlated and, therefore, that the estimates of the coefficients b_1 and b_2 may lack precision, making it difficult to clearly assess the degree of nonlinearity in a data set (i.e., to assess $\hat{b}_2$).

The data given in Table 4.5 and Figure 4.7 serve to illustrate. The least squares estimated coefficients for the second-degree polynomial model are $\hat{a} = 82.150$, $\hat{b}_1 = -21.008$, and $\hat{b}_2 = 4.504$, giving the estimated regression equation $\hat{y}_j = 82.150 - 21.008x_j + 4.504x_j^2$. Also shown in Table 4.5 are the estimated values $\hat{y}^{(i)}$ and their corresponding residual values. The notation (i) indicates two different models: (1) the model excluding and (2) the model including the second-order term (Figure 4.7). Identical to the previous regression analyses, the squared multiple correlation coefficient is the ratio of the regression sum of squares to the total sum of squares. Specifically, $R^2 = 0.963$ reflects the high degree of accuracy of the quadratic model applied to the data in Table 4.5.

A Model with Higher-Order Terms

A statistical test of the coefficients that determine the nonlinearity ($\hat{b}_2$ in the illustration case) indicates whether the data support a model with higher-order terms. A t-test or F-to-remove test are the usual approach to evaluating the gains from representing the data by a quadratic model. The residual values from the illustrative data are substantially reduced by including the x^2-term (the residual sums of squares $= 13,710.53$ versus $3,000.53$). Also note the pattern of the residual values (residuals$^{(1)}$) shows five consecutive negative values). The t-test applied to $\hat{b}_2$ ($\hat{b}_2 = 4.504$ with a standard error of 0.901 yields a t-value of 4.999 and a p-value < 0.001) also gives a strong indication that the x^2-term is an important part of the description of the data.

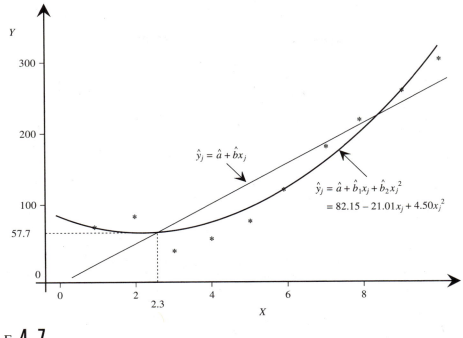

The figure shows a scatter of data points with two fitted curves. The linear fit is labeled $\hat{y}_j = \hat{a} + \hat{b}x_j$ and the quadratic fit is labeled:

$$\hat{y}_j = \hat{a} + \hat{b}_1 x_j + \hat{b}_2 x_j^2$$
$$= 82.15 - 21.01 x_j + 4.50 x_j^2$$

The Y-axis shows values 0, 57.7, 100, 200, 300 and the X-axis shows values 0, 2, 2.3, 4, 6, 8.

FIGURE 4.7

Least squares fit of a second-degree polynomial equation

For the quadratic model, the maximum or minimum occurs at $x_m = -b_1/(2\,b_2)$ and can be estimated by using the estimated coefficients. From the example, $\hat{x}_m = 21.008/(2 \times 4.504) = 2.332$. The corresponding estimated minimum value of the dependent variable is $\hat{y}_m = \hat{a} + \hat{b}_1 \hat{x}_m + \hat{b}_2 \hat{x}_m^2$ and, continuing the example, is $\hat{y}_m = 57.651$.

When higher-order terms are included in a linear model, they are usually correlated with the lower-order terms. For the example data in Table 4.5, the correlation between x and x^2 is 0.975. As already noted, high correlations among independent variables reduce the reliability of the estimates of the corresponding coefficients. A simple transformation sometimes reduces this correlation. If the analysis is carried out using the variable $x - \bar{x}$ instead of x, a more precise analysis can be achieved because the correlation between $x - \bar{x}$ and $(x - \bar{x})^2$ is frequently smaller than the correlation between x and x^2. For the data in Table 4.5, the *correlation*$[(x - \bar{x}), (x - \bar{x})^2]$ is 0.0 (an extreme case). Such a transformation centers the data and often has only a small effect on the estimated regression coefficients while reducing the variability associated with their estimation.

In general, a polynomial equation of any degree $(k < n)$ can be estimated and evaluated as a possible analytic model to represent the relationships within a data set. The "data" are values constructed from the observed values (e.g., $y, x, x^2, x^3, x^4, \cdots, x^k$). The usual least squares method is applied to estimate the coefficients and the associated sums of squares provide a statistical evaluation of the usefulness of the higher-order terms. An *F*-to-remove strategy is helpful in deciding how many terms should be included in a polynomial regression equation, or

$$F = \frac{(SS_{\text{lower-order}} - SS_{\text{higher-order}})/(df_0 - df_1)}{(SS_{\text{higher-order}})/df_1}$$

where df_1 are the degrees of freedom associated with the higher-order model and df_0 are the degrees of freedom associated with the lower-order model (i.e., $n - (k + 1) =$ degrees of freedom for a k-degree polynomial). The test statistic F, as before, has an F-distribution with degrees of freedom $df_0 - df_1$ and df_1 when the higher-order terms add no systematic information to the description of the data.

Including Interaction Effects

A special type of polynomial regression model allows for interaction effects. An interaction is the failure of one variable to have the same relationship to another variable for all values of a third variable. It is often necessary to describe data using specialized polynomial models to reflect interactions. For example, if the relationship between an independent variable x_1 and a dependent variable Y depends on the level of an another variable x_2, then x_1 and x_2 are said to interact. A polynomial regression model that includes an interaction is

$$E(Y \mid x_1, x_2) = a + b_1 x_1 + b_2 x_2 + b_3 x_1 x_2 . \tag{4.26}$$

This polynomial form of a regression equation allows the influence of x_1 on the dependent variable to depend on specific values of x_2. This dependency is most easily seen when x_2 is a binary variable coded as 0 or 1, although x_2 can be a continuous measure and the same principles apply. In the binary case, the relationship between x_1 and the dependent variable Y is described by two different simple linear regression equations depending on the value of x_2; namely

$$E(Y \mid x_1, x_2) = a + b_1 x_1 + b_2 x_2 + b_3 x_1 x_2 .$$

becomes

$$x_2 = 0: \ E(Y \mid x_1, x_2 = 0) = a + b_1 x_1, \ \text{and}$$
$$x_2 = 1: \ E(Y \mid x_1, x_2 = 1) = (a + b_2) + (b_1 + b_3)x_1 = A + Bx_1 .$$

If b_2 and b_3 are not zero, different intercepts and different slopes are necessary to describe the response in Y associated with x_1 for each level of x_2 (an interaction). That is, the relationship between Y and x_1 is different depending on the value of x_2, two straight lines with different slopes and intercepts. If $b_3 = 0$, then two parallel lines (same slope but different intercepts; no interaction) describe the relationship between Y and x_1, or

$$x_2 = 0: \ E(Y \mid x_1, x_2 = 0) = a + b_1 x_1 , \ \text{and}$$
$$x_2 = 1: \ E(Y \mid x_1, x_2 = 1) = (a + b_2) + b_1 x_1 = A + b_1 x_1 .$$

Because the relationship between Y and x_1 is the same for both values of x_2, no interaction is present. That is, the relationship between x_1 and Y can be described without reference to variable x_2. Furthermore, if b_2 is also 0, then the variable x_2 plays no role in the description of the dependent variable. Therefore, a series of hypotheses can be evaluated concerning the relationship of the independent variables x_1 and x_2 to the outcome Y by estimating and assessing a series of nested regression models. To summarize, the four nested regression models are

Regression Coefficients			Model	Expression
$b_1 \neq 0$	$b_2 \neq 0$	$b_3 \neq 0$	Two nonparallel lines	$a + b_1 x_1 + b_2 x_2 + b_3 x_1 x_2$
$b_1 \neq 0$	$b_2 \neq 0$	$b_3 = 0$	Two parallel lines	$a + b_1 x_1 + b_2 x_2$
$b_1 \neq 0$	$b_2 = 0$	$b_3 = 0$	One line, nonzero slope	$a + b_1 x_1$
$b_1 = 0$	$b_2 = 0$	$b_3 = 0$	One horizontal line	a

A Regression Model Example with an Interaction Term

To illustrate a regression model with an interaction term, data on body weights and cholesterol levels of $n = 30$ men age 40 classified as type A (coronary-prone behavior) and type B (noncoronary-prone behavior) are given in Table 4.6 and plotted in Figure 4.8.

A statistical model describing the relationship between body weight and cholesterol while accounting for the two behavior-type classifications is given by a polynomial equation with an interaction term included (expression 4.26). The value x_1 rep-

T A B L E **4.6**

Body weight, cholesterol (mg/100 ml), and behavior type (A and B) for 30 men (age = 40).

Obs.	Chol. (y)	Weight (x_1)	A/B (x_2)	Obs.	Chol. (y)	Weight (x_1)	A/B (x_2)
1	289	231	A=1	16	261	210	A=1
2	206	136	B=0	17	247	222	A=1
3	200	140	B=0	18	197	185	B=0
4	319	192	A=1	19	179	135	B=0
5	263	165	A=1	20	191	147	B=0
6	240	158	B=0	21	194	150	B=0
7	181	135	B=0	22	337	155	B=0
8	181	135	B=0	23	256	160	A=1
9	275	192	A=1	24	256	170	B=0
10	135	133	B=0	25	315	228	A=1
11	293	191	A=1	26	283	250	A=1
12	233	150	B=0	27	232	160	B=0
13	176	106	B=0	28	317	191	A=1
14	263	182	A=1	29	193	178	B=0
15	282	262	A=1	30	140	138	B=0

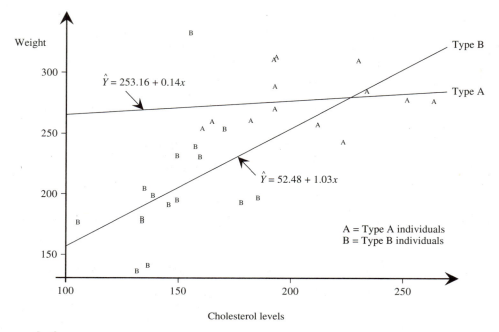

FIGURE **4.8**

Data on body weight, cholesterol, and behavior type (A and B) for 30 men (age = 40)

resents weight and x_2 is a binary variable representing behavior type (type A = 1, aggressive personalities prone to coronary disease, and type B = 0, less aggressive personalities who are less coronary prone). The dependent variable is an individual's cholesterol level. The term $x_1 x_2$ allows for the possibility that body weight and behavior type interact. The resulting estimated parameters are $\hat{a} = 52.477$, $\hat{b}_1 = 1.027$, $\hat{b}_2 = 200.684$, and $\hat{b}_3 = -0.888$, yielding the estimated regression equation

$$\hat{y}_i = 52.477 + 1.027x_{1j} + 200.684x_{2j} - 0.888x_{1j}\,x_{2j}$$

as a representation of the relationship of cholesterol to body weight for type A and type B individuals. Using the standard error of $\hat{b}_3$ ($S_{\hat{b}_3} = 0.593$) and a t-test ($t = -0.888/0.593 = -1.497$ with degrees of freedom = 26) indicates (p-value = 0.073) that weight and cholesterol possibly interact ($b_3 \neq 0$) with regard to behavior type. That is, it is probably necessary to describe the relationship between weight and cholesterol by two straight lines with different intercepts and slopes, one for type B individuals and one for type A individuals. Specifically,

$$\text{type B } (x_2 = 0): \ \hat{y}_i = \hat{a} + \hat{b}_1 x_{1i} = 52.477 + 1.027x_{1i}, \quad \text{and}$$
$$\text{type A } (x_2 = 1): \ \hat{y}_i = (\hat{a} + \hat{b}_2) + (\hat{b}_1 + \hat{b}_3)x_{1i} = 253.161 + 0.139x_{1i}.$$

Test Case | **Path Coefficients**

Path Coefficient

Path coefficients using the data from the test case in Chapter 3: $Y = CO$, $x_1 =$ cigarettes smoked per day, and $x_2 =$ time since smoked last cigarette

Standardized data (from previous test case, Chapter 3):

	1	2	3	4	5	6	7	8	9	10	Mean
$CO = y_j^*$	−2.178	−0.854	−0.412	−0.118	0.177	0.177	0.177	0.765	0.912	1.354	0.0
Cigs./day = z_{1j}	−0.671	0.156	.615	−0.671	−1.130	−0.992	−0.303	0.156	2.223	0.615	0.0
Time = z_{2j}	0.415	1.674	−0.741	0.796	−0.452	−1.029	1.281	0.022	−0.937	−1.029	0.0

where

$$y_j^* = (y_j - \overline{y})/S_Y \qquad z_{1j} = (x_{1j} - \overline{x}_1)/S_1 \qquad z_{2j} = (x_{2j} - \overline{x}_2)/S_2 .$$

Standardized variables:

$$S_{z_1}^2 = S_{z_2}^2 = S_Y^2 = 1.0 \qquad \text{and} \qquad \overline{y}^* = \overline{z}_1 = \overline{z}_2 = 0$$

$$S_{12} = r_{12} = -0.286 \qquad S_{1Y^*} = r_{Y1} = 0.412 \qquad S_{2Y^*} = r_{Y2} = -0.488 .$$

Normal equations:

$$\hat{p}_1 + \hat{p}_2 r_{12} = r_{Y1} \qquad \hat{p}_1 + \hat{p}_2(-0.286) = 0.412$$

and

$$\hat{p}_1 r_{12} + \hat{p}_2 = r_{Y2} \qquad \hat{p}_1(-0.286) + \hat{p}_2 = -0.488 .$$

Estimates of path coefficients:

$$\hat{p}_1 = (r_{Y1} - r_{Y2} r_{12})/(1 - r_{12}^2) = 0.296 \text{ or}$$
$$\hat{p}_1 = \hat{b}_1 S_1/S_Y = 0.296$$

and

$$\hat{p}_2 = (r_{Y2} - r_{Y1} r_{12})/(1 - r_{12}^2) = -0.403 \text{ or}$$
$$\hat{p}_2 = \hat{b}_2 S_2/S_Y = -0.403 .$$

The regression equation is

$$\hat{y}_j^* = \hat{p}_1 z_{1j} + \hat{p}_2 z_{2j}$$
$$= 0.296 z_{1j} - 0.403 z_{2j} .$$

Relevant sums of squares:

$$\Sigma\{\hat{y}_j^*\}^2 = (n-1)(\hat{p}_1 r_{Y1} + \hat{p}_2 r_{Y2}) = (n-1)R^2 = 9(.318) = 2.865$$

$$\Sigma(y_j^* - \hat{y}_j)^2 = (n-1)(1 - \hat{p}_1 r_{Y1} - \hat{p}_2 r_{Y2}) = (n-1)(1-R^2) = 9(.682) = 6.135$$

$$\Sigma\{y_j^*\}^2 = n-1 = 9$$

Analysis of Variance Table

	Sum of Squares	Degrees of Freedom	Mean Square
Regression	2.865	$k = 2$	1.433
Residual	6.135	$n - (k+1) = 7$	0.876
Total	9.000	$n - 1 = 9$	

H_0: $\hat{p}_1$ and $\hat{p}_2$ are simultaneously equal to 0

$$F = 1.433/0.876 = 1.635 \approx F_{2,7} \quad p = 0.261$$

$$\text{and} \quad R^2 = 2.865/9 = 0.318$$

Squared part correlations are

$$C_{(1)}^2 = \hat{p}_1^2(1 - r_{12}^2) = (0.088)(0.918) = 0.080 \quad \text{influence from number of cigs./day}$$

$$C_{(2)}^2 = \hat{p}_2^2(1 - r_{12}^2) = (0.162)(0.918) = 0.149 \quad \text{influence from time last smoked}$$

Linear Regression

The following STATA code produces a regression analysis using the test data from Chapter 3 but standardized to illustrate path coefficients. As in Chapter 3, Y = CO levels (ppm), x_1 = amount reported smoked per day, and x_2 = time since last smoked (hours). Note that the path coefficients are also easily produced by adding the word *beta* to the usual STATA regression command (e.g., "regress co time cigs, beta"). As noted, a path coefficient is sometimes referred to as a beta coefficient.

```
example.path

. infile co time cigs using co2.dat
(10 observations read)

. correlate co time cigs
(obs=10)

         |      co      time      cigs
---------+---------------------------------
      co |   1.0000
    time |  -0.4875    1.0000
    cigs |   0.4117   -0.2862    1.0000
```

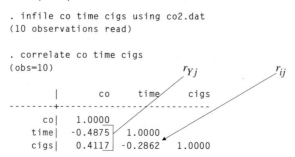

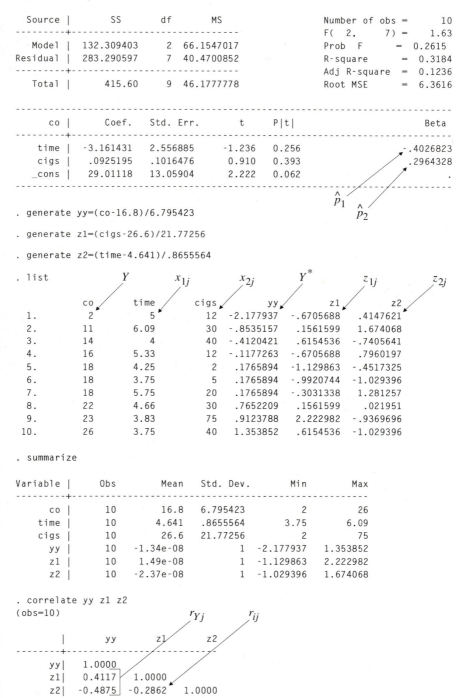

. regress co time cigs, beta

Source	SS	df	MS
Model	132.309403	2	66.1547017
Residual	283.290597	7	40.4700852
Total	415.60	9	46.1777778

Number of obs = 10
F(2, 7) = 1.63
Prob F = 0.2615
R-square = 0.3184
Adj R-square = 0.1236
Root MSE = 6.3616

co	Coef.	Std. Err.	t	P|t|	Beta
time	-3.161431	2.556885	-1.236	0.256	-.4026823
cigs	.0925195	.1016476	0.910	0.393	.2964328
_cons	29.01118	13.05904	2.222	0.062	.

$\hat{p}_1$ $\hat{p}_2$

. generate yy=(co-16.8)/6.795423

. generate z1=(cigs-26.6)/21.77256

. generate z2=(time-4.641)/.8655564

. list

Y x_{1j} x_{2j} Y^* z_{1j} z_{2j}

	co	time	cigs	yy	z1	z2
1.	2	5	12	-2.177937	-.6705688	.4147621
2.	11	6.09	30	-.8535157	.1561599	1.674068
3.	14	4	40	-.4120421	.6154536	-.7405641
4.	16	5.33	12	-.1177263	-.6705688	.7960197
5.	18	4.25	2	.1765894	-1.129863	-.4517325
6.	18	3.75	5	.1765894	-.9920744	-1.029396
7.	18	5.75	20	.1765894	-.3031338	1.281257
8.	22	4.66	30	.7652209	.1561599	.021951
9.	23	3.83	75	.9123788	2.222982	-.9369696
10.	26	3.75	40	1.353852	.6154536	-1.029396

. summarize

Variable	Obs	Mean	Std. Dev.	Min	Max
co	10	16.8	6.795423	2	26
time	10	4.641	.8655564	3.75	6.09
cigs	10	26.6	21.77256	2	75
yy	10	-1.34e-08	1	-2.177937	1.353852
z1	10	1.49e-08	1	-1.129863	2.222982
z2	10	-2.37e-08	1	-1.029396	1.674068

. correlate yy z1 z2
(obs=10)

r_{Yj} r_{ij}

	yy	z1	z2
yy	1.0000		
z1	0.4117	1.0000	
z2	-0.4875	-0.2862	1.0000

```
. regress yy z1 z2
```

Source	SS	df	MS
Model	2.86521824	2	1.43260912
Residual	6.13478248	7	.876397498
Total	9.00000072	9	1.00000008

Number of obs = 10
$F(2, 7)$ = 1.63
Prob > F = 0.2615
R-square = 0.3184
Adj R-square = 0.1236
Root MSE = .93616

n

F

p-value

R^2

yy	Coef.	Std. Err.	t	P|t|	[95% Conf. Interval]
z1	.2964328	.3256792	0.910	0.393	-.4736761 1.066542
z2	-.4026823	.3256792	-1.236	0.256	-1.172791 .3674266
_cons	-2.74e-08	.2960401	-0.000	1.000	-.7000237 .7000236

$\hat{p}_j$ $S_{\hat{p}_j}$ *p*-value

```
. exit
```

Applied Example

Multiple Linear Regression

Nine variables, selected from the Child Health and Development Study data, illustrate a multiple regression analysis. The first step in attempting to understanding the multivariable relationships among these nine variables is to examine the array of all possible pairwise correlation coefficients (see following computer output). A number of associations emerge:

1   The highest correlation ($r = 0.817$) is, not surprisingly, between age of mother (MAGE) and age of father (FAGE).

2   Length of gestation (GESTWKS) is essentially uncorrelated with the parental variables ($|r| < 0.07$) and is strongly correlated with birth weight ($r = 0.426$).

3   A number of variables show rather expected associations—variables related to adult size (mother's prepregnancy weight (MPPWT), mother's height (MHEIGHT), and father's height (FHEIGHT)) are all moderately correlated with each other ($r > 0.166$). Also maternal (MNOCIG) and paternal (FNOCIG) smoking habits are correlated ($r = 0.262$).

4   At least one variable shows an unexpected correlation—father's age and father's height have a negative correlation of $r = -0.134$, whereas mother's age and mother's height show almost no correlation ($r = 0.018$).

Exploring a correlation array is an unstructured but a productive way to begin an analysis. The process ultimately becomes unsatisfactory because correlation coefficients only indicate pairwise linear associations when typically more complicated relationships exist.

A multiple regression analysis with birth weight (BWT) as the dependent variable and nine independent variables is one approach to exploring the entire multivariate structure. The regression equation explains a significant 26.2% of the total variation ($F = 26.42$ with p-value < 0.001). The squared multiple correlation coefficient adjusted for shrinkage ($R_{adj}^2 = 0.252$) is not very different from the unadjusted value ($R^2 = 0.262$). Shrinkage is not much of an issue when the number of independent variables ($k = 9$) is much less than the number of observations ($n = 680$).

The comparison of the nine independent variables by means of individual t-statistics indicates the relative magnitude of the unique contribution of each variable to the overall variability in birth weight. Comparing the t-values is essentially the same as comparing the part correlation coefficients because these two statistics always give the same relative magnitude; they differ by a constant factor. Gestation period, as expected from the pairwise correlation structure, is the largest contributor to the explained variation in birth weight. Maternal smoking, maternal prepregnancy weight, paternal height, and maternal height are the next four most important contributors. The remaining four variables (mother's age, father's age, father's smoking, and father's education) have significance probabilities greater than 0.05, which is generally, but somewhat arbitrarily, the point above which variables are considered nonsignificant contributors.

A question of interest is: What is the relative influence on birth weight from the maternal variables compared to the paternal variables? This question is addressed by contrasting two separate regression analyses based on reduced models with the full model employing all nine variables. To investigate the impact of the paternal variables, for example, the regression equation with all the four paternal variables (FHEIGHT, FNOCIG, FEDYRS, and FAGE) removed yields a residual sum of squares 605.071. The removal of the paternal variables shows an increase in the residual sum of squares of 7.100 over the full model (residual sum of squares = 597.971). A formal assessment of this increase comes from generating an F-statistic:

full model—all nine variables:

$$SS_1 = 597.971 \text{ with degrees of freedom} = 670$$

paternal variables removed:

$$SS_0' = 605.071 \text{ with degrees of freedom} = 674$$

yielding

$$F = \frac{(605.071 - 597.971)/(674 - 670)}{597.971/670} = \frac{1.775}{0.892} = 1.989 \ .$$

Under the hypothesis that the paternal variables have only a random influence on birth weight, $F = 1.989$ comes from an F-distribution with 4 and 670 degrees of freedom producing a significance probability of 0.095.

Similarly, removal of the four variables that directly involve maternal effects (MHEIGHT, MNOCIG, MPPWT, and MAGE) also allows evaluation of the relative impact of maternal variables and produces a statistical assessment of the maternal influence on birth weight:

full model—all nine variables:

$$SS_1 = 597.971 \text{ with degrees of freedom} = 670$$

maternal variables removed:

$$SS_0'' = 645.652 \text{ with degrees of freedom} = 674$$

yielding

$$F = \frac{(645.652 - 597.971)/(674 - 670)}{597.971/670} = \frac{11.920}{0.892} = 13.356 .$$

The likelihood of $F = 13.356$ occurring by chance when the maternal variables have no role in determining birth weight is less than 0.001, from an F-distribution with 4 and 670 degrees of freedom.

Contrasting sums of squares also provides a commensurate way of expressing the relative influences of the independent variables on the dependent variable. A direct measure of the relative impact on birth weight from the maternal and paternal variables is the decrease in goodness-of-fit incurred by deleting each set of parental variables (maternal: $SS_0'' - SS_1 = 47.7$; paternal: $SS_0' - SS_1 = 7.1$). The 6.7-fold increase in difference shows the considerably stronger influence of the maternal variables on an infant's birth weight.

Notice that the estimates of the coefficients that remain in the regression equation after parental variables are removed change in value, but not dramatically. For example, the coefficient associated with length of gestation in the full model is $\hat{b}_1 = 0.233$, $\hat{b}_1 = 0.246$ in the model with maternal variables removed, and $\hat{b}_1 = 0.234$ with paternal variables removed. These estimates illustrate a slight confounding bias associated with gestation time from the maternal variables and almost no bias associated with the paternal variables.

Linear Regression: *k*-Variable Regression

The CHDS data ("use chds") is used to analyze the influence of 9 parental variables on the birth weight of a newborn infant. A description of these 10 variables (the 9 independent variables and the 1 constant term) is given in Chapter 2. Of particular interest is an evaluation of the relative influences of the paternal versus the maternal variables on birth weight.

chds.multivariable

```
. use chds

. correlate bwt gestwks mage mnocig mheight mppwt fage fedyrs fnocig fheigh
(obs=680)
```

```
       |      bwt  gestwks     mage   mnocig  mheight    mppwt     fage
-------+-------------------------------------------------------------------
   bwt|   1.0000
gestwks|   0.4259   1.0000
  mage|   0.0013   0.0034   1.0000
mnocig|  -0.1794  -0.0708   0.0450   1.0000
mheight|   0.2025   0.0476   0.0175   0.0259   1.0000
 mppwt|   0.2216   0.0517   0.1157  -0.0258   0.4942   1.0000
  fage|   0.0165   0.0422   0.8171   0.0277   0.0180   0.1240   1.0000
fedyrs|   0.0330   0.0354   0.2406   0.0237   0.1080   0.0013   0.2204
fnocig|  -0.0234  -0.0025   0.0166   0.2617  -0.0147  -0.0275   0.0397
fheight|   0.1542   0.0240  -0.0711   0.0108   0.3033   0.1664  -0.1344


       |   fedyrs   fnocig  fheight
-------+---------------------------
fedyrs|   1.0000
fnocig|  -0.1823   1.0000
fheight|   0.1078   0.0136   1.0000
```

. regress bwt gestwks mage mnocig mheight mppwt fage fedyrs fnocig fheigh

```
  Source |       SS       df       MS              Number of obs =     680
---------+------------------------------              F(  9,   670) =   26.42
   Model | 212.224792        9  23.5805324           Prob  F       =  0.0000
Residual | 597.970754      670  .892493663           R-square      =  0.2619
---------+------------------------------              Adj R-square  =  0.2520
   Total | 810.195546      679  1.19321877           Root MSE      = .94472
```

```
-------------------------------------------------------------------------------
     bwt |     Coef.   Std. Err.      t     P|t|     [95% Conf. Interval]
---------+---------------------------------------------------------------------
 gestwks |  .2334471    .019471    11.989   0.000     .1952156    .2716787
    mage | -.0013084   .0116336    -0.112   0.910    -.0241511    .0215344
  mnocig | -.0152209   .0033571    -4.534   0.000    -.0218127   -.0086291
 mheight |  .0397564   .0175108     2.270   0.024     .0053737    .0741391
   mppwt |  .0084209   .0023736     3.548   0.000     .0037604    .0130814
    fage |  .0000664   .0104467     0.006   0.995    -.0204459    .0205786
  fedyrs |  .0041575   .0176683     0.235   0.814    -.0305345    .0388495
  fnocig |  .0018638   .0027182     0.686   0.493    -.0034733    .007201
 fheight |  .0390176   .0147421     2.647   0.008     .0100713    .067964
   _cons | -8.090991   1.439882    -5.619   0.000    -10.91821   -5.263768
-------------------------------------------------------------------------------
```

. regress bwt gestwks mage mnocig mheight mppwt

```
  Source |       SS       df       MS              Number of obs =     680
---------+------------------------------              F(  5,   674) =   45.70
   Model | 205.124818        5  41.0249636           Prob  F       =  0.0000
Residual | 605.070729      674  .897731051           R-square      =  0.2532
---------+------------------------------              Adj R-square  =  0.2476
   Total | 810.195546      679  1.19321877           Root MSE      = .94749
```

```
     bwt |      Coef.   Std. Err.        t    P|t|     [95% Conf. Interval]
---------+--------------------------------------------------------------------
 gestwks |    .234383    .0194694    12.039   0.000      .196155     .272611
    mage |   -.002312    .0067161    -0.344   0.731     -.015499    .0108751
  mnocig |  -.0145217    .0032423    -4.479   0.000     -.020888   -.0081555
 mheight |   .0520824    .0168865     3.084   0.002      .018926    .0852389
   mppwt |   .0085377    .0023615     3.615   0.000      .003901    .0131745
   _cons |   -6.07666    1.235067    -4.920   0.000    -8.501702   -3.651619
---------+--------------------------------------------------------------------
```

. regress bwt gestwks fage fedyrs fnocig fheigh

```
  Source |       SS       df       MS                  Number of obs =     680
---------+------------------------------              F(  5,   674) =   34.35
   Model | 164.543495     5   32.908699               Prob  F       =  0.0000
Residual | 645.652051    674  .957940729              R-square      =  0.2031
---------+------------------------------              Adj R-square  =  0.1972
   Total | 810.195546    679  1.19321877              Root MSE      =  .97874
```

```
     bwt |      Coef.   Std. Err.        t    P|t|     [95% Conf. Interval]
---------+--------------------------------------------------------------------
 gestwks |   .2455619    .0200598    12.241   0.000     .2061746    .2849492
    fage |     .00378    .0063947     0.591   0.555     -.008776    .0163359
   fedyrs |  -.0036479    .0180339    -0.202   0.840    -.0390574    .0317615
   fnocig |  -.0020451    .0027084    -0.755   0.450     -.007363    .0032728
  fheight |   .0612989    .0145365     4.217   0.000     .0327566    .0898412
    _cons |  -6.609072    1.299398    -5.086   0.000    -9.160427   -4.057717
---------+--------------------------------------------------------------------
```

. exit

Notation

Summary values:

$$k = \text{number of independent variables:}$$

$$n = \text{number of sets of observations } (y_j , x_{1j} , x_{2j} , \cdots , x_{kj})$$

$$S_i^2 = \frac{\Sigma(x_{ij} - \overline{x}_i)^2}{n - 1}$$

$$S_{ik} = \frac{\Sigma(x_{ij} - \overline{x}_i)\,(x_{kj} - \overline{x}_k)}{n - 1}$$

$$S_{Yi} = \frac{\Sigma(x_{ij} - \overline{x}_i)\,(y_j - \overline{y})}{n - 1}$$

Regression equation:

$$E(Y \mid x_1 , x_2 , \cdots , x_k) = a + b_1 x_1 + b_2 x_2 + \cdots + b_k x_k$$

Estimated regression equation:

$$\hat{y}_j = \hat{a} + \hat{b}_1 x_{1j} + \hat{b}_2 x_{2j} + \cdots + \hat{b}_k x_{kj}$$

for the individual observation indexed by j.

Estimated variance:

$$S^2_{Y|x_1, x_2, \cdots, x_k} = \frac{\Sigma(y_j - \hat{y}_j)^2}{n - (k + 1)}$$

The F-test for contribution from $b_1, b_2, \cdots, b_q$ (to remove q variables):

SS_1 = residual sum of squares for the full model; degrees of freedom = $n - (k + 1)$

SS_0 = residual sum of squares for the model with q variables removed; degrees of
degrees of freedom = $n - (k - q + 1)$

$$F = \frac{(SS_0' - SS_1)/[(n - (k - q + 1)) - (n - (k + 1))]}{SS_1/(n - (k + 1))} = \frac{(SS_0' - SS_1)/q}{SS_1/(n - (k + 1))}$$

where F has an F-distribution with q and $n - (k + 1)$ degrees of freedom when the q variables removed do not have a systematic influence.

Problems

1. Continuing the regression analysis (Table 1.4) concerning an infant's birth weight and the amount of weight gained by the mother during pregnancy, conduct a multi-variable regression analysis $k = 4$ where:

   Y = birth weight, x_1 = net weight gain ($netgain = weight_1 - weight_0 - y$), x_2 = height, x_3 = gestation, and x_4 = cigarettes smoked per day.

   Rank these four variables in terms of their importance in influencing an infant's birth weight. Calculate the part correlation associated with each of these four independent variables.

2. Calculate the 48 residual values r_i associated with $\hat{y}_i$ for the regression analysis in problem 1 and plot these values against each of the four independent variables. Plot y_i against $\hat{y}_i$. Calculate directly the correlation between $\hat{y}_i$ and y_i using all $n = 48$ observations and verify that this value equals the multiple correlation coefficient R.

3. Why not conduct the same analysis as in problem 1 but also include the mother's prepregnancy weight ($weight_0$) and last measured weight ($weight_1$)?

4. What transformation linearizes the relationship

   $$y_j = a x_{1j}^{b_1} x_{2j}^{b_2} ?$$

   Estimate a, b_1, and b_2 where x_1 = height, x_2 = weight, and Y = birth weight from Table 1.4. Why is it inappropriate to compare the residual sum of squares from this transformed data to the residual sum of squares generated from a bivariate analysis of the untransformed data to evaluate which model best represents of the relationship between dependent and independent variables?

5   If $E(Y | x) = a + bx$, show that the estimate of b is simply divided by 10 when $x' = 10x + 3$ is used instead of x. If $E(Y | x) = a + b_1x + b_2x^2$, show how the estimate of b_1 is affected when $x' = 10x + 3$ is used instead of x.

6   Construct a diagram analogous to Figure 4.2 and calculate the values a, b, c, and d for the data on five-year-olds given in problem 6, Chapter 3.

7   Fit a quadratic regression equation to the data from Table 1.4 where

$$Y = \text{birth weight}, x = \textit{netgain} \text{ (from problem 1), and}$$
$$x^2 = \text{the squared value of } \textit{netgain} .$$

Is there evidence that a "squared"-term helps describe the relationship between mother's weight gain and the weight of her infant? Repeat the same analysis using $x - \bar{x}$, and describe the differences from the analysis using x.

8   For the four independent variables in problem 1, compute $R_{(i)}^2$ for $i = 1, 2, 3$, and 4. Using these values, calculate the variance for $\hat{b}_i$ for the same four variables and compare the results to the variances estimated in problem 1.

9   Show that the mean of the n estimated values $\hat{y}_i$ from a linear regression equation is identical to the mean of the n observed values (i.e., $\hat{\bar{y}} = \bar{y}$).

10   Consider the following data where Y = length of stay of patients in a intensive care unit (hours), x_1 = severity index, and x_2 = hospital size classified as large = 0 and small = 1.

Obs.	Y	x_1	x_2
1	24	22	0
2	48	80	0
3	37	55	0
4	81	140	1
5	48	90	0
6	2	10	1
7	24	40	1
8	64	120	1
9	12	30	1
10	8	25	1
11	4	10	1
12	36	77	0
13	12	32	0
14	8	14	1
15	4	15	1
16	88	200	0
17	54	120	0

Conduct a bivariate linear regression analysis employing an interaction term to assess whether the relationship between length of stay and the severity index differs in the two types of hospitals. Test for the presence of an interaction two ways: using the coefficient $\hat{b}_3$ (associated with the interaction term) and contrasting two nested models (one containing the interaction term and one with the interaction term deleted).

References

1 Vogt., T. M., Selvin, S., and Hulley, S. B. Comparison of Biochemical and Questionnaire Estimates of Tobacco Exposure. *Preventive Medicine* 8 (1979): 23–33.
2 Cohen, J., and Cohen, P. *Applied Multiple Regression/Correlation Analysis for the Behavioral Sciences.* Hillsdale, N.J.: Lawrence Erlbaum Associates, 1975.
3 Li, C. C. *Path Analysis—A Primer.* Pacific Grove, Calif.: Boxwood Press, 1975.
4 Blalock, H. M. (editor). *Causal Models in Social Science Research.* Chicago: Aldine-Atherton, 1971.
5 Neter, J., Wasserman, W., and Kutner, M. H. *Applied Linear Statistical Models: Regression, Analysis of Variance and Experimental Designs.* Homewood, Ill.: Irwin, 1990.
6 Belsley, D. A., Kuh, E. K., and Welsch, R. E. *Regression Diagnosis.* New York: John Wiley, 1980.
7 David, H. A. *Order Statistics.* New York: John Wiley, 1980.
8 Dixon, W. J., and Massey, F. J. *Introduction to Statistical Analysis.* New York: McGraw-Hill, 1969.
9 Conover, W. J. *Practical Nonparametric Statistics.* New York: John Wiley, 1971.
10 Chambers, J. M., Cleveland, W. S., Kleiner, B., and Tukey, P. A. *Graphical Methods for Data Analysis.* Belmont, Calif.: Wadsworth International Group, 1983.
11 Afifi, A. A., and Clark, V. *Computer-Aided Multivariate Analysis,* 2nd ed. New York: Chapman and Hall, 1990.

5

Analysis of Covariance

Background

The analysis of covariance, a special case of regression analysis (Chapters 2–4), involves a continuous dependent variable and both categorical and continuous independent variables. The central purpose of an analysis of covariance is to compensate for influences on the dependent variable that interfere with direct comparisons among a series of categories. Situations arise, for example, where mean values differ to some extent because of the influence of one or more extraneous continuous variables. The analysis of covariance seeks to remove the influence of these variables, called covariables, that bias direct comparisons among a series of categories. For example, the mean birth weights of a sample of infants differ when classified by mother's smoking habits (never smoked, past smokers, and present smokers), but the three groups also differ by maternal age (covariable). The contributions to the observed differences caused by maternal age, though of little interest, bias the evaluation of possible effects of maternal smoking exposure on birth weight. The analysis adjusts the mean birth weights among the three smoking categories so that comparisons are "free" from the influences of maternal age. The word *free* should not be taken too literally. Statistical adjustment, which is at the center of the analysis of covariance, depends on the validity (or, at least, the goodness-of-fit) of a specific statistical model. When this underlying model is appropriate, it is possible to remove effectively the influence of a covariable by regression techniques. Differences observed among the adjusted mean values are then no longer influenced by the effects of one or more extraneous variables. Similar to regression analysis, goodness-of-fit is addressed as part of the analysis. The following sections concentrate on the simplest analysis of covariance followed by a less detailed description of more general and more complicated approaches.

Method: Criterion

The basic analysis of covariance combines simple linear regression (Chapter 2) with one-way analysis of variance (reviewed in Chapter 1). The data are collected in pairs (a covariable and a dependent variable) and classified into a series of categories, illustrated in Table 5.1. The symbol g represents the number of categories or groups (independent variable). The symbol x represents a continuous covariable and the sampled value of the dependent variable is denoted, as before, by y. The notation (y_{ij}, x_{ij}) indicates the jth observation from the ith group. Furthermore, the covariable x is related to the dependent variable and, on the average, differs among the g categories. The direct comparison of the sample mean values $\bar{y}_1, \bar{y}_2, \cdots, \bar{y}_g$, therefore, reflects any effects from the categorical classification as well as influences from the covariable, making direct evaluation of the role of the categorical variable difficult without compensating for the influence of the covariable x. If the variable x is unrelated to the dependent variable or does not differ among the g categories, then the comparison of the $\bar{y}$-values is not affected by x and an analysis of covariance is unnecessary.

An analysis of covariance can be viewed as a sequential investigation of the fit of three nested statistical models. As before, the criterion to examine the goodness-of-fit of these competing models is an F-statistic, used to contrast residual sums of squares.

Model I: Interaction

The key to an analysis of covariance is the feasibility of a specific statistical model, called for simplicity, model II. The effectiveness of model II is evaluated by a comparison with a more general statistical structure called model I (depicted in Figure

T A B L E **5.1**

Notation for the analysis for covariance

Categories	1	2	$\cdots$	g
Observations	(x_{11}, y_{11})	(x_{21}, y_{21})	$\cdots$	(x_{g1}, y_{g1})
	(x_{12}, y_{12})	(x_{22}, y_{22})	$\cdots$	(x_{g2}, y_{g2})
	.			
	.			
	.			
	(x_{1n_1}, y_{1n_1})	(x_{2n_2}, y_{2n_2})	$\cdots$	(x_{gn_g}, y_{gn_g})
Sample size per group	n_1	n_2	$\cdots$	n_g
Mean values per group	$(\bar{x}_1, \bar{y}_1)$	$(\bar{x}_2, \bar{y}_2)$	$\cdots$	$(\bar{x}_g, \bar{y}_g)$

FIGURE **5.1**

Analysis of covariance,
interaction model
(model I)

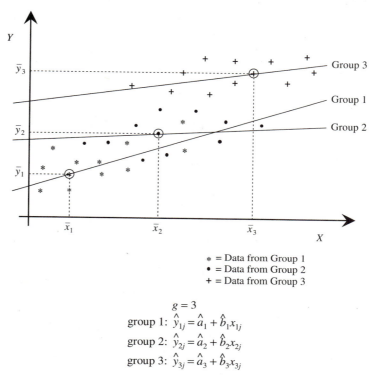

* = Data from Group 1
• = Data from Group 2
+ = Data from Group 3

$$g = 3$$
$$\text{group 1: } \hat{y}_{1j} = \hat{a}_1 + \hat{b}_1 x_{1j}$$
$$\text{group 2: } \hat{y}_{2j} = \hat{a}_2 + \hat{b}_2 x_{2j}$$
$$\text{group 3: } \hat{y}_{3j} = \hat{a}_3 + \hat{b}_3 x_{3j}$$

5.1). Model I postulates that the covariable x is related to the dependent variable Y by a series of simple linear regression equations that differ among some or all of the g groups. The property that the relationship between the covariable and the dependent variable differs among groups is an interaction. Like the previous description of an interaction (expression 4.26), the relationship between variables x and Y depends on the value of a third variable (categories).

Specifically, a series of straight lines with different intercepts (a_i) and different slopes (b_i) describes the sample data within each group. In symbols, the expected values of the dependent variable for each group is given by the simple linear equation

$$E(Y_{ij} \mid x_{ij}) = a_i + b_i x_{ij} \qquad (5.1)$$

where i indicates group membership and j indicates the specific observation within the group. This statistical model allows the parameters of each regression line (a_i and b_i) to differ among the g categories ($i = 1, 2, \cdots, g$).

A simple linear regression analysis applied to the n_i observed pairs (x_{ij}, y_{ij}) in each of the g categories yields an estimate of a_i and b_i denoted, as before, $\hat{a}_i$ and $\hat{b}_i$. The process involves g separate regression analyses, requires $2g$ estimates, and produces g estimated linear regression lines. Furthermore, each estimated regression line certainly fails to fit the data perfectly and the lack of fit is measured by the residual sum of squares within each group. The residual sum of squares for the ith category is the sum of squared deviations of the points on the estimated regression line from each corresponding observation or, for the ith category,

$$\hat{y}_{ij} = \hat{a}_i + \hat{b}_i x_{ij} .$$

Then,

$$Res_i = \sum_j (y_{ij} - \hat{y}_{ij})^2 \tag{5.2}$$

is the residual sum of squares for a specific group. This residual sum of squares divided by the variance of the dependent variable has a chi-square distribution with $n_i - 2$ degrees of freedom when the requirements hold for a regression analysis (independence, normality, equal variance, and linearity). The estimation and evaluation of each of these g regression equations is identical to the simple linear regression analysis described in Chapter 2; the process is just repeated g times, once for each group.

An overall assessment of model I comes from summarizing the total fit of the g regression equations by

$$Res(I) = Res_1 + Res_2 + \cdots + Res_g = \sum Res_i$$
$$= \sum_i \sum_j (y_{ij} - \hat{y}_{ij})^2 . \tag{5.3}$$

The value $Res(I)$ divided by the variance of Y also has a chi-square distribution under model I with $\sum(n_i - 2) = N - 2g$ degrees of freedom where $N = \sum n_i$ is the total number of observations. It is necessary to make $2g$ independent estimates to establish the N estimated values $\hat{y}_{ij}$ under model I, yielding $N - 2g$ degrees of freedom.

Model II: No Interaction

Model II allows the expected values among the g categories to differ but the relationship between covariable x and the dependent variable Y within each category is the same, namely linear with equal slopes (Figure 5.2). In symbols, model II is

$$E(Y \mid x_{ij}) = a_i' + b' x_{ij} . \tag{5.4}$$

Model II is a special case of model I and provides the specific structure to statistically remove the influence of the covariable. A set of $g + 1$ parameters $a_1', a_2',$ $\cdots, a_g',$ and b' defines a series of parallel regression lines. These lines have g intercepts that differ among categories but have the same slope, denoted b'. Because the relationship between the independent and dependent variable is the same in each group, no interaction exists (no-interaction model). In other words, the relationship between variables x and Y does not depend on the values of a third variable (categories).

The fit of model II is also measured by a residual sum of squares. For each of the g groups, the estimated linear model for the ith group is

$$\hat{y}_{ij}' = \hat{a}_i' + \hat{b}' x_{ij}$$

FIGURE **5.2**

Analysis of covariance,
no interaction (model II)

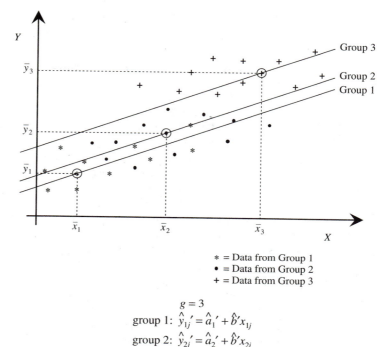

* = Data from Group 1
• = Data from Group 2
+ = Data from Group 3

$$g = 3$$

group 1: $\hat{y}_{1j}' = \hat{a}_1' + \hat{b}'x_{1j}$

group 2: $\hat{y}_{2j}' = \hat{a}_2' + \hat{b}'x_{2j}$

group 3: $\hat{y}_{3j}' = \hat{a}_3' + \hat{b}'x_{3j}$

and

$$Res_i' = \sum_j (y_{ij} - \hat{y}_{ij}')^2 ,$$

(5.5)

producing the summary measure of fit of the g regression equations

$$Res(II) = Res_1' + Res_2' + \cdots + Res_g' = \Sigma Res_i'$$

(5.6)

or, similar to model I,

$$Res(II) = \sum_i \sum_j (y_{ij} - \hat{y}_{ij}')^2 .$$

(5.7)

The quantity represented by $Res(II)$ divided by the variance of Y has a chi-square distribution under model II with $N - (g + 1)$ degrees of freedom. Because $g + 1$ estimates are required to produce the N estimated values $\hat{y}_{ij}'$, the degrees of freedom are $N - (g + 1)$. The residual sum of squares is related to a chi-square distribution when, in addition to the four usual regression analysis requirements, the slopes of the regression lines within each group are identical. Note that the intercept values a_i' generated by model II are not equal to the a_i-values used to define model I.

Because model II ($b_1 = b_2 = \cdots = b_g = b'$) is a restriction of model I, an F-statistic serves to assess whether model II is consistent with the data. The formal F-statistic is

$$F = \frac{(Res(II) - Res(I))/([N - (g + 1)] - [N - 2g])}{Res(I)/(N - 2g)}$$

$$= \frac{(Res(II) - Res(I))/(g - 1)}{Res(I)/(N - 2g)} \tag{5.8}$$

which has an F-distribution with $g - 1$ and $N - 2g$ degrees of freedom when the g slopes are equal, no interaction.

If the value F is unlikely to have occurred by chance, then model II is not a tractable representation of the data and adjustment by linear regression techniques is not meaningful. The rejection of model II indicates a statistical interaction—the relationship between x and Y differs among some or all of the g categories. Geometrically, a large difference between the fit of model I and model II indicates that the relationship between x and Y is not adequately represented by a set of parallel straight lines. A direct consequence of rejecting model II (accepting model I) is that a comparison of the mean values $\bar{y}_1, \bar{y}_2, \cdots, \bar{y}_g$ depends on the value of x (Figure 5.1). A comparison of the mean values at one choice of x produces a difference $\bar{y}_i - \bar{y}_j$ that differs for the same comparison for other choices of x; therefore, no general comparison can be made among the categories free of the influence of the covariable x.

When an F-statistic indicates that model II differs from model I by no more than chance variation, the no-interaction model provides a basis to remove the influence of the covariable x, yielding a set of adjusted mean values. Model II requires the differences between mean values to be the same for any choice of x. Adjusted mean values based on model II, free from the influence of the x-variate, are then directly compared to detect possible differences associated with the g categories. Clearly, the validity or the close fit of model II is crucial because the computation of the adjusted mean values depends on the requirement that the regression lines be parallel within all groups compared. Failure to reject model II does not unequivocally imply that the model represents the relationships within the data. It simply means that in face of no evidence to the contrary, the analysis proceeds, often somewhat pragmatically, as though no interaction exists.

Adjustment of the Mean Values

When model II adequately represents the data, adjusted mean values are estimated for each group. Under the structure specified by model II, the distances between between the g categories can be examined at any value of the covariable because the distance between parallel lines is the same for all choices of the x-values. A traditional point of comparison is the overall mean of the covariable ($\bar{x}$).

Model II states that

$$\hat{b}' = \frac{\bar{y}_i - \bar{y}_i{}'}{\bar{x}_i - \bar{x}} \quad \text{(definition of a slope)} . \tag{5.9}$$

The value $\bar{y}_i$ is the mean of the dependent variable in group i at the mean of the covariable $\bar{x}_i$ (i.e., $\bar{y}_i = \hat{a}_i' + \hat{b}' \bar{x}_i$). The value $\bar{y}_i{}'$ is the predicted mean of the dependent variable in the ith group at the overall mean of the covariable, $\bar{x}$. An adjusted mean follows directly as

$$\bar{y}_i' = \bar{y}_i - \hat{b}'(\bar{x}_i - \bar{x}) . \qquad (5.10)$$

The adjusted mean results from applying the definition of a slope (change in the y-variable divided by the change in the x-variable) under the parallel slope requirement of model II. The estimated mean $\bar{y}_i'$ is the adjusted mean for i^{th} group and the influence of the covariable is no longer an issue. Comparisons among the adjusted means $\bar{y}_1', \bar{y}_2', \cdots, \bar{y}_g'$ are free from the effects of the covariable x because the associated value of the covariable for each group is identical, namely $\bar{x}$. The distance between pairs of adjusted means $\bar{y}_i' - \bar{y}_j'$ is not due to differences in values of the covariable; the value of the covariable is the same for all comparisons of the adjusted mean values. Calculating adjusted mean values at other values of the covariable yields different estimates but the differences between these adjusted values are the same. Another point of comparison, for example, is the differences among the categories when $x = 0$ (i.e., the y-intercept), giving

$$\bar{y}_i^* = \hat{a}_i' = \bar{y}_i - \hat{b}' \bar{x}_i \qquad (5.11)$$

but $\bar{y}_i' - \bar{y}_j' = \bar{y}_i^* - \bar{y}_j^*$. Geometrically, adjustment amounts to shifting the observed mean values $\bar{y}_i$ along the line $\hat{y}_{ij} = \hat{a}_i' + \hat{b}'x_{ij}$ to a common value of the x-variable (Figure 5.3), producing a series of artificial mean values for each group but with identical values of the covariable. Or simply, the adjusted mean $\bar{y}_i'$ is the value of the ith regression equation evaluated at a specific value, such as $\bar{x}$, under model II, or $\bar{y}_i' = \hat{a}_i' + \hat{b}' \bar{x}$.

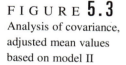

FIGURE 5.3

Analysis of covariance, adjusted mean values based on model II

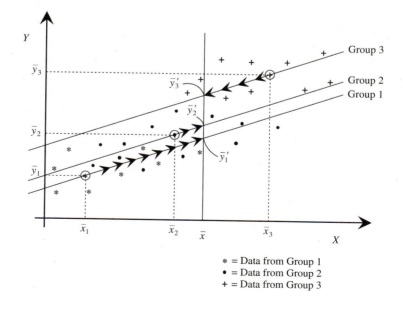

* = Data from Group 1
• = Data from Group 2
+ = Data from Group 3

Covariables

Model III: Main Effects

When the data are consistent with model II, interest is then focused on evaluating the degree of influence of the g-level categorical variable. The primary question is: After adjustment, do the mean values $\bar{y}_i'$ provide evidence of a systematic difference among categories? The analysis once again takes the form of comparing two models. A model III is postulated to represent the data as though no difference exists among the g groups. This requirement is identical to postulating that the observed differences among the categories are due strictly to the influence of the covariable and the random nature of the dependent variable and, therefore, that the data are no more than random deviations from a single straight line (Figure 5.4). Model III is

$$E(Y \mid x_{ij}) = a'' + b''x_{ij} .$$

(5.12)

That is, the original division of the data into categories is not relevant ($a_1 = a_2 = \cdots = a_g = a$) and is, therefore, not taken into account by the statistical model.

The parameters $\hat{a}''$ and $\hat{b}''$ are estimated by considering all $N = \Sigma\, n_i$ observations and not distinguishing among the categories. The estimated regression equation is

$$\hat{y}_{ij}'' = \hat{a}'' + \hat{b}''x_{ij}$$

(5.13)

and the fit of the data to model III is measured by the residual sum of squares

$$Res(III) = \sum_i \sum_j (y_{ij} - \hat{y}_{ij}'')^2 .$$

(5.14)

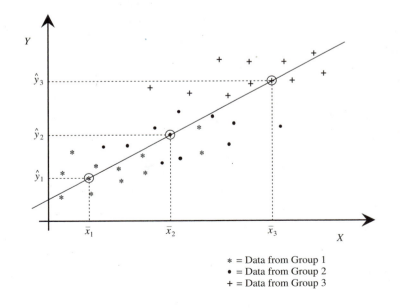

F I G U R E 5.4

Analysis of covariance, regression model (model III)

* = Data from Group 1
• = Data from Group 2
+ = Data from Group 3

T A B L E **5.2**

Analysis of variance table for model III

Source	Sum of Squares	Degrees of Freedom
Regression	Total − *Res(III)*	1
Residual	*Res(III)*	$N - 2$
Total	$N - 1$

As before, *Res(III)* divided by the variance of *Y* has a chi-square distribution with $N - 2$ degrees of freedom when the data deviate from model III only because of random variation. Two independent estimates establish the *N* estimated values $\hat{y}_{ij}''$, producing $N - 2$ degrees of freedom.

Model III is the same linear regression equation described in Chapter 2 with slightly different notation but involves the same statistical requirements (expression 2.1). The analysis of variance summaries shown in Table 5.2 result from a regression analysis based on model III ignoring the *g*-level categorical variable.

To evaluate whether the covariable exerts a significant influence on *Y* (usually not an important issue), an assessment of the hypothesis that the slope *b″* equals zero is carried out using the test statistic

$$F = \frac{(Total - Res(III))/([N-1]-[N-2])}{Res(III)/(N-2)}$$

(5.15)

which is the same *F*-statistic discussed in Chapter 2 (expression 2.13). The existence of a relationship between *x* and *Y* is usually a foregone conclusion, because *x* is measured in conjunction with the dependent variable precisely because it is thought to influence with the comparison among the *g* groups.

Using Model III to Measure the Influence of Categorical Classification

The principal function of model III is to provide a contrast to model II to measure the influence of the categorical classification. The difference between model II and model III reflects the impact of the *g* categories on the dependent variable *Y*. The residual sum of squares calculated from model III divides into two pieces: the residual from model II (*Res(II)*) and a piece that measures the influence of the categorical variable, called the main effects sum of squares. The difference in fit between model II and model III directly indicates the magnitude of the main effects (i.e., the influence of the *g*-level categorical variable). Specifically,

$$main\ effects = Res(III) - Res(II)$$

or

$$main\ effects = \Sigma\Sigma(y_{ij} - \hat{y}_{ij}'')^2 - \Sigma\Sigma(y_{ij} - \hat{y}_{ij}')^2.$$

(5.16)

FIGURE **5.5**

Main effects sum of
squares = $Res(III) -$
$Res(II)$

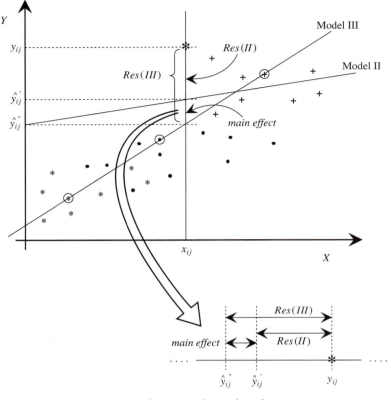

Note: $y_{ij} - \hat{y}_{ij}'' = (y_{ij} - \hat{y}_{ij}') + (\hat{y}_{ij}' - \hat{y}_{ij}'')$

Sums of squares: $Res(III) = Res(II) +$ main effects

Degrees of freedom: $(N - 2) = [N - (g + 1)] + [g - 1]$

The values $\hat{y}_{ij}''$ are estimated from a single line (model III) and the values $\hat{y}_{ij}'$ are estimated from a series of g parallel lines (model II).

To summarize, note that $y_{ij} - \hat{y}_{ij}'' = (y_{ij} - \hat{y}_{ij}') + (\hat{y}_{ij}' - \hat{y}_{ij}'')$, which implies that $Res(III) = Res(II) +$ main effects (expression 5.16). The size of the main effects sum of squares is dictated by the difference in fit between using the model generated values $\hat{y}_{ij}'$ (g categories have a systematic influence) instead of $\hat{y}_{ij}''$ (g categories have no systematic influence) as estimates of the dependent variable y_{ij}. The degrees of freedom follow the same pattern as the sum of squares. The degrees of freedom associated with the main effects sum of squares are $(N - 2) - [N - (g + 1)] = g - 1$.

The geometry of the analysis of covariance is displayed in Figure 5.5 and the analysis of variance is given in Table 5.3. Because model III (no influence of the categorical variable) is a special case of model II (the categorical variable has a systematic influence), an F-statistic allows a formal comparison of these two nested models. The F-statistic to examine the influence of the categorical variable is

TABLE 5.3

Analysis of covariance contrasting model II with model III

Source	Sum of Squares	Degrees of Freedom
Covariable—regression	Total − *Res(III)*	1
Main effects	*Res(III)* − *Res(II)*	$g - 1$
"Explained" variation	Total − *Res(II)*	g
"Unexplained" variation	*Res(II)*	$N - (g + 1)$
Total		$N - 1$

$$F = \frac{(Res(III) - Res(II))/(g - 1)}{Res(II)/[N - (g + 1)]} \qquad (5.17)$$

where F has an F-distribution with $g - 1$ and $N - (g + 1)$ degrees of freedom when the categorical classification is not relevant and only the covariable along with random variation causes the differences observed among the g groups.

Another View of the Main Effects

An alternative view of the main effects comes from directly evaluating the observed differences among the g adjusted mean values, $\bar{y}_i'$. A sum of squares summarizing the differences among categories calculated directly from the adjusted mean values is

$$\text{"main-effects" sum of squares} = \Sigma n_i(\bar{y}_i' - \bar{y})^2 \qquad (5.18)$$

where $\bar{y}_i'$ is the adjusted mean from the ith category at $\bar{x}$ and $\bar{y}' = \bar{y}$ is the overall mean of the N dependent variables. The "main-effects" sum of squares reflects the variation among the adjusted mean values and is analogous to the between sum of squares used in a one-way analysis of variance (expression 1.18). If the sum of squares $\Sigma n_i(\bar{y}_i' - \bar{y})^2$ is small, the categorical classification is likely unimportant, as the g adjusted mean values are similar to the overall mean, $\bar{y}$. Conversely, when $\Sigma n_i(\bar{y}_i' - \bar{y})^2$ is large, then one or more of the adjusted mean values differ among the categories under study. This directly calculated "main-effect" sum of squares is not identical to the main-effect sum of squares calculated from the differences between the residual sums of squares from model II and model III. In symbols this bias is

$$\Sigma n_i(\bar{y}_i' - \bar{y})^2 - bias = Res(III) - Res(II) . \qquad (5.19)$$

However, the bias is often small and both quantities measure the same phenomenon—the influence of the g categories. Assessing the bias is not important because the exact main-effects sum of squares is calculated from $Res(III) - Res(II)$. The fundamental fact to notice is that the main-effects sum of squares has essentially the same interpretation as the usual analysis of variance between groups sum of squares (expression 1.18) but applied to series of the adjusted mean values.

Extension of Analysis of Covariance

An extension of the three previous analysis of covariance models allows adjustment for the presence of k covariables. Model I is extended by postulating that the dependent variable is related to the k covariables by a different multivariable linear regression equation for each of g groups (interaction model), or

$$E(Y \mid x_{1j}, x_{2j}, \cdots, x_{kj}) = a_i + b_1^{(i)}x_{1j} + b_2^{(i)}x_{2j} + \cdots + b_k^{(i)}x_{kj} \tag{5.20}$$

where i denotes one of the g categories. A total of $g + gk$ parameters are estimated to completely specify the estimated dependent variable $\hat{y}_{ij}$—that is, g separate k-variate multivariable regression equations requiring a total of $g(k + 1)$ parameters.

Model II has the same structure as model I except the k regression coefficients are identical for all g regression equations (groups) and only the intercepts a_i differ (no interaction model), or

$$E(Y \mid x_{1j}, x_{2j}, \cdots, x_{kj}) = a_i' + b_1'x_{1j} + b_2'x_{2j} + \cdots + b_k'x_{kj} . \tag{5.21}$$

Model II requires $g + k$ parameters to estimate the dependent variables $\hat{y}_{ij}'$. Analogous to adjusting for a single covariable, model II is the statistical structure that allows the adjustment of the mean values within each group for the influence of the k variables. Sums of squares based on $\hat{y}_{ij}$ and $\hat{y}_{ij}'$ from models I and II are compared using an F-statistic to examine whether the data support adjustment. Because model II is a special case of model I, then

$$F = \frac{(Res(II) - Res(I))/k(g - 1)}{Res(I)/(N - g[k + 1])} \tag{5.22}$$

where F has an F-distribution with $k(g - 1)$ and $N - g(k + 1)$ degrees of freedom when the data differ from model II only because of random variation. If $k = 1$, expression 5.22 is identical to the F-statistic from the single covariable analysis (expression 5.8).

If model II (parallel k-dimensional planes) is tenable (F is small—probable), then model III (a single k-dimensional plane) is postulated so the comparison of sums of squares from model II and model III indicates differences among the categories associated with the g groups (called again main effects). The differences in these two sums of squares are again directly related to the differences among the adjusted mean values. Model III involves $k + 1$ parameters and is exactly the multiple regression model examined in Chapter 4 (expression 4.1). That is, model III is

$$E(Y \mid x_{1j}, x_{2j}, \cdots, x_{kj}) = a'' + b_1''x_{1j} + b_2''x_{2j} + \cdots + b_k''x_{kj} . \tag{5.23}$$

Model III requires $k + 1$ parameters to estimate the dependent variables $\hat{y}_{ij}''$, which represents the data ignoring the g categories. The comparison of model II with model

T A B L E **5.4**

Analysis of covariance table, model II versus model III

Source	Sum of Squares	Degrees of Freedom
Covariable	Total − Res(III)	k
Main effects	Res(III) − Res(II)	$g - 1$
"Explained"	Total − Res(II)	$k + g - 1$
"Unexplained"	Res(II)	$N - (g + k)$
Total	$N - 1$

III again measures the influence from the categorical classification (main effects) and is a key element in an analysis of covariance table. Again, the regression sum of squares measures the covariable effects (model III: total − *Res(III)*) and the measure of the main effects is, as before, the difference between the residual sums of squares from model III and from model II (*Res(III) − Res(II)*). The analysis of covariance for k covariates yields an analysis of variance table almost identical to the single co-variable analysis; only the degrees of freedom differ. Of course, the estimated regression equations producing $\hat{y}_{ij}$, $\hat{y}_{ij}'$, and $\hat{y}_{ij}''$ are more complicated, but the three fundamental sums of squares remain as

$$Residual\ model\ I = Res(I) = \Sigma\Sigma(y_{ij} - \hat{y}_{ij})^2$$

$$Residual\ model\ II = Res(III) = \Sigma\Sigma(y_{ij} - \hat{y}_{ij}')^2$$

$$Residual\ model\ III = Res(III) = \Sigma\Sigma(y_{ij} - \hat{y}_{ij}'')^2 .$$

The comparison of model II with model III is again the central issue but only makes sense if model II is a useful description of the data (no important difference between model I and model II—no interaction). The components of the k-variate analysis of covariance are summarized in Table 5.4.

The main-effects sum of squares can be further investigated by employing analysis of variance techniques (available in many computer programs such as SAS, BMDP, and SPSS). The total influence associated with the categorical classification (main effects) can be partitioned into meaningful sums of squares based on the adjusted mean values. For example, if the main-effects sum of squares measures both the influence of smoking and racial categories, then a two-way analysis can separate the impact of these two variables based on the adjusted mean values [1]. The assessment is free from the effects of the k covariables, which are removed by use of the no-interaction linear regression model (model II). A series of F-tests allows statistical inferences about these variables but, of course, the conclusions depend on the validity of model II and the basic requirements underlying a regression analysis. Analysis of covariance is a topic that can be extended to adjust and compare mean values under a large number of conditions ([1], [2], and [3]).

An Alternate View of the Analysis of Covariance

The properties of multiple linear regression analysis discussed in Chapters 2–4 equally apply to the analysis of covariance, which is a special case. Using interaction terms in a regression equation allows differences observed among the dependent variables associated with a categorical variable to be evaluated while accounting for the influences of one or more covariables. Each category is characterized by a binary variable defined as one when an observation belongs to a specific category and zero otherwise. Only $g - 1$ of these variables are necessary to identify the role of g groups in an analysis. The binary variables constructed to represent the g groups along with the values of the covariable and the dependent variable are entered into a regression analysis to produce estimated coefficients. These estimated coefficients lead to exactly the same sums of squares and, therefore, the same conclusions as the previously discussed approach of comparing three nested models.

To illustrate, consider the comparison of the mean values from three groups adjusted for a single covariable variable (x). The regression model that allows the data from each group to be represented by a different straight line (interaction model—model I) is

$$E(Y \mid x_{ij}) = A + B_1 x_{ij} + B_2 g_{1j} + B_3 g_{2j} + B_4 x_{ij} g_{1j} + B_5 x_{ij} g_{2j} \tag{5.24}$$

where the variable g_{ij} is constructed so $g_{ij} = 1$ if the jth observation belongs to the ith group and is 0 otherwise. Therefore, $g_{1j} = 1$ when an observation belongs to the group labeled 1 and 0 otherwise, and $g_{2j} = 1$ when the observation belongs to group 2 and is 0 otherwise. A similar variable for group 3 is not necessary. The binary variables used to indicate group membership are called design variables or dummy variables. These specialized variables play a wide variety of roles in statistical analyses. Design variables are discussed in more detail in other texts (e.g., [4] or [5]) and are further developed in Chapter 10.

A regression model (expression 5.24) employing two design variables to represent three groups is no more than a compact expression of the previous model I. That is,

if $g_{1j} = 1$ and $g_{2j} = 0$, then $\hat{y}_{1j} = (\hat{A} + \hat{B}_2) + (\hat{B}_1 + \hat{B}_4) x_{1j} = \hat{a}_1 + \hat{b}_1 x_{1j}$;

if $g_{1j} = 0$ and $g_{2j} = 1$, then $\hat{y}_{2j} = (\hat{A} + \hat{B}_3) + (\hat{B}_1 + \hat{B}_5) x_{2j} = \hat{a}_2 + \hat{b}_2 x_{2j}$; and

if $g_{1j} = 0$ and $g_{2j} = 0$, then $\hat{y}_{3j} = \hat{A} + \hat{B}_1 x_{3j} = \hat{a}_3 + \hat{b}_3 x_{3j}$

represents the data by three straight lines with different intercepts and different slopes for each group (six parameters). The coefficients $\hat{a}_i$ and $\hat{b}_i$ are identical to those derived from model I (expression 5.1).

The No-Interaction Model (Model II)

The no-interaction model, model II, is constructed by setting B_4 and B_5 equal to 0 in expression 5.24, yielding the model

$$E(Y \mid x_{ij}) = A' + B_1'x_{ij} + B_2'g_{1j} + B_3'g_{2j}$$

where

if $g_{1j} = 1$ and $g_{2j} = 0$, then $\hat{y}_{1j}' = (\hat{A}' + \hat{B}_2') + \hat{B}_1'x_{1j} = \hat{a}_1' + \hat{b}'x_{1j}$;

if $g_{1j} = 0$ and $g_{2j} = 1$, then $\hat{y}_{2j}' = (\hat{A}' + \hat{B}_3') + \hat{B}_1'x_{2j} = \hat{a}_2' + \hat{b}'x_{2j}$; and

if $g_{1j} = 0$ and $g_{2j} = 0$, then $\hat{y}_{3j}' = \hat{A}' + \hat{B}_1'x_{3j} = \hat{a}_3' + \hat{b}'x_{3j}$

represents the data as three parallel lines, which is model II (four parameters).

Finally, the no-main-effects model, model III, is constructed by additionally setting B_2 and B_3 equal to 0 in expression 5.24 (i.e., deleting the design variables completely from the model), giving

$$E(Y \mid x_{ij}) = \hat{A}'' + \hat{B}_1''x_{ij} = \hat{a}'' + \hat{b}''x_{ij} \, ,$$

which represents the data as a single straight line ignoring the categorical variable (two parameters). These three models demonstrate what is true in general; that the analysis of covariance is a special case of multiple linear regression analysis in which design variables and interaction terms are used to indicate categorical classification.

The test case with 18 observed birth weights and maternal ages for three categories of smoking, analyzed as part of the following test case, demonstrates a multiple regression approach to analysis of covariance. The data, the design variables, and the interaction terms are given in Table 5.5.

Using the full regression model (six terms) yields estimates of the constant term and five regression coefficients, given in Table 5.6. The six estimated coefficients describe three regression lines:

$\hat{A} + \hat{B}_2 = 4.210 - 0.678$ and $\hat{B}_1 + \hat{B}_4 = 0.096 + 0.0514$,
  giving $\hat{a}_1 = 3.531$ and $\hat{b}_1 = 0.148$;

$\hat{A} + \hat{B}_3 = 4.210 + 0.219$ and $\hat{B}_1 + \hat{B}_5 = 0.096 - 0.006$,
  giving $\hat{a}_2 = 4.429$ and $\hat{b}_2 = 0.090$; and

$\hat{A} = 4.210$ and $\hat{B}_1 = 0.096$,
  giving $\hat{a}_3 = 4.210$ and $\hat{b}_3 = 0.096$.

These coefficients and, therefore, regression lines are the same as those previously calculated directly from model I.

To describe the data in terms of three parallel lines (model II), the estimated regression coefficients are given in Table 5.7, where $B_4 = B_5 = 0$. Four estimated coefficients are necessary to describe the three parallel regression lines, or

$\hat{A}' + \hat{B}_2' = 3.785 + 0.839$, giving $\hat{a}_1' = 4.624$ and $\hat{b}' = 0.113$;

$\hat{A}' + \hat{B}_3' = 3.785 - 0.003$, giving $\hat{a}_2' = 3.782$ and $\hat{b}' = 0.113$; and

$\hat{A}' = 3.785$ and $\hat{B}_1' = 0.113$, giving $\hat{a}_3' = 3.785$ and $\hat{b}' = 0.113$.

Again, these estimates are identical to those calculated directly from model II.

T A B L E   **5.5**
Birth-weight test-case data and constructed design variables

i	j	y_{ij}	x_{ij}	g_{1j}	g_{2j}	$x_{ij}g_{1j}$	$x_{ij}g_{2j}$
1	1	9.1	35	1	0	35	0
1	2	8.9	29	1	0	29	0
1	3	8.5	34	1	0	34	0
1	4	7.4	32	1	0	32	0
1	5	7.5	28	1	0	28	0
1	6	7.3	28	1	0	28	0
2	1	7.2	32	0	1	0	32
2	2	7.7	30	0	1	0	30
2	3	6.8	26	0	1	0	26
2	4	7.0	33	0	1	0	33
2	5	7.4	28	0	1	0	28
2	6	6.2	25	0	1	0	25
3	1	6.7	24	0	0	0	0
3	2	6.5	23	0	0	0	0
3	3	7.2	28	0	0	0	0
3	4	6.4	26	0	0	0	0
3	5	6.5	26	0	0	0	0
3	6	7.1	30	0	0	0	0

T A B L E   **5.6**
Estimated coefficients for the regression approach, model I

	Coefficient	Standard Error
$\hat{A}$	4.210	2.453
$\hat{B}_1$	0.096	0.093
$\hat{B}_2$	−0.678	3.435
$\hat{B}_3$	0.219	3.271
$\hat{B}_4$	0.051	0.121
$\hat{B}_5$	−0.006	0.119

T A B L E   **5.7**
Estimated coefficients for the regression approach, model II

	Coefficient	Standard Error
$\hat{A}$	3.785	1.158
$\hat{B}_1$	0.113	0.043
$\hat{B}_2$	0.839	0.358
$\hat{B}_3$	−0.003	0.315

No-Effect Model (Model III)

The regression approach with $B_2 = B_3 = B_4 = B_5 = 0$ (model III) produces an estimated simple linear equation with coefficients

$$\hat{A}'' = \hat{a}'' = 2.447 \text{ and } \hat{B}_1'' = \hat{b}'' = 0.169 .$$

Because the comparison of a series of nested models and using multiple regression equations with design variables are only different representations of an identical process, the sums of squares and other summaries are identical (e.g., $Res(I) = 3.436$, $Res(II) = 3.530$, and $Res(III) = 5.708$, for the test data). Both the regression approach and the comparison of a series of nested models (models I, II, and III) produce the same inferences.

Test Case

Computer Implementation

Analysis of Covariance (One-Way Classification with One Covariable)

A set of test-case data consists of Y = infant birth weight, x = covariable = maternal age with categories nonsmokers, quitters, and smokers—$n = 6$, $g = 3$, and $N = 18$

Data

Nonsmokers

						Mean	s^2	n
Birth weight=Y	9.1	8.9	8.5	7.4	7.5	7.3	8.117	0.658	6
Maternal age=X	35	29	34	32	28	28	31.0	9.600

Quitters

Birth weight=Y	7.2	7.7	6.8	7.0	7.4	6.2	7.050	0.271	6
Maternal age=X	32	30	26	33	28	25	29.0	10.400

Smokers

Birth weight=Y	6.7	6.5	7.2	6.4	6.5	7.1	6.733	0.115	6
Maternal age=X	24	23	28	26	26	30	26.167	6.567

$$W_{yy}^{(1)} = 3.288 \qquad W_{yy}^{(2)} = 1.355 \qquad W_{yy}^{(3)} = 0.573 \qquad W_{yy} = \Sigma W_{yy}^{(i)} = 5.217$$
$$W_{xy}^{(1)} = 7.100 \qquad W_{xy}^{(2)} = 4.700 \qquad W_{xy}^{(3)} = 3.167 \qquad W_{xy} = \Sigma W_{xy}^{(i)} = 14.967$$
$$W_{xx}^{(1)} = 48.000 \qquad W_{xx}^{(2)} = 52.000 \qquad W_{xx}^{(3)} = 32.833 \qquad W_{xx} = \Sigma W_{xx}^{(i)} = 132.833$$

$$B_{yy} = 6.303 \qquad\qquad T_{yy} = 5.216 + 6.303 = 11.520$$
$$B_{xy} = 19.433 \qquad\qquad T_{xy} = 14.967 + 19.433 = 34.400$$
$$B_{xx} = 70.778 \qquad\qquad T_{xx} = 132.833 + 70.778 = 203.611$$

Complete definitions of these sums of squares are found in the notation section at the end of the chapter.

Note: $\hat{b}_i = W_{xy}^{(i)}/W_{xx}^{(i)}$, giving

$$\hat{b}_1 = 7.100/48.000 = 0.148, \hat{b}_2 = 4.700/52.000 = 0.090,$$
$$\text{and } \hat{b}_3 = 3.167/32.833 = 0.096 .$$

Also

$$\hat{a}_1 = 3.531, \hat{a}_2 = 4.429, \text{ and } \hat{a}_3 = 4.210 \text{ where } \hat{a}_i = \bar{y}_i - \hat{b}_i\bar{x}_i.$$

Computation for an Analysis of Covariance

Sums of Squares

$$\text{Residual model II} = Res(II) = \Sigma \, \Sigma(y_{ij} - \hat{y}_{ij}')^2 = W_{yy} - \hat{b}'W_{xy}$$
$$= 5.217 - (0.113)(14.967)$$
$$= 3.530$$

where $\hat{b}' = W_{xy}/W_{xx} = 14.967/132.833 = 0.113$

$$\text{Residual model III} = Res(III) = \Sigma \, \Sigma(y_{ij} - \hat{y}_{ij}'')^2 = T_{yy} - \hat{b}''T_{xy}$$
$$= 11.520 - (0.169)(34.400)$$
$$= 5.708$$

where $\hat{b}'' = T_{xy}/T_{xx} = 34.400/203.611 = 0.169$

$$\text{Total} = \Sigma \, \Sigma(y_{ij} - \bar{y})^2 = T_{yy} = 11.520$$

Analysis of covariance table:

Source	Sum of Squares	Degrees of Freedom
Covariables	Total − $Res(III)$ = 5.812	1
Main effects	$Res(III) - Res(II)$ = 2.178	$g - 1 = 2$
"Explained"	Total − $Res(II)$ = 7.990	$g = 3$
"Unexplained"	$Res(II)$ = 3.530	$N - (g + 1) = 14$
	Total = 11.520	$N - 1 = 17$

Adjusted Mean Values (Model II)

$$\bar{y}_i' = \bar{y}_i - \hat{b}'(\bar{x}_i - \bar{x})$$

group 1: $\bar{y}_1' = 8.117 - 0.113(31.000 - 28.722) = 7.860$

group 2: $\bar{y}_2' = 7.050 - 0.113(29.000 - 28.722) = 7.019$

group 3: $\bar{y}_3' = 6.733 - 0.113(26.167 - 28.722) = 7.021$

Test of Main Effects

$$H_0: \text{ no difference exists between adjusted mean values}$$

$$F = \frac{(Res(III) - Res(II))/([N-2] - [N-(g+1)])}{Res(II)/[N-(g+1)]} = \frac{2.178/2}{3.530/14} = 4.318$$

$$p\text{-value} = 0.035 \quad \text{degrees of freedom} = 2 \text{ and } 14$$

Note:

$$\text{"main effects"} = \Sigma n_i (\bar{y}_i' - \bar{y})^2 = 2.823$$

and the bias is ("main effects" − main effects) = 2.823 − 2.178 = 0.645.

Analysis of Covariance: Test Case

The following test data and analysis of covariance illustrate the STATA implementation ("anova"). The dependent variable is Y = newborn birth weight, the covariable is X = maternal age, and the data are classified into $g = 3$ smoking exposure groups: nonsmokers, quitters, and smokers (i.e., $n = 6$, $g = 3$, and $N = 18$).

```
example.covariance

. infile g y x using cov.dat
(18 observations read)

. list
```

	g	y	x
1.	1	9.1	35
2.	1	8.9	29
3.	1	8.5	34
4.	1	7.4	32
5.	1	7.5	28
6.	1	7.3	28
7.	2	7.2	32
8.	2	7.7	30
9.	2	6.8	26
10.	2	7	33
11.	2	7.4	28
12.	2	6.2	25
13.	3	6.7	24
14.	3	6.5	23
15.	3	7.2	28
16.	3	6.4	26
17.	3	6.5	26
18.	3	7.1	30

With the annotations: Y and X_{ij} labeling the y and x columns; y_{1j} and x_{1j} for observations 1–6; y_{2j} and x_{2j} for observations 7–12; y_{3j} and x_{3j} for observations 13–18.

```
. sort g
```

```
. by g: summarize y x
```

$\overline{y}_i$
$\overline{x}_i$

```
- g=      1
Variable |     Obs        Mean    Std. Dev.        Min        Max
---------+---------------------------------------------------------
       y |       6    8.116667    .8109665        7.3        9.1
       x |       6          31    3.098387         28         35

- g=      2
Variable |     Obs        Mean    Std. Dev.        Min        Max
---------+---------------------------------------------------------
       y |       6        7.05    .5205766        6.2        7.7
       x |       6          29    3.224903         25         33

- g=      3
Variable |     Obs        Mean    Std. Dev.        Min        Max
---------+---------------------------------------------------------
       y |       6    6.733333    .3386246        6.4        7.2
       x |       6    26.16667    2.562551         23         30
```

```
. anova y x g g*x, continuous(x)
```

```
                      Number of obs =      18      R-square     = 0.7017
                      Root MSE      = .53512      Adj R-square = 0.5774

           Source |  Partial SS      df       MS            F     Prob  F
      ------------+----------------------------------------------------------
            Model |  8.08376531       5   1.61675306        5.65    0.0066
                  |
                x |  1.58897765       1   1.58897765        5.55    0.0363
                g |   .02316316       2    .01158158        0.04    0.9605
              g*x |  .094098666       2   .047049333        0.16    0.8504
                  |
         Residual |  3.43623513      12   .286352928
      ------------+----------------------------------------------------------
            Total |  11.5200004      17   .677647085
```

Res(I)

```
. anova y x g, continuous(x)
```

```
                      Number of obs =      18      R-square     = 0.6935
                      Root MSE      = .502162     Adj R-square = 0.6279

           Source |  Partial SS      df       MS            F     Prob  F
      ------------+----------------------------------------------------------
            Model |  7.98966664       3   2.66322221       10.56    0.0007
                  |
                x |  1.68633226       1   1.68633226        6.69    0.0216
                g |  2.17780226       2   1.08890113        4.32    0.0346
                  |
         Residual |  3.5303338       14   .2521667
      ------------+----------------------------------------------------------
            Total |  11.5200004      17   .677647085
```

Res(II)   *F*   *p*-value

```
. anova y x, continuous(x)
```

| | | Number of obs = | 18 | R-square | = 0.5045 |
| | | Root MSE | = .597293 | Adj R-square = | 0.4735 |

Source	Partial SS	df	MS	F	Prob F
Model	5.81186438	1	5.81186438	16.29	0.0010
x	5.81186438	1	5.81186438	16.29	0.0010
Residual	5.70813606	16	.356758504		
Total	11.5200004	17	.677647085		

Res(III)

Analysis of Covariance: Test Case Using Design Variables

The following computer output uses the standard STATA linear regression analysis command ("regress") to produce an analysis of covariance using design variables. Again, the dependent variable is Y = newborn birth weight, the covariable is X = maternal age, and the data are classified into $g = 3$ smoking exposure groups (i.e., $n = 6$, $g = 3$, and $N = 18$).

example.covariance

```
. infile g y x using cov.dat
(18 observations read)

. generate g1=g==1

. generate g2=g==2

. generate gg1=g1*x

. generate gg2=g2*x

. list y x g g1 g2 gg1 gg2 , nodisplay
```

Group indicator

g_{1j} g_{2j} $x_{ij}g_{1j}$ $x_{ij}g_{2j}$

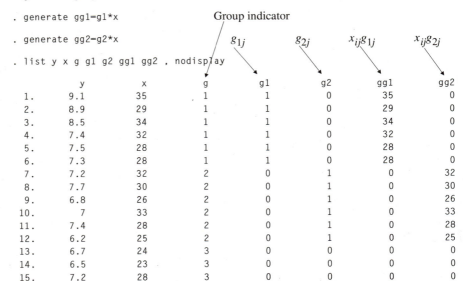

	y	x	g	g1	g2	gg1	gg2
1.	9.1	35	1	1	0	35	0
2.	8.9	29	1	1	0	29	0
3.	8.5	34	1	1	0	34	0
4.	7.4	32	1	1	0	32	0
5.	7.5	28	1	1	0	28	0
6.	7.3	28	1	1	0	28	0
7.	7.2	32	2	0	1	0	32
8.	7.7	30	2	0	1	0	30
9.	6.8	26	2	0	1	0	26
10.	7	33	2	0	1	0	33
11.	7.4	28	2	0	1	0	28
12.	6.2	25	2	0	1	0	25
13.	6.7	24	3	0	0	0	0
14.	6.5	23	3	0	0	0	0
15.	7.2	28	3	0	0	0	0
16.	6.4	26	3	0	0	0	0
17.	6.5	26	3	0	0	0	0
18.	7.1	30	3	0	0	0	0

```
. regress y x g1 g2 gg1 gg2        Res(I)

   Source |      SS       df       MS                    Number of obs =      18
----------+----------------------------                  F( 5,    12) =    5.65
    Model | 8.08376531      5  1.61675306                Prob  F      =  0.0066
 Residual | 3.43623513     12  .286352928                R-square     =  0.7017
----------+----------------------------                  Adj R-square =  0.5774
    Total | 11.5200004     17  .677647085                Root MSE     =  .53512

------------------------------------------------------------------------------
        y |    Coef.    Std. Err.      t     P|t|     [95% Conf. Interval]
----------+-------------------------------------------------------------------
        x |  .0964467   .0933885     1.033   0.322    -.1070293    .2999227
       g1 |  -.678396   3.435109    -0.197   0.847    -8.162856    6.806064
       g2 |  .2192016   3.270804     0.067   0.948    -6.907268    7.345671
      gg1 |   .05147    .1211903     0.425   0.679    -.212581     .315521
      gg2 | -.0060621   .119282     -0.051   0.960    -.2659552    .2538311
    _cons |  4.209645   2.453411     1.716   0.112    -1.135878    9.555168
------------------------------------------------------------------------------

. regress y x g1 g2            Res(II)

   Source |      SS       df       MS                    Number of obs =      18
----------+----------------------------                  F( 3,    14) =   10.56
    Model | 7.98966664      3  2.66322221                Prob  F      =  0.0007
 Residual |  3.5303338     14  .2521667                  R-square     =  0.6935
----------+----------------------------                  Adj R-square =  0.6279
    Total | 11.5200004     17  .677647085                Root MSE     =  .50216
                                              ^
                                              b'
------------------------------------------------------------------------------
        y |    Coef.    Std. Err.      t     P|t|     [95% Conf. Interval]
----------+-------------------------------------------------------------------
        x |  .1126725   .0435703     2.586   0.022     .0192236    .2061215
       g1 |  .8387496   .3583345     2.341   0.035     .0701986    1.607301
       g2 | -.0025721   .3151115    -0.008   0.994    -.6784191    .6732748
    _cons |  3.785069   1.158374     3.268   0.006     1.300603    6.269534
------------------------------------------------------------------------------

. regress y x                 Res(III)

   Source |      SS       df       MS                    Number of obs =      18
----------+----------------------------                  F( 1,    16) =   16.29
    Model | 5.81186438      1  5.81186438                Prob  F      =  0.0010
 Residual | 5.70813606     16  .356758504                R-square     =  0.5045
----------+----------------------------                  Adj R-square =  0.4735
    Total | 11.5200004     17  .677647085                Root MSE     =  .59729

------------------------------------------------------------------------------
        y |    Coef.    Std. Err.      t     P|t|     [95% Conf. Interval]
----------+-------------------------------------------------------------------
        x |  .1689495   .0418588     4.036   0.001     .0802129    .2576862
    _cons |  2.447394   1.210491     2.022   0.060    -.1187332    5.013521
------------------------------------------------------------------------------

. exit,clear
```

Applied Example

Analysis of Covariance

The effects from maternal smoking on birth weight can be directly investigated by comparing the mean birth weight of infants classified by maternal smoking exposure. The 680 white male infants from the Child Health and Development Study produce the following mean birth weights ($\bar{y}_i$):

Group	Sample Size	Mean Value (pounds)
Never smoked	266	7.71
Past smoker	115	7.78
Smoked < pack/day	169	7.22
Smoked ≥ pack/day	130	7.27
Total	680	7.52

A one-way analysis of variance to compare statistically these four mean values produces an F-value of 12.049 with a p-value < 0.001, indicating a likely systematic difference among the four mean birth weights.

Birth weight is strongly influenced by length of gestation ($r = 0.426$). Furthermore, increases in smoking exposure tend to be associated with shorter gestation (i.e., 39.92 weeks for past smokers, 39.48 weeks for <1 pack/day, 39.67 weeks for ≥1 pack/day, and 39.94 weeks for never smoked). That is, average length of gestation differs among the exposure categories. Although these differences are not large, it is of interest to adjust the mean levels of birth weight so the influence from differing lengths of gestation is removed from the comparisons of newborn birth weights among the four categories of smoking exposure.

The first step in an analysis of covariance compares the relative feasibility of the two analytic models constructed to summarize the relationship of birth weight and length of gestation (model I versus model II). The first model specifies that birth weight relates to length of gestation by a different simple linear equation for at least some of the four smoking exposure categories (model I, interaction). The second model restricts these four linear relationships to have the same slope but different intercepts (model II, no interaction).

Under the hypothesis that each smoking category generates four different linear relationships between birth weight and gestation, the summaries in Table 5.8 emerge. The overall fit of these four simple regression analyses (model I) is measured by

$$Res(I) = 633.217 \text{ with degrees of freedom} = 672 \,.$$

Restricting the analytic structure to four straight lines with the same slope (model II) for each smoking category gives a measure of fit of

$$Res(II) = 635.775 \text{ with degrees of freedom} = 675.$$

The F-statistic to evaluate the fit of model II is

$$F = \frac{(635.775 - 633.217)/(675 - 672)}{633.217/672} = 0.905 \qquad p\text{-value} = 0.438) \,,$$

T A B L E   **5.8**

Smoking and birth weight, model I (interaction)

	Sample Size	$\hat{R}_i^2$	$\hat{b}_i$	Res_i	df_i
Past smoker	115	0.093	0.165	104.551	113
<1 pack/day	169	0.186	0.260	158.810	167
≥1 pack/day	130	0.214	0.256	120.666	128
Never smoked	266	0.183	0.245	249.189	264
Total	680	—	—	633.217	672

showing no evidence to reject the description of the data in terms of four straight lines with a common slope, no interaction. The validity of this inference, as usual, rests on the applicability of the four requirements of a regression analysis. The estimate of the common slope is $\hat{b}' = 0.238$, producing the adjusted mean values given in Table 5.9. Although the adjusted mean birth weights show no dramatic shifts, comparisons among the $\bar{y}_i'$ values are no longer affected by differing lengths of gestation among the four smoking categories. The mean gestation associated with each adjusted mean birth weight is $\bar{x} = 39.77$ weeks.

To compare the four adjusted mean values, a third analytic structure is proposed (model III, no main effects). The possible influences from the four smoking exposure categories are ignored and the relationship between birth weight and gestation is describe by a single regression equation. Regression analysis based on this structure (model III) produces the analysis of variance summaries given in Table 5.10.

The comparison of the residual sum of squares calculated ignoring differences among smoking categories (model III) to the residual sum of squares calculated including differences among smoking categories (model II) measures the influence

T A B L E   **5.9**

Smoking and birth weight, adjusted mean values

	$\bar{y}_i - \hat{b}'(\bar{x}_i - \bar{x})$	$= \bar{y}_i'$
Past smoker	7.78 − 0.238(39.92 − 39.77)	= 7.75
<1 pack/day	7.22 − 0.238(39.48 − 39.77)	= 7.29
≥1 pack/day	7.27 − 0.238(39.67 − 39.77)	= 7.29
Never smoked	7.71 − 0.238(39.94 − 39.77)	= 7.67

T A B L E **5.10**

Smoking and birth weight, analysis of variance (model III)

Source	Sum of Squares	Degrees of Freedom
Regression	146.931	1
Residual	663.265	678
Total	810.196	679

of smoking on birth weights (main effects) free of influences from differences associated with gestation time. That is,

$$Res(II) = 635.775 \quad \text{with degrees of freedom} = 675$$

and

$$Res(III) = 663.265 \quad \text{with degrees of freedom} = 678 \,,$$

giving

$$F = \frac{(663.265 - 635.775)/(678 - 675)}{635.775/675} = 9.73 \,,$$

which has an F-distribution with 3 and 675 degrees of freedom when the four smoking exposure categories have no influence on birth weight. An F-value of 9.73 has an associated significance probability (p-value) of less than 0.001, indicating that birth weight remains significantly associated with mother's smoking exposure status after adjustment for the differences in lengths of gestation. A summary of this analysis of covariance is given in Table 5.11.

T A B L E **5.11**

Smoking and birth weight, model II versus model III

Source	Sum of Squares	Degrees of Freedom
Covariable—gestation	Total – $Res(III)$ = 146.931	$k = 1$
Main effects—smoking	$Res(III) - Res(II)$ = 27.490	$g - 1 = 3$
"Explained" variation	Total – $Res(II)$ = 174.420	$k + g - 1 = 4$
"Unexplained" variation	$Res(II)$ = 635.775	$N - (g + k) = 675$
	Total = 810.196	$N - 1 = 679$

Analysis of Covariance

The CHDS data and the following analysis of covariance, using a regression analysis format, illustrates the STATA implementation. The dependent variable is Y = newborn birth weight in pounds, the covariable is X = gestational age in weeks, and the data are classified into four smoking exposure groups (0 = past smoker, 1 = smokes < 20 cigarettes per day, 2 = smokes $\geq$ 20 cigarettes per day, and 3 = never smoked). Also, $n_1 = 115$, $n_2 = 169$, $n_3 = 130$, $n_4 = 266$, $g = 4$, and $N = 680$.

```
chds.covariance

. use chds

. oneway bwt status

                                Analysis of Variance
        Source              SS         df      MS            F      Prob  F
    - - - - - - - - - - - - - - - - - - - - - - - - - - - - - - - - - - - - - - -
    Between groups      41.1241493      3   13.7080498     12.05    0.0000
    Within groups      769.071397     676    1.13767958
    - - - - - - - - - - - - - - - - - - - - - - - - - - - - - - - - - - - - - - -
        Total          810.195546     679    1.19321877

Bartlett's test for equal variances:  chi2(3) =   0.9523  Probchi2 = 0.813

. sort status

. by status: summarize bwt gestwks

- status=        0
Variable |    Obs        Mean   Std. Dev.       Min        Max
- - - - - - - - -+- - - - - - - - - - - - - - - - - - - - - - - - - - - - - - - - - -
     bwt |    115    7.783478   1.005418        5.2       11.4
 gestwks |    115    39.92174   1.855044         34         45

- status=        1
Variable |    Obs        Mean   Std. Dev.       Min        Max
- - - - - - - - -+- - - - - - - - - - - - - - - - - - - - - - - - - - - - - - - - -
     bwt |    169    7.221302    1.07776        5.2         10
 gestwks |    169    39.47929   1.786486         33         46

- status=        2
Variable |    Obs        Mean   Std. Dev.       Min        Max
- - - - - - - - -+- - - - - - - - - - - - - - - - - - - - - - - - - - - - - - - - -
     bwt |    130    7.266154   1.090946        4.4        9.4
 gestwks |    130    39.66923   1.974208         34         45

- status=        3
Variable |    Obs        Mean   Std. Dev.       Min        Max
- - - - - - - - -+- - - - - - - - - - - - - - - - - - - - - - - - - - - - - - - - -
     bwt |    266    7.710902   1.073099        3.3         11
 gestwks |    266    39.93985   1.874393         29         48
```

```
. generate g1=status==1

. generate g2=status==2

. generate g3=status==3

. generate gg1=g1*gestwks

. generate gg2=g2*gestwks

. generate gg3=g3*gestwks

. regress bwt gestwks g1 g2 g3 gg1 gg2 gg3
```

Source	SS	df	MS		
Model	176.978934	7	25.2827049		
Residual	633.216612	672	.942286625		
Total	810.195546	679	1.19321877		

```
Number of obs =      680
F(  7,   672) =    26.83
Prob  F       =   0.0000
R-square      =   0.2184
Adj R-square  =   0.2103
Root MSE      =  .97071
```

bwt	Coef.	Std. Err.	t	P|t|	[95% Conf. Interval]	
gestwks	.1650574	.04901	3.368	0.001	.0688262	.2612886
g1	-4.249813	2.565355	-1.657	0.098	-9.286889	.7872632
g2	-4.070149	2.606312	-1.562	0.119	-9.187644	1.047346
g3	-3.275763	2.335454	-1.403	0.161	-7.861428	1.309903
gg1	.0952567	.0644934	1.477	0.140	-.0313761	.2218895
gg2	.0906119	.0653922	1.386	0.166	-.0377858	.2190096
gg3	.0801254	.05843	1.371	0.171	-.0346019	.1948527
_cons	1.1941	1.958658	0.610	0.542	-2.651725	5.039924

```
. regress bwt gestwks g1 g2 g3
```

Source	SS	df	MS		
Model	174.420194	4	43.6050486		
Residual	635.775352	675	.941889411		
Total	810.195546	679	1.19321877		

```
Number of obs =      680
F(  4,   675) =    46.30
Prob  F       =   0.0000
R-square      =   0.2153
Adj R-square  =   0.2106
Root MSE      =  .97051
```

bwt	Coef.	Std. Err.	t	P|t|	[95% Conf. Interval]	
gestwks	.237543	.019968	11.896	0.000	.1983362	.2767498
g1	-.4570758	.1176508	-3.885	0.000	-.6880813	-.2260703
g2	-.4573428	.1243425	-3.678	0.000	-.7014875	-.2131982
g3	-.076878	.1083116	-0.710	0.478	-.2895462	.1357902
_cons	-1.699651	.8022762	-2.119	0.034	-3.274908	-.1243944

```
. regress bwt gestwks
```

Source	SS	df	MS
Model	146.930653	1	146.930653
Residual	663.264893	678	.978266805
Total	810.195546	679	1.19321877

```
Number of obs =     680
F( 1,   678) =  150.19
Prob  F      =  0.0000
R-square     =  0.1814
Adj R-square =  0.1801
Root MSE     =  .98907
```

bwt	Coef.	Std. Err.	t	P|t|	[95% Conf. Interval]	
gestwks	.2480389	.0202391	12.255	0.000	.2083	.2877778
_cons	-2.348182	.805816	-2.914	0.004	-3.930377	-.7659874

```
. exit,clear
```

Notation

Summary values:

$$n_i = \text{total number of observations per group}$$

$$(\text{special case of equal-size groups: } n_1 = n_2 = \cdots = n_g = n)$$

$$g = \text{the number of comparison groups (levels of the categorical variable)}$$

$$N = \text{the total number of observations} = \Sigma n_i \text{ (special case: } N = gn)$$

Sum of squares:

Within i^{th} group:

$$W_{xx}^{(i)} = \Sigma(x_{ij} - \overline{x}_i)^2 \,, \; W_{xy}^{(i)} = \Sigma(x_{ij} - \overline{x}_i)(y_{ij} - \overline{y}_i) \,, \; W_{yy}^{(i)} = \Sigma(y_{ij} - \overline{y}_i)^2$$

Within sums of squares:

$$W_{xx} = \Sigma W_{xx}^{(i)} \,, \; W_{xy} = \Sigma W_{xy}^{(i)} \,, \; W_{yy} = \Sigma W_{yy}^{(i)}$$

Between sums of squares:

$$B_{xx} = \Sigma n_i(\overline{x}_i - \overline{x})^2 \,, \; B_{xy} = \Sigma n_i(\overline{x}_i - \overline{x})(\overline{y}_i - \overline{y}) \,, \; B_{yy} = \Sigma n_i(\overline{y}_i - \overline{y})^2$$

Total sums of squares:

$$T_{xx} = W_{xx} + B_{xx} = \Sigma\Sigma(x_{ij} - \overline{x})^2$$
$$T_{xy} = W_{xy} + B_{xy} = \Sigma\Sigma(x_{ij} - \overline{x})(y_{ij} - \overline{y})$$
$$T_{yy} = W_{yy} + B_{yy} = \Sigma\Sigma(y_{ij} - \overline{y})^2$$

Slopes:

Within each group (model I):

$$\hat{b}_i = \frac{W_{xy}^{(i)}}{W_{xx}^{(i)}}$$

Combined among groups (model II):

$$\hat{b}' = \frac{\Sigma W_{xx}^{(i)}\hat{b}_i}{\Sigma W_{xx}^{(i)}} = \frac{\Sigma W_{xy}^{(i)}}{\Sigma W_{xx}^{(i)}} = \frac{W_{xy}}{W_{xx}}$$

Ignoring groups (model III):

$$\hat{b}'' = \frac{T_{xy}}{T_{xx}}$$

Problems

1   Conduct an analysis of covariance using the data in Table 1.4 where $Y =$ birth weight and $x =$ maternal prepregnancy weight for two groups, women who gained more than 13 kilograms during their pregnancy and those who did not. Calculate the residual sum of squares for model I, calculate the residual sum of squares for model II, and calculate the residual sum of squares for model III. Test the fit of model II as a representation of the data.

2   Repeat the analysis in problem 1 using the linear regression approach and the single design variable $g = 1$ if weight gain > 13 kilograms and $g = 0$ if weight gain ≤ 13 kilograms.

3   Compute the adjusted mean values for each weight-gain group based on the estimates from the analysis of covariance in problem 1. Compute directly the between sum of squares; that is, *between sum of squares* $= \Sigma n_i(\bar{y}_i' - \bar{y})^2$ where $\bar{y}_i'$ is an adjusted mean value. Compare the *between sum of squares* to the corresponding difference in the residual sum of squares from model II and model III (i.e., *Res(III)* − *Res(II)* = main-effects sum of squares).

4   Based on the analysis of covariance in problem 1, calculate the estimated birth weights $\hat{y}_{ij}$, $\hat{y}_{ij}'$, and $\hat{y}_{ij}''$ for models I, II, and III for all 48 observations. (Hint: Use the regression approach.) Demonstrate that $Res(III) - Res(II) = \Sigma\Sigma(\hat{y}_{ij}' - \hat{y}_{ij}'')^2$.

5   Two adjusted mean values for group i and group j are $\bar{y}_i'$ and $\bar{y}_j'$. Also, the intercept of two regression lines with the same slope are $\hat{a}_i'$ and $\hat{a}_j'$. Show that $\bar{y}_i' - \bar{y}_j' = \hat{a}_i' - \hat{a}_j'$.

6   Show that the overall mean of the adjusted values $\bar{y}'$ is equal to the overall mean of the observed values $\bar{y}$ (i.e., $\bar{y}' = \bar{y}$).

7   Estimate the parameters and sums of squares for models I, II, and III using the following data (repeated from Chapter 4) where $Y =$ length of stay of patients in a critical care unit (hours), $x_1 =$ severity index, and $x_2 =$ hospital size classified as large $= 0$ and small $= 1$:

Obs.	Y	x_1	$x_2 =$ group
1	24	22	0
2	48	80	0
3	37	55	0
4	81	140	1

Obs.	Y	x_1	x_2 = group
5	48	90	0
6	2	10	1
7	24	40	1
8	64	120	1
9	12	30	1
10	8	25	1
11	4	10	1
12	36	77	0
13	12	32	0
14	8	14	1
15	4	15	1
16	88	200	0
17	54	120	0

Contrast the results of applying the three analysis of covariance models with the analysis from Chapter 4 (problem 10) where regression techniques are employed.

References

1 Dunn, O. J., and Clark, V. A. *Applied Statistics: Analysis of Variance and Regression*. New York: John Wiley, 1974.

2 Sokal, R. R., and Rohlf, F. J. *Biometry*. San Francisco: W. H. Freeman and Co., 1969.

3 Winer, B. J., Brown, D. R., and Michels, K. M. *Statistical Principles in Experimental Design*, 3rd ed. New York: McGraw-Hill, 1991.

4 Hosmer, D. W., and Lemeshow, S. *Applied Logistic Regression*. New York: John Wiley, 1989.

5 Selvin, S. *Statistical Analysis of Epidemiologic Data*. New York: Oxford University Press, 1991.

6

Linear Discriminant Analysis

Multivariate Distance

Statistical distance, like all distance measures, addresses the question: How far is one location from another? The simplest distance is the difference between two observations, $|x_1 - x_2|$. Complexity increases as more than one variable per observation is considered. To compare two observations each made up of two variables, the distance between the observations (x_1, y_1) and (x_2, y_2) is found by applying the Pythagorean theorem. Therefore, the distance between observations 1 and 2 (d_{12}) is

$$d_{12} = \sqrt{(x_1 - x_2)^2 + (y_1 - y_2)^2} .$$

For three variables x, y, and z measured on each observation, a natural extension of the two-variable distance is

$$d_{12} = \sqrt{(x_1 - x_2)^2 + (y_1 - y_2)^2 + (z_1 - z_2)^2}$$

where d_{12} is again the distance between observations 1 and 2 (Figure 6.1). Distance based on a sum of squared differences is called the Euclidean distance. In general, Euclidean distance between two observations for k measured variables per observation is

$$d_{12} = \sqrt{\sum_{i=1}^{k} (x_{i1} - x_{i2})^2} .$$

(6.1)

Euclidean distance depends on the units used to measure the k variables. If the variables that make up an observation are measured in different units (e.g., feet, pounds, and dollars), the distance calculated by combining these variables is not easily interpreted. Furthermore, Euclidean distance is dominated by those variables with the greatest variability even when the components are measured in the same units. For example, when a bivariate observation is characterized by a mother's prepregnancy weight and her infant's birth weight, maternal weight dominates any conclusions

FIGURE **6.1**

Illustration of distance in three dimensions

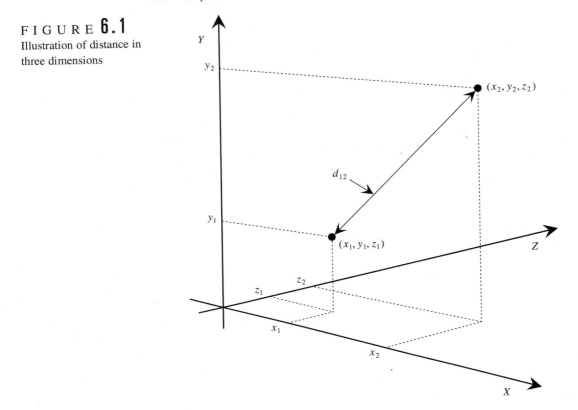

based on Euclidean distance. Infant birth weights are rarely less than five pounds or greater than nine pounds, whereas maternal weight varies over a much broader range. A difference of one or two pounds in infant weight (which is considerable) will be lost in a Euclidean distance summary containing maternal weight even though both measurements are in the same units.

One approach to studying variables that differ in units and/or in variability is to divide each variable by its standard deviation. This process produces a series of variables in units of standard deviations, giving each variable a commensurate role in the makeup of the Euclidean distance.

To illustrate Euclidean distance, four groups of males are considered where each group is characterized by the mean of four variables (body weight (pounds), systolic blood pressure (mm), cholesterol level (mg/100ml) and amount smoked (cigarettes/day); $k = 4$). The four groups are type A individuals who have coronary heart disease (C), type A individuals who do not have coronary heart disease ($\overline{C}$), type B individuals who have coronary disease (C), and type B individuals who do not have coronary disease ($\overline{C}$). The standardized Euclidean distances among the four behavior/coronary groups are given in Table 6.1, constructed from four standardized vari-

T A B L E **6.1**

Mean values and standardized distances among four groups of individuals measured for four characteristics

Means Groups	Weight	Blood Pressure	Cholesterol	Smoking
$B\mid\bar{C}$	168.76	127.07	222.36	9.92
$A\mid\bar{C}$	170.39	129.05	226.27	12.45
$B\mid C$	174.49	134.96	252.43	16.41
$A\mid C$	174.45	135.57	249.02	16.78

Distances Groups	$B\mid\bar{C}$	$A\mid\bar{C}$	$B\mid C$	$A\mid C$
$B\mid\bar{C}$	0.000	0.569	1.781	1.679
$A\mid\bar{C}$	0.569	0.000	1.372	1.532
$B\mid C$	1.781	1.372	0.000	2.635
$A\mid C$	1.679	1.532	2.635	0.000

ables: weight (lbs.), systolic blood pressure (mm), cholesterol (mg%), and smoking (cigarettes per day). The standardized Euclidean distances, for example, show that type A and type B individuals are relatively close when both groups do not have coronary heart disease ($\bar{C}$; 0.569) and further apart when both groups are made up of individuals with coronary heart disease (C; 2.635).

The influence of a single variable or several variables on the Euclidean distance is simply isolated. Distance is calculated with the variable(s) included and then calculated with the variable(s) excluded. The difference between these two distances reflects the direct influence of the variable(s) removed. For example, the standardized distance between type A and type B groups with coronary disease ($A\mid C$ versus $B\mid C$) is 2.635; with cholesterol removed the distance is 2.491, showing the influence exclusively attributable to standardized cholesterol levels.

Penrose [1] suggests a similar standardized measure that also accounts for differences in variability where the distance between observations 1 and 2 based on k variables is

$$d_{12}' = \sqrt{\sum_{i=1}^{k} \frac{(x_{i1} - x_{i2})^2}{k\,S_i^2}}$$

(6.2)

where S_i^2 is the pooled variance of the groups 1 and 2 from variable i. A property of Penrose's suggestion, as well as Euclidean distance in general, is that correlations among the variables used in the distance measure are not taken into account. If two variables are highly correlated, they essentially measure the same thing, adding a redundancy to the Euclidean and Penrose measures of distance. Measures exist that account for the correlations among variables, producing a single multivariate distance. One such multivariate distance, sometimes called Mahalanobis distance, emerges in the study of the linear discriminant function. (Mahalanobis distance is named after Prasanta Chandra Mahalanobis [1893–1972], who was trained in mathematical physics but became a preeminent contributor to the field of statistics and a pioneer in establishing statistical research in India.)

Background

Situations arise in which it is desirable to classify, as well as possible, an observation into one of several categories. A statistical method designed for this purpose is a linear discriminant function. Suppose a limited number of individuals can be accepted into a program to help people stop smoking cigarettes. Accurate classification of individuals as to whether they are likely to continue or likely to stop smoking would be useful. Assume specific information is available on each person, such as the number of cigarettes smoked each day, the number of years each individual smoked, blood pressure, residual lung capacity, and several smoking-related biochemical determinations. A linear discriminant function provides a statistical basis for using this complex multivariate information to identify individuals likely to be the "best" candidates for the smoking control program. A discriminant function is relatively simple (linear), combines multivariate measurements into a single value (discriminant score), and allows classification of observations into a series of categories with minimum misclassification. The word *best* means the linear discriminant function is superior to all other possibilities but in a limited sense. Like a regression analysis, a useful but somewhat arbitrary criterion is at the foundation for constructing the "best" statistical tool for classifying observations.

Method: Criterion

The linear discriminant function, invented by the British statistician-geneticist R. A. Fisher [2], fundamentally serves as a tool for classification. However, the process of classifying observations into a series of categories also leads to insights into the structure of multivariate data. Like many multivariable methods, the linear discriminant function arises from employing a mathematical criterion to produce a statistical procedure that is optimum in a specific sense. For regression analysis, the criterion is the squared distance from the data points to the values from a regression equation and is optimum in the sense that this distance is minimum. The linear discriminant function involves a seemingly different criterion, but similarities to regression analysis appear.

The simplest linear discriminant function involves two variables measured per observation (x and y) and has the form

$$l_{ij} = \alpha_x x_{ij} + \alpha_y y_{ij} \qquad (6.3)$$

where l_{ij} is a discriminant score, i denotes the group from which the observation was sampled, and j indicates a specific member of that group. The values represented by α_x and α_y are weights assigned to each variable called the linear discriminant coefficients. This linear discriminant score l_{ij} reduces a bivariate measurement to a single number, which can be evaluated with a variety of elementary statistical techniques.

Specifically, consider the situation where the number of groups to which an observation can possibly belong is two ($g = 2$), the number of measurements per observation is also two ($k = 2$), and the number of sampled observations in each group is denoted by n_i. The notation is displayed in Table 6.2. The mean discriminant scores ($\bar{l}_1$ and $\bar{l}_2$) summarize much of the bivariate information contained in the variables x and y. The group means ($\bar{x}_1$, $\bar{y}_1$) and ($\bar{x}_2$, $\bar{y}_2$) indicate the location of two bivariate distributions (e.g., Figure 1.5). A discriminant function translates these bivariate mean values into univariate measures of location ($\bar{l}_1$ and $\bar{l}_2$).

Discriminant scores are not restricted to just two variables and are readily extended to summarize information contained in k-variate measurements (k variables measured per observation) for any number of groups. The mean discriminant scores from g groups and k variables are

$$\bar{l}_1 = \alpha_1 \bar{x}_{11} + \alpha_2 \bar{x}_{21} + \alpha_3 \bar{x}_{31} + \cdots + \alpha_k \bar{x}_{k1}$$
$$\bar{l}_2 = \alpha_1 \bar{x}_{12} + \alpha_2 \bar{x}_{22} + \alpha_3 \bar{x}_{32} + \cdots + \alpha_k \bar{x}_{k2}$$

$$.$$
$$.$$
$$.$$

$$\bar{l}_g = \alpha_1 \bar{x}_{1g} + \alpha_2 \bar{x}_{2g} + \alpha_3 \bar{x}_{3g} + \cdots + \alpha_k \bar{x}_{kg} \qquad (6.4)$$

where $\bar{x}_{ij}$ represents the mean value of the ith variable from group j and α_i is the discriminant coefficient associated with the ith variable. For example, the value $\bar{x}_{32}$ is the

T A B L E **6.2**

Notation for a discriminant function (groups = g = 2 and variables = k = 2)

	Group 1 ($i = 1$)			Group 2 ($i = 2$)	
	Observations	Discriminant Scores		Observations	Discriminant Scores
1	x_{11} y_{11}	$l_{11} = \alpha_x x_{11} + \alpha_y y_{11}$	1	x_{21} y_{21}	$l_{21} = \alpha_x x_{21} + \alpha_y y_{21}$
2	x_{12} y_{12}	$l_{12} = \alpha_x x_{12} + \alpha_y y_{12}$	2	x_{22} y_{22}	$l_{22} = \alpha_x x_{22} + \alpha_y y_{22}$
.			.	,	
.			.	,	
.			.	,	
n_1	x_{1n_1} y_{1n_1}	$l_{1n_1} = \alpha_x x_{1n_1} + \alpha_y y_{1n_1}$	n_2	x_{2n_2} y_{2n_2}	$l_{2n_2} = \alpha_x x_{2n_2} + \alpha_y y_{2n_2}$
Means	$\bar{x}_1$ $\bar{y}_1$	$\bar{l}_1 = \alpha_x \bar{x}_1 + \alpha_y \bar{y}_1$	Means	$\bar{x}_2$ $\bar{y}_2$	$\bar{l}_2 = \alpha_x \bar{x}_2 + \alpha_y \bar{y}_2$

mean of the third variable measured among the k possible from group 2 and has associated coefficient α_3. Again, a multivariate mean ($\bar{x}_{1j}, \bar{x}_{2j}, \cdots, \bar{x}_{kj}$) is translated into a far simpler univariate measurement of location ($\bar{l}_j$), one for each group.

Normal Equations

Fundamental to any estimated value is its variance. For the bivariate linear discriminant function, the variance of the discriminant score l_{ij}, like any linear function, is estimated by (expression 1.11 shows the general case)

$$variance(l_{ij}) = S_L^2 = \alpha_x^2 S_X^2 + \alpha_y^2 S_Y^2 + 2\alpha_x \alpha_y S_{XY} \quad \text{(for } g = 2 \text{ and } k = 2\text{)}.$$

The construction of a linear discriminant function comes from optimizing a criterion to generate specific discriminant function coefficients. The criterion dictates that the α-coefficients are chosen so the squared distance between the mean discriminant scores is maximum relative to the variance of the discriminant function. For $g = 2$ and $k = 2$, the criterion is Q:

$$Q = \frac{(\bar{l}_1 - \bar{l}_2)^2}{S_L^2} = \frac{[(\alpha_x \bar{x}_1 + \alpha_y \bar{y}_1) - (\alpha_x \bar{x}_2 + \alpha_y \bar{y}_2)]^2}{\alpha_x^2 S_X^2 + \alpha_y^2 S_Y^2 + 2\alpha_x \alpha_y S_{XY}}. \tag{6.5}$$

Specific values of the coefficients α_x and α_y are found that maximize Q—that is, $(\bar{l}_1 - \bar{l}_2)^2$ is made as large as possible when divided by S_L^2. To find α_x and α_y so Q is maximum requires calculus and produces the normal equations

$$\alpha_x S_X^2 + \alpha_y S_{XY} = (\bar{x}_2 - \bar{x}_1)$$
$$\alpha_x S_{XY} + \alpha_y S_Y^2 = (\bar{y}_2 - \bar{y}_1) \tag{6.6}$$

which are solved to produce the linear discriminant coefficients.

These normal equations are similar to the ones used to estimate the bivariate regression coefficients. The values on the left side of the equation are identical to the left side of the reduced normal equations associated with bivariate regression analysis (expression 3.4). The values on the right side, $\bar{x}_2 - \bar{x}_1$ and $\bar{y}_2 - \bar{y}_1$, play the roles of S_{Y1} and S_{Y2}. This similarity illustrates that discriminant analysis can be viewed as a special case of regression analysis.

The equivalence of the discriminant function and multiple regression normal equations means that many of the approaches developed in the context of regression analysis (Chapters 2–5) apply to discriminant analysis. For example, the procedure to formally assess the components of a discriminant function relies on an analysis of variance strategy (F-to-remove), as will be seen. A novel feature (also described later) is that computer programs that produce a linear regression analysis can be used to produce a linear discriminant analysis.

The solution to the normal equations gives specific coefficients α_x and α_y that maximize Q, or

$$\alpha_x = \frac{(\bar{x}_2 - \bar{x}_1)S_Y^2 - (\bar{y}_2 - \bar{y}_1)S_{XY}}{S_X^2 S_Y^2 - S_{XY}^2} \quad \text{and}$$

$$\alpha_y = \frac{(\bar{y}_2 - \bar{y}_1)S_X^2 - (\bar{x}_2 - \bar{x}_1)S_{XY}}{S_X^2 S_Y^2 - S_{XY}^2}. \tag{6.7}$$

In general, the k coefficients of a linear discriminant function that maximizes the distance between mean discriminant scores relative to the variance come from solving k normal equations. These equations are

$$\alpha_1 S_1^2 + \alpha_2 S_{21} + \alpha_3 S_{31} + \cdots + \alpha_k S_{k1} = (\bar{x}_{12} - \bar{x}_{11})$$

$$\alpha_1 S_{12} + \alpha_2 S_2^2 + \alpha_3 S_{32} + \cdots + \alpha_k S_{k2} = (\bar{x}_{22} - \bar{x}_{21})$$

$$\alpha_1 S_{13} + \alpha_2 S_{23} + \alpha_3 S_3^2 + \cdots + \alpha_k S_{k3} = (\bar{x}_{32} - \bar{x}_{31})$$

$$\vdots$$

$$\alpha_1 S_{1k} + \alpha_2 S_{2k} + \alpha_3 S_{3k} + \cdots + \alpha_k S_k^2 = (\bar{x}_{k2} - \bar{x}_{k1}). \tag{6.8}$$

The solution $\alpha_1, \alpha_2, \cdots, \alpha_k$ is usually calculated by a computer program. It should be noted that the development of the linear discriminant function presented here assumes the sample sizes are large enough so that occasionally ignoring the distinction between sample estimates and population values has negligible effects.

Invariance Property

If a constant value is added to discriminant scores l_{ij}, then it does not affect the criterion Q (i.e., $S_{L+C}^2 = S_L^2$ and $[(\bar{l}_1 + C) - (\bar{l}_2 + C)] = [\bar{l}_1 - \bar{l}_2]$). Also, if the discriminant scores are multiplied by a constant value, the criterion Q is unchanged because

$$Q = \frac{(B\bar{l}_1 - B\bar{l}_2)^2}{B^2 S_L^2} = \frac{B^2(\bar{l}_1 - \bar{l}_2)^2}{B^2 S_L^2} = \frac{(\bar{l}_1 - \bar{l}_2)^2}{S_L^2}. \tag{6.9}$$

Because Q remains unchanged when the discriminant scores l_{ij} are transformed by multiplication or addition, the opportunity arises to define a discriminant function with particularly convenient properties. In other words, the scores associated with a linear discriminant function are not unique and a variety of valid choices exist for the coefficients.

A constant value typically subtracted from l_{ij} is

$$\bar{l} = \alpha_1 \bar{x}_1 + \alpha_2 \bar{x}_2 + \cdots + \alpha_k \bar{x}_k, \tag{6.10}$$

where $\bar{x}_i$ is the mean value of the ith variable disregarding the classification of observations into groups (the overall mean for each of the k variables). Also, each discriminant score is often divided by the estimated standard deviation of the discriminant scores, S_L. The transformed discriminant scores (denoted d_{ij}) then have mean = 0. More important, the variance of these transformed discriminant scores is exactly 1.0 ($S_D^2 = 1.0$).

Specifically, the transformed discriminant score d_{ij}, when $g = 2$ and $k = 2$, is

$$d_{ij} = \alpha_x(x_{ij}/S_L) + \alpha_y(y_{ij}/S_L) - \bar{l}/S_L$$
$$= a_x x_{ij} + a_y y_{ij} - (a_x \bar{x} + a_y \bar{y}) \quad \text{where} \quad a_x = \alpha_x/S_L \text{ and } a_y = \alpha_y/S_L .$$

Therefore,

$$d_{ij} = a_x(x_{ij} - \bar{x}) + a_y(y_{ij} - \bar{y}) .$$

The "new" mean discriminant scores are

$$\bar{d}_1 = a_x(\bar{x}_1 - \bar{x}) + a_y(\bar{y}_1 - \bar{y}) \quad \text{and} \quad \bar{d}_2 = a_x(\bar{x}_2 - \bar{x}) + a_y(\bar{y}_2 - \bar{y}) .$$

The overall mean discriminant score is zero (i.e., $\bar{d} = 0$). Additionally, the variance is

$$S_D^2 = a_x^2 S_X^2 + a_y^2 S_Y^2 + 2a_x a_y S_{XY} = (\alpha_x^2 S_X^2 + \alpha_y^2 S_Y^2 + 2\alpha_x \alpha_y S_{XY})/S_L^2 = 1.0 .$$

However, the criterion Q remains maximized for the transformed discriminant scores d_{ij}.

To summarize, this section is a somewhat complicated demonstration of the simple property that Q remains unchanged for multiplicative and additive transformations of the linear discriminant scores. Because these scores are not unique, they are often transformed to have specific properties. In the following discussion, discriminant scores (l_{ij}) are transformed (d_{ij}) to have mean value of zero ($\bar{d} = 0$) and a variance of 1.0 ($S_D^2 = 1.0$), allowing some simplification in presentation and interpretation. The transformed values d_{ij} are also often the discriminant scores that appear in the output from linear discriminant computer programs (e.g., STATA and SPSS).

Underlying Structure for a Linear Discriminant Analysis

The four basic requirements for a discriminant analysis are independence of observations (not variables), equal variances among groups, equal covariances among groups, and normally distributed sample observations.

For linear discriminant analysis, independence means that each of N observations is independently selected, producing a sample of independent multivariate observations. Independence is certainly not necessary among the k variables measured on each observation. In fact, the opposite is usually true. A multivariate analysis is conducted because the k measurements made on a single observation are almost always related.

The property of equal variances and covariances is similar to the property of equal variances required for regression analysis. All variables must have the same variances and the same covariances in every group considered. Estimates of the variances and covariances come, therefore, from pooling the estimates made from within each group (weighted averages, where the weights are the degrees of freedom). The pooled estimates are

$$\text{variances: } S_{X_i}^2 = \frac{\Sigma \, (n_j - 1) S_{X_{ij}}^2}{N - g} \tag{6.11}$$

and

$$\text{covariances: } S_{X_i X_m} = \frac{\Sigma \, (n_j - 1) S_{X_{ij} X_{mj}}}{N - g} \tag{6.12}$$

where i and m denote specific variables, j represents groups, g = the number of groups, and $\Sigma n_j = N$ is the total number of observations.

The requirement of normally distributed data applies to the k variables measured on each observation. Each k-variate observation is assumed to come from a multivariate normal distribution. Multivariate normally distributed variables form linear discriminant scores that have normal distributions (for more detail see [3] or [4]).

These four requirements produce transformed linear discriminant scores with normal distributions and variance = $S_D^2 = 1.0$ that summarize values of multivariate measurements. Experience and some mathematics show that discriminant analysis works well when the requirements of normality and equal variance/covariance only approximately describe the structure underlying the data to be analyzed; however, like regression analysis, the accuracy of the analysis is sensitive to the requirement of independence of the sampled observations.

Geometric Interpretation

The result of maximizing the squared distance between mean discriminant scores relative to the variance (maximizing Q) has a simple geometric interpretation for two groups ($g = 2$) with two variables ($k = 2$). Ellipses that represent probability contours of the two bivariate normal distributions of X and Y surrounding the means ($\bar{x}_i$, $\bar{y}_i$) are shown in Figure 6.2 (dashed lines). These probability ellipses are described in Chapter 1. A line is constructed that passes through the intersection of the probability contours generated by the two distributions (line A in Figure 6.2). Then, a line perpendicular to A represents the possible values of the discriminant scores (line B in Figure 6.2). A discriminant score d_{ij} computed from a data point (x_{ij}, y_{ij}) is a point on line B. More technically, the two-dimensional data are projected onto a one-dimensional line. The characteristics of these univariate discriminant scores along line B is the primary focus of a discriminant analysis. For example, the point on line B corresponding to $\bar{d}_i$ is the summary of the location of the bivariate mean values ($\bar{x}_i$, $\bar{y}_i$) for each group. The process is, in principle, the same for the k-variable case. The k-dimensional measurements are projected onto a line, producing a distribution of one-dimensional discriminant scores that are easily displayed and analyzed.

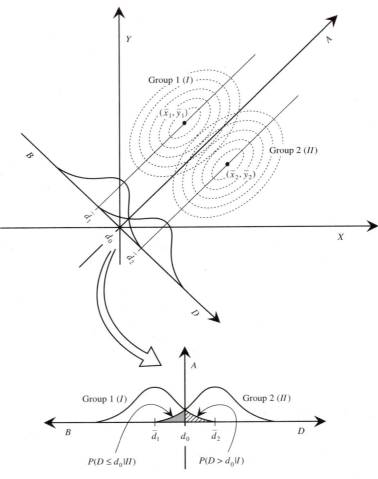

F I G U R E **6.2**
Discriminant analysis based on $d_{ij} = a_x(x_{ij} - \overline{x}) + a_y(y_{ij} - \overline{y})$

Relationships

The following four statistical expressions begin to show how and why a linear discriminant function works when applied to classifying observations into two groups. Several of these expressions serve as computational forms but, more important, they express in mathematical terms fundamental issues involved in discriminant analysis. These results provide some foundation for the discussion contained in the following sections. This section, however, can be omitted without loss because less technical discussions of the same issues follow. As before, the primary focus is on the relatively simple two-variable analysis with less detailed reference to the higher-dimensional cases.

1 The standard deviation of the discriminant function (Box 6.1) is expressed as

$$S_L = a_x(\bar{x}_2 - \bar{x}_1) + a_y(\bar{y}_2 - \bar{y}_1)$$

(6.13)

where a_x and a_y are linear discriminant coefficients calculated from the previous normal equations (expression 6.6).

B O X **6.1**   VARIABILITY OF THE DISCRIMINANT SCORE S_L^2 *

The normal equations are

$$\alpha_x S_X^2 + \alpha_y S_{XY} = (\bar{x}_2 - \bar{x}_1)$$
$$\alpha_x S_{XY} + \alpha_y S_Y^2 = (\bar{y}_2 - \bar{y}_1)$$

and α_x times the first equation and α_y times the second equation yields

$$\alpha_x^2 S_X^2 + \alpha_x \alpha_y S_{XY} = \alpha_x(\bar{x}_2 - \bar{x}_1)$$
$$\alpha_x \alpha_y S_{XY} + \alpha_y^2 S_Y^2 = \alpha_y(\bar{y}_2 - \bar{y}_1) \, .$$

Adding the two equations gives

$$\alpha_x^2 S_X^2 + \alpha_y^2 S_Y^2 + 2\alpha_x \alpha_y S_{XY} = \alpha_x(\bar{x}_2 - \bar{x}_1) + \alpha_y(\bar{y}_2 - \bar{y}_1)$$

or

$$S_L^2 = \alpha_x(\bar{x}_2 - \bar{x}_1) + \alpha_y(\bar{y}_2 - \bar{y}_1) \, .$$

Because

$$a_x = \alpha_x / S_L \quad \text{and} \quad a_y = \alpha_y / S_L \, ,$$

then

$$S_L^2 = a_x S_L(\bar{x}_2 - \bar{x}_1) + a_y S_L(\bar{y}_2 - \bar{y}_1)$$

and

$$S_L = a_x(\bar{x}_2 - \bar{x}_1) + a_y(\bar{y}_2 - \bar{y}_1) \, .$$

In general, for $g = 2$ groups and any number of variables measured per observation (k), then

$$S_L = \Sigma a_i(\bar{x}_{i2} - \bar{x}_{i1}) \, .$$

*($g = 2$ and $k = 2$)

2 The estimated variance of the discriminant score (Box 6.2) is the multivariate distance between two groups and is sometimes referred to as Mahalanobis distance. That is, $S_L^2 = (\bar{d}_1 - \bar{d}_2)^2$.

3 Maximizing the criterion of Q means that $(\bar{d}_1 - \bar{d}_2)^2$ is the maximum squared distance between the measures of location for $g = 2$ groups because $Q_{max} = (\bar{d}_1 - \bar{d}_2)^2$ where, as before, $\bar{d}_i$ is the estimated mean discriminant score of the ith group (see Box 6.3).

To summarize, $(\bar{d}_1 - \bar{d}_2)^2 = S_L^2 = Q_{max}$ and is maximized for a given data set by the choice of the discriminant function coefficients. The distance/variance is maximized in a relative sense. The multivariate distance between the mean discriminant scores is not an absolute maximum but is the maximum distance between means among all possible distances that could result from applying the criterion Q. That is, it is the maximum distance between the discriminant scores among all possible projections of the multivariate data onto a line. The same is true with respect to S_L^2—it is also the maximum variance among all possible variances resulting from projecting the multivariate data on a line.

4 If again two groups are considered, labeled group *I* and group *II* for convenience, then

$$d_0 = -\frac{1}{S_L} \log\left[\frac{P(II)}{P(I)}\right] + \tfrac{1}{2}(\bar{d}_1 + \bar{d}_2)$$

(6.14)

B O X **6.2**   RELATIONSHIP BETWEEN THE $\bar{d}_i\,s$-VALUES AND THE VARIANCE OF $L\,(S_L^2)$ *

$$\bar{d}_1 = a_x(\bar{x}_1 - \bar{x}) + a_y(\bar{y}_1 - \bar{y})$$
$$\bar{d}_2 = a_x(\bar{x}_2 - \bar{x}) + a_y(\bar{y}_2 - \bar{y})$$
$$(\bar{d}_1 - \bar{d}_2) = a_x(\bar{x}_1 - \bar{x}_2) + a_y(\bar{y}_1 - \bar{y}_2)$$

Therefore,

$$(\bar{d}_1 - \bar{d}_2)^2 = [a_x(\bar{x}_1 - \bar{x}_2) + a_y(\bar{y}_1 - \bar{y}_2)]^2 = S_L^2$$

because

$$S_L = a_x(\bar{x}_2 - \bar{x}_1) + a_y(\bar{y}_2 - \bar{y}_1) \,.$$

In general, for $g = 2$ groups and any number of variables measured per observation (k), then

$$(\bar{d}_1 - \bar{d}_2)^2 = S_L^2 \,.$$

*($k = 2$ and $g = 2$)

where d_0 is a classification point—it divides the discriminant scores into those observations assigned to *I* and those assigned to *II*. The symbol $P(I)$ represents the probability that a random observation belongs to *I*, and $P(II) = 1 - P(I)$ represents the complimentary probability that a random observation belongs to *II*. A rigorous justification of the value of d_0 is complex and found elsewhere [5]; however, a description of the properties of the classification point d_0 follows.

If *D* is a new observation (i.e., an observation not used to construct the discriminant function) and if $D \leq d_0$, then the observation is classified as belonging to *I*; if $D > d_0$, the observation is classified as belonging to *II*. When $D \leq d_0$ for an observation belonging to *II*, one kind of misclassification occurs, and when $D > d_0$ for an observation belonging to *I*, another type of misclassification occurs. The magnitude of these misclassification errors is expressed in terms of probabilities: $P(D \leq d_0 \mid II)$ and $P(D > d_0 \mid I)$.

The cut-point d_0 (expression 6.14) is designed to give the least possible misclassification for the analyzed data set. If it is equally likely that an observation comes from *I* or *II* ($P(I) = P(II) = 0.5$), then the probability of misclassifying a new observation *D* is $P(\text{misclassification}) = \frac{1}{2}P(D < d_0 \mid II) + \frac{1}{2}P(D > d_0 \mid I)$. Using d_0 guarantees that no other linear combination of the observations has a smaller unconditional probability of misclassification. Misclassification when $P(I) \neq P(II)$ is discussed in a following section on prior and posterior probabilities. Not surprisingly, the point d_0 "best" classifies observations only when the data meet the requirements for a linear discriminant analysis.

B O X **6.3**   RELATIONSHIP BETWEEN $\bar{d_i}$'s-VALUES AND CRITERION Q*

$$Q_{max} = \frac{(\bar{l}_1 - \bar{l}_2)^2}{S_L^2}$$

$$\bar{l}_i = \alpha_x \bar{x}_i + \alpha_y \bar{y}_i = a_x S_L \bar{x}_i + a_y S_L \bar{y}_i$$

Therefore,

$$Q_{max} = \frac{[(a_x S_L \bar{x}_1 + a_y S_L \bar{y}_1) - (a_x S_L \bar{x}_2 + a_y S_L \bar{y}_2)]^2}{S_L^2} = \frac{S_L^2 [a_x(\bar{x}_1 - \bar{x}_2) + a_y(\bar{y}_1 - \bar{y}_2)]^2}{S_L^2}$$

$$= \frac{S_L^2 (\bar{d}_1 - \bar{d}_2)^2}{S_L^2} = (\bar{d}_1 - \bar{d}_2)^2 .$$

In general, for $g = 2$ groups and any number of variables measured per observation (k), then

$$Q_{max} = (\bar{d}_1 - \bar{d}_2)^2 .$$

*($k = 2$ and $g = 2$)

Classification

Once a discriminant function is established, assessing its performance is the next step. Consider a person reporting the number of cigarettes smoked per day, the number of years smoked, blood pressure, level of carbon monoxide, level of thiocyanate, and residual lung capacity; a discriminant score combines these six measurements into a single value that may indicate whether this person is likely to give up smoking when enrolled in a smoking cessation program. At the same time, the misclassification probability associated with the predictive process is estimated. It is this probability of misclassification that is primarily used to evaluate the performance of a discriminant function. Two methods of calculating the misclassification probability, among the more than half dozen available, are described.

Parametric Estimate of the Probability of Misclassification

To classify an observation into one of two groups based on a discriminant score, a decision rule is necessary. The rule that produces the least misclassification error is:

If $D > d_0$, assign the new observation D to one group and

if $D \leq d_0$, assign the new observation D to the other group.

The symbol D represents a discriminant score calculated from an observation not used in the estimation of the discriminant function, and the classification point d_0 is defined in the previous section (expression 6.14). Assume that any new observation to be classified is equally likely to come from either group *I* or *II* (i.e., $P(I) = P(II) = 0.5$). These quantities are called prior probabilities. The word *prior* refers to the property that these probabilities are known before any specific measurements are made on the sampled observations. The assumption or knowledge of equal prior probabilities makes $d_0 = \frac{1}{2}(\bar{d}_1 + \bar{d}_2)$. When, for example, $d_0 = \frac{1}{2}(\bar{d}_1 + \bar{d}_2) = 0$, then values $D > 0$ are classified as one group and values $D \leq 0$ are classified as the other group. In general, classification decisions are made on the basis of both the prior probabilities and a discriminant score. When the prior probabilities are equal, however, classification is based strictly on the discriminant score. Schematically, the decision rule is illustrated in Figure 6.3 (repeated from Figure 6.2).

An alternative and perhaps helpful view of the classification rule based on d_0 is an equivalent rule given by

$$\text{if } (D - \bar{d}_1)^2 > (D - \bar{d}_2)^2, \text{ classify the new observation } D \text{ as belonging to } II$$

and

$$\text{if } (D - \bar{d}_1)^2 \leq (D - \bar{d}_2)^2, \text{ classify the new observation } D \text{ as belonging to } I . \quad \text{(6.15)}$$

This means a new observation D is naturally classified as belonging to the group with estimated mean discriminant score ($\bar{d}_1$ or $\bar{d}_2$) closest to the new observed value. A little algebra shows that this classification rule is identical to comparing D to d_0 ($P(I) = P(II) = 0.5$).

FIGURE **6.3**
Classification based on
discriminant scores
$(P(I) = P(II) = 0.5)$

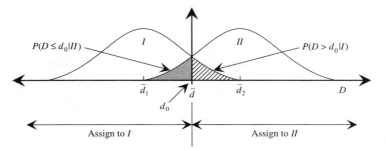

The shaded area to the right of d_0 in Figure 6.3 represents the probability that an error occurs when classifying an observation as a member of group *II*. This error occurs when the decision is made on the basis of the discriminant score *D* to assign an observation to group *II* $(D > d_0)$ when it actually belongs to group *I*. When the discriminant scores have a normal distribution, the probability of misclassifying a member of group *I* as belonging to group *II* is estimated by

$$P(D > d_0 \mid I) = P(D > \tfrac{1}{2}[\bar{d}_1 + \bar{d}_2] \mid I) = P\left(Z > \frac{\tfrac{1}{2}[\bar{d}_2 + \bar{d}_1] - \bar{d}_1}{S_D}\right)$$
$$= P(Z > \tfrac{1}{2}[\bar{d}_2 - \bar{d}_1]) = P(Z > \tfrac{1}{2} S_L) \tag{6.16}$$

where *Z* denotes a standard normal variable. Similarly, the probability of misclassifying a member of group *II* as a member of group *I* is estimated by

$$P(D \leq d_0 \mid II) = P(Z \leq -\tfrac{1}{2} S_L) \tag{6.17}$$

which is represented as the shaded area to the left of d_0 in Figure 6.3. Therefore, the overall (unconditional) estimated probability of a misclassification is

$$P(misclassification) = P(Z > \tfrac{1}{2} S_L)P(I) + P(Z \leq -\tfrac{1}{2} S_L)P(II) . \tag{6.18}$$

Because $P(I) = P(II) = 0.5$ and $P(Z > \tfrac{1}{2} S_L) = P(Z \leq -\tfrac{1}{2} S_L)$, the probability of misclassifying a new observation *D* is

$$P(misclassification) = P(Z > \tfrac{1}{2} S_L) . \tag{6.19}$$

The unconditional probability of misclassification corresponds to the total shaded area in Figure 6.3. The classification rule and description applies when $\bar{d}_2 > \bar{d}_1$ and is easily modified when $\bar{d}_2 < \bar{d}_1$.

The probability of misclassification is inversely related to the variance of the discriminant function S_L^2. The larger the variance S_L^2, the smaller the likelihood that a new observation is misclassified. Increased variability produces a more useful statistical summary because differences among observations are seen more clearly across a wide range of values. The opposite is clearly true. If a discriminant function has little or no variability, it is useless as a classification tool. A discriminant function is constructed to be as variable as possible, because $Q = S_L^2$ is maximized. Maximum variability guarantees the minimum probability of misclassification for the analyzed data set.

Classification with a Single Variable

At this point it is useful to digress to consider classification based on a single variable in contrast to using a discriminant function. When two variables X and Y are available for classifying an observation, assume X alone is chosen as a basis for deciding group membership. Similar to the structure underlying a discriminant function, these variables are assumed to come from normal distributions that have the same variance in both groups. A decision rule is (see Figure 6.4):

If a new observation $X > x_0 = \frac{1}{2}(\bar{x}_1 + \bar{x}_2)$, assign the observation to group *II*

and

if a new observation $X \leq x_0 = \frac{1}{2}(\bar{x}_1 + \bar{x}_2)$, assign the observation to group *I*.

The probability of misclassification is then

$$P(misclassification) = P(X > x_0 \mid I)P(I) + P(X \leq x_0 \mid II)P(II)$$

$$= P\left[Z > (\frac{1}{2}) \frac{\bar{x}_2 - \bar{x}_1}{S_X} \right] \text{ when } P(I) = P(II) = 0.5$$

where S_X^2 is the pooled estimate of the common variance. Again, Z is a standard normal variate, allowing the probabilities associated with misclassification to be calculated. The same considerations hold when the variable Y is chosen for classification and the decision point $y_0 = \frac{1}{2}(\bar{y}_1 + \bar{y}_2)$ is used.

Classification based on a single variable is geometrically equivalent to projecting the data onto a coordinate axis (Figure 6.4) and disregarding the other variables. The misclassification probability is the area of overlap between groups *I* and *II* measured along the axis associated with the variable used to discriminate. The amount of overlap decreases with increases in the distance $|\bar{x}_1 - \bar{x}_2|$ and decreases with increases in S_X. The effectiveness of employing X or Y, like the discriminant function, relates to the overlap of the distribution of the projected variables. The discriminant function is constructed to minimize the amount of overlap between the distributions of the discriminant scores and, as will be seen from several points of view, produces the smallest probability of misclassification for a linear combination (i.e., geometrically, the smallest overlap).

To illustrate classification based on a single variable, if

$$\bar{x}_1 = 5, \bar{x}_2 = 10, \text{ and } S_X^2 = 5.111$$

then, based on X,

$$P(misclassification) = P(Z > \frac{1}{2}(\bar{x}_2 - \bar{x}_1)/S_X)$$

$$= P(Z > 1.106) = 0.134.$$

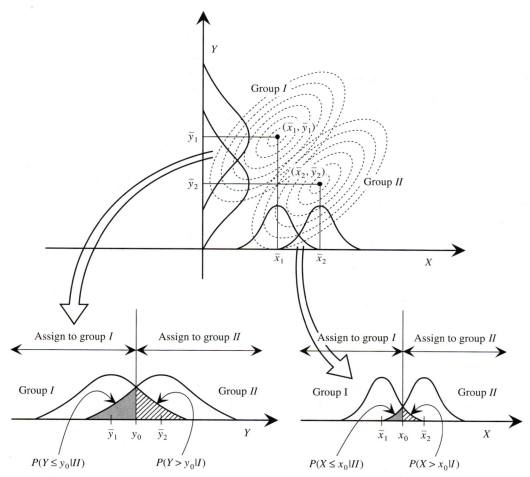

FIGURE **6.4**
Discrimination based on one variable X or Y

Also, if

$$\bar{y}_1 = 10 \, , \bar{y}_2 = 8 \, , \text{ and } S_Y^2 = 13.556$$

then, based on Y,

$$P(misclassification) = P(Z > \tfrac{1}{2}(\bar{y}_1 - \bar{y}_2)/S_Y)$$
$$= P(Z > 0.272) = 0.393 \, .$$

Finally, for the discriminant function based on both X and Y,

$$P(misclassification) = P(Z > 1.244) = 0.107$$

(see test case at the end of this chapter). The smaller probability is associated with the discriminant function as expected because a discriminant function is designed so that among all possible projections of the data, the probability of misclassification is the smallest.

Geometry of Misclassification

The geometry of misclassification for the two-dimensional case is not complicated. The observations could be projected on any line. The choice of the discriminant coefficients is equivalent to selecting a specific line onto which the data are projected. The projection on the x-axis, on the y-axis, and on the line constructed for the discriminant scores have been discussed, but it is certainly possible to choose coefficients to project the data onto any line. Among all possible choices of a line, the line selected to produce the discriminant scores is the one that produces the minimum overlap between the distributions of the projected data. This choice assures that the probability of misclassification is the smallest possible (Figure 6.5 displays the two-variable case). Maximum distance between mean discriminant scores, maximum variability of the discriminant scores, and minimum probability of misclassification are no more than different ways of expressing the geometric fact that the overlap between the distributions of the discriminant scores associated with groups *I* and *II* is the smallest possible for the sampled data.

Nonparametric Estimation of the Probability of Misclassification

The performance of a linear discriminant function can also be evaluated regardless of the distribution of the discriminant scores (i.e., nonparametrically). The derived discriminant function is evaluated by testing it with the same data used to produce the linear function. Applying the classification rule to the same d_{ij}-scores used to calculate the point d_0 produces the elements of Table 6.3.

T A B L E **6.3**

Discriminant scores classified by actual and predicted group

		Predicted Group		
		Group *I*	Group *II*	
Actual Group	Group *I*	a	b	n_1
	Group *II*	c	d	n_2

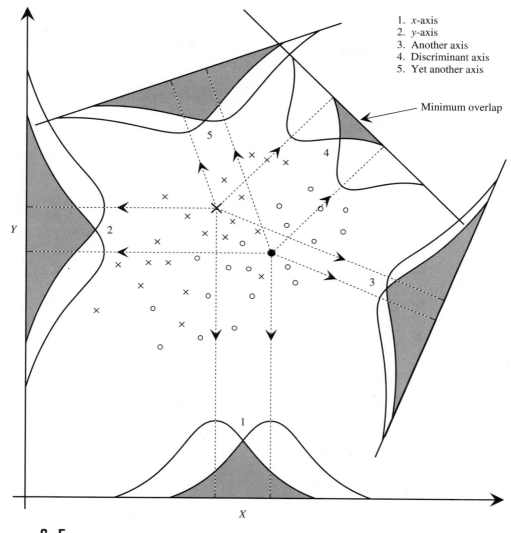

1. *x*-axis
2. *y*-axis
3. Another axis
4. Discriminant axis
5. Yet another axis

Minimum overlap

F I G U R E **6.5**

Five different distributions arising from projections of (x , y)-data on selected axes

The classification into the actual groups is based directly on the data, where classification into the predicted group is based on the derived discriminant scores and d_0. The values b and c represent the number of misclassified observations within the data set, and the probability of misclassification is estimated by

$$P(misclassification) = \left(\frac{b}{n_1}\right)P(I) + \left(\frac{c}{n_2}\right)P(II) \,.$$

(6.20)

If $P(I) = P(II) = 0.5$, then this estimate is

$$P(misclassification) = (\tfrac{1}{2}) \left(\frac{b}{n_1} + \frac{c}{n_2} \right).$$

(6.21)

A special case arises when the sample of the $n_1 + n_2$ observations is a random sample with respect to the membership in groups *I* and *II*. The prior probabilities $P(I)$ and $P(II)$ can then be estimated from the observed data by $\hat{P}(I) = n_1/(n_1 + n_2)$ and $\hat{P}(II) = n_2/(n_1 + n_2)$, giving an estimated probability of misclassification of

$$P(misclassification) = \frac{b + c}{n_1 + n_2}.$$

(6.22)

The nonparametric estimate of the probability of misclassification involves a certain amount of bias, because the observations are first used to produce an optimum linear discriminant function that is then applied to the same sample of data. Therefore, assessing a discriminant function with the data that produced the coefficients gives the best possible results for that data set. The misclassification rate is likely to be greater when the derived linear discriminant function is applied to new data. In other words, the estimated probability of misclassification is downwardly biased, much like the upward bias of the squared multiple correlation coefficient R^2 encountered in regression analysis. The potential overstatement of the performance of the discriminant function is the principal disadvantage of a nonparametric estimate of the probability of misclassification. However, for large sample sizes n_1 and n_2 (say, $n_i > 10k$), this overstatement is small.

Another version of the nonparametric approach to evaluating a linear discriminant function occasionally used when the data are plentiful involves splitting the data set into two sets. The discriminant function is developed on the first half of the data ("training sample") and then tested on the second half. Again, this approach to evaluating a discriminant function requires no assumptions about the distribution of the sampled data (nonparametric). Furthermore, no bias is incurred because the observations used to estimate the probability of misclassification are not used in the development of the discriminant function. This method is obviously practical only when sufficient numbers of observations exist to divide the data in half and still have an adequate number of observations to estimate and evaluate accurately the discriminant function.

If a nonparametric approach to calculating the probability of misclassification is used, no restrictions on the variables employed are necessary. Specifically, the observed values do not need to be normally distributed or even continuous. A function of variables is constructed and assessed using the data that produced the function or a second data set, providing a measure of the effectiveness of the process. The observed probability of misclassification reflects the likelihood that new data will be misclassified. However, this approach is not guaranteed to be optimal and requires moderate or large numbers of observations to be effective.

The nonparametric estimate of P (misclassification) is called the apparent probability of misclassification. The parametric calculation based on the estimates $\bar{d}_1$, $\bar{d}_2$, and S_L^2 along with the assumption of normally distributed data is called the actual probability of misclassification. Other methods for estimating the misclassification probabilities are available, as mentioned, but the two described are representative and readily calculated.

To summarize, the linear discriminant function is designed to minimize the probability of misclassification ($Q_{max} = S_L^2 = (\bar{d}_1 - \bar{d}_2)^2$ is maximized) among all sets of values resulting from projecting the observed data on a straight line and is, in this sense, the "best" possible statistical tool for classification. The word *best* has several limitations. First, the linear discriminant function is "best" only among linear functions. It is certainly possible that more complicated functions will lower the misclassification probability. Second, the estimated distance between mean discriminant scores is subject to a bias. Totally analogous to the squared multiple correlation coefficient R^2, the discriminant function gives the largest distance between mean discriminant scores $S_L^2 = (\bar{d}_2 - \bar{d}_1)^2$, producing the smallest possible probability of misclassification for the data set from which the linear discriminant coefficients are derived. If a discriminant function calculated from one data set is applied to other observations, the misclassification probability is likely to be higher. Third, the parametric assessment of a discriminant function, needless to say, is most meaningful when the data meet the requirements of being at least approximately normally distributed, being independently sampled, and having close to equal variances and covariances among all groups analyzed.

Wilks' Lambda

The previous section explored the probability of misclassification as a way to evaluate the performance of a discriminant function. Wilks' lambda (denoted λ), like a misclassification probability, indicates the overall worth of a discriminant function. (Samuel S. Wilks [1906–1964] was a major contributor in a number of areas of statistical research, particularly in the development of multivariate methods.) Estimated misclassification probabilities are an excellent descriptive tool but are not easily assessed statistically (i.e., tested against random variation). Wilks' lambda is not the most natural descriptive statistic but is readily assessed with a testing procedure. For a discriminant analysis with $g = 2$ groups, Wilks' lambda summary has a simple interpretation in terms of sums of squares from a one-way analysis of variance (Table 6.4).

Using standard one-way analysis of variance techniques (Chapter 1) in which the discriminant scores are treated as "data" produces a sum of squares for the difference between two groups (expression 1.18) of

$$between \; sum \; of \; squares = \Sigma n_i (\bar{d}_i - \bar{d})^2 = \frac{n_1 n_2}{n_1 + n_2} (\bar{d}_1 - \bar{d}_2)^2$$

and a within-groups sum of squares of $n_1 + n_2 - 2$. The within-groups sum of squares does not involve the values d_{ij} because the discriminant scores have been standardized so the estimated within-group variation S_D^2 equals 1.0. The degrees of freedom

T A B L E   **6.4**

Analysis of variance table using discriminant scores

Source	Sum of Squares	Degrees of Freedom
Between groups	$\Sigma n_i(\bar{d}_i - \bar{d})^2 = \dfrac{n_1 n_2}{n_1 + n_2}(\bar{d}_1 - \bar{d}_2)^2$	k
Within groups	$\Sigma\Sigma(d_{ij} - \bar{d}_i)^2 = n_1 + n_2 - 2$	$n_1 + n_2 - (k + 1)$
Total	$\Sigma\Sigma(d_{ij} - \bar{d})^2$	$n_1 + n_2 - 1$

are the same as those used in a regression analysis with $n_1 + n_2$ total observations and have the same justification (i.e., k coefficients and one constant are estimated from the data). Wilks' summary value λ is defined as the within sum of squares divided by the total sum of squares. That is, for $g = 2$ groups with k variables measured per observation, Wilks' lambda is

$$\lambda = \frac{within\ sum\ of\ squares}{total\ sum\ of\ squares} = \frac{n_1 + n_2 - 2}{\Sigma\Sigma(d_{ij} - \bar{d})^2}.$$

(6.23)

Wilks' lambda measures the proportion of total variation among discriminant scores not summarized by the linear discriminant function; in other words, the proportion of total variation not related to the distance between mean discriminant scores $\bar{d}_1$ and $\bar{d}_2$. Wilks' lambda decreases as the distance between mean discriminant scores increases and, additionally, λ is a summary measure between 0 and 1 (i.e., $0 \leq \lambda \leq 1$). The value $1 - \lambda$, therefore, is a single quantitative evaluation of the effectiveness of a discriminant function to correctly classify observations.

Adding Variables

Parallel to the properties of the squared multiple correlation coefficient R^2, the quantity $1 - \lambda$ increases as the number of variables used in a discriminant function increases. Because the addition of new variables to the discriminant analysis always increases $(\bar{d}_2 - \bar{d}_1)^2$ (perhaps only a small increase but nevertheless an increase), the actual probability of misclassification necessarily decreases as variables are added to the discriminant function. However, increases in the performance of the discriminant function from adding variables are balanced to some degree by a reduction in the reliability of the estimates of the discriminant coefficient. Like a regression analysis, a tradeoff exists between the precision of the discriminant coefficients and the addition of correlated variables to the analysis.

Not all summary statistics calculated from a discriminant analysis follow an easily predicted pattern when the number of variables employed in the analysis is increased. For example, the apparent probability of misclassification (nonparametric assessment of the misclassification probability) generally decreases but can increase when variables are added to the analysis.

At the very least, a discriminant function should classify new observations into the specified groups more accurately than would be expected by random assignment. A formal statistical test exists to investigate whether the multivariate distance between mean discriminant scores meets this minimum predictive value. However, a discriminant function that classifies observations better than would be expected by chance is not necessarily useful.

An exact approach to evaluating a discriminant function applied to a sample generated from two groups ($g = 2$) involves the analysis of variance F-test (Table 6.4) of the hypothesis that no difference exists between mean discriminant scores. The ratio of the between mean square to the within mean square (Table 6.4) is

$$F = \frac{mean\ square\ between}{mean\ square\ within} = \frac{n_1 + n_2 - (k + 1)}{n_1 + n_2 - 2} \times \frac{1}{k} \times \frac{n_1 n_2}{n_1 + n_2} (\bar{d}_1 - \bar{d}_2)^2$$

$$= \frac{n_1 + n_2 - (k + 1)}{k} \times \frac{1 - \lambda}{\lambda}.$$

(6.24)

The value F has an F-distribution with k and $n_1 + n_2 - (k + 1)$ degrees of freedom when the two mean discriminant scores differ only because of random variation. This F-test used in a different context is called the multivariate t-test and will be discussed (see "Two-Sample Multivariable t-Test").

When a discriminant function is used with more than two groups ($g \geq 3$), the analysis of variance approach no longer accurately applies. Alternatively, an approximate test of the overall effectiveness of a linear discriminant function is possible using Wilks' λ measure and a chi-square statistic. A logarithmic transformation of Wilks' λ has an approximate chi-square distribution when no differences among the g groups can be identified with a discriminant function, or, specifically,

$$X^2 = -m \log(\lambda)$$

(6.25)

has an approximate chi-square distribution with $k(g - 1)$ degrees of freedom where $m = (N - \frac{1}{2}(k + g) - 1)$. The value N is, as before, the total number of observations in all groups ($N = \Sigma n_i$). When λ is small, X^2 is large (logarithms of small fractions produce large negative values), leading to the inference that the discriminant function is potentially an effective analytic tool for classification. Remember, λ is the proportion of the variation in the discriminant scores not accounted for by the discriminant function. Conversely, if λ is large (close to 1.0), then the multivariate distance between mean discriminant scores is not sufficient to allow a discriminant function to work well.

Assessment of the Discriminant Function Variables

One approach to evaluating the relative contribution of each variable to the discriminant function involves computing and comparing standardized discriminant coefficients. The magnitude of unstandardized coefficients depends on the units of measurement, which means that direct comparisons among coefficients measured in different units do not cleanly indicate their relative contribution to the discriminant score. The same situation is encountered in the comparison of regression coefficients.

A standardized linear discriminant coefficient is

$$a_i^* = a_i S_i^*$$ (6.26)

where S_i^* is the standard deviation of the ith variable disregarding the classification of the data into groups. Other versions of a standardized discriminant coefficient exist. For example, $a_i' = a_i S_i$ where S_i is the estimated pooled variance associated with the ith variable. In the sense that standardized coefficients are expressed in units of standard deviations, the discriminant coefficients are commensurate. Large (in absolute value) standardized coefficients indicate those variables likely to contribute most influentially to detecting differences among groups.

F-Test

A basic difference between discriminant and regression coefficients stems from the fact that discriminant coefficients are not uniquely determined. Because a unique set of discriminant coefficients is not produced from a data set, the variance of a discriminant function coefficient is not easily interpreted [6]. However, an F-statistic provides an evaluation of the specific contributions of each variable.

Similar to the F-to-remove procedure encountered in regression analysis, an F-statistic produces an assessment of the impact of removing a subset of variables from the discriminant function. For a discriminant function based on k variables per observation, say a subset of $k - q$ variables remains in the discriminant function after q variables are removed; then two multivariate distances are relevant. They are

all k variables included in the discriminant function: $D_k^2 = (\bar{d}_1 - \bar{d}_2)^2$

and

$k - q$ variables remain in the discriminant function: $D_{k-q}^2 = (\bar{d}_1^* - \bar{d}_2^*)^2$.

To test formally the decrease in multivariate distance incurred by removing q variables $(D_k^2 - D_{k-q}^2)$,

$$F = \frac{n_1 + n_2 - (k + 1)}{q} \times \frac{n_1 n_2 (D_k^2 - D_{k-q}^2)}{(n_1 + n_2)(n_1 + n_2 - 2) + n_1 n_2 D_{k-q}^2}$$ (6.27)

has an F-distribution with q and $n_1 + n_2 - (k + 1)$ degrees of freedom when the q variables removed have no systematic influence on the discriminant function [6]. Unlikely values of F (large F) indicate that the variable or variables removed are probably important contributors to the classification process. Note that if $k = q$ (remove all variables), $D_{k-q}^2 = 0$ and expression 6.27 is the same as expression 6.24.

To assess the role of a single variable, such as x_i, the discriminant function is calculated with the variable included and then calculated with the ith variable excluded

($q = 1$). The quantity $D_k^2 - D_{k-1}^2$ represents the decrease in multivariate distance from deleting x_i from the discriminant equation and is analogous to the squared part correlation from a linear regression analysis. The difference in multivariate distances $D_k^2 - D_{k-1}^2$ is due only to the influence of the variable x_i. An F-statistic provides an evaluation of the differences between the two discriminant functions (distances). Like the previous F-statistics, it is used to compare two statistical models (two discriminant functions—a k-variate versus $(k-1)$-variate discriminant function).

Consider the following multivariate distances between two groups based on 10 observations per group where two variables are measured; X = years smoked and Y = CO ppm measured in a person's lungs (see the test case for complete details):

$$D_{X+Y}^2 = 6.187 ;$$

if X is removed from the analysis, then based on Y alone

$$D_Y^2 = 0.295 ,$$

and if Y is removed, then based on X alone

$$D_X^2 = 4.891 .$$

The distances D_X^2 and D_Y^2 immediately suggest that removing variable X has a much greater impact than removing Y. The formal evaluation of the individual contributions of X and Y to the probability of discriminating between two groups is given by

$$\text{assessment of } X: \quad F = \frac{17}{1} \frac{(10)\,(10)\,(6.187 - 0.295)}{(20)\,(18) + (10)\,(10)\,(0.295)} = 25.716$$

with p-value < 0.001, and

$$\text{assessment of } Y: \quad F = \frac{17}{1} \frac{(10)\,(10)\,(6.187 - 4.891)}{(20)\,(18) + (10)\,(10)\,(4.891)} = 2.595$$

with p-value $= 0.126$

where $n_1 = n_2 = 10$, $k = 2$, and $q = 1$, giving 1 and 17 degrees of freedom. As suspected, the variable represented by X is an important contributor (p-value < 0.001), whereas Y is not an important contributor to a discriminant function based on X and Y.

Correlation with a Discriminant Function

The correlation between the discriminant function scores and the observed values of a variable used in the construction of the discriminant scores has a direct interpretation in terms of multivariate distance. These k correlations provide yet another view of the way each variable contributes to the distance between mean discriminant scores. The product-moment correlation coefficient relating the discriminant score d and a specific variable x is

$$r_{XD} = \frac{S_{XD}}{S_X S_D} = \sqrt{\frac{(\bar{x}_2 - \bar{x}_1)^2 / S_X^2}{(\bar{d}_2 - \bar{d}_1)^2}} = \sqrt{\frac{distance\ based\ on\ x}{distance\ based\ on\ all\ k\ variables}} . \qquad (6.28)$$

The squared standardized distance between two groups based on a specific variable (x) ignoring all other variables is $(\bar{x}_2 - \bar{x}_1)^2 / S_X^2$, and the total squared multivariate distance between two groups is $(\bar{d}_1 - \bar{d}_2)^2$. Therefore, r_{XD}^2 is the proportion of the total multivariate distance explainable in terms of the standardized distance between groups 1 and 2 based on the single variable x (details in Box 6.4).

For the data set concerning the discrimination between individuals who quit and individuals who failed to quit smoking (details in the test case) with variables $X =$ years smoked and $Y = CO$ ppm, the following are calculated in the usual way:

$$S_{XD} = 2.010 ,\ S_D = 1.0,\ and\ S_X = 2.261\ (pooled\ covariance\ and\ variances)$$

and applying the definition of a correlation coefficient gives

$$r_{XD}^2 = \left[\frac{2.010}{2.261(1.0)} \right]^2 = (0.889)^2 = 0.790 .$$

B O X **6.4**   RELATIONSHIP BETWEEN CORRELATION COEFFICIENT (r_{XD}) AND DISTANCE

The usual definition of a correlation coefficient (expression 1.7) is

$$r_{XD} = \frac{covariance(X, D)}{\sqrt{variance(X)\ variance(D)}} = \frac{S_{XD}}{S_X S_D}$$

where D is the value from a discriminant function and X represents a specific variable $(g = 2)$. Because

$$S_{XD} = \frac{\Sigma\Sigma(d_{ij} - \bar{d}_i)(x_{ij} - \bar{x}_i)}{n_1 + n_2 - 2} \quad (expression\ 1.6\ for\ both\ groups\ pooled)$$

$$= \frac{\Sigma\Sigma[a_x(x_{ij} - \bar{x}_i) + a_y(y_{ij} - \bar{y}_i)]\ (x_{ij} - \bar{x}_i)}{n_1 + n_2 - 2}$$

$$= a_x S_X^2 + a_y S_{XY}$$

$$= \frac{(\bar{x}_2 - \bar{x}_1)}{S_L} \quad from\ the\ normal\ equations\ (expression\ 6.6)$$

then, because $S_D^2 = 1$,

$$r_{XD}^2 = \left[\frac{(\bar{x}_1 - \bar{x}_2)/S_L}{S_X S_D} \right]^2 = \frac{(\bar{x}_1 - \bar{x}_2)^2 / S_X^2}{(\bar{d}_1 - \bar{d}_2)^2} = \sqrt{\frac{distance\ based\ on\ x}{distance\ based\ on\ all\ k\ variables}} .$$

Or, alternatively, because $\bar{x}_2 = 10$, $\bar{x}_1 = 5$, and the multivariate distance based on X and Y between the two mean discriminant scores is $(\bar{d}_2 - \bar{d}_1)^2 = 6.187$, then

$$r^2_{XD} = \frac{(10-5)^2/2.261^2}{6.187} = \frac{4.891}{6.187} = 0.790 \ .$$

That is, $(\bar{x}_2 - \bar{x}_1)^2/S_X^2 = 4.891$ reflects the distance between groups 1 and 2 based on the single variable $X = $ years smoked, which yields $4.891/6.187 = 0.790$ as the proportion of the overall multivariate distance predicted by "years smoked." Similarly, the correlation between the discriminant scores and the CO measurements is -0.218, implying that $r^2_{YD} = (-0.218)^2 = 0.048$ or that 4.8% of the total distance between mean discriminant scores is predicted by CO measurements.

The proportion of multivariate distance attributed to a specific variable ($r^2_{X_i D}$) reflects the direct influences on the discriminant score D as well as indirect influences from the other $k-1$ variables through their association with x_i. The correlation $r_{X_i D}$ does not measure the influence of X isolated from the other variables in the discriminant function but rather its relative predictive worth. The value $D_k^2 - D_{k-1}^2$ is an assessment of the unique influence of a specific variable.

Prior and Posterior Probabilities

Information from sources other than the k measured variables improves the classification accuracy of a discriminant function. Knowledge about the sampled population incorporated into a decision rule lowers the probability of misclassifying an observation. This knowledge takes the form of prior probabilities. A prior probability (denoted, as before, as $P(I)$ and $1 - P(I) = P(II)$) is defined as the probability that an observation belongs to one of the classification groups prior to any measurements or knowledge specific to that observation. For example, if a discriminant function is used to classify mothers who are likely or unlikely to have a low birth-weight infant (fewer than 6.0 pounds) and 15% of pregnancies result in low birth weight, then the probabilities 0.15 and 0.85 can serve as prior probabilities. Such prior probabilities become part of the classification process. Another example arises when a discriminant function is used to classify a series of individuals into racial categories based on genetic information. If California is the sampled population, it is known that the racial composition is 75% white, 7% black, 15% Hispanic, 1% Asian, and 2% other. These four proportions could serve as prior probabilities.

Two related quantities that employ prior probabilities in conjunction with a discriminant function are

$$d_0 = -\frac{1}{S_L} \log\left[\frac{P(II)}{P(I)}\right] + \tfrac{1}{2}(\bar{d}_1 + \bar{d}_2)$$

(6.29)

and

$$P(I \mid D = d_{ij}) = \frac{P(I)}{P(I) + P(II) \ e^{(\bar{d}_2 - \bar{d}_1)(d_{ij} - d_c)}}$$

(6.30)

where $d_c = (\bar{d}_1 + \bar{d}_2)/2$ is the point midway between the means $\bar{d}_1$ and $\bar{d}_2$ (c for center). The value d_0 is the critical point used to construct a classification rule (expression 6.14, repeated) and $P(I \mid D = d_{ij})$ is called the posterior probability. A posterior probability is the probability an observation belongs to a specific group when the discriminant score associated with that observation is known. Contrasting a prior probability with a posterior probability shows the impact of information from a discriminant function on a specific observation. The posterior probability is an "update" of the prior probability based on specific measurements summarized by the discriminant score. The two posterior probabilities $P(I \mid D = d_{ij})$ and $P(II \mid D = d_{ij}) = 1 - P(I \mid D = d_{ij})$ are a combination of the prior probabilities and specific quantities from the discriminant function. The derivation of the expression for the posterior probabilities involves knowledge from mathematical statistics and is not present here (see [5]). Expression 6.30, however, identifies an extremely important concept, namely the role of the prior probabilities and the discriminant score in determining the likelihood that an observation is correctly classified.

A classification rule that incorporates prior probabilities into the discriminant analysis is the same as previously discussed; that is,

if $D > d_0$, assign the new observation D to one group and

if $D \leq d_0$, assign the new observation D to the other group.

This rule employs the classification cut-point d_0, which involves prior probabilities. The decision rule based on equal prior probabilities is a special case where the prior probabilities do not play a role in the classification process (i.e., $P(I) = P(II) = 0.5$; then

$$\log\left[\frac{P(II)}{P(I)}\right] = \log(1.0) = 0 \text{ and } d_0 = (\bar{d}_1 + \bar{d}_2)/2 \, ,$$

as before).

Probability of Misclassification

The actual probability of misclassification for unequal prior probabilities is estimated in the same way as the equal prior probabilities case. The estimated probability of misclassification is

$$P(misclassification) = P_I P(I) + P_{II} P(II) \tag{6.31}$$

with

$$P_I = P(D > d_0 \mid I) = P\left(Z > -\frac{1}{S_L} \log\left[\frac{P(II)}{P(I)}\right] + \tfrac{1}{2} S_L\right) \tag{6.32}$$

and

$$P_{II} = P(D \leq d_0 \mid II) = P\left(Z \leq -\frac{1}{S_L} \log\left[\frac{P(II)}{P(I)}\right] - \tfrac{1}{2} S_L\right) \tag{6.33}$$

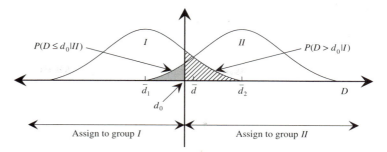

FIGURE 6.6

Classification based on discriminant scores with unequal prior probabilities ($P(I) < P(II)$)

where Z has a standard normal distribution. Notice that when $P(I) = P(II) = 0.5$, $P_I = P(Z > \frac{1}{2} S_L)$ and $P_{II} = P(Z \leq -\frac{1}{2} S_L)$, which agrees with the previous equal prior probabilities case.

Prior Probabilities

The use of prior probabilities shifts the value d_0 toward the population with the lower prior probability, producing a reduction in the posterior probability. For example, when $P(I)$ is less than $P(II)$, d_0 shifts toward group I, increasing the misclassification error associated with group I and decreasing the misclassification error associated with group II, because it is less likely an observation comes from population I or, conversely, it is more likely an observation comes from population II, illustrated in Figure 6.6. The net effect is to incorporate prior knowledge into the classification process, lowering the overall probability of misclassification (an example is given in Box 6.5).

BOX **6.5**   CLASSIFICATION WITH UNEQUAL PRIOR PROBABILITIES *

Letting $P(I) = 0.3$ and $P(II) = 0.7$ with $\bar{d}_1 = -1.244$ and $\bar{d}_2 = 1.244$ gives

$$S_L^2 = (\bar{d}_2 - \bar{d}_1)^2 = (1.244 - [-1.244])^2 = 6.187$$

$$d_0 = -\frac{1}{S_L} \log [P(II)/P(I)] + \frac{1}{2}(\bar{d}_1 + \bar{d}_2)$$

$$= -\frac{1}{\sqrt{6.187}} \log[0.7/0.3] + \frac{1}{2}(-1.244 + 1.244)$$

$$= -0.341$$

$$P(D > d_0 \mid I) = P(D > -0.341 \mid \bar{d}_1 = -1.244)$$

$$= P(Z > 0.903) = 0.183$$

$$P(D \leq d_0 \mid II) = P(D \leq -0.341 \mid \bar{d}_2 = 1.244)$$

$$= P(Z \leq -1.585) = 0.057$$

$$P(misclassification) = 0.183(0.3) + 0.057(0.7) = 0.095 \ .$$

*($\bar{d}_1$ and $\bar{d}_2$ come from the test case.)

If a pregnant woman has a probability of 0.85 of having an infant with "normal" birth weight prior to any information about her specific characteristics, then this information should be taken into account when she is classified as to whether she is likely or unlikely to have a low birth-weight child. For a discriminant function to be of any use in classifying mothers, the misclassification rate must be less than 15%. The simple rule of classifying all pregnant mothers into the "normal" category has a misclassification rate of 15%, as this is the proportion of low birth-weight infants and, therefore, the proportion of misclassifications. The use of prior probabilities $P(I) = 0.15$ for low birth weight and $P(II) = 0.85$ for "normal" birth weight in the computation of d_0 means the actual misclassification probability based on the discriminant function is improved; that is, more than 0.15 (classifying all observations into the "normal" group). Using prior probabilities, however, does not imply that a discriminant function is automatically useful. It implies only that the discriminant function adds to already known information—perhaps effectively or perhaps not.

Posterior probabilities can also be used to classify new observations. A decision rule is

if $P(II \mid D = d_{ij}) > P(I \mid D = d_{ij})$, assign the new observation D to group II and

if $P(II \mid D = d_{ij}) \leq P(I \mid D = d_{ij})$, assign the new observation D to group I.

This rule assigns the new observation D to the group with the highest posterior probability. It is the same as employing d_0 as a cut-point for classifying observations. Any observation such that $D \leq d_0$ will yield $P(II \mid d_{ij}) \leq P(I \mid d_{ij})$ and vice versa (a bit of algebra will show these rules to be identical). An example of calculating a posterior probability is given in Box 6.6.

Relationships between Posterior Probabilities and the Linear Discriminant Function

One final expression shows the relationships among the posterior probabilities, the prior probabilities, and the linear discriminant function (Box 6.7). Namely,

$$\log\left[\frac{P(I \mid d_{ij})}{P(II \mid d_{ij})}\right] = \log\left[\frac{P(I)}{P(II)}\right] + (\bar{d}_2 - \bar{d}_1)(d_c - d_{ij}).$$

(6.34)

The logarithm of the posterior odds (i.e., odds $= p/(1 - p)$ where p is a probability) is equal to the logarithm of the prior odds plus a contribution from the linear discriminant function. The contribution from the linear discriminant function consists of the square root of the multivariate distance between mean discriminant scores $(\bar{d}_2 - \bar{d}_1)$ and the distance of a specific discriminant score from the midvalue $(d_c - d)$. Therefore, the discriminant function, as noted, is a statistical tool that "updates" the prior probability based on a discriminant score. That is,

$$\log\{\text{posterior odds}\} \leftarrow [\text{prior knowledge}] +$$
$$[\text{group properties}] \times [\text{specific measurements}],$$

B O X   **6.6**   CALCULATION OF POSTERIOR PROBABILITIES

For example, consider the mean discriminant scores from the test case, with

$$\bar{d}_1 = -1.244, \text{ and } \bar{d}_2 = 1.244, \text{ making } (\bar{d}_2 - \bar{d}_1) = 2.487 \text{ and } d_c = 0 .$$

If $P(I) = P(II) = 0.5$, and the discriminant score associated with the 10th observation in group I is $d_{1,10} = -0.528$ (from the test case), then

$$P(I \mid -0.528) = \frac{0.5}{0.5 + 0.5 e^{(2.487)\,(-0.528\,-\,0)}} = 0.788$$

and

$$P(II \mid -0.528) = 0.212 .$$

For $P(I) = 0.3$ and $P(II) = 0.7$, then

$$P(I \mid -0.528) = \frac{0.3}{0.3 + 0.7 e^{(2.487)\,(-\,0.528\,-\,0)}} = 0.614$$

and

$$P(II \mid -0.528) = 0.386 .$$

Employing $P(I) = 0.3$ and $P(II) = 0.7$ decreases the posterior probability that the observation generating the discriminant score -0.528 belongs to group I when compared to the equal prior case (i.e., 0.614 versus 0.788), consistent with the fact that observations from group I are less likely than those from group II.

B O X   **6.7**   CHARACTERIZATION OF A DISCRIMINANT FUNCTION

Because

$$P(group\ I \mid d) = \frac{P(I)}{P(I) + P(II) e^{(\bar{d}_2 - \bar{d}_1)\,(d - d_c)}}$$

and

$$P(group\ II \mid d) = 1 - P(group\ I \mid d) ,$$

then

$$\frac{P(group\ I \mid d)}{P(group\ II \mid d)} = \frac{P(I)}{P(II) e^{(\bar{d}_2 - \bar{d}_1)\,(d - d_c)}}$$

and taking the logarithm gives

$$\log\left[\frac{P(group\ I \mid d)}{P(group\ II \mid d)} \right] = \log\left[\frac{P(I)}{P(II)} \right] + (\bar{d}_2 - \bar{d}_1)\,(d_c - d) .$$

where prior knowledge, group properties, and individual measurements are combined to predict the group membership of a specific observation in terms of a posterior probability (more formally described in Box 6.7). The discriminant function is related to the logistic regression model (Chapter 10), in which again the logarithm of the odds is also described by a linear combination of variables.

Analysis of Multivariate Mean Values

One-sample and two-sample t-tests are common univariate statistical procedures [3] used to address questions about differences in mean values. For multivariate measurements, the mean value is not a single quantity but rather a set of related mean values. For example, a sample of individuals might be measured for a series of characteristics such as heart disease severity, weight, blood pressure, and cholesterol level. As noted earlier, it is not satisfactory to assess each mean value separately. A multivariate approach, parallel to the univariate t-test, is necessary to analyze effectively a set of correlated means. Two specific approaches to the analysis of multivariate means are equivalent to a specialize application of a discriminant function. The general subject of multivariate analysis of variance is left to other texts (e.g., [8]) but these two methods are briefly discussed.

Plausibility of a Set of Postulated Mean Values

Occasionally an observed multivariate mean value ($\bar{x}_1, \bar{x}_2, \cdots, \bar{x}_k$) derived from a sample of multivariate observations is compared to a known or postulated multivariate mean ($\mu_1, \mu_2, \cdots, \mu_k$), parallel to the univariate one-sample t-test ($\bar{x}$ versus μ_0; expression 1.4). For example, the data in Table 6.5 come from measuring heart disease using a severity index (a clinical measurement), weight, blood pressure, and cholesterol level of 12 men who adopted a special diet to reverse severe heart disease.

Suppose it is known that the population mean values (μ_i) of these four variables are μ_1 = severity index = 70, μ_2 = weight = 170 pounds, μ_3 = blood pressure = 128 mm, and μ_4 = cholesterol = 220 mg/100 ml. The estimated means ($\bar{x}_i$) for the 12 men are $\bar{x}_1$ = severity = 73.42, $\bar{x}_2$ = weight = 182.25 pounds, $\bar{x}_3$ = blood pressure = 138.92 mm, and $\bar{x}_4$ = cholesterol = 252.17 mg/100 ml. A question naturally arises: Are the four differences between the sample means and the population means likely to be due exclusively to random variation or is a systematic difference involved? That is, a null hypothesis postulates that the population means from which the sample was drawn are 70, 170, 128, and 220 (H_0), respectively. To assess this hypothesis, a discriminant function can be used. A discriminant function computed from the data and the null-hypothesis-generated values (where $g = 2$, $n_1 = 12$ = observations, and $n_2 = 1$ = postulated set of means) applied to the mean values ($\bar{x}_i$) produces a mean discriminant score, $\bar{d}_x$. The mean discriminant score, as before, effectively reflects the information contained in the k measured variables. For the example data, the dis-

TA B L E   **6.5**

Four values measured on 12 men, illustrative data

	Severity	Weight	Blood Pressure	Cholesterol
1	84	155	132	258
2	79	162	160	237
3	65	186	120	185
4	65	169	102	277
5	61	178	152	236
6	79	208	146	238
7	83	231	150	276
8	76	171	131	278
9	62	188	141	265
10	83	182	154	283
11	67	163	121	267
12	77	194	158	226

criminant function coefficients associated with the four variables (severity index, weight, blood pressure, and cholesterol) are −0.0294, 0.0149, 0.0288, and 0.0319. The mean discriminate score derived from the observed and postulated means using these coefficients is $\bar{d}_x = 12.602$ (i.e., $\bar{d}_x = -0.0294(73.42) + 0.0149(182.25) + 0.0288(138.92) + 0.0319(252.17) = 12.602$). The same coefficients applied to the postulated means yield a "null" discriminant score of d_0. For the diet data, $\bar{d}_0 = 11.179$ (i.e., $\bar{d}_0 = -0.0294(70) + 0.0149(170) + 0.0288(128) + 0.0319(220) = 11.179$). The squared difference $(\bar{d}_x - \bar{d}_0)^2$ is the multivariate distance between the sample mean values and the postulated mean values. The "size" of this distance is evaluated with an F-statistic. Specifically, the test statistic

$$F = \frac{n-k}{n-1}\frac{1}{k}\,n(\bar{d}_x - \bar{d}_0)^2$$

(6.35)

is calculated where n is the number of observations and k is the number of variables (the number of observed mean values). The test statistic F has an F-distribution with k and $n-k$ degrees of freedom when all k observed mean values $\bar{x}_i$ differ from the population values μ_i by chance alone. From the example data, the multivariate distance between the mean discriminant score based on the 12 observed men and the "null" discriminant score is $(\bar{d}_x - \bar{d}_0)^2 = (12.602 - 11.179)^2 = 2.024$, giving a value for F of 4.415 with 4 and 8 degrees of freedom, yielding an associated p-value of 0.035. The multivariate analysis of the four variables shows some evidence that the 12 individuals differ from the postulated population mean values.

Two-Sample Multivariate *t*-Test

The classic univariate two-sample *t*-test is used to examine whether two mean values differ because of random variation or whether evidence exists of a systematic difference between the two sampled populations. The analogous multivariate procedure is called Hotelling's T^2-test. (Harold Hotelling [1895–1973] was a leader in the field of multivariate statistics and played a prominent part in the development of mathematical statistics.)

Hotelling's T^2-test is designed to compare sets of multivariate measurements arising from two sources. The issue to be examined is the same as the univariate case (random variation versus systematic difference). For example, a multivariate investigation of the morphological constitutions of smokers and nonsmokers might consist of comparing a variety of characteristics, such as height, weight, chest breadth, head circumference, hand length, and hand width [9]. Hotelling's T^2-test takes into account the correlated, multivariate nature of the variables and simultaneously takes advantage of the independent information contained in each measurement to evaluate differences between two sources of data. The T^2-test is a function of a multivariate distance—a multivariate distance between two sets of sample means, each selected from a different source.

The multivariate T^2-statistic is no more than a comparison of two mean discriminant scores. To compare the morphology of smokers and nonsmokers, for example, a discriminant function yields a mean discriminant score for smokers ($\bar{d}_1$ based on n_1 observations) and a mean discriminant score for nonsmokers ($\bar{d}_2$ based on n_2 observations). The squared difference between mean discriminant scores $\bar{d}_1$ and $\bar{d}_2$ is the multivariate distance between two groups and, similar to comparing $\bar{x}_1$ and $\bar{x}_2$ with the univariate *t*-test, serves to indicate systematic differences between the two sampled populations. Discriminant scores effectively combine the information from multivariate measurements.

A formal test of the hypothesis that no difference exists between the two populations sampled as reflected by the multivariate distance comes from calculating the value *F*, where

$$F = \frac{n_1 + n_2 - (k + 1)}{n_1 + n_2 - 2} \times \frac{1}{k} \times \frac{n_1 n_2 (\bar{d}_1 - \bar{d}_2)^2}{n_1 + n_2} \quad \text{(expression 6.24, repeated)} ,$$

which has an *F*-distribution with k and $n_1 + n_2 - (k + 1)$ degrees of freedom when the mean discriminant scores differ only because of random variation. Again, k is the number of variables that make up the discriminant function. This test statistic was previously discussed as an assessment of the overall worth of a discriminant function.

Continuing the previous example, suppose that a control group (nondiet) was selected and the same four variables measured on 20 individuals, giving the data in Table 6.6. The mean values for the control group are $\bar{y}_1$ = severity index = 68.35, $\bar{y}_2$ = weight = 176.15 pounds, $\bar{y}_3$ = blood pressure = 136.05 mm, and $\bar{y}_4$ = cholesterol =

247.00 mg/100 ml. To contrast the previous group of 12 men with this control group, a discriminant function is again used. The discriminant coefficients based on all 32 observations ($n_1 = 12$ = cases and $n_2 = 20$ = controls) of the severity index, weight, blood pressure, and cholesterol are 0.0769, 0.0004, 0.0190, and 0.0096, respectively. The two mean discriminant scores calculated using these coefficients applied to the four mean values are $\bar{d}_1 = 10.779$ (diet group) and $\bar{d}_2 = 10.283$ (control group). The estimated multivariate distance between these two groups is then $(\bar{d}_1 - \bar{d}_2)^2 = (10.779 - 10.283)^2 = 0.247$. To evaluate whether it is likely that this distance results from some nonrandom influence, an F-statistic is calculated (expression 6.24). For the example data, $F = 0.416$ with 4 and 27 degrees of freedom and the associated p-value is 0.795, indicating that these two groups are not likely to differ in any systematic way based on evidence from the four variables used to construct the discriminant function. One last note: The assumption that the data are independently sampled from multivariate normally distributed populations with the same variance/covariance structure for each group is still required, as it is for discriminant analysis.

T A B L E **6.6**

Four values measured on 20 men (control group)

	Severity	Weight	Blood Pressure	Cholesterol
1	81	145	112	228
2	29	132	150	232
3	64	176	120	175
4	61	163	101	267
5	69	172	155	233
6	81	208	142	228
7	86	211	140	266
8	72	171	131	275
9	65	185	131	265
10	63	172	152	273
11	67	153	121	247
12	67	174	153	223
13	61	175	142	233
14	69	204	136	236
15	81	231	150	256
16	66	151	131	262
17	62	172	141	266
18	83	181	134	272
19	68	163	131	267
20	72	184	148	236

Test Case   **Computer Implementation**

Discriminant Analysis ($g = 2$)

Twenty smokers are classified into two groups: those who gave up smoking (quitters) and those who did not quit after attending a one-year smoking cessation clinic (non-quitters), where X = number of years smoking cigarettes and Y = levels of measured carbon monoxide (CO ppm).

Population *I:* Non-quitters ($n_1 = 10$):

Sample											Mean
X = years	2	2	3	4	5	5	6	7	8	8	$\bar{x}_1 = 5$
Y = CO	16	12	9	12	14	6	8	11	8	4	$\bar{y}_1 = 10$

Population *II:* Quitters ($n_2 = 10$):

Sample											Mean
X = years	7	7	8	9	10	10	11	12	13	13	$\bar{x}_2 = 10$
Y = CO	14	10	7	10	12	4	6	9	6	2	$\bar{y}_2 = 8$

Summary Statistics

I	*II*	Pooled (within)
$S_X^2 = 5.111$	$S_X^2 = 5.111$	$S_X^2 = 5.111$
$S_{Y_1}^2 = 13.556$	$S_{Y_2}^2 = 13.556$	$S_Y^2 = 13.556$
$S_{X_1Y_1} = -5.333$	$S_{X_2Y_2} = -5.333$	$S_{XY} = -5.333$
$r_{X_1Y_1} = -0.641$	$r_{X_2Y_2} = -0.641$	$r_{XY} = -0.641$

Univariate F-Test for $\mu_1 = \mu_2$

Variable *X:*

$$F = n\Sigma(\bar{x}_i - \bar{x})^2/S_X^2 = 125/5.111 = 24.457 \text{ (expression 1.20)}$$

degrees of freedom 1 and 18, $p < 0.001$

Variable *Y:*

$$F = n\Sigma(\bar{y}_i - \bar{y})^2/S_Y^2 = 20/13.556 = 1.475$$

degrees of freedom 1 and 18, $p = 0.240$

Discriminant Coefficients

$$\alpha_x = [(\bar{x}_2 - \bar{x}_1)S_Y^2 - (\bar{y}_2 - \bar{y}_1)S_{XY}]/[S_X^2 S_Y^2 - S_{XY}^2]$$

$$= [(5)(13.556) - (-2)(-5.333)]/[(5.111)(13.556) - (-5.333)^2]$$

$$= 57.111/40.840 = 1.398$$

$$\alpha_y = [(\bar{y}_2 - \bar{y}_1)S_X^2 - (\bar{x}_2 - \bar{x}_1)S_{XY}]/[S_X^2 S_Y^2 - S_{XY}^2]$$

$$= [(-2)(5.111) - (5)(-5.333)]/[(5.111)(13.556) - (-5.333)^2]$$

$$= 16.443/40.840 = 0.403$$

$$S_L^2 = (\alpha_x)^2 S_X^2 + (\alpha_y)^2 S_Y^2 + 2\alpha_x \alpha_y S_{XY}$$

$$= \alpha_x(\bar{x}_2 - \bar{x}_1) + \alpha_y(\bar{y}_2 - \bar{y}_1)$$

$$= (1.398)(5) + (0.403)(-2) = 6.187$$

and the linear discriminant function is

$$l_{ij} = \alpha_x x_{ij} + \alpha_y y_{ij} = 1.398x_{ij} + 0.403y_{ij}.$$

Transformed Discriminant Coefficients

X	$a_x = \alpha_x/S_L = 1.398/\sqrt{6.187} = 0.562$
Y	$a_y = \alpha_y/S_L = 0.403/\sqrt{6.187} = 0.162$

$$mean = a_x\bar{x} + a_y\bar{y} = 0.562(7.5) + 0.162(9) = 5.674$$

and

$$d_{ij} = a_x(x_{ij} - \bar{x}) + a_y(y_{ij} - \bar{y}) = 0.562(x_{ij} - 7.5) + 0.162(y_{ij} - 9.0).$$

Now,

$$\bar{d} = 0.0 \quad \text{and} \quad S_D^2 = 1.0.$$

Standardized Discriminant Coefficients

X	$a'_x = a_x S_X^* = 0.562\sqrt{11.421} = 1.900$
Y	$a'_y = a_y S_Y^* = 0.162\sqrt{13.895} = 0.603$

where S_X^* and S_Y^* are the standard deviations estimated from the data ignoring the classification into groups.

Mean Discriminant Scores for the Two Groups

X	$\bar{d}_1 = a_x(\bar{x}_1 - \bar{x}) + a_y(\bar{y}_1 - \bar{y}) = 0.562(5 - 7.5) + 0.162(10 - 9.0) = -1.244$
Y	$\bar{d}_2 = a_x(\bar{x}_2 - \bar{x}) + a_y(\bar{y}_2 - \bar{y}) = 0.562(10 - 7.5) + 0.162(8 - 9.0) = 1.244$

Discriminant Scores

From sample *I* where, for example, $x_{11} = 2$ and $y_{11} = 16$, then

$$d_{11} = a_x(x_{11} - \bar{x}) + a_y(y_{11} - \bar{y}) = 0.562(2 - 7.5) + 0.162(16 - 9.0) = -1.959$$

Summary

Sample I (X, Y)	(2, 16)	(2, 12)	(3, 9)	(4, 12)	(5, 14)	(5, 6)	(6, 8)	(7, 11)	(8, 8)	(8, 4)
d_{1j}	−1.959	−2.607	−2.530	−1.482	−0.596	−1.891	−1.005	0.043	0.119	−0.528
Classified* as	I	I	I	I	I	I	I	II	II	I
$P(I \mid D = d_{ij})$	0.992	0.998	0.998	0.976	0.815	0.991	0.924	0.473	0.426	0.788

Sample II (X, Y)	(7, 14)	(7, 10)	(8, 7)	(9, 10)	(10, 12)	(10, 4)	(11, 6)	(12, 9)	(13, 6)	(13, 2)
d_{2j}	0.528	−0.119	−0.043	1.005	1.891	0.596	1.482	2.530	2.607	1.959
Classified* as	II	I	I	II	II	II	II	II	II	II
$P(II \mid D = d_{ij})$	0.788	0.426	0.473	0.924	0.991	0.815	0.976	0.998	0.998	0.992

*Classification rule: If $D > 0$ assign to II and if $D \leq 0$ assign to I Note: $P(I) = P(II) = 0.50$ and $d_0 = d_c = 0$.

Summary Table

		Classified (predicted)		
		I	II	Total
Observed (actual)	I	8	2	10
	II	2	8	10
	Total	10	10	20

Apparent Misclassification Probability

$$P(misclassification) = (2/10)(1/2) + (2/10)(1/2) = 1/5 = 0.20$$

Actual Misclassification Probability

$$P(misclassification) = P(D > 0 \mid I)P(I) + P(D \leq 0 \mid II)P(II)$$
$$= P(Z > [0 - (-1.244)]/1.0)(1/2) + P(Z \leq [0 - (1.244)]/1.0)(1/2)$$
$$= 0.107(1/2) + 0.107(1/2) = 0.107$$

Multivariate (Mahalanobis) Distance

$$S_L^2 = (\bar{d}_1 - \bar{d}_2)^2 = (1.244 - [-1.244])^2 = 6.187 = \text{Mahalanobis distance}$$

Analysis of Variance Table

Source	Sum of Squares	Degrees of Freedom
Between	$\Sigma n_i(\bar{d}_i - \bar{d})^2 = n_1 n_2(\bar{d}_1 - \bar{d}_2)^2/(n_1 + n_2) = 30.934$	$k = 2$
Within	$\Sigma\Sigma(d_{ij} - \bar{d}_i)^2 = n_1 + n_2 - 2 = 18.000$	$n_1 + n_2 - (k + 1) = 17$
Total	$\Sigma\Sigma(d_{ij} - \bar{d})^2 = 48.934$	$n_1 + n_2 - 1 = 19$

$$\text{Wilks' } \lambda : \quad \lambda = 18.000/48.934 = 0.368$$

The F-test of overall effectiveness of the discriminant function is

$$F = \frac{30.934/2}{18/17} = 14.608$$

and F has an F-distribution with $k = 2$ and $n_1 + n_2 - (k + 1) = 17$ degrees of freedom, giving p-value = 0.0002. Alternatively,

$$X^2 = -[n_1 + n_2 - \tfrac{1}{2}(g + k) - 1]\log(\lambda)$$
$$= -17\log(0.368) = 17.002$$

has an approximate chi-square distribution with $k(g - 1) = 2(2 - 1) = 2$ degrees of freedom, giving p-value = 0.0002.

Discriminant Analysis

The following STATA code implements a discriminant analysis using the previous test data ($n = 10$) with X = the number of years an individual smoked and Y = the levels of carbon monoxide measured in that person's lungs. A test case is a necessity for the use of any computer program, not to check the accuracy of the results (accuracy is not generally a problem) but to check whether the labeled values are what the user thinks they are.

A linear regression computer program can also be used to produce the statistics for a discriminant analysis ([5] or [10]). The dependent variable is defined as

$$f_i = \frac{n_2}{n_1 + n_2}$$

when the ith observation belongs to group 1 and

$$f_i = -\frac{n_1}{n_1 + n_2}$$

when the ith observation belongs to group 2, and the k measured variables play the role of the independent variables. Using these pseudo-dependent variables, a typical multiple regression program produces a set of discriminant coefficients (regression coefficients). To illustrate the computer implementation both the STATA discriminant ("discrim") and the STATA regression ("regress") programs are used to demonstrate the equivalence of these two approaches.

example.discriminant

```
. infile g x y using quit.dat
(20 observations read)

. list g x y
```

	g	x	y
1.	1	2	16
2.	1	2	12
3.	1	3	9
4.	1	4	12
5.	1	5	14
6.	1	5	6
7.	1	6	8
8.	1	7	11
9.	1	8	8
10.	1	8	4
11.	0	7	14
12.	0	7	10
13.	0	8	7
14.	0	9	10
15.	0	10	12
16.	0	10	4
17.	0	11	6
18.	0	12	9
19.	0	13	6
20.	0	13	2

The x column rows 1–10 are labeled x_{1j} and the y column rows 1–10 are labeled y_{1j}; the x column rows 11–20 are labeled x_{2j} and the y column rows 11–20 are labeled y_{2j}.

```
. discrim g x y, anova detail
```

Dichotomous Discriminant Analysis

Observations	= 20		Obs Group 0 =	10
Indep variables	= 2		Obs Group 1 =	10
Centroid 0	=	-1.2437	R-square	= 0.6322
Centroid 1	=	1.2437	Mahalanobis =	6.1868
Grand Cntd	=	0.0000		
Eigenvalue	=	1.7186	Wilk's Lambda =	0.3678
Canon. Corr.	=	0.7951	Chi-square	= 17.0017
Eta Squared	=	0.6322	Sign Chi2	= 0.0002

Annotations: Centroid 0 is $\bar{d}_1$ and Centroid 1 is $\bar{d}_2$; the difference relates to $(\bar{d}_1 - \bar{d}_2)^2$. Obs Group 0 = n_1, Obs Group 1 = n_2. Wilk's Lambda is λ, Chi-square is X^2, and Sign Chi2 is the p-value.

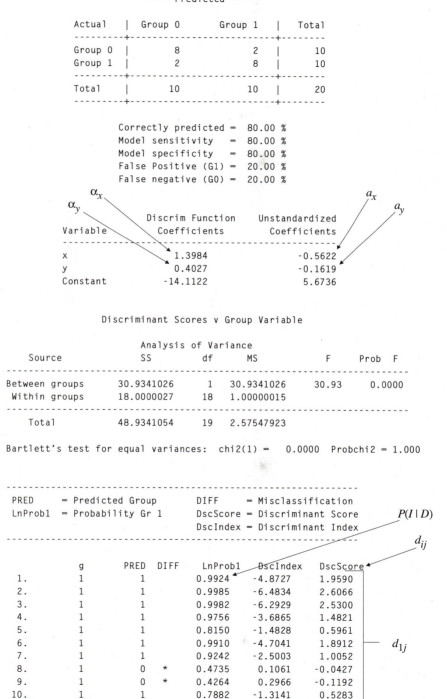

```
                    ----- Predicted -----

    Actual   | Group 0      Group 1   | Total
    ---------+-------------------------+--------
    Group 0  |     8            2      |   10
    Group 1  |     2            8      |   10
    ---------+-------------------------+--------
    Total    |    10           10      |   20
    ---------+-------------------------+--------


            Correctly predicted =  80.00 %
            Model sensitivity   =  80.00 %
            Model specificity   =  80.00 %
            False Positive (G1) =  20.00 %
            False negative (G0) =  20.00 %
```

α_x α_y a_x a_y

```
                    Discrim Function   Unstandardized
    Variable        Coefficients       Coefficients
    --------------------------------------------------
    x                   1.3984             -0.5622
    y                   0.4027             -0.1619
    Constant          -14.1122              5.6736
```

```
            Discriminant Scores v Group Variable

                    Analysis of Variance
    Source          SS          df      MS          F       Prob  F
    ----------------------------------------------------------------
    Between groups  30.9341026   1   30.9341026    30.93    0.0000
    Within groups   18.0000027  18    1.00000015
    ----------------------------------------------------------------
    Total           48.9341054  19    2.57547923
```

Bartlett's test for equal variances: chi2(1) =   0.0000 Probchi2 = 1.000

```
    ----------------------------------------------------------------
    PRED    = Predicted Group       DIFF    = Misclassification
    LnProb1 = Probability Gr 1       DscScore = Discriminant Score
                                     DscIndex = Discriminant Index
    ----------------------------------------------------------------
```

$P(I \mid D)$ d_{ij}

```
            g    PRED  DIFF   LnProb1   DscIndex   DscScore
    1.      1     1           0.9924    -4.8727    1.9590
    2.      1     1           0.9985    -6.4834    2.6066
    3.      1     1           0.9982    -6.2929    2.5300
    4.      1     1           0.9756    -3.6865    1.4821
    5.      1     1           0.8150    -1.4828    0.5961
    6.      1     1           0.9910    -4.7041    1.8912
    7.      1     1           0.9242    -2.5003    1.0052
    8.      1     0    *      0.4735     0.1061   -0.0427
    9.      1     0    *      0.4264     0.2966   -0.1192
    10.     1     1           0.7882    -1.3141    0.5283
```

d_{1j}

```
  11.      0        0         0.2118    1.3141   -0.5283 ⎤
  12.      0        1   *     0.5736   -0.2966    0.1192 │
  13.      0        1   *     0.5265   -0.1061    0.0427 │
  14.      0        0         0.0758    2.5003   -1.0052 │
  15.      0        0         0.0090    4.7041   -1.8912 │
  16.      0        0         0.1850    1.4828   -0.5961 ⎬── d₂ⱼ
  17.      0        0         0.0244    3.6865   -1.4821 │
  18.      0        0         0.0018    6.2929   -2.5300 │
  19.      0        0         0.0015    6.4834   -2.6066 │
  20.      0        0         0.0076    4.8727   -1.9590 ⎦
```

. generate f=-((g-1)*0.5+g*0.5)

. regress f x y

$$(\bar{d}_2 - \bar{d}_1)\,(d - d_c)$$

```
    Source |       SS         df       MS              Number of obs =      20
-----------+------------------------------             F( 2,    17) =   14.61
     Model | 3.16079173        2  1.58039586           Prob  F       =  0.0002
  Residual | 1.83920827       17  .108188722           R-square      =  0.6322
-----------+------------------------------             Adj R-square  =  0.5889
     Total |      5.00        19  .263157895           Root MSE      =  .32892
```

```
------------------------------------------------------------------------------
         f |      Coef.   Std. Err.       t    P|t|     [95% Conf. Interval]
-----------+------------------------------------------------------------------
         x |   .1428889    .028178     5.071   0.000     .0834385    .2023394
         y |   .0411431   .0255469     1.610   0.126    -.0127562    .0950424
     _cons |  -1.441955   .4027601    -3.580   0.002    -2.291704   -.5922053
------------------------------------------------------------------------------
```

. predict l

. list g l

```
              g          l
   1.         1   -.4978873
   2.         1   -.6624597
   3.         1   -.6430001
   4.         1   -.3766819
   5.         1   -.1515067
   6.         1   -.4806516
   7.         1   -.2554765
   8.         1    .0108418
   9.         1    .0303013
  10.         1   -.1342711
  11.         0    .1342711
  12.         0   -.0303013
  13.         0   -.0108418
  14.         0    .2554765
  15.         0    .4806516
  16.         0    .1515067
  17.         0    .3766819
  18.         0    .6430001
  19.         0    .6624597
  20.         0    .4978873
```

```
. sort g

. by g: summarize 1

- g=        0
Variable |     Obs        Mean    Std. Dev.        Min        Max
---------+-----------------------------------------------------
       1 |      10     .3160792    .2541512   -.0303013    .6624597

- g=        1
Variable |     Obs        Mean    Std. Dev.        Min        Max
---------+-----------------------------------------------------
       1 |      10    -.3160792    .2541512   -.6624597    .0303013


. generate d1=1/.2541512 if g == 0

. generate d2=1/.2541512 if g == 1

. list d1 d2
```

d_{1j}

```
              d1          d2
 1.     1.95902           .
 2.    2.606557           .
 3.     2.52999           .
 4.    1.482117           .
 5.    .5961283           .
 6.    1.891203           .
 7.    1.005215           .
 8.   -.0426587           .
 9.   -.1192257           .
10.    .5283119           .
```

d_{2j}

```
11.           .    -.5283119
12.           .     .1192257
13.           .     .0426587
14.           .    -1.005215
15.           .    -1.891203
16.           .    -.5961283
17.           .    -1.482117
18.           .     -2.52999
19.           .    -2.606557
20.           .     -1.95902

. summarize d1 d2
```

$\bar{d}_1$ $\bar{d}_2$

```
Variable |     Obs        Mean    Std. Dev.        Min        Max
---------+-----------------------------------------------------
      d1 |      10     1.243666    .9999999   -.1192257    2.606557
      d2 |      10    -1.243666    .9999999   -2.606557    .1192257

. generate temp=1/.2541512
```

```
. oneway temp g
                           Analysis of Variance
       Source            SS          df       MS           F      Prob  F
------------------------------------------------------------------------
Between groups       30.9340957       1    30.9340957    30.93    0.0000
Within groups        17.9999979      18    .999999881
------------------------------------------------------------------------
       Total         48.9340936      19    2.57547861

. generate dtemp= d1  0 & d2  0

. tabulate dtemp g
          | g
   dtemp|        0            1  |    Total
--------+--------------------------+----------
      0 |        2            8  |       10
      1 |        8            2  |       10
--------+--------------------------+----------
   Total|       10           10  |       20

. exit,clear
```

Applied Example

Discriminant Analysis

Four maternal variables from the Child Health and Development Study allow a discriminant function to be created to classify mothers into two categories to predict those mothers who are at high risk of having a low birth-weight infant (birth weight < 6 pounds). Using $n = 680$ births, a discriminant function based on maternal measurements of age, smoking exposure, height, and prepregnancy weight is estimated. The data consist of 41 infants with a birth weight of fewer than 6 pounds and 639 infants whose birth weight is greater than or equal to 6 pounds. Comparison of the maternal mean values between these two groups begins to indicate which variables might be important predictors of low birth-weight pregnancies. Because these variables are measured in different units, come from distributions with different variances, and are interrelated, the inspection of the differences between the four mean values is, at best, a start.

A linear discriminant analysis of the four maternal variables with prior probabilities of 0.15 (low birth weight) and 0.85 ("normal" birth weight) allows a multivariate assessment of these measures as predictors of the occurrence of low birth-weight infants. Using all four variables, the estimated mean discriminant scores are −0.474 for group *I* (< 6 pounds) and 0.030 for group *II* (≥ 6 pounds). To statistically assess whether the multivariate distance $(\bar{d}_1 - \bar{d}_2)^2 = (0.030 - (-0.474))^2 = 0.255$ is sufficiently large to classify new observations more accurately than random assignment, an *F*-test or Wilks' λ summary can be used. The Wilks' λ approach yields

$$\lambda = 678.000/687.814 = 0.986 \quad \text{with} \quad m = 680 - (4 + 2)/2 - 1 = 676; \quad \text{then}$$

$$X^2 = -m \log(\lambda) = -676 \log(0.986) = 9.714 \, .$$

Because X^2 has an approximate chi-square distribution with four degrees of freedom when the four maternal variables have no relationship to birth weight, a *p*-value = 0.046 is observed. The discriminant function shows that the four maternal variables are not strong indicators of infant low birth weight because the difference between the mean discriminant scores is not large and somewhat likely to have occurred by chance variation.

The nonparametric approach to calculating the misclassification probability involves simply evaluating the discriminant function with the data used to generate the coefficients and yields a 32% error rate. Because 19 infants of less than 6 pounds are misclassified as more than 6 pounds and 185 infants more than 6 pounds are misclassified as less than 6 pounds, the apparent probability of misclassification is (19/41) (0.15) + (185/639) (0.85) = 0.316, which further indicates that the discriminant function is not effective in predicting low birth-weight infants.

```
chds.discriminant

. use chds

. generate bwt0=bwt = 6

. tabulate bwt0
        bwt0|      Freq.      Percent        Cum.
------------+-----------------------------------
          0 |         41         6.03        6.03
          1 |        639        93.97      100.00
------------+-----------------------------------
       Total|        680       100.00

. sort bwt0

. by bwt0: summarize mage mnocig mheight mppwt

- bwt0=         0
Variable |       Obs        Mean    Std. Dev.         Min         Max
---------+-----------------------------------------------------------
    mage |        41    26.60976     6.752326          18          41
  mnocig |        41    12.31707     12.32363           0          50
 mheight |        41    64.12195     2.471792          58          70
   mppwt |        41    125.2683     18.53648          99         200

- bwt0=         1
Variable |       Obs        Mean    Std. Dev.         Min         Max
---------+-----------------------------------------------------------
    mage |       639    25.80908      5.37303          15          42
  mnocig |       639    7.117371     11.13864           0          50
 mheight |       639    64.45383     2.484557          57          71
   mppwt |       639         127      17.8445          85         246

. discrim bwt0 mage mnocig mheight mppwt, a
```

```
                    Dichotomous Discriminant Analysis

Observations   = 680                     Obs Group 0 =        41
Indep variables = 4                      Obs Group 1 =       639


Centroid 0   =   -0.4743                 R-square    =    0.0143
Centroid 1   =    0.0304                 Mahalanobis =    0.2547
Grand Cntd   =   -0.4438


Eigenvalue    =   0.0145                 Wilks' lambda =   0.9857
Canon. Corr. =    0.1194                 Chi-square   =    9.7143
Eta Squared  =    0.0143                 Sign Chi2    =    0.0455
                    ----- Predicted -----

        Actual  | Group 0      Group 1  |  Total
        --------+------------------------+--------
        Group 0 |     22           19   |    41
        Group 1 |    185          454   |   639
        --------+------------------------+--------
        Total   |    207          473   |   680
        --------+------------------------+--------


               Correctly predicted =  70.00 %
               Model sensitivity   =  53.66 %
               Model specificity   =  71.05 %
               False Positive (G1) =  89.37 %
               False negative (G0) =   4.02 %


                     Discrim Function    Unstandardized
        Variable      Coefficients        Coefficients
        - - - - - - - - - - - - - - - - - - - - - - - -
        mage             0.0246             -0.0486
        mnocig           0.0411             -0.0815
        mheight         -0.0530              0.1050
        mppwt           -0.0020              0.0041
        Constant         2.6235             -5.4203


end of do-file

. exit,clear
```

Multivariate T^2-Test

It is possible that maternal age is related to the size of a newborn infant. To evaluate this possibility 680 CHDS pregnant women are classified into two age categories (≤ 37 years and > 37 years of age) and their live-born male infants are measured for head circumference, length, and birth weight with the following results:

	Group 1 (age ≤ 37 years)	Group 2 (age > 37 years)
Head circumference	$\bar{x}_1 = 13.219$	$\bar{y}_1 = 13.222$
Length	$\bar{x}_2 = 20.288$	$\bar{y}_2 = 20.074$
Birth weight	$\bar{x}_3 = 7.530$	$\bar{y}_3 = 7.181$
Sample size	$n_1 = 653$	$n_2 = 27$

Correlations (r_{ij}) ignoring group status:

$$r_{12} = \text{head-length} = 0.456$$

$$r_{13} = \text{head–birth weight} = 0.625$$

$$r_{23} = \text{length–birth weight} = 0.711$$

The linear discriminant function based on these 680 observations gives mean discriminant scores of $\bar{d}_1 = 0.016$ (mothers ≤ 37 years of age) and $\bar{d}_2 = -0.398$ (mothers > 37 years of age), producing a multivariate distance of $(-0.398 - 0.016)^2 = 0.172$ and an F-statistic of

$$F = \frac{653 + 27 - 3 - 1}{653 + 27 - 2} \times \frac{1}{3} \times \frac{653(27)}{653 + 27} (-0.398 - 0.016)^2 = 1.484 \ .$$

The value $F = 1.484$ comes from an F-distribution with 3 and 676 degrees of freedom when $\bar{d}_1$ and $\bar{d}_2$ differ only because of random variation. The F-statistic shows no evidence (p-value = 0.218) of a difference between mothers age ≤ 37 and mothers age > 37 with respect to infant "size" as measured by the three-variate variable (head circumference, length, and birth weight).

chds.t-test

```
. use chds

. generate g = mage > 37

. sort g

. by g: summarize headcir length bwt

- g=        0
Variable |     Obs        Mean    Std. Dev.        Min        Max
---------+------------------------------------------------------------
 headcir |      27    13.22222    .6405126         12         15
  length |      27    20.07407     1.14105         18         23
     bwt |      27    7.181482    1.126614        5.3        9.2
```

```
- g=            1
Variable |    Obs       Mean    Std. Dev.      Min       Max
---------+-------------------------------------------------------
 headcir |    653    13.21899   .6262222        11        15
  length |    653    20.2879    .9750526        17        23
     bwt |    653    7.530322   1.089577        3.3       11.4

. correlate headcir length bwt
(obs=680)

         |  headcir    length      bwt
---------+-----------------------------
 headcir|   1.0000
  length|   0.4558    1.0000
     bwt|   0.6246    0.7114    1.0000

. discrim g headcir length bwt, a
```

Dichotomous Discriminant Analysis

```
Observations    = 680          Obs Group 0 =          27
Indep variables = 3            Obs Group 1 =         653


Centroid 0   =   -0.3978       R-square     =    0.0065
Centroid 1   =    0.0164       Mahalanobis  =    0.1716
Grand Cntd   =   -0.3813


Eigenvalue   =    0.0066       Wilks' lambda =   0.9935
Canon. Corr. =    0.0807       Chi-square    =   4.4241
Eta Squared  =    0.0065       Sign Chi2     =   0.2192
```

----- Predicted -----

Actual	Group 0	Group 1	Total
Group 0	17	10	27
Group 1	292	361	653
Total	309	371	680

```
              Correctly predicted  =  55.59 %
              Model sensitivity    =  62.96 %
              Model specificity    =  55.28 %
              False Positive (G1)  =  94.50 %
              False negative (G0)  =   2.70 %
```

	Discrim Function	Unstandardized
Variable	Coefficients	Coefficients
headcir	0.5381	-1.2990
length	0.0116	-0.0280
bwt	-0.4939	1.1925
Constant	-3.7140	8.7759

```
. exit,clear
```

Notation

Summary values:

$$g = \text{number of groups}$$
$$k = \text{number of measurements on each member of a group}$$
$$n_i = \text{number of observations in the } i^{th} \text{ group}$$
$$N = \Sigma n_i = \text{total number of observations}$$

Untransformed:

The discriminant scores for the jth group for the ith variable are

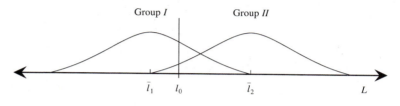

where $\bar{l}_1 = \alpha_1 \bar{x}_{11} + \alpha_2 \bar{x}_{21} + \cdots + \alpha_k \bar{x}_{k1}$

$$\bar{l}_2 = \alpha_1 \bar{x}_{12} + \alpha_2 \bar{x}_{22} + \cdots + \alpha_k \bar{x}_{k2}$$

$$\bar{l} = \alpha_1 \bar{x}_1 + \alpha_2 \bar{x}_2 + \cdots + \alpha_k \bar{x}_k$$

and k = number of variables employed to discriminate.

Transformed:

The discriminant scores for the jth group for the ith variable are

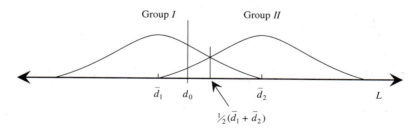

where $a_i = \alpha_i / S_L$ and, therefore,

$$\bar{d}_1 = a_1(\bar{x}_{11} - \bar{x}_1) + a_2(\bar{x}_{21} - \bar{x}_2) + \cdots + a_k(\bar{x}_{k1} - \bar{x}_k)$$

$$\bar{d}_2 = a_1(\bar{x}_{12} - \bar{x}_1) + a_2(\bar{x}_{22} - \bar{x}_2) + \cdots + a_k(\bar{x}_{k2} - \bar{x}_k) .$$

Variance of discriminant score:

$$S_D^2 = 1.0$$

$$d_0 = \text{classification cut-point}$$

That is,

if $D > d_0$, assign the new observation D to one group and

if $D \leq d_0$, assign the new observation D to the other group,

where

$$d_0 = -\frac{1}{S_L} \log\,[P(II)/P(I)] + \tfrac{1}{2}(\bar{d}_1 + \bar{d}_2)$$

Problems

T A B L E   **6.7**

Carapace dimensions of painted turtles (*Chrysemys picta marginata*)

24 Males			24 Females		
Length	Width	Height	Length	Width	Height
93	74	37	98	81	38
94	78	35	103	84	38
96	80	35	103	86	42
101	84	39	105	86	40
102	85	38	109	88	44
103	81	37	123	92	50
104	83	39	123	95	46
106	83	39	133	99	51
107	82	38	133	102	51
112	89	40	133	102	51
113	88	40	134	100	48
114	86	40	136	102	49
116	90	43	137	98	51
117	90	41	138	99	51
117	91	41	141	105	53
119	93	41	147	108	57
120	89	40	149	107	55
120	93	44	153	107	56
121	95	42	155	115	63
125	93	45	155	117	60
127	96	45	158	115	62
128	95	45	159	118	63
131	95	46	162	124	61
135	106	47	177	132	67

Note: The data in Table 6.7 consist of three measurements on the carapace (shell) of 24 male and 24 female painted turtles [11].

1 Using the data from Table 6.7, conduct a discriminant analysis to classify turtles into male and female categories based on the three measures of their carapace (length, width, and height). Compute and test Wilks' λ-statistic. Estimate the misclassification probability associated with the discriminant function both parametrically (assuming that discriminant scores have at least an approximate normal distribution) and nonparametrically when $P(I) = P(male) = P(II) = P(female) = 0.5$.

2 Using the data in Table 6.7, classify the observations into male and female categories based on each variable separately (e.g., classify the turtles as males or females based on carapace length). Plot length against width, length against height, and height against width. Contrast these univariate approaches to the discriminant analysis in problem 1.

3 Show that $P(II \mid D) < P(I \mid D)$ occurs only when $D < d_0$ occurs.

4 If $P(I) = P(II) = 0.5$, demonstrate that if $(D - \bar{d}_1)^2 > (D - \bar{d}_2)^2$, then $D > d_0$.

5 If X and Y are uncorrelated, find an expression for the discriminant coefficients α_x and α_y. Using these estimates, write an expression for S_L^2. Show that $r_{XD}^2 + r_{YD}^2 = 1$, for this case where $S_{XY} = 0$.

6 If $\bar{d}_1 = -1.6$ and $\bar{d}_2 = 1.6$ with $P(I) = P(II) = 0.5$, compute d_0. If $\bar{d}_1 = -1.6$ and $\bar{d}_2 = 1.6$ with $P(I) = 0.2$ and $P(II) = 0.8$, compute d_0. If a specific discriminant score is $d = -0.4$, calculate $P(I \mid d)$ and $P(II \mid d)$ when the prior probabilities are $P(I) = P(II) = 0.5$. If a specific discriminant score is $d = -0.4$, calculate $P(I \mid d)$ and $P(II \mid d)$ when the prior probabilities are $P(I) = 0.2$ and $P(II) = 0.8$.

7 Compute the mean lengths, heights, and widths for the male and female carapace measurements (Table 6.7). Perform a Hotelling's T^2-test to evaluate the multivariate distance between these two sets of mean values.

8 Using the data below on length of stay in an intensive care unit, can one discriminate between large and small hospitals using the variables x_1 = length of stay in an intensive care unit and x_2 = severity of the patient's condition?

Obs.	x_1	x_2	Group
1	24	22	0
2	48	80	0
3	37	55	0
4	81	140	1
5	48	90	0
6	2	10	1
7	24	40	1
8	64	120	1
9	12	30	1
10	8	25	1
11	4	10	1
12	36	77	0
13	12	32	0
14	8	14	1

Obs.	x_1	x_2	Group
15	4	15	1
16	88	200	0
17	54	120	0

Compute a discriminant function based on variables x_1 and x_2. Assess the effectiveness of this discriminant function by computing and testing the value of Wilks' lambda. Also use an F-test to assess the same issue.

References

1 Manly, B. F. J. *Multivariate Statistical Methods—A Primer*. New York: Chapman and Hall, 1986.

2 Fisher, R. A. The Use of Multiple Measurement in Taxonomic Problems. *Annals of Eugenics* 7 (1936): 179–188.

3 Winer, B. J., Brown, D. R., and Michels, K. M. *Statistical Principles in Experimental Design*, 3rd ed. New York: McGraw-Hill, 1991.

4 Mood, A. M., Graybill, F. A., and Boes, D. C. *Introduction to the Theory of Statistics*, 3rd ed. New York: McGraw-Hill, 1974.

5 Lachenbruch, P. A. *Discriminant Analysis*. New York: Hafner Press, 1975.

6 Rao, C. R. *Linear Statistical Inference and Its Applications*. New York: John Wiley, 1965.

7 Lindgren, B. L. *Statistical Theory*. New York: Macmillan, 1976.

8 Morrison, D. L. *Multivariate Statistical Methods*, 3rd ed. New York: McGraw-Hill, 1990.

9 Seltzer, C. C. Morphologic Constitution and Smoking. *Journal of the American Medical Association* 183 (1963): 639–645.

10 Flury, B., and Riedwyl, H. *Multivariate Statistics—A Practical Approach*. New York: Chapman and Hall, 1988.

11 Bryant, E. H., and Atchley, W. R. (editors). *Multivariate Statistics Methods: Within-Groups Covariation*. New York: Halsted Press, 1975.

7

Principal Components

Background

Usually, most expense and effort is spent collecting a sample of observations. Once multivariate observations are collected, recording a number of characteristics on each observation is often easy and relatively inexpensive. Multiple measurements on a single observation are undoubtedly related, complicated to analyze, and difficult to interpret, as already noted. For example, a simple plot often reveals outlying observations in univariate data. However, the complexity of multivariate data is such that detection of outliers usually requires special tools. One approach to understanding interrelated measurements is to create a new set of variables from the original data, simplifying the analysis and, perhaps, the interpretation. For example, levels of several biochemical substances related to smoking exposure, such as carbon monoxide and thiocyanate, are measured on each of a series of smokers. These variables overlap in terms of information on cigarette smoking exposure but, at the same time, make independent contributions to defining exposure levels. Forming a weighted sum (linear combination) of these measurements produces a single summary containing information from all variables. A variety of choices exist for selecting the weights. Discriminant scores require the calculation of one set of weights, producing summary linear combinations with specific properties. Principal component analysis involves yet another choice of weights, producing quite different and potentially valuable summary statistics.

Principal components are linear combinations that not only summarize information by combining multivariate measurements, but also constitute a set of new observations with a simplified structure. This procedure, like discriminant analysis, reduces a multivariate observation to one or a few summary values, called canonical variables. In some situations, a few canonical variables are all that is needed to conduct an effective analysis. In the next sections, the theory (mostly for the two-variable case) and computation of principal components are discussed, followed by a description of several ways these canonical summaries are used to explore multivariate data.

Method: Criterion

Unlike the previously discussed techniques, principal component analysis requires no assumptions about the population from which the data are sampled. In particular, no assumption is necessary about the distribution of the sampled observations. These summary components, however, do take on additional meaning when distributional assumptions are appropriate.

Formally, a principal component is a linear combination of k multivariate measurements. For the jth observation, symbolically the first principal component is

$$P_j^{(1)} = a_1 x_{1j} + a_2 x_{2j} + \cdots + a_k x_{kj} \tag{7.1}$$

where x_{ij} represents a specific variable. Notationally, x_{ij} is the jth observation on the ith variable and the coefficients a_i are the associated weights, producing principal component $P_j^{(1)}$. This linear combination is the first of k different principal components that can be constructed from n observations each with k measurements ($k < n$). However, because a basic feature of principal component analysis is a reduction in the number of variables, interest is generally focused on only a few principal component summaries. The second and third principal components are similar linear combinations each defined by k coefficients, or

$$P_j^{(2)} = b_1 x_{1j} + b_2 x_{2j} + \cdots + b_k x_{kj} \text{ and}$$

$$P_j^{(3)} = c_1 x_{1j} + c_2 x_{2j} + \cdots + c_k x_{kj} . \tag{7.2}$$

Different principal components (each consisting of a different set of coefficients) are denoted by superscripts (1), (2), $\cdots$, up to (k).

The properties that make principal components potentially useful summaries of k related variables result from the way the coefficients are chosen to construct each linear combination. Principal components are generated by two requirements placed on the linear combinations. They are:

1 The variance of the first principal component is greatest; the second principal component has the next largest variance; the third, the next largest variance; and so forth.

2 At the same time, each principal component generates transformed values that are uncorrelated with the values generated by any other principal component. That is, *correlations*$(P^{(i)}, P^{(j)}) = 0$ for all pairs of principal components.

Two-Variable Case

Consider the simplest case of two ($k = 2$) variables measured on each observation, represented by x and y. (Note: The symbol y simply represents one of two variables and should not be confused with a dependent Y used in the context of regression

T A B L E **7.1**

Notation for *n* observed bivariate observations ($k = 2$)

Observation	1	2	3	...	*n*
Variable *x*	x_1	x_2	x_3	...	x_n
Variable *y*	y_1	y_2	y_3	...	y_n

analysis; also a_x and a_y represent two coefficients unrelated to the coefficients used in a discriminant analysis.) For a sample size *n*, the sampled values are represented in Table 7.1. When

$$P_j^{(1)} = a_x x_j + a_y y_j \tag{7.3}$$

and

$$P_j^{(2)} = b_x x_j + b_y y_j \tag{7.4}$$

then the transformed values $P_j^{(1)}$ and $P_j^{(2)}$ are represented by the symbols given in Table 7.2 (see the test case for a numerical illustration.)

Principal components transform the natural measurements (e.g., x_j and y_j) into alternative values (e.g., $P_j^{(1)}$ and $P_j^{(2)}$). These alternative values have two special properties. The principal component coefficients a_x, a_y, b_x, and b_y are chosen so that:

1 The estimated variance of $P^{(1)}$ is greater than the estimated variance of $P^{(2)}$, or $S_{P^{(1)}}^2 > S_{P^{(2)}}^2$.

2 The correlation between the two principal component values is zero, or $correlation(P^{(1)}, P^{(2)}) = 0$.

The analogous situation holds for the *k*-variate case or, in general,

1 $S_{P^{(1)}}^2 > S_{P^{(2)}}^2 > S_{P^{(3)}}^2 > \cdots > S_{P^{(k)}}^2$

2 $correlation(P^{(i)}, P^{(j)}) = 0 \ (i \neq j)$

for all pairs of principal components, as mentioned.

T A B L E **7.2**

Notation of *n* bivariate principal components ($k = 2$)

Observation	1	2	3	...	*n*
1st principal component, $P^{(1)}$	$P_1^{(1)}$	$P_2^{(1)}$	$P_3^{(1)}$	...	$P_n^{(1)}$
2nd principal component, $P^{(2)}$	$P_1^{(2)}$	$P_2^{(2)}$	$P_3^{(2)}$	...	$P_n^{(2)}$

Estimation of Principal Components (Two-Variable Case)

The criteria of maximum variability and zero correlation produce the following equations for two measurements per observation ($k = 2$):

$$a_x S_X^2 + a_y S_{XY} = a_x \lambda_1$$
$$a_x S_{XY} + a_y S_Y^2 = a_y \lambda_1 \tag{7.5}$$

and the solution yields values of the coefficients of the first principal component. There are two equations, but three values are unknown (a_x, a_y, and λ_1). A constraint is added so these equations can be uniquely solved—namely, $a_x^2 + a_y^2 = 1$. Using the sample estimates S_X^2, S_Y^2, and S_{XY} and the three equations, values a_x and a_y are calculated.

The solution to these three equations is

$$\lambda_1 = (S_X^2 + S_Y^2 + F)/2 \quad \text{where} \quad F^2 = (S_X^2 - S_Y^2)^2 + 4S_{XY}^2 . \tag{7.6}$$

Then

$$a_x = \frac{-S_{XY}}{\sqrt{S_{XY}^2 + (S_X^2 - \lambda_1)^2}} \tag{7.7}$$

and

$$a_y = \frac{S_X^2 - \lambda_1}{\sqrt{S_{XY}^2 + (S_X^2 - \lambda_1)^2}} \quad (\text{Note: } a_x^2 + a_y^2 = 1) . \tag{7.8}$$

The coefficients (a_x and a_y) make it possible to generate the n values of the first principal component.

Two similar equations produce the coefficients for the second principal component, and the solution is

$$\lambda_2 = (S_X^2 + S_Y^2 - F)/2 , \tag{7.9}$$

giving

$$b_x = \frac{S_Y^2 - \lambda_2}{\sqrt{S_{XY}^2 + (S_Y^2 - \lambda_2)^2}} = -a_y \tag{7.10}$$

and

$$b_y = \frac{-S_{XY}}{\sqrt{S_{XY}^2 + (S_Y^2 - \lambda_2)^2}} = a_x . \tag{7.11}$$

The coefficients (a_x, a_y) and (b_x, b_y) produce n principal component pairs ($P_j^{(1)}$, $P_j^{(2)}$) from n observed pairs (x_j, y_j). Two properties of principal components with $k = 2$ variables, which apply to the general case of k measurements per observation, are

1 $a_x b_x + a_y b_y = 0$ and, more important,
2 $S_{P^{(1)}}^2 = \lambda_1$ and $S_{P^{(2)}}^2 = \lambda_2$ (see Box 7.1). $\tag{7.12}$

B O X **7.1**   VARIANCE OF A PRINCIPAL COMPONENT ($P^{(i)}$)

The following shows that the variance $S_{P^{(1)}}^2 = \lambda_1$.

From the principal component equations, (expression 7.5), the expressions

$$a_x S_X^2 + a_y S_{XY} = a_x \lambda_1$$

$$a_x S_{XY} + a_y S_Y^2 = a_y \lambda_1$$

give

$$
\begin{aligned}
S_{P^{(1)}}^2 = S_{a_x X + a_y Y}^2 &= a_x^2 S_X^2 + a_y^2 S_Y^2 + 2 a_x a_y S_{XY} \\
&= a_x (a_x S_X^2 + a_y S_{XY}) + a_y (a_y S_Y^2 + a_x S_{XY}) \\
&= a_x \lambda_1 a_x + a_y \lambda_1 a_y = \lambda_1 (a_x^2 + a_y^2) = \lambda_1
\end{aligned}
$$

because $a_x^2 + a_y^2 = 1$.

In general, for k variables per observation,

$$S_{P^{(i)}}^2 = \lambda_i \quad i = 1, 2, \cdots, k .$$

Additionally, the principal component transformation redistributes the original variability but maintains the same total variance. In symbols, this property of the variance means that

$$\lambda_1 + \lambda_2 = S_{P^{(1)}}^2 + S_{P^{(2)}}^2 = S_X^2 + S_Y^2 . \tag{7.13}$$

B O X **7.2**   PRINCIPAL COMPONENTS PARTITION THE TOTAL VARIANCE

To show that $\lambda_1 + \lambda_2 = S_X^2 + S_Y^2$, begin with

$$\lambda_1 = [S_X^2 + S_Y^2 + F]/2$$

$$\lambda_2 = [S_X^2 + S_Y^2 - F]/2 \quad \text{(expressions 7.6 and 7.9)} .$$

Adding the two equations gives

$$\lambda_1 + \lambda_2 = S_{P^{(1)}}^2 + S_{P^{(2)}}^2 = S_X^2 + S_Y^2 .$$

In general, for k variables per observation,

$$\Sigma \lambda_i = \Sigma S_{P^{(i)}}^2 = \Sigma S_{X_i}^2 .$$

k-Variable Case

For an observation consisting of k variables, the criteria of maximum variability and zero correlation yield a set of equations for the first principal component as

$$a_1 S_1^2 + a_2 S_{21} + a_3 S_{31} + \cdots + a_k S_{k1} = a_1 \lambda_1$$
$$a_1 S_{12} + a_2 S_2^2 + a_3 S_{32} + \cdots + a_k S_{k2} = a_2 \lambda_1$$

$$\vdots$$

$$a_1 S_{1k} + a_2 S_{2k} + a_3 S_{3k} + \cdots + a_k S_k^2 = a_k \lambda_1 \,. \qquad (7.14)$$

These equations are usually solved by computer techniques and produce values for the coefficients $a_1, a_2, a_3, \cdots, a_k$, making it possible to calculate the first principal component $P_j^{(1)}$ for each k-variate observation in the data set. Similar equations are solved for the coefficients $b_1, b_2, \cdots, b_k$, producing the second principal component $P_j^{(2)}$ so $S_{P^{(1)}}^2 > S_{P^{(2)}}^2$ and the correlation between $P^{(1)}$ and $P^{(2)}$ is zero. A third set of equations generates the coefficients of the third principal component $P_j^{(3)}$ where $S_{P^{(1)}}^2 > S_{P^{(2)}}^2 > S_{P^{(3)}}^2$ and all pairs of components have zero correlation. This process is repeated to produce any number of principal components up to a maximum of k, the number of variables per observation. As part of the calculation, the values $\lambda_1, \lambda_2, \cdots, \lambda_k$ are also produced and play an important role because these values indicate the variance associated with each principal component (i.e., $\lambda_i = S_{P^{(i)}}^2$).

Individual principal components independently summarize some but not all of the information contained in the original multivariate measurement. The worth of each summary is directly related to its variability. A discriminant function increases in classification accuracy as the variability of the discriminant scores increases. Similarly, the degree of effectiveness of a principal component as a summary is defined in terms of its variability. The first component has the maximum possible variance among all linear combinations of the original observed values, making it the most effective principal component summary. However, "the most effective" may fall short of being useful. Its worth is usually judged by the proportion of the total variation "explained." The maximized variance of the first principal component is $S_{P^{(1)}}^2 = \lambda_1$ and the percentage of the total variation associated with this summary is

$$\% \ total \ variation = \frac{\lambda_1}{\lambda_1 + \lambda_2 + \cdots + \lambda_k} \times 100$$

$$= \frac{\lambda_1}{S_P^{2(1)} + S_P^{2(2)} + \cdots + S_P^{2(k)}} \times 100$$

$$= \frac{\lambda_1}{S_{X_1}^2 + S_{X_2}^2 + \cdots + S_{X_k}^2} \times 100 \,. \qquad (7.15)$$

Again note that $\Sigma \lambda_i = \Sigma S_{P^{(i)}}^2 = \Sigma S_{X_i}^2$, or the total variance is not changed but redistributed among the transformed values.

The relative effectiveness of any particular principal component employed as a summary is judged by

$$\% \text{ total variation of principal component } i = \frac{\lambda_i}{\lambda_1 + \lambda_2 + \cdots + \lambda_k} \times 100$$

(7.16)

where i represents any one of the k principal components. The question of what percentage of the total variation makes a principal component an adequate summary of the data is not easily answered. An often quoted rule of thumb, which has little basis in statistical theory or practical experience, is that a principal component is an effective summary of a k-variable measurement only when the variance is greater than one ($\lambda_i > 1$). The adequacy of a principal component as a summary is best decided in a nonstructured manner based on subject-matter considerations.

A principal component can be an effective way to reduce the complexity of a data set when the original multivariate measurements are correlated. When the measured variables are uncorrelated the principal component transformation has little effect. Each principal component essentially corresponds to one of the original uncorrelated variables but is arranged in descending order of variability.

Like regression analysis, principal components can be derived from standardized data. The data are standardized in the usual way so all mean values are zero and all variances are 1.0. The standardization produces "data" expressed in units of standard deviations and only the correlation structure remains to define the principal components (i.e., all variances = 1.0).

The calculated values and effectiveness of a principal component differ depending on whether it is derived from data in natural units or from standardized data. Many applications employ standardized data in keeping with patterns established largely in the social sciences. When the units of measurements differ (miles, pounds, years, dollars, etc.), a principal component derived from unstandardized data is a weighted sum of noncommensurate values (combining "apples and oranges"). Employing standardized data improves the interpretability of the summary values, particularly the coefficients associated with each variable. When all measurements are made in the same units (often the case for a series of morphological measurements on a biological specimen), the interpretation of a principal component derived from unstandardized data is relatively straightforward. Such a principal component is a more sensitive statistical tool because it uses information from variances as well as covariances.

Geometric Description

For principal components constructed from two variables per observation ($k = 2$), the first principal component is related to the projection of the data points on the major axis of an ellipse (Figure 7.1) describing the data centered at ($\overline{x}$, $\overline{y}$). Formally, a projection is the shortest distance from a point to a line—the perpendicular distance.

FIGURE **7.1**

Major axis and the first
principal component

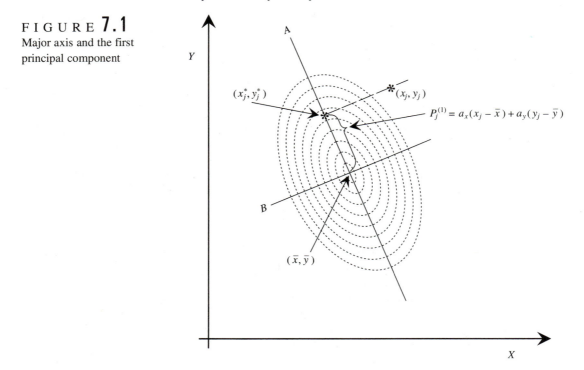

The major axis of the data ellipse is established by finding that line that minimizes the squared perpendicular distance between the observed points (x_j, y_j) and the corresponding points (x_j^*, y_j^*) on the major axis (labeled A in Figure 7.1). That is,

$$Q = \Sigma \, d_j^2 = \Sigma[(x_j - x_j^*)^2 + (y_j - y_j^*)^2] \tag{7.17}$$

is minimized, producing the equations in expression 7.5, which are solved to produce the principal component coefficients.

The distance between the point on the estimated major axis (x_j^*, y_j^*) and the point $(\overline{x}, \overline{y})$ on the major axis is the first principal component associated with the point (x_j, y_j). Algebraically, the distance between (x_j^*, y_j^*) and $(\overline{x}, \overline{y})$, given by

$$D_j = \sqrt{(x_j^* - \overline{x})^2 + (y_j^* - \overline{y})^2} \,, \tag{7.18}$$

and is the jth value of the first principal component $P_j^{(1)}$, or

$$D_j = P_j^{(1)} = a_x x_j + a_y y_j - (a_x \overline{x} + a_y \overline{y}) = a_x(x_j - \overline{x}) + a_y(y_j - \overline{y}) \tag{7.19}$$

where the value of a_x and a_y are calculated from expression 7.5 and the overall mean $a_x \overline{x} + a_y \overline{y}$ is subtracted so the mean overall principal component $\overline{P}^{(1)} = 0$, for convenience.

The slope of the major axis is estimated by

$$\hat{\beta} = \frac{a_y}{a_x} = \frac{\lambda_1 - S_X^2}{S_{XY}} \quad [1].$$

(7.20)

The estimated major axis can be plotted by constructing a line with slope $\hat{\beta}$ through the point $(\overline{x}, \overline{y})$.

The major axis (line A, Figure 7.1) is the "best" estimated line describing a bivariate data set when x and y play symmetric roles. That is, variables x and y often have no specific directional relationship; changes in x do not produce changes in y. This situation differs from linear regression analysis, in which the x-values are considered to be antecedent to the dependent variable Y; changes in the independent variables are viewed as producing changes in the dependent variable. For example, if x is the temperature of a chemical reaction and y is the quantity of chemical produced, changes in temperature cause changes in yield. However, if x is the birth weight of one twin and y is the birth weight of the other, no ordering is apparent and minimizing the perpendicular distance is a more appropriate criterion to generate a summary line for a scatter of data points. Geometrically, the difference between the two approaches to representing the relationship between x and y depends on how distance to the summary line is measured. A regression line is estimated by measuring distance parallel to the y-coordinate axis, whereas the major axis of a data ellipse is estimated by measuring the perpendicular distance between the data and the line.

Second Principal Component

The second principal component is related to the minor axis of a data ellipse in the same way the first principal component is related to the major axis. An estimate of the minor axis is established by a line perpendicular to the major axis that also passes through the point $(\overline{x}, \overline{y})$. Like the first principal component, the second principal component is the distance between $(\overline{x}, \overline{y})$ and the point $(\tilde{x}_j, \tilde{y}_j)$ on the minor axis. The point $(\tilde{x}_j, \tilde{y}_j)$ is the point closest to (x_j, y_j) on the estimated minor axis of the ellipse (line B, Figure 7.2). That is, $(\tilde{x}_j, \tilde{y}_j)$ results from projecting the data point (x_j, y_j) onto the minor axis. The fact that the minor axis (line B) is perpendicular to the major axis (line A) guarantees that the values $P_j^{(1)}$ are uncorrelated with the values $P_j^{(2)}$. Because the major and minor axes are perpendicular, the estimated slope of the minor axis is necessarily $-1/\hat{\beta}$.

The technical details tend to obscure the fact that the principal components are no more than a redefinition of the coordinate axes. The major axis of the data ellipses centered at $(\overline{x}, \overline{y})$ becomes the new "y-axis" and the minor axis becomes the new "x-axis" (Figure 7.3). The principal components are then the "x/y"-values associated with each point $(P_j^{(1)}, P_j^{(2)})$. This redefinition of the coordinate axes produces the properties of the principal components (maximum variability and uncorrelated values).

FIGURE **7.2**
Minor axis and the second principal component

FIGURE **7.2**
Minor axis and the second principal component

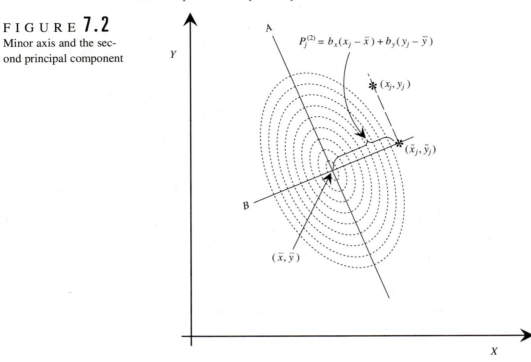

$$P_j^{(2)} = b_x(x_j - \bar{x}) + b_y(y_j - \bar{y})$$

(x_j, y_j)

$(\tilde{x}_j, \tilde{y}_j)$

$(\bar{x}, \bar{y})$

FIGURE **7.3**
The rotation of the *x/y* coordinate axes to form the principal components $(k = 2)$

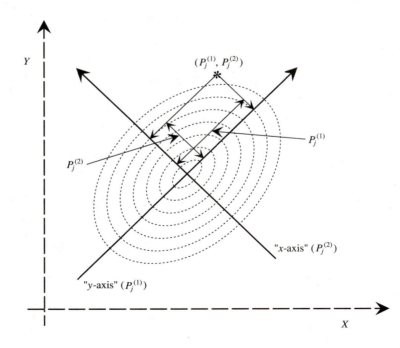

$(P_j^{(1)}, P_j^{(2)})$

$P_j^{(1)}$

$P_j^{(2)}$

"x-axis" $(P_j^{(2)})$

"y-axis" $(P_j^{(1)})$

Applications of Principal Components

Among the variety of ways principal components are used, four specific possibilities are:

1 The first principal component can be used as a univariate summary of original multivariate measurements—an index.

2 The principal component coefficients can detect meaningful combinations of the original multivariate measurements.

3 The first several principal components can be employed as a graphical tool to search for "natural" clustering of multivariate observations or, similarly, as a way to explore k-variable data for extreme values.

4 Principal components, because they are uncorrelated, can be used in a linear regression or discriminant analysis in place of highly correlated (collinear) independent variables.

Index

Employing the first principal component as a univariate canonical variable in place of k variables has the advantage of simplicity but incurs a loss of efficiency. Using all k variables makes the analysis 100% efficient (i.e., "explaining" all of the variation), whereas combining the original variables into one or a few summary numbers by means of a principal component index is expectedly less efficient. However, occasions arise when the first principal component effectively summarizes a large proportion of the variability contained in the original multivariate values. Typically the first principal component summarizes the original measurements with high efficiency when a series of correlated measurements are made on a single observation. A classic example ([2] or [3]) is the measurement of turtle shells in terms of length, height, and width (see following application). The first principal component relates to turtle shell "size" and is 98% as efficient as employing three separate correlated variables.

A principal component employed as a summary reduces complex multivariate data to a univariate canonical variable. Univariate statistical procedures such as plots, correlations, t-tests, and analysis of variance techniques can then be applied to the principal component values to describe the various relationships under study. Furthermore, the second and, perhaps, the third principal component often summarize other aspects of the multivariate observation (i.e., they are uncorrelated).

Application

Index

Turtle Carapace

Variables:

$$\text{turtle carapace length} = \text{length}$$
$$\text{turtle carapace width} = \text{width}$$
$$\text{turtle carapace height} = \text{height}$$

	Covariance Array				Correlation Array		
	Length	Width	Height		Length	Width	Height
Length	451.39	271.17	168.70	Length	1.0	0.974	0.973
Width		171.73	103.29	Width		1.0	0.965
Height			66.65	Height			1.0

Summary:

Component	Variances	Percent	Cumulative Percent
$P^{(1)}$	$\lambda_1 = 680.40$	98.64	98.64
$P^{(2)}$	$\lambda_2 = 6.50$	0.94	99.59
$P^{(3)}$	$\lambda_3 = 2.86$	0.41	100.00
Total	689.76	100.00	—

Principal component coefficients, a_i, b_i, and c_i:

Variables	$P^{(1)}$	$P^{(2)}$	$P^{(3)}$
Coefficients	a_i	b_i	c_i
Length ($i = 1$)	0.813	−0.545	−0.205
Width ($i = 2$)	0.496	0.832	−0.249
Height ($i = 3$)	0.307	0.101	0.947

Therefore,

$$P_j^{(1)} = a_1 x_{1j} + a_2 x_{2j} + a_3 x_{3j}$$

$$P_j^{(1)} = a_1 x_{1j} + a_2 x_{2j} + a_3 x_{3j} = 0.813 x_{1j} + 0.496 x_{2j} + 0.307 x_{3j}$$

$$P_j^{(2)} = b_1 x_{1j} + b_2 x_{2j} + b_3 x_{3j} = -0.545 x_{1j} + 0.832 x_{2j} + 0.101 x_{3j}$$

$$P_j^{(3)} = c_1 x_{1j} + c_2 x_{2j} + c_3 x_{3j} = -0.205 x_{1j} - 0.249 x_{2j} + 0.947 x_{3j}$$

with

$$S_{P^{(1)}}^2 = \lambda_1 = 680.40, \quad S_{P^{(2)}}^2 = \lambda_2 = 6.50, \quad \text{and} \quad S_{P^{(3)}}^2 = \lambda_3 = 2.86 .$$

Coefficients

The coefficients associated with the variables that make up a principal component can reflect interesting and sometimes important interpretations of the entire component. If the variables are not commensurate, then the relative sizes of the coefficients are meaningless because their relative magnitude depends on the units of measurement. Typically variables measured in differing units are made commensurate by standardizing the data (means = 0 and variances = 1) used to generate the principal components. For commensurate variables, the larger coefficients (in absolute value) in each component are associated with those variables that dominate the linear combination. Occasionally, the joint contribution from the larger coefficients gives the principal component a general interpretation. For example, in a study of wooden supports [4] used in the construction of mines, a series of standardized physical measurements ($[x_{ij} - \bar{x}_j]/S_j$) were made on a sample of supports (see following application). The largest coefficients from the first component are associated with the size of the support (top-diameter, length, number of rings at the buse, and whorls). The second principal component is dominated by the coefficients associated with seasoning (specific gravity and moisture content). The third component emphasizes variables related to growth (oven-dry specific gravity and number of annual rings at the top). Note that "size," "seasoning," and "growth" are certainly properties of wooden supports but were not directly measured. Not all principal components have coefficients with neat and compact interpretations, but examining the coefficients sometimes leads to general ways of viewing a summary of a multivariate observation by identifying the important elements.

Application   ## Coefficients

Properties of Wooden Supports

Variables:

top diameter of prop = top

length of prop = length

moisture content = moist

specific gravity of timber at time of test = spg.

oven-dry specific gravity = oven spg.

number of annual rings at top = rg. top

number of annual rings at base = rg. base

maximum bow = max. bow

distance of maximum bow from top = bow dist.

number of knot whorls = whorls

length of clear prop (without knots) = clear

average number of knots per whorl = knots

average diameter of knots = avg. knots

Descriptive statistics:

Variable	Mean	Std. Deviation	Variable	Mean	Std. Deviation
top	4.31	0.98	max. bow	0.65	0.43
length	46.88	11.29	bow dist.	23.40	9.06
moist	114.60	57.10	whorls	2.49	1.17
spg.	0.88	0.23	clear	10.70	6.27
oven spg.	0.42	0.07	knots	5.45	1.65
rg. top	13.30	3.24	avg. knots	0.82	0.33
rg. base	16.30	3.95			

Variances:

Component	Variances	Percent	Cumulative Percent
$P^{(1)}$	$\lambda_1 = 4.219$	32.4	32.4
$P^{(2)}$	$\lambda_2 = 2.378$	18.3	50.7
$P^{(3)}$	$\lambda_3 = 1.878$	14.4	65.1
$P^{(4)}$	$\lambda_4 = 1.109$	8.5	73.6
$P^{(5)}$	$\lambda_5 = 0.910$	7.0	80.6

Principal component coefficients:

Variables	$P^{(1)}$	$P^{(2)}$	$P^{(3)}$	$P^{(4)}$	$P^{(5)}$
top	0.83	0.34	−0.28	−0.10	0.08
length	0.83	0.29	−0.32	−0.11	0.11
moist	0.25	0.83	0.19	0.08	−0.33
spg.	0.36	0.70	0.48	0.06	−0.34
oven spg.	0.12	−0.26	0.66	0.05	−0.17
rg. top	0.58	−0.02	0.65	−0.07	0.30
rg. base	0.82	−0.29	0.35	−0.07	0.21
max. bow	0.60	−0.29	−0.33	0.30	−0.18
bow dist.	0.73	0.02	−0.28	0.10	0.10
whorls	0.78	−0.38	−0.16	−0.22	−0.15
clear	−0.02	0.32	−0.10	0.85	0.33
knots	−0.24	0.53	0.13	−0.32	0.57
avg. knots	−0.23	0.48	0.45	−0.32	0.08

Graphical

The first few principal components can serve as a graphical tool to explore possible clustering among k-variate observations or to identify differing subgroups within a data set. A plot of the first and second principal components may detect clusters or

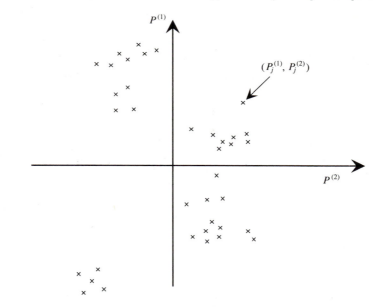

FIGURE **7.4**
Plot of values of first
and second principal
components for the
winged aphids data

outlier observations that are difficult to discover when the data are not summarized in some way (number of variables not reduced). Biologists, for example, often make a large number of measurements on a particular insect or animal. The first few principal components can possibly separate these multivariate observations into morphological types or even species. In a study of winged aphids [4], plots of the first two principal components each based on 19 variables clearly show distinct morphological types (see following application and Figure 7.4).

In another example, the technique of plotting the values of the first two principal components worked well to identify an outlier when k measurements of automobile safety features were studied (e.g., stopping distance, turning radius, center of gravity, density, etc.). A principal component plot showed two distinct clusters and a single point: American cars, Japanese cars, and the Mercedes-Benz.

Application   ## Graphical

Winged Aphids

Variables:

x_1	body length	x_2	number of antennal spines
x_3	body width	x_4	leg length, tarsus III
x_5	fore-wing length	x_6	leg length, tibia III
x_7	hind-wing length	x_8	leg length, femur III

x_9	number of spiracles	x_{10}	rostrum
x_{11}	length of antennal segment I	x_{12}	ovipositor
x_{13}	length of antennal segment II	x_{14}	number of ovipositor spines
x_{15}	length of antennal segment III	x_{16}	anal folub
x_{17}	length of antennal segment IV	x_{18}	number of hind-wing hooks
x_{19}	length of antennal segment V		

	$S^2_{P^{(i)}}$	Percent	Cumulative Percent
$P_j^{(1)}$	13.86	73.0	73.0
$P_j^{(2)}$	2.37	12.5	85.5

Regression

As noted previously, highly correlated independent variables used to estimate regression or discriminant coefficients adversely affect the precision of the estimated values. One way to improve the precision of these estimates is to exclude from the analysis the variable or variables that cause collinearity. Alternatively, principal component summaries can be generated and these uncorrelated components used in place of the highly correlated variables.

For two correlated variables ($k = 2$), the first principal component is a weighted sum ($a_x x_1 + a_y x_2$) related to the major axis of the probability ellipse. The second principal component is a weighted difference ($a_y x_1 - a_x x_2$) related to the minor axis of the probability ellipse. These uncorrelated transformed variables can be employed in a regression analysis and treated as independent variables. For example, age and years of employment are often part of measuring cumulative "exposure" to toxic material in the workplace. Entering both variables into the regression equation decreases the reliability of the estimated regression coefficients associated with both age and years of employment due to the typically high correlation between these two variables. Two principal components formed from these variables yield two uncorrelated measures of "exposure" and summarize 100% of the variation of the original correlated variables.

The process of replacing a series of collinear variables with uncorrelated principal component values applies to any number of independent variables. The set of transformed variables can be more awkward to interpret than the original measurements, but the "new" variables are uncorrelated and, therefore, improve the precision of the estimated regression coefficients. The relative merits of interpretability versus improved estimation precision is a typical tradeoff in statistical approaches.

Computer Implementation

Principal Components ($k = 2$)

Continuing the data from the test case in Chapter 6: For $n = 10$ observations, $x =$ number of years smoked, and $y =$ level of carbon monoxide (CO) for those individuals who failed to quit smoking, then

Variable	1	2	3	4	5	6	7	8	9	10	Mean
x = years	2	2	3	4	5	5	6	7	8	8	5
y = CO	16	12	9	12	14	6	8	11	8	4	10

$$S_X^2 = 5.111 \qquad S_Y^2 = 13.556 \qquad S_{XY} = -5.333$$

λ_j-values (variances):

$$F = \sqrt{(S_X^2 - S_Y^2)^2 + 4S_{XY}^2}$$
$$= \sqrt{(5.111 - 13.556)^2 + 4(-5.333)^2} = 13.605$$
$$\lambda_1 = (S_X^2 + S_Y^2 + F)/2 = (5.111 + 13.556 + 13.605)/2 = 16.136$$
$$\lambda_2 = (S_X^2 + S_Y^2 - F)/2 = (5.111 + 13.556 - 13.605)/2 = 2.531$$

Principal Component Coefficients

Major axis:

$$a_x = -S_{XY}/\sqrt{S_{XY}^2 + (S_X^2 - \lambda_1)^2}$$
$$= 5.333/\sqrt{(-5.333)^2 + (5.111 - 16.136)^2} = 0.435$$
$$a_y = (S_X^2 - \lambda_1)/\sqrt{S_{XY}^2 + (S_X^2 - \lambda_1)^2}$$
$$= (5.111 - 16.136)/\sqrt{(-5.333)^2 + (5.111 - 16.136)^2} = -0.900$$
$$P_j^{(1)} = a_x(x_j - \bar{x}) + a_y(y_j - \bar{y})$$
$$= 0.435(x_j - 5) - 0.900(y_j - 10)$$

Minor axis:

$$b_x = (S_Y^2 - \lambda_2)/\sqrt{S_{XY}^2 + (S_Y^2 - \lambda_2)^2}$$
$$= (13.556 - 2.531)/\sqrt{(-5.333)^2 + (13.556 - 2.531)^2} = 0.900$$
$$\text{or} \quad b_x = -a_y = 0.900$$
$$b_y = -S_{XY}/\sqrt{S_{XY}^2 + (S_Y^2 - \lambda_2)^2}$$
$$= 5.333/\sqrt{(-5.333)^2 + (13.556 - 2.531)^2} = 0.435$$
$$\text{or} \quad b_y = a_x = 0.435$$

$$P^{(2)}_j = b_x(x_j - \overline{x}) + b_y(y_j - \overline{y})$$
$$= 0.900(x - 5) + 0.435(y_j - 10)$$

Values of $P^{(1)}_j$ and $P^{(2)}_j$:

Variable	1	2	3	4	5	6	7	8	9	10	Mean
$P^{(1)}$	−6.708	−3.107	0.029	−2.236	−3.601	3.601	2.236	−0.029	3.107	6.708	0.0
$P^{(2)}$	−0.088	−1.830	−2.236	−0.029	1.742	−1.742	0.029	2.236	1.830	0.088	0.0

Variances:

$$S^2_{P^{(1)}} = 145.317/9 = 16.136 = \lambda_1$$
$$S^2_{P^{(2)}} = 22.779/9 = 2.531 = \lambda_2$$
$$correlation(P^{(1)}, P^{(2)}) = 0.0$$
$$\text{Note: } S^2_{P^{(1)}} + S^2_{P^{(2)}} = 18.667 = S^2_X + S^2_Y$$

Slopes of the major and minor axes:

$$major\ axis = \hat{\beta} = (\lambda_1 - S^2_X)/S_{XY} = (16.136 - 5.111)/-5.333 = -2.067$$
$$minor\ axis = -1/\hat{\beta} = 0.484$$

Proportion variation explained:

$$P^{(1)}: \lambda_1/(\lambda_1 + \lambda_2) = 16.136/18.667 = 0.864$$
$$P^{(2)}: \lambda_2/(\lambda_1 + \lambda_2) = 2.531/18.667 = 0.136$$

Principal Components: Test Case

The principal component analysis implemented by the STATA system ("factor") using both natural units and standardized values is contained in the following output. The variable X = years smoked cigarettes and Y = amount of measured carbon monoxide exposure among smokers who were unable to quit smoking after attending a smoking control program (data repeated from Chapter 6).

```
example.principal

. infile x y using princ.dat
(10 observations read)

. list

              x           y
    1.        2          16
    2.        2          12
    3.        3           9
    4.        4          12
    5.        5          14
    6.        5           6
    7.        6           8
    8.        7          11
    9.        8           8
   10.        8           4
```

```
. factor x y, pc covariance
(obs=10)
                (principal components; 2 components retained)
Component    Eigenvalue    Difference    Proportion    Cumulative
-----------------------------------------------------------------
        1      16.13566      13.60465        0.8644        0.8644
        2       2.53101             .        0.1356        1.0000
```

$$\lambda_1$$
$$\lambda_2$$
$$\frac{\lambda_1}{\lambda_1 + \lambda_2}$$
$$\frac{\lambda_2}{\lambda_1 + \lambda_2}$$

```
              Eigenvectors
  Variable |     1          2
-----------+---------------------
        x |  -0.43549    0.90020
        y |   0.90020    0.43549
```

a_x a_y b_x b_y

```
. generate p1=0.43549*(x-5)-0.90020*(y-10)

. generate p2=0.90020*(x-5)+0.43549*(y-10)

. list p1 p2
```

$P_j^{(1)}$ $P_j^{(2)}$

```
            p1          p2
  1.   -6.70767     -.08766
  2.   -3.10687    -1.82962
  3.     .02922    -2.23589
  4.   -2.23589     -.02922
  5.    -3.6008     1.74196
  6.     3.6008    -1.74196
  7.    2.23589      .02922
  8.    -.02922     2.23589
  9.    3.10687     1.82962
 10.    6.70767      .08766


. summarize p1 p2

  Variable |      Obs        Mean    Std. Dev.        Min         Max
-----------+-----------------------------------------------------------
        p1 |       10           0    4.016945    -6.70767     6.70767
        p2 |       10           0    1.590924    -2.23589     2.23589

. correlate p1 p2
(obs=10)

           |       p1        p2
-----------+------------------
        p1 |   1.0000
        p2 |  -0.0000    1.0000


. factor x y, pc means
(obs=10)
  Variable |      Mean    Std. Dev.         Min         Max
-----------+----------------------------------------------------
        x |         5     2.260777           2           8
        y |        10     3.681787           4          16
```

```
            (principal components; 2 components retained)
Component    Eigenvalue    Difference    Proportion    Cumulative
- - - - - - - - - - - - - - - - - - - - - - - - - - - - - - - - - - - - -
    1          1.64074      1.28148        0.8204        0.8204
    2          0.35926         .           0.1796        1.0000


             Eigenvectors
Variable |      1           2
- - - - - - - - +- - - - - - - - - - - - - - - - - - - - -
       x |   -0.70711     0.70711
       y |    0.70711     0.70711

. exit,clear
```

Applied Example

Principal Components

Three variables from the Child Health and Development Study related to the size of a newborn infant are head circumference, body length, and birth weight. Treating these variables in three individual analyses is to some extent redundant and makes the interpretation of the results difficult because overlapping and nonoverlapping influences are not separated. Creating a principal component index provides a summary of the three measurements, reducing each trivariate observation to a single number. The first principal component is

$$P_j^{(1)} = 0.539(\text{standardized head circumference}_j) + 0.571(\text{standardized length}_j)$$
$$+ 0.619(\text{standardized birth weight}_j) \, .$$

This principal component is derived from standardized data because of the differing measurement units (inches and pounds). The values $P_j^{(1)}$ summarize infant "size" and form a parsimonious description that can be analyzed in a variety of ways. The efficiency associated with $P_j^{(1)}$ is $(2.201/3) \times 100 = 73.4\%$. (Note: For standardized data, $\Sigma \lambda_i = k$.) Although some loss of efficiency is incurred by employing this univariate canonical summary of size, compensation occurs in the form of simplification of analytic procedures and ease of interpretation of a univariate summary.

The separate components (head circumference, length, and birth weight) are related to parental measurements. It is, therefore, not surprising that the variable "size" (i.e., first principal component) is also associated with these parental variables. For example, "size" is positively correlated with length of gestation ($r = 0.404$) and negatively associated with amount of maternal smoking ($r = -0.195$).

The variable "size" considered as a summary measure of intrauterine infant growth can be used to examine the influences of maternal smoking. Four categories of maternal smoking exposure yield the mean values of the first principal component summary given in Table 7.3. The "size" of the infant as measured by three characteristics and summarized by the first principal component shows that mothers who smoke more than a pack of cigarettes per day have smallest infants. The four mean values of the principal component can be formally assessed with an F-statistic generated from a one-way analysis of variance (Chapter 1). The F-value is 12.77, produc-

T A B L E **7.3**

The mean values of the first principal component

Group		Sample size	Mean Value ($\bar{P}_j^{(1)}$)
1	Never smoker	266	0.292
2	Past smoker	115	0.315
3	Smoked < pack/day	169	−0.360
4	Smoked ≥ pack/day	130	−0.409
	Total	680	0.000

ing a p-value < 0.001, implying that the observed differences are unlikely to have occurred by chance variation. Implicit in this test is the assumption that the three variables (head circumference, length, and weight) have normal or at least approximately normal distributions. Also, the fact that the principal components are calculated quantities and not sampled data causes the F-test to be an approximate but generally reliable procedure, particularly for large sample sizes.

A regression analysis included in the following computer output illustrates another possible application of principal component summaries. The variable "size" is the dependent variable and maternal smoking and infant gestation (standardized) are employed as independent variables. The overall F-statistic 79.74 with a p-value < 0.001 indicates that these two variables are likely related to infant "size." Both independent variables also show individual associations with "size" (gestational time: p-value < 0.001, and maternal smoking: p-value < 0.001). These results are similar to those observed in the parallel regression analysis of birth weight (Chapter 3).

Principal Components

The CHDS data are used to analyze three measurements made on each newborn infant: head circumference, body length, and birth weight. A brief description of these three variables is given in Chapter 2. A principal component summary combines the three measurements and the resulting "size" variable is then used to further analyze the relationship between infant characteristics and two maternal variables.

chds.principal

```
. use chds


. factor headcir length bwt, pc mean
(obs=680)
 Variable |      Mean    Std. Dev.          Min         Max
----------+------------------------------------------------
  headcir |  13.21912    .6263148           11          15
   length |  20.27941    .9821018           17          23
      bwt |  7.516471    1.092346          3.3        11.4
```

```
                  (principal components; 3 components retained)
Component     Eigenvalue     Difference     Proportion     Cumulative
----------------------------------------------------------------------
    1           2.20053        1.64933         0.7335         0.7335
    2           0.55120        0.30293         0.1837         0.9172
    3           0.24827           .            0.0828         1.0000

                  Eigenvectors
Variable |      1           2           3
---------+----------------------------------
headcir  |   0.53894     0.78081     0.31604
length   |   0.57136    -0.61454     0.54395
   bwt   |   0.61894    -0.11258    -0.77733
```

. generate size1=.53894*(headcir-13.21912)/0.6263148+.57136*(length-20.27941)/0
.9821018+0.61894*(bwt-7.516471)/1.092346

. summarize size1 mnocig gestwks

```
Variable |     Obs        Mean      Std. Dev.      Min         Max
---------+-----------------------------------------------------------
  size1  |     680      -1.21e-06    1.483414    -5.583251    4.507249
 mnocig  |     680       7.430882   11.27202         0          50
 gestwks |     680       39.77059    1.875433        29          48
```

. generate smnocig=(mnocig-7.430882)/11.27202

. generate sgestwks=(gestwks-39.77059)/1.875433

. correlate size1 smnocig sgestwks
(obs=680)

```
         |   size1    smnocig   sgestwks
---------+------------------------------
   size1 |  1.0000
 smnocig | -0.1950    1.0000
sgestwks |  0.4035   -0.0708    1.0000
```

. gen status=mnocig

. recode status 0=0 1/20=1 21/30=2 31/max=4
(299 changes made)

. sort status

. by status: summarize size1

```
- status=          0
Variable |     Obs        Mean      Std. Dev.      Min         Max
---------+-----------------------------------------------------------
  size1  |     381      .2990908    1.394788    -5.006064    4.507249

- status=          1
Variable |     Obs        Mean      Std. Dev.      Min         Max
---------+-----------------------------------------------------------
  size1  |     169     -.3595021    1.495067    -3.63103     3.940634
```

```
- status=          2
Variable |      Obs          Mean    Std. Dev.          Min          Max
---------+------------------------------------------------------------
   size1 |      105    -.4764622    1.570766    -5.583251     2.456863

- status=          4
Variable |      Obs          Mean    Std. Dev.          Min          Max
---------+------------------------------------------------------------
   size1 |       25    -.1268018    1.332366    -2.487229     1.818429

. oneway size1 status

                          Analysis of Variance
        Source              SS         df        MS          F     Prob  F
------------------------------------------------------------------------
Between groups       80.1630029        3   26.721001      12.77    0.0000
Within groups        1413.98732      676   2.09169722
------------------------------------------------------------------------
     Total           1494.15032      679   2.20051594

Bartlett's test for equal variances: chi2(3) =    3.1384   Probchi2 = 0.371

. regress size1 smnocig sgestwks

   Source |      SS         df        MS                Number of obs =      680
---------+------------------------------              F( 2,   677) =    79.74
    Model |  284.881123       2   142.440561            Prob  F       =   0.0000
 Residual |   1209.2692     677   1.78621743            R-square      =   0.1907
---------+------------------------------              Adj R-square  =   0.1883
    Total |  1494.15032     679   2.20051594            Root MSE      =   1.3365

------------------------------------------------------------------------
   size1 |      Coef.    Std. Err.        t      P|t|    [95% Conf. Interval]
---------+--------------------------------------------------------------
 smnocig |   -.2480575    .0514191     -4.824    0.000    -.3490176   -.1470974
 sgestwks |    .5810403    .0514191     11.300    0.000     .4800802    .6820004
    _cons |   -6.63e-07    .0512522     -0.000    1.000    -.1006331    .1006318
------------------------------------------------------------------------

. exit,clear
```

Notation

First three principal components:

$$P_j^{(1)} = a_1 x_{1j} + a_2 x_{2j} + \cdots + a_k x_{kj}$$

$$P_j^{(2)} = b_1 x_{1j} + b_2 x_{2j} + \cdots + b_k x_{kj}$$

$$P_j^{(3)} = c_1 x_{1j} + c_2 x_{2j} + \cdots + c_k x_{kj}$$

Estimated variance of $P^{(i)}$ or λ_i-value:

$$S_{P^{(i)}}^2 = \frac{\sum_j (P_j^{(i)} - \overline{P}^{(i)})^2}{n-1} = \frac{\sum_j [P_j^{(i)}]^2}{n-1} = \lambda_i$$

Distance:

$$D_j = \sqrt{(x_j^* - \bar{x})^2 + (y_j^* - \bar{y})^2}$$
$$D_j = a_x(x_j - \bar{x}) + a_y(y_j - \bar{y}) = P_j^{(1)}$$

Slope of the major axis:

$$\hat{\beta} = \frac{a_y}{a_x} = \frac{\lambda_1 - S_X^2}{S_{XY}}$$

Problems

1 Using the turtle data in Table 6.7, calculate the first principal component based on the turtle carapace measurements, ignoring the male and female classification. What is the percent variation accounted for by $P_i^{(1)}$?

2 Do the values the $P_i^{(1)}$ differ between the male and female turtles (e.g., plot the values and calculate the mean values and variances to assess the difference in these two mean values)?

3 Consider the three points (x_i, y_i): (3, 2), (9, 10), and (2, 9). Find the coefficients for the first and second principal component based on these three observations: $P_i^{(1)} = a_x x_i + a_y y_i$ and $P_i^{(2)} = b_x x_i + b_y y_i$.

4 Plot the points in problem 3. Calculate and draw the major and minor axes associated with $P_i^{(1)}$ and $P_i^{(2)}$. Show numerically that $P_2^{(1)} = 5.0$ corresponds to the point (9, 10). Show that the point $(x_2^*, y_2^*) = (7.666, 11)$ is that point on the major axis perpendicularly below $(x_2, y_2) = (9, 10)$. Compute the distance between (x_2^*, y_2^*) and $(\bar{x}, \bar{y})$.

5 Show that the coefficients a_x and a_y (expressions 7.7 and 7.8) are the solutions to the equations that generate the first principal component (expression 7.5).

6 If

$$\hat{\beta} = \frac{\lambda_1 - S_X^2}{S_{XY}} \quad \text{and} \quad \hat{\beta}' = \frac{S_{XY}}{\lambda_1 - S_Y^2},$$

show that $\hat{\beta} = \hat{\beta}'$.

7 If $S_X^2 = S_Y^2$, show that $P_j^{(1)}$ is proportional to $x_j + y_j$ and $P_j^{(2)}$ is proportional to $x_j - y_j$.

8 If $S_{XY} = 0$, show that $\lambda_1 = S_X^2$ and $\lambda_2 = S_Y^2$.

9 Consider the following "data":

Obs.	x_1	x_2
1	3	2
3	5	4
4	5	6
5	4	7
6	7	4
7	7	6

Obs.	x_1	x_2
8	8	3
9	9	2
10	9	8

Conduct a principal component analysis of these data. Sketch the transformed data ellipses and indicate the location of the principal component pairs $(P_j^{(1)}, P_j^{(2)})$. Why is this principal component analysis not very useful?

References

1 Sokal, R. R., and Rohlf, F. J. *Biometry*. San Francisco: W. H. Freeman and Co., 1969.

2 Morrison, D. F. *Multivariate Statistical Methods*, 3rd ed. New York: McGraw-Hill, 1990.

3 Jolicoeur, P., and Masimann, J. E. Size and Shape Variation in the Painted Turtle. *Growth* 24 (1960): 339–354.

4 Jeffers, J. N. R. Two Case Studies in the Application of Principal Component Analysis. *Applied Statistics* (1960): 225–236.

8

Contingency Table Analysis I

Background

The past seven chapters deal with techniques primarily designed to analyze continuous multivariable data. The next two chapters describe statistical methods that apply to multivariate but discrete data, distributed into tables—referred to as multidimensional contingency tables. When each sampled observation is classified by one or more characteristics, the total count of observations with the same characteristics make up the cell frequencies of a contingency table. Individuals classified by race, sex and blood-pressure status illustrate a contingency table, shown in Table 8.1 and Figure 8.1.

Tables by themselves are important summary descriptions of collected data. However, further analysis of contingency tables often identifies interesting features of the data and helps answer specific questions concerning the relationships among the variables used to form the categories. For example, is hypertension associated

FIGURE 8.1

A graphical representation of a three-dimensional contingency table

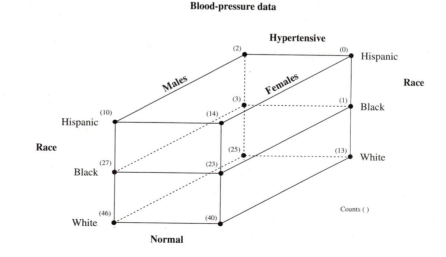

Blood-pressure data

T A B L E **8.1**

Data on 204 individuals classified by race, sex and blood-pressure status

Males	Normal	Hypertensive	Total
Hispanic	10	2	12
Black	27	3	30
White	46	25	71
Total	83	30	113

Females	Normal	Hypertensive	Total
Hispanic	14	0	14
Black	23	1	24
White	40	13	53
Total	77	14	91

with race? Or, does the association between hypertension and gender differ among race categories? Or, is the relationship between race and hypertension the same for both males and females?

The number of categorical variables used to classify a sample of observations is called the dimensionality of the table (e.g., Table 8.1 has dimensionality = 3). Not surprisingly, the complexity of the relationships within a contingency table increases as the dimensionality increases. This chapter describes methods most appropriate for the analysis of one- and two-dimensional tables. The following chapter explores extensions of these methods and introduces the log-linear model as a method for investigating and interpreting multivariable discrete data from higher-dimensional contingency tables.

Contingency table analysis involves a large number of techniques for exploring tabular data. The one-way table serves as a starting point where the Pearson chi-square and the likelihood ratio approaches are developed. For the two-way table, several general techniques are discussed, with particular emphasis on identifying associations between the classification variables. Also included is a description of the restricted chi-square test, which is a basic statistical tool used in the following chapter.

One-Way Tables

The essence of a statistical analysis of contingency table data is the comparison of the observed values to theoretically generated expected values. The comparison of the observed count x_i to the expected value $E(X_i)$ takes primarily two forms. A difference $[x_i - E(X_i)]$ is certainly a relevant form of comparison, as is the ratio $[x_i/E(X_i)]$. The issue of which measure is superior (difference versus ratio) is not settled here but, for most data, both measures produce similar results.

T A B L E   **8.2**

The components of a one-way table with two categories

	Category 1	Category 2	Total
Probability	p	$1 - p = q$	1.0
Expected probability	p_0	$1 - p_0 = q_0$	1.0
Estimated probability	$\hat{p} = x_1/n$	$1 - \hat{p} = \hat{q} = x_2/n$	1.0
Expected counts	$E(X_1) = np_0$	$E(X_2) = nq_0$	n
Observed counts	x_1	$n - x_1 = x_2$	n

Pearson Chi-Square Statistic

To understand how these two measures lead to effective analytic tools, consider the simplest contingency table—one-way classification with two categories. A set of n observations (x_1 of one kind and x_2 of another) classified into two cells produces the components displayed in Table 8.2.

The symbol p represents an unknown probability that a random observation belongs to category 1, whereas the symbol p_0 represents a postulated probability that a random observation belongs to category 1. The value p_0 is generated from theoretical considerations. The symbol $\hat{p}$ represents the observed proportion of observations falling into category 1, an estimate of p. The quantities $[x_1 - E(X_1)] = [x_1 - np_0]$ and $[x_2 - E(X_2)] = [x_2 - nq_0]$ indicate the magnitude of the differences between the counts of sample observations and theoretically derived counts (based on p_0 and q_0). Or, viewed slightly differently, the difference between $\hat{p}$ and p_0 is measured by $[x_1 - E(X_1)] = n[\hat{p} - p_0]$ and $[x_2 - E(X_2)] = n[\hat{q} - q_0]$. A summary that combines these two differences is the Pearson chi-square statistic, which, for two categories, is

$$X^2 = \frac{[x_1 - E(X_1)]^2}{E(X_1)} + \frac{[x_2 - E(X_2)]^2}{E(X_2)}.$$

(8.1)

This statistic has an approximate chi-square distribution with one degree of freedom when the theory generating p_0 is correct (when the hypothesis that produced p_0 is true; i.e., $p = p_0$). The interplay of the population parameter p, the theoretical value p_0, and the estimated value based on the data $\hat{p}$ is a fundamental part of a contingency table analysis.

Alternatively, a seemingly different approach, employing the test statistic

$$z = \frac{x_1 - E(X_1)}{\sqrt{np_0(1 - p_0)}} = \frac{\hat{p} - p_0}{\sqrt{p_0(1 - p_0)/n}},$$

(8.2)

is also used to assess data arising from observations classified into two categories. The value z has an approximate standard normal distribution under the hypothesis that $p = p_0$. However, z^2 is identical to X^2 (expression 8.1), as a little algebra will show (Box 8.1), and, therefore, analyses based on z do not differ from those based on X^2.

An important feature of the expression symbolized by X^2 is that it directly generalizes to the analyses of tables with more than two cells. That is,

$$X^2 = \sum_{all\ cells} \frac{[x_i - E(X_i)]^2}{E(X_i)}$$

(8.3)

has an approximate chi-square distribution when the hypothesis that generates the expected values accurately reflects the observed data. The degrees of freedom depend on the size and configuration of the table under consideration.

Likelihood Chi-Square Statistic

A ratio measure to compare observed and expected cell frequencies also applies to data classified into a one-way table with two cells (Table 8.2). Two probabilities are central, denoted $\hat{P}$ and P_0. The probability $\hat{P}$ is derived directly from the observed data whereas P_0 depends on postulating relationships within the data.

When observations have a binomial distribution (reviewed as part of the appendix to Chapter 12), then for a one-way classification the likelihood of observing x_1 values of one kind and $n - x_1 = x_2$ observations of another kind is estimated by

$$\hat{P} = \binom{n}{x_1} \hat{p}^{x_1} (1 - \hat{p})^{n - x_1}$$

(8.4)

B O X **8.1**   EQUIVALENCE OF X^2 AND z^2

The following demonstrates that $X^2 = z^2$.

$$X^2 = \frac{[x_1 - E(X_1)]^2}{E(X_1)} + \frac{[x_2 - E(X_2)]^2}{E(X_2)}$$

$$= \frac{[x_1 - np_0]^2}{np_0} + \frac{[x_2 - n(1 - p_0)]^2}{n(1 - p_0)}$$

$$= \frac{[x_1 - np_0]^2}{np_0} + \frac{[(n - x_1) - n(1 - p_0)]^2}{n(1 - p_0)}$$

$$= [x_1 - np_0]^2 \left[\frac{1}{np_0} + \frac{1}{n(1 - p_0)} \right]$$

$$= \frac{[x_1 - np_0]^2}{np_0(1 - p_0)} = \frac{[x_1 - E(X_1)]^2}{np_0(1 - p_0)} = z^2$$

where $\hat{p} = x_1/n$ estimates the probability that a random observation belongs to category 1. The estimate $\hat{P}$ is called the likelihood associated with a specific set of observations, namely x_1 and x_2. The probability $\hat{P}$ is generated entirely by the observed data.

Another likelihood value P_0 is also calculated based on the binomial distribution associated with x_1 and x_2 but derived under the specific hypothesis that the probability that a random observation belongs to category 1 is p_0 (i.e., null hypothesis; $H_0: p = p_0$), or

$$P_0 = \binom{n}{x_1} p_0^{x_1} (1 - p_0)^{n - x_1} .$$

$$(8.5)$$

The probability P_0 is generated from the observed data as well as a specific postulated parameter, p_0. The ratio of $\hat{P}$ to P_0 is likely close to 1 when the null hypothesis is true ($p = p_0$) and likely differs from 1 when the null hypothesis is false ($p \neq p_0$). Ratios contrasting such observed and expected probabilities are called likelihood ratios and occur in a variety of statistical contexts. An effective summary of the likelihood ratio ($\hat{P}/P_0$) is

$$Y^2 = 2 \log\left(\frac{\hat{P}}{P_0}\right).$$

$$(8.6)$$

The value Y^2 has an approximate chi-square distribution with one degree of freedom when the hypothesis generating P_0 is true (i.e., $H_0 : p = p_0$). Some algebraic manipulations yield the same quantity in a more generalizable form, or

$$Y^2 = 2 \left[x_1 \log\left(\frac{x_1}{E(X_1)}\right) + x_2 \log\left(\frac{x_2}{E(X_2)}\right) \right],$$

$$(8.7)$$

called the likelihood ratio chi-square. Similar to the X^2-statistic, the expression for Y^2 applies to tables with any number of cells and has the general form

$$Y^2 = 2 \sum_{\text{all cells}} x_i \log\left(\frac{x_i}{E(X_i)}\right).$$

$$(8.8)$$

The general likelihood ratio Y^2-statistic also has an approximate chi-square distribution when the data differ only by chance from the hypothesis that generates the expected values. Both X^2– and Y^2–statistics have the same associated degrees of freedom, which are occasionally tricky to calculate.

Generally, the degrees of freedom for these two chi-square measures from contingency table data are

degrees of freedom = (the number of cells in a table minus one) – (the number of independent estimated parameters necessary to define the expected values).

Calculations surrounding the degrees of freedom will be developed further. For a table with two cells, no estimates are made from the data to establish H_0, so the degrees of freedom for both X^2 and Y^2 are $(2 - 1) - 0 = 1$, as noted.

Applying X^2- and Y^2-Statistics to the Same Data (An Example)

At this point, it is instructive to apply X^2- and Y^2-statistics to the same data. When two heterozygotic individuals are mated, dominant Mendelian traits in the offspring are three times more likely than recessive traits. Data from $n = 40$ offspring of heterozygous matings collected to test the conjecture that a specific trait follows the Mendelian ratio (3:1) produced Table 8.3, where 24 dominant and 16 recessive offspring were observed. The hypothesis to be evaluated is

$$H_0: \ p = p_0 = \frac{3}{4}, \ \text{or} \ E(X_1) = np_0 = 30,$$

with the alternative hypothesis

$$H_1: \ p \neq p_0 = \frac{3}{4}, \ \text{or} \ E(X_1) \neq np_0,$$

giving a Pearson test statistic of

$$X^2 = \frac{(24-30)^2}{30} + \frac{(16-10)^2}{10} = 4.800 \ .$$

Also, the likelihood ratio chi-square is

$$Y^2 = 2\left[24 \log\left(\frac{24}{30}\right) + 16 \log\left(\frac{16}{10}\right)\right] = 4.329 \ .$$

Note that the alternative form,

$$Y^2 = 2 \log\left(\frac{\hat{P}}{P_0}\right) = 2 \log\left(\frac{0.1279}{0.0147}\right) = 4.329 \ ,$$

gives the identical result.

Both statistics, the Pearson X^2 and the likelihood ratio statistic Y^2, have similar values, which is typical. In fact, as the number of observations increases, the differ-

T A B L E **8.3**

Data from heterozygotic matings producing offspring classified by dominant and recessive genetic traits

	Dominant	Recessive	Total
Observed counts	$x_1 = 24$	$x_2 = 16$	40
Estimated probability	$\hat{p} = 24/40 = 0.6$	$\hat{q} = 16/40 = 0.4$	1.0
Expected probability	$p_0 = 3/4$	$q_0 = 1/4$	1.0
Expected counts	$E(X_1) = 30$	$E(X_2) = 10$	40

T A B L E **8.4**

Analysis of 60 individuals classified by 6 electrophoretic categories

Categories	1	2	3	4	5	6	Total
Observed numbers	4	10	6	10	14	16	60
Hypothesis *I*							
Expected frequencies I	0.167	0.167	0.167	0.167	0.167	0.167	1.0
Expected values *I*	10.0	10.0	10.0	10.0	10.0	10.0	60.0
Hypothesis *II*							
Types	*AA*	*AB*	*BB*	*BC*	*AC*	*CC*	
Expected frequencies *II*	0.071	0.142	0.071	0.249	0.249	0.218	1.0
Expected values *II*	4.267	8.533	4.267	14.933	14.933	13.067	60.0

ence between these two approaches disappears. In this chapter the Pearson chi-square statistic, employing differences and represented by X^2, is most often used, but no strong reason exists to prefer one measure over the other. In the next chapter, the likelihood ratio statistic Y^2 is used because this test statistic fits more naturally with the log-linear model approach. The important point again is that both methods usually produce similar results, particularly when the frequencies in all cells of a contingency table are large. The illustrative data show that the probability of a more extreme result under the hypothesis that the ratio of dominant to recessive traits equals 3:1 (H_0: $p = 3/4$) is p-value = 0.028 using X^2 and p-value = 0.037 using Y^2. These test statistics indicate that simple Mendelian inheritance is not a likely explanation for the observed distribution of the trait.

One-Way Contingency Tables (An Example)

Analysis of data distributed into one-way contingency tables with more than two cells follows the same pattern. The chi-square statistic consists of a sum of values $[x_i - E(X_i)]^2/E(X_i)$—one for each of the k cells in the table—and has an approximate chi-square distribution when the hypothesis generating the expected values is true. For example, 60 individuals biochemically analyzed for a serum protein called erthrocyte acid phosphatase (EAP) yield six distinct electrophoretic types classified as shown in Table 8.4 (top row).

A possible distribution underlying these data is that the six electrophoretic types are equally frequent in the population sampled (hypothesis *I*). Hypothesis *I* generates an expected frequency of $E(X_i) = 10$ individuals for each cell. The chi-square test statistic to evaluate this hypothesis is

$$X^2 = \frac{(4-10)^2}{10} + \frac{(10-10)^2}{10} + \cdots + \frac{(16-10)^2}{10} = 10.400$$

or, based on the likelihood ratio chi-square, the test statistic is

$$Y^2 = 2\left[4 \log\left(\frac{4}{10}\right) + 10 \log\left(\frac{10}{10}\right) + \cdots + 16 \log\left(\frac{16}{10}\right) \right] = 11.001 .$$

The degrees of freedom are $(6 - 1) - 0 = 5$ in both cases, and more extreme values than the ones observed occur with probability 0.065 (for X^2) and 0.051 (for Y^2) when the expected frequency in each of the six categories is $p_0 = 1/6$ ($E(X_i) = np_0 = 60(1/6) = 10$). These significance probabilities lead to the inference that these six types of individuals are not likely to occur with equal frequency in the population sampled.

A more sophisticated conjecture based on Mendelian patterns of inheritance is that three genes independently determine the six different types of individuals (hypothesis *II*). These genes, called *A*, *B*, and *C* for convenience, have population frequencies denoted as denoted as P_A, P_B, and P_C. Based on these gene frequencies and statistical independence, the six electrophoretic types would have frequencies P_A^2, $2P_A P_B$, P_B^2, $2P_B P_C$, $2P_A P_C$, and P_C^2 for types AA, AB, BB, BC, AC, and CC, respectively. To generate the expected values to test hypothesis *II*, specific values for P_A, P_B, and P_C are necessary. The usual source of values is the sampled data. For the EAP data, the proportion of *A* genes is $\hat{P}_A = (2(4) + 10 + 14)/2(60) = 0.267$, and the proportion of *B* genes is $\hat{P}_B = (2(6) + 10 + 10)/2(60) = 0.267$; $\hat{P}_C = 1 - \hat{P}_A - \hat{P}_B = 0.466$ is necessarily the proportion of *C* genes. Then, the estimated proportions expected in each cell of the table are $\hat{P}(AA) = \hat{P}_A^2 = (0.267)^2 = 0.071$, $\hat{P}(AB) = 2\hat{P}_A\hat{P}_B = 2(0.267)(0.267) = 0.142$, $\hat{P}(AC) = 2\hat{P}_A\hat{P}_C = 2(0.267)(0.466) = 0.249$, and so on (Table 8.4). The chi-square statistics to evaluate hypothesis *II* using estimated values based on these estimated frequencies (Table 8.4) are

$$X^2 = \frac{(4 - 4.267)^2}{4.267} + \cdots + \frac{(16 - 13.067)^2}{13.067} = 3.320$$

or

$$Y^2 = 2\left[4 \log\left(\frac{4}{4.267}\right) + \cdots + 16 \log\left(\frac{16}{13.067}\right) \right] = 3.400 .$$

The degrees of freedom in both cases are $(k - 1) - s$ where k is the number of cells in the table and s is the number of independent estimates necessary to define the hypothesis that generated the estimated values—namely $\hat{P}_A$ and $\hat{P}_B$—giving $(k - 1) - s = (6 - 1) - 2 = 3$ degrees of freedom. Because the mean of a chi-square distribution equals its associated degrees of freedom, a value of 3.320 is likely to occur by chance variation under hypothesis *II*; in fact, *p*-value = 0.345 for X^2 and *p*-value = 0.334 for Y^2.

Using the observed data to help establish the expected values involves a certain amount of circular reasoning. Values estimated from the observed data will more closely resemble the data than values generated from other considerations. Using these estimated values then reduces the chi-square measure of fit, producing an apparent improvement in correspondence between the observed and the estimated values. To compensate for this artificial increase in goodness-of-fit, statistical theory dictates that the degrees of freedom $(k-1)$ be correspondingly reduced. One degree of freedom is subtracted for each independent estimate made to specify the hypothesis generated values. For hypothesis *I* no estimates were made ($k = 6$ and $s = 0$), and for hypothesis *II* two estimates ($\hat{P}_A$ and $\hat{P}_B$) were necessary ($k = 6$ and $s = 2$), giving $(6-1) - 0 = 5$ and $(6-1) - 2 = 3$ degrees of freedom, respectively.

A General Consideration

A chi-square statistic is essentially made up of three components: degrees of freedom, sample size (n), and a measure of lack of fit. A bit of algebraic manipulation [1] shows an approximate expression that identifies these components of a Pearson chi-square statistic. That is,

$$\text{the expected chi-square value} \approx \text{degrees of freedom} + n(\text{lack of fit})^2 . \quad (8.9)$$

Several issues are direct consequences of viewing the chi-square statistic in this form. If the lack-of-fit component is zero or is small, the chi-square statistic is expected to be close to its associated degrees of freedom. Conversely, if the lack-of-fit component is not zero, then the magnitude of the chi-square statistic is a function of sample size. Large sample sizes produce large chi-square statistics. Therefore, for sufficiently large n, any lack of fit is reflected by elevated values of chi-square statistic and clearly, the greater the lack-of-fit component, the larger the chi-square value for any sample size. Last, expression 8.9 shows that direct comparisons of chi-square values are useless unless the chi-square values are derived from tables with the same dimensions and sample sizes.

Two-Way Tables: 2-by-2 Case

The simplest two-way contingency table consists of four cells formed by classifying observations by two binary variables. A great deal can be learned by careful study of this fundamental 2-by-2 table.

Hypergeometric Distribution

When a population of N individuals measured for the presence or absence of a specific characteristic is independently sampled without replacement, four possibilities occur with respect to the classification of each member of the population. Each individual possesses or does not possess the characteristic and each individual is either included or not included in the sample. "Without replacement" means that once an individual is sampled, that person cannot be selected again. For N individuals, m with the characteristic and $(N-m)$ without the characteristic, a sample of size n produces a description of the population in terms of a 2-by-2 contingency table, given in Table 8.5.

The variable X represents the number of individuals in the sample with the characteristic where (to repeat) each sample observation is selected independently, without replacement. The word *independently* in this context means the probability that

T A B L E **8.5**

The notation for a sample of size n from a population of N individuals

	With the Characteristic	Without the Characteristic	Total
Sampled	X	$n - X$	n
Not sampled	$m - X$	$(N - m) - (n - X)$	$N - n$
Total	m	$N - m$	N

any individual is selected depends only on the composition of the population at the time of selection and does not depend in any way on whether the individual possesses or does not possess the characteristic. The minimum value of X is the larger of 0 or $n + m - N$, and the maximum value is the smaller of the values m or n. The probability distribution describing X, called the hypergeometric distribution, is given by

$$P(X = x) = \frac{\binom{m}{x}\binom{N - m}{n - x}}{\binom{N}{n}} = \frac{\binom{n}{x}\binom{N - n}{m - x}}{\binom{N}{m}}.$$

(8.10)

Hypergeometric probabilities can be calculated or found in tables [2] when population parameters n, m, and N are known. Using these probabilities, the expected value and variance of the hypergeometric distribution are (derived in [3] or [4])

$$expected\ value = E(X) = n\left(\frac{m}{N}\right) \text{ and}$$

(8.11)

$$variance = \sigma_X^2 = n\left(\frac{m}{N}\right)\left(1 - \frac{m}{N}\right)\left(\frac{N - n}{N - 1}\right).$$

(8.12)

The term $(N - n)/(N - 1)$ is called the finite population correction factor. The variance of X decreases as the proportion of the population sampled increases (n/N becomes closer to 1.0). Ultimately, when $n = N$, the variance is zero because X must equal exactly m for any sample ($\sigma_X^2 = 0$). The hypergeometric probabilities $P(X = x)$, expected value, and variance require that the marginal frequencies (i.e., the values m, $N - m$, n, and $N - n$) of the 2-by-2 table are fixed in advance of selecting the n sample observations; that is, not subject to random variation.

Fisher's exact test is an application of these hypergeometric probabilities. A concrete example of a hypergeometric distribution comes from a casino lottery game called Keno. A player chooses n different numbers from a possible $N = 80$ consecutive numbers; the casino independently selects without replacement $m = 20$ numbers and X represents the number of selections chosen by both the player and the casino. Suppose a player selects $n = 8$ values; then the 2-by-2 table describing this Keno population is given by Table 8.6.

T A B L E **8.6**

Sampling without replacement illustrated by the game Keno

		Casino		Total
		Selected	Not Selected	
Player	Selected	X	$8 - X$	8
	Not Selected	$20 - X$	$52 + X$	72
	Total	20	60	80

The probabilities generated by this specific version of Keno are given by

$$P(X = x) = \frac{\binom{20}{x}\binom{60}{8-x}}{\binom{80}{8}}$$

(8.13)

for $x = 0, 1, 2, \cdots, 8$. The payoffs are based on values of the hypergeometric variable X with probability $X = 8$ paying \$25,000 with probability $P(X = 8) = 0.00000435$. By contrast, the probability of choosing six correct values in the California state lottery "Pick-6" is also given by the hypergeometric distribution. Because N equals 51, $m = 6$, and $n = 6$, then

$$P(X = 6) = \frac{\binom{6}{6}\binom{45}{0}}{\binom{51}{6}} = \frac{1}{18,009,460} = 0.000000056$$

but pays millions to a winner. For Keno and other lotteries, the marginal frequencies are unequivocally fixed by the rules of the game. Outside the artificial world of gambling, realistic situations rarely exist in which both marginal frequencies are fixed and known in advance of selecting a sample.

Fisher's Exact Test

Fisher's exact test employs hypergeometric probabilities to find the probability of the occurrence of the observed 2-by-2 table and all more extreme results, when the sampling is unrelated to the characteristic under study. The sum of the probabilities associated with "more extreme results" quantifies the likelihood that the observed values represent a random sample from a hypergeometric distribution.

The following tables and probabilities clarify the phrase *more extreme results*. If $N = 20$, $m = 10$, and $n = 6$, then $x = 0, 1, 2, 3, 4, 5$, and 6 gives seven possible 2-by-2 tables. They are:

	B_1	B_2
A_1	0	6
A_2	10	4

	B_1	B_2
A_1	1	5
A_2	9	5

	B_1	B_2
A_1	2	4
A_2	8	6

	B_1	B_2
A_1	3	3
A_2	7	7

	$x = 0$	$x = 1$	$x = 2$	$x = 3$
$P(X = x)$	0.005	0.065	0.244	0.372

	B_1	B_2
A_1	4	2
A_2	6	8

	B_1	B_2
A_1	5	1
A_2	5	9

	B_1	B_2
A_1	6	0
A_2	4	10

	$x = 4$	$x = 5$	$x = 6$
$P(X = x)$	0.244	0.065	0.005

where B_1 means possesses the characteristic (B_2 = not) and A_1 indicates an observation is included among the sampled values (A_2 = not). The variable X is the count of occurrences in which the characteristic is observed in the sample and A_1 and B_1 occur. When the count of simultaneous occurrences of A_1 and B_1 is large or small, either a rare event has occurred by chance or the sampling of the observations is related to whether the observations possess or do not possess the characteristic, causing the extreme value of X. The probability of observing $X = 5$, from the example, is 0.065, and the probability of an equal or larger value (i.e., $X = 5$ or 6) is 0.065 + 0.005 = 0.070, giving the value $p = 0.070 + 0.070 = 0.140$ as the probability of a more extreme result ($X \geq 5$ or $X \leq 1$) when the sampling is unrelated to the characteristic, sometimes called a two-sided test. Here *more extreme* is defined as results with associated probabilities equal to or smaller than the probability associated with the observed table. Such a p-value indicates the likelihood that the observed association between the two binary categorical variables arose by chance, but it should be noted that the way tables are determined to be extreme is not uniform and often depends on the applied situation. Conventionally, the hypothesis of independence is rejected when the p-value falls below 0.05.

Identification tests provide a situation in which the marginal frequencies are fixed, making Fisher's exact test appropriate. For example, a medical student is asked to visually differentiate between two tumor types, A and B. A test is designed with 30 specimens classified by microscopic criteria which determines the type. Fifteen of each type of tumor are prepared and the student declares 15 type A and the remaining 15 type B producing the data in Table 8.7.

For this particular data set, five A-type tumors were correctly identified. To assess whether this result is better than chance, the Fisher's exact test applies (hypergeometric probabilities). The probability that less than six A-type tumors are identified when the person doing the identification has no ability to differentiate A-type from B-type is $P(X = 0) + P(X = 1) + P(X = 2) + P(X = 3) + P(X = 4) + P(X = 5) = P(X \leq 5) = 0.072$ (expression 8.10).

T A B L E **8.7**

Student identification of tumor types labeled A and B

	A	B	Total
A	5	10	15
B	10	5	15
Total	15	15	30

Note: A and B are the determination by the student where it is
known that there are 15 A and 15 B type tumors.

A natural question arises: What if the marginal frequencies of a 2-by-2 table are not fixed? Do the hypergeometric probabilities apply? The answer to this question is not simple and the complex issues surrounding the application of Fisher's exact test when the marginal frequencies are not fixed is not discussed [5]. Nevertheless, the application of Fisher's exact test is justified, produces reliable results, and is often recommended when sample sizes are small.

Chi-Square Test: 2-by-2 Tables

Another approach to identifying the presence of an association in a 2-by-2 table uses a chi-square statistic. A chi-square test applies to tables with fixed as well as random marginal frequencies. A general notation for a 2-by-2 table with binary variables A (A_1 and A_2) and B (B_1 and B_2) is given in Table 8.8. The symbol f_{ij} represents the frequency observed in the ith, j^{th} cell; $f_{i.} = f_{i1} + f_{i2}$ represents the ith row marginal frequency; $f_{.j} = f_{1j} + f_{2j}$ represents the jth column marginal frequency; and $f_{11} + f_{12} + f_{21} + f_{22} = f_{..} = n$ represents the total number of observations. The position of the "dot" indicates the sum over either a row or a column in the table.

The hypothesis of independence of categorical variables A and B is expressed as

$$H_0: \ p_{ij} = P(A_iB_j) = P(A_i)P(B_j) = p_iq_j$$

where p_{ij} represents the probability that an observation falls into a specific cell (i, j), p_i represents the probability an observation falls into category A_i (row i), and q_j represents the probability an observation falls into category B_j (column j). When probabilities p_i and q_j are unknown, estimates are made to evaluate the hypothesis of independence H_0. These probabilities are estimated from the marginal frequencies where $\hat{p}_i = f_{i.}/n$ and $\hat{q}_j = f_{.j}/n$ are derived from the observed data. When variables A and B are independent, the estimated count for a specific cell is given by

$$\hat{f}_{ij} = n\hat{p}_i\hat{q}_j = n\left(\frac{f_{i.}}{n}\right)\left(\frac{f_{.j}}{n}\right) = \frac{f_{i.}f_{.j}}{n}.$$

(8.14)

T A B L E **8.8**

Notation for a 2-by-2 table

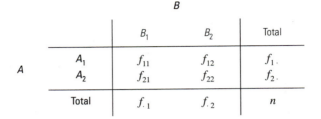

Using these estimated values, a test statistic for the hypothesis of independence between categorical variables A and B is

$$X^2 = \sum_i \sum_j \frac{(f_{ij} - \hat{f}_{ij})^2}{\hat{f}_{ij}}$$

$$= \frac{\left(f_{11} - \frac{f_{1.}f_{.1}}{n}\right)^2}{\frac{f_{1.}f_{.1}}{n}} + \frac{\left(f_{12} - \frac{f_{1.}f_{.2}}{n}\right)^2}{\frac{f_{1.}f_{.2}}{n}} + \frac{\left(f_{21} - \frac{f_{2.}f_{.1}}{n}\right)^2}{\frac{f_{2.}f_{.1}}{n}} + \frac{\left(f_{22} - \frac{f_{2.}f_{.2}}{n}\right)^2}{\frac{f_{2.}f_{.2}}{n}}$$

(8.15)

and X^2 has an approximate chi-square distribution with $(k-1)-s = (4-1)-2 = 1$ degree of freedom. The value s is two because $\hat{p}_1$ and $\hat{q}_1$ are estimated from the data, whereas $\hat{p}_2 = 1 - \hat{p}_1$ and $\hat{q}_2 = 1 - \hat{q}_1$ follow automatically.

The X^2-statistic applied to a 2-by-2 table (expression 8.15) algebraically simplifies to a "shortcut" formula of

$$X^2 = \frac{n(f_{11}f_{22} - f_{12}f_{21})^2}{f_{1.}f_{.1}f_{2.}f_{.2}},$$

(8.16)

which saves some computation.

Test Based on Proportions

Another approach to analyzing data classified into a 2-by-2 table consists of testing the hypothesis that the probability an observation belongs to a particular A_i-category is the same for both B-categories (columns; B_1 and B_2), or $P_1 = P(A_1 | B_1) = P(A_1 | B_2) = P_2$. A similar hypothesis can be developed in terms of rows where $P_1' = P(B_1 | A_1) = P(B_1 | A_2) = P_2'$. To test the hypothesis that $P_1 = P_2$, the value

$$z = \frac{\hat{P}_1 - \hat{P}_2}{\sqrt{\hat{P}(1 - \hat{P}) \left[\frac{1}{f_{.1}} + \frac{1}{f_{.2}} \right]}}$$

(8.17)

T A B L E **8.9**

Cases of ankylosing spondylitis and controls distributed into a 2-by-2 table

	Cases	Controls	Total
Trait	$f_{11} = 16$	$f_{12} = 6$	$f_{1.} = 22$
No trait	$f_{21} = 26$	$f_{22} = 36$	$f_{2.} = 62$
Total	$f_{.1} = 42$	$f_{.2} = 42$	$n = 84$

is calculated where $\hat{P}_1 = f_{11}/f_{.1}$, $\hat{P}_2 = f_{12}/f_{.2}$, and $\hat{P} = f_{1.}/n$. The value z has an approximate standard normal distribution when $P_1 = P_2$. This approach offers nothing new. The hypothesis of equality of column probabilities is a special case of statistical independence and z^2 is identical to X^2, for the 2-by-2 table (expression 8.15). The test analogous to expression 8.17 for $P_1' = P_2'$ also yields the same value of z^2. The test of $P_1 = P_2$ or $P_1' = P_2'$ is referred to as a test of homogeneity. As shown, the test for independence is the same as the test of homogeneity but the inference is focused on a somewhat different issue.

The data in Table 8.9 illustrate the chi-square analysis (expression 8.15) of a 2-by-2 contingency table. A disease called ankylosing spondylitis is thought to be associated with a specific genetic trait (in the HLA region). A sample of 42 cases (diseased) and 42 control individuals analyzed for the presence or absence of this trait results in Table 8.9. Under the hypothesis that the genetic trait is not associated with the case/control status, the estimated values are given in Table 8.10. The chi-square statistic is

$$X^2 = \frac{(16-11)^2}{11} + \frac{(6-11)^2}{11} + \frac{(26-31)^2}{31} + \frac{(36-31)^2}{31} = 6.158$$

or, equally,

$$\frac{84[(16)\,(36) - (6)\,(26)]^2}{(42)\,(22)\,(62)\,(42)} = 6.158 .$$

T A B L E **8.10**

Estimated numbers of cases of ankylosing spondylitis under the hypothesis of independence distributed into a 2-by-2 table

	Cases	Controls	Total
Trait	$\hat{f}_{11} = 11$	$\hat{f}_{12} = 11$	$f_{1.} = 22$
No trait	$\hat{f}_{21} = 31$	$\hat{f}_{22} = 31$	$f_{2.} = 62$
Total	$f_{.1} = 42$	$f_{.2} = 42$	$n = 84$

Because X^2 has an approximate chi-square distribution with one degree of freedom when no relationship exists between case/control status and presence or absence of the trait, the probability of observing a result more extreme than 6.158 is p-value = 0.013. The small significance probability indicates that the trait is likely associated with the disease.

The likelihood ratio chi-square approach gives

$$Y^2 = 2\left[16 \log\left(\frac{16}{11}\right) + 6 \log\left(\frac{6}{11}\right) + 26 \log\left(\frac{26}{31}\right) + 36 \log\left(\frac{36}{31}\right)\right] = 6.337 \ .$$

Because Y^2 also has an approximate chi-square distribution with one degree of freedom when no association exists, the probability of observing a result more extreme than 6.337 is p-value = 0.012.

Also note that $\hat{P}(trait \mid case) = \hat{P}_1 = 16/42 = 0.381$, $\hat{P}(trait \mid control) = \hat{P}_2 = 6/42 = 0.143$, and $\hat{P} = 22/84 = 0.262$, giving

$$z = \frac{.381 - 0.143}{\sqrt{(0.262)\,(0.738)\left[\frac{1}{42} + \frac{1}{42}\right]}} = 2.482$$

and $z^2 = (2.482)^2 = 6.158 = X^2$.

Adjusted Chi-Square Statistic

The application of a chi-square analysis to a 2-by-2 table involves a degree of approximation but generally produces accurate values of the associated significance probabilities (p-values). The chi-square statistic works well for large sample sizes but works less well as the sample size decreases. A fairly conservative rule of thumb states that a chi-square statistic gives an accurate result (p-value) as long as all four null-hypothesis-generated frequencies exceed five.

Another version of the chi-square statistic occasionally recommended for small sample situations—represented as X_a^2, a for adjusted—is

$$X_a^2 = \frac{n(\,|\,f_{11}f_{22} - f_{12}f_{21}\,| - n/2)^2}{f_{1\cdot}\,f_{\cdot 1}\,f_{2\cdot}\,f_{\cdot 2}} \ . \tag{8.18}$$

The value X_a^2 also has an approximate chi-square distribution with one degree of freedom. The value X_a^2 for the ankylosing spondylitis data is

$$X_a^2 = 84(\,|\,(16)\,(36) - (6)\,(26)\,| - 84/2)^2/(42)\,(22)\,(62)\,(42) = 4.988 \ ,$$

producing a p-value = 0.026.

The value X_a^2 is necessarily smaller than X^2, producing a conservative test of the hypothesis of no association; conservative in the sense that if a hypothesis is rejected using X_a^2 (expression 8.18), then it will also be rejected using X^2 unadjusted (expression 8.15). Some argue that X_a^2 is too conservative and should not be used (e.g., [5]). The unadjusted chi-square statistic is used in the following analyses and examples.

Matched Pairs

Matched pairs is a name given to a specific pattern of collecting data and produces a specific type of 2-by-2 contingency table. Matched pair observations are correlated within a pair where the chi-square analysis of two binary variables (expression 8.15) requires observations to be independent. The relationship of sampled observations should not be confused with independence or nonindependence of the categorical variables. An analysis of a contingency table explores the relationship between the categorical variables and the data collection pattern dictates the analytic pattern. Matched pair designs apply to a large number of situations but are discussed in terms of a clinical trial to simplify terminology.

Suppose a clinical trial consists of testing two drugs (symbolized by S and T) where patients take S on one occasion and T on another. For each occasion the patient reports whether the drug was efficacious or not, producing matched pair data consisting of two correlated observations from the same individual. Such data are displayed in Table 8.11, where the entries in the table, represented by a, b, c and d, are the numbers of pairs classified by the four possible responses. Here the observations are likely related within patient (same individual) but the matched pair analysis focuses on the question of association between drugs S and T.

McNemar's Test

The number of times a study participant reported both drugs to be effective, labeled a in Table 8.11, does not provide information about the relative merits of the two treatments. In other words, differences between drugs S and T cannot be investigated in a group of individuals where no differences are perceived. Similarly, the pairs, represented by d, are also noninformative because all patients declared drugs S and T identical (neither efficacious). For the two categories where one or the other of the two drugs is declared efficacious, the expected values are estimated by $(b + c)/2$, under the hypothesis that drugs S and T have the same impact (i.e., H_0: $E(b) = E(c)$). That is, when treatments S and T do not differ, then S is declared superior to T or T is

T A B L E **8.11**

Notation for data collected in a matched pair design

		Drug S		
		Efficacious	Not Efficacious	Total
Drug T	Efficacious	a	b	$a + b$
	Not efficacious	c	d	$c + d$
	Total	$a + c$	$b + d$	n

T A B L E   **8.12**

Data from a matched pair design to study the risk of lung cancer among beauticians

		Lung Cancer Cases		
		Beautician	Nonbeautician	Total
Controls (without cancer)	Beautician	0	1	1
	Nonbeautician	6	169	175
	Total	6	170	176

declared superior to S only by chance. In the long run, the numbers of such pairs are expected to be equally frequent. Using these estimates, a chi-square statistic X^2 (expression 8.1) produces for a matched pair analysis

$$X^2 = \frac{(b - (b+c)/2)^2}{(b+c)/2} + \frac{(c - (b+c)/2)^2}{(b+c)/2} = \frac{(b-c)^2}{b+c}.$$
(8.19)

The accuracy of this chi-square statistic, referred to as McNemar's test, is improved by using the adjusted form of

$$X_a^2 = \frac{(|b - c| - 1)^2}{b+c}.$$
(8.20)

Both forms have approximate chi-square distributions with one degree of freedom when $E(b) = E(c)$; that is, no difference between drugs S and T.

To illustrate McNemar's test, consider data collected to study the risk of lung cancer among beauticians. A series of 176 lung cancer deaths were matched to individuals of the same sex, race, and age who died of causes other than cancer during the same month [6]. The number of beauticians and nonbeauticians among the 176 pairs of individuals sampled is given in Table 8.12. Under the hypothesis that beauticians are not at excess risk for lung cancer, the chi-square statistic is

$$X^2 = \frac{(6-1)^2}{(6+1)} = 3.571 \quad \text{with } p\text{-value} = 0.059$$

or

$$X_a^2 = \frac{(|6-1|-1)^2}{(6+1)} = 2.286 \ p\text{-value} = 0.131.$$

Unadjusted and adjusted chi-square statistics differ most dramatically when the sample size b and c are small (e.g., $b + c = 7$). The matched pair data give some indication of an association between lung cancer and being employed as a beautician.

Confidence Interval

The evaluation of a set of matched pairs with a chi-square test is not influenced by the frequency of like pairs, as noted. The number of like pairs, a and d, is not included in the calculations and, therefore, does not play a role in assessing the hypothesis of no association. However, the number of like pairs is relevant to the analysis of matched pair data when confidence intervals are estimated or power calculations are made. To illustrate, using the notation in Table 8.11, a natural measure of the impact of two drugs is the difference in frequency of efficacious outcomes, or

$$\hat{p}_2 - \hat{p}_1 = \frac{b - c}{n} \quad \text{where} \quad \hat{p}_1 = \frac{a + c}{n} \quad \text{and} \quad \hat{p}_2 = \frac{a + b}{n} .$$

(8.21)

An estimate of the variance of $\hat{p}_2 - \hat{p}_1$ is

$$variance(\hat{p}_2 - \hat{p}_1) = \frac{(a + d)(b + c) + 4bc}{n^3} .$$

(8.22)

Although the estimated difference in proportions $\hat{p}_2 - \hat{p}_1$ is not directly affected by the values a and d, the variance of the difference clearly depends on the numbers of all types of matched pairs (like and unlike). An approximate confidence interval, for example, $\hat{p}_2 - \hat{p}_1 \pm z_{1-\alpha/2}\sqrt{variance(\hat{p}_2 - \hat{p}_1)}$ depends on the values of a and d. For the lung cancer matched pair data, the difference in frequency of beauticians between cases and controls is $\hat{p}_2 - \hat{p}_1 = 1/176 - 6/176 = -5/176 = -0.028$ with standard error = 0.015, giving an approximate 95% confidence interval of $(-0.058, 0.001)$. This interval will differ if the 176 observations are distributed differently in the 2-by-2 table. It is reasonable that a data set with 1000 like pairs would produce a more precise confidence interval than a data set of 10 like pairs, regardless of the pattern observed in the unlike pairs. This influence is reflected by the estimated variance.

Measuring Association in 2-by-2 Tables

A chi-square statistic is chiefly used to evaluate the likelihood an association arose by chance. More appropriate statistics measure the magnitude of an association, because an X^2-value does not directly indicate the strength of an association without knowledge of the sample size and the degrees of freedom, as noted (expression 8.9). The following are a few of the many possible ways to describe an association in 2-by-2 tables:

Pearson's Coefficient

The Pearson's coefficient of contingency is defined as

$$\hat{P} = \sqrt{\frac{X^2}{X^2 + n}}$$

(8.23)

where X^2 is the previously discussed chi-square statistic. The range of $\hat{P}$ is between 0 and 1. For the genetic case/control data (Table 8.9), $\hat{P} = 0.261$.

Yule's Measure of Association

Yule's measure of association is

$$\hat{Y} = \frac{f_{11}f_{22} - f_{12}f_{21}}{f_{11}f_{22} + f_{12}f_{21}}.$$

(8.24)

The range of $\hat{Y}$ is $-1 \le \hat{Y} \le 1$, taking on the value 1 when f_{12} or f_{21} is zero and the value -1 when f_{11} or f_{22} is zero. A Yule measure of association equal to 1 or -1 is generally an overstatement of the degree of association because $\hat{Y} = 1$ or -1 does not imply perfect association (e.g., $\hat{Y} = 1$ while $f_{12} \ne 0$ but $f_{21} = 0$). When $f_{11}f_{22} = f_{12}f_{21}$, $\hat{Y} = 0$, which occurs when the data reflect perfect independence (i.e., $f_{ij} = f_{i.}f_{.j}/n$). Again using the ankylosing spondylitis data (Table 8.9) $\hat{Y} = 0.574$. The estimated variance of $\hat{Y}$ is

$$variance(\hat{Y}) = \frac{1}{4}\left(1 - \hat{Y}^2\right)^2\left(\frac{1}{f_{11}} + \frac{1}{f_{12}} + \frac{1}{f_{21}} + \frac{1}{f_{22}}\right).$$

(8.25)

An approximate 95% confidence interval based on $\hat{Y} = 0.574$ is

$$\hat{Y} \pm 1.96\sqrt{variance(\hat{Y})},$$

yielding (0.216, 0.931).

Correlation Coefficient

A 2-by-2 contingency table describes a series of n pairs of observations. Each pair consists of two values, which are coded zero or one producing four possible pairs—namely, (0, 0), (0, 1), (1, 0), and (1, 1). There are f_{11} pairs (0, 0), f_{12} pairs (0, 1), f_{21} pairs (1, 0), and f_{22} pairs (1, 1), summarized in a 2-by-2 table. Entering each of the n pairs into the usual expression for the product-moment correlation coefficient (expression 1.7) yields, as always, a measure of association between 1 and -1. Because of the simple nature of the data (zeros and ones), the product-moment correlation coefficient formula simplifies to

$$\hat{\varphi} = \frac{f_{11}f_{22} - f_{12}f_{21}}{\sqrt{f_{1.}f_{.1}f_{2.}f_{.2}}} = \sqrt{\frac{X^2}{n}}.$$

(8.26)

Employing the expression for product-moment correlation guarantees that $\hat{\varphi}$ is between -1 and 1. However, $\hat{\varphi}$ is not related to a linear regression analysis like the Pearson's product-moment correlation coefficient r_{XY} applied to two continuous variables x and y. For the case/control data (Table 8.9), $\hat{\varphi} = 0.271$.

Odds Ratio

The odds ratio measure of association is estimated by

$$\hat{OR} = \frac{f_{11}/f_{21}}{f_{12}/f_{22}} = \frac{f_{11}f_{22}}{f_{12}f_{21}}$$

(8.27)

because an estimate of the odds is the number of times an event occurs divided by the number of times it does not occur (e.g., f_{11}/f_{21} or f_{12}/f_{22}). The value $\hat{OR}$ ranges from 0 to infinity, with $\hat{OR} = 1.0$ indicating perfect statistical independence. Because the distribution of $\hat{OR}$ is asymmetric, confidence intervals are typically constructed in terms of $\log(\hat{OR})$ rather than directly from $\hat{OR}$. The value $\log(\hat{OR})$ ranges from $-\infty$ to ∞ and has a more symmetric distribution that is more accurately approximated by a normal distribution. An estimate of the variance of $\log(\hat{OR})$ is

$$variance[\log(\hat{OR})] = \frac{1}{f_{11}} + \frac{1}{f_{12}} + \frac{1}{f_{21}} + \frac{1}{f_{22}},$$

(8.28)

giving an approximate 95% confidence interval for $\log(OR)$ of $\log(\hat{OR}) \pm 1.96\sqrt{variance[\log(\hat{OR})]}$. For the genetic case/control data (Table 8.9);

$$\hat{OR} = 3.692$$
$$\log(\hat{OR}) = 1.306$$

and

$$variance[\log(\hat{OR})] = 0.295 .$$

The approximate 95% confidence interval for $\log(OR)$ based on these estimates is $(0.241, 2.372)$ and an approximate 95% confidence interval for OR is $(e^{0.241}, e^{2.372}) = (1.272, 10.714)$.

The odds ratio has four notable properties:

1　A probabilistic interpretation—the odds ratio is an estimate of the ratio of probabilities from a 2-by-2 table, $OR = (p_{11}/p_{21})/(p_{12}/p_{22})$.

2　Multiplying rows or columns of a 2-by-2 table by a constant value does not change the value of the odds ratio.

3　Reversing the coding of a binary variable gives the reciprocal of the odds ratio.

4　The estimated odds ratio $\hat{OR}$ is symmetric in the sense that $\hat{OR}$ and $1/\hat{OR}$ ($\log(\hat{OR})$ and $-\log(\hat{OR})$) measure the same degree of association in the opposite directions.

Also note that Yule's $\hat{Y}$ (expression 8.24) and the odds ratio are related—$\hat{Y} = (\hat{OR} - 1)/(\hat{OR} + 1)$.

Another estimate of the odds ratio is

$$\hat{OR}^* = \frac{(f_{11} + \frac{1}{2})\,(f_{22} + \frac{1}{2})}{(f_{12} + \frac{1}{2})\,(f_{21} + \frac{1}{2})} .$$

(8.29)

The value $\hat{OR}^*$ measures association in the same way as $\hat{OR}$ but remains defined when either f_{12} or f_{21} is zero. For the genetic case/control data (Table 8.9), $\hat{OR}^* = 3.496$. Adding $\frac{1}{2}$ to each cell improves the accuracy of the estimate (reduces bias) and can be justified by a rather technical argument [7]. The odds ratio relates in a natural way to the log-linear and logistic approaches to analyzing multivariate data used in the next two chapters.

Two-Way Tables: *r-by-c* Case

Tables created from two variables, represented as *A* and *B*, constructed from categories labeled $A_1, A_2, \cdots, A_r$ and $B_1, B_2, \cdots, B_c$ (*r* for rows and *c* for columns), are analyzed in much the same manner as a 2-by-2 table. The notation for a general *r*-by-*c* contingency table is displayed in Table 8.13.

Like the 2-by-2 table, $\hat{p}_i = f_i/n$ is an estimate of the probability that a random observation falls into category A_i, and similarly $\hat{q}_j = f_j/n$ is an estimate of the probability that a random observation falls into category B_j. When variables *A* and *B* are independent, the expected frequencies associated with each cell are again estimated by

$$\hat{f}_{ij} = n\hat{p}_i\hat{q}_j = \frac{f_i \cdot f_j}{n}.$$

(8.30)

A chi-square statistic calculated to assess the hypothesis of no association—each cell contributes a value of $(f_{ij} - \hat{f}_{ij})^2/\hat{f}_{ij}$—is

$$X^2 = \sum_{all\ cells} \frac{(f_{ij} - [f_i \cdot f_j/n])^2}{f_i \cdot f_j/n}.$$

(8.31)

The value X^2 has an approximate chi-square distribution with $(r-1)(c-1)$ degrees of freedom when variables *A* and *B* are independent. The degrees of freedom are $(rc-1)-s$ where $s=(r-1)+(c-1)$, because $r-1$ values $\hat{p}_1, \hat{p}_2, \cdots, \hat{p}_{r-1}$, $(\hat{p}_r = 1 - \hat{p}_1 - \hat{p}_2 - \cdots - \hat{p}_{r-1}$ follows automatically) and $c-1$ values $\hat{q}_1, \hat{q}_2, \cdots, \hat{q}_{c-1}$ $(\hat{q}_c = 1 - \hat{q}_1 - \hat{q}_2 - \cdots - \hat{q}_{c-1}$ follows automatically) must be estimated to specify the

T A B L E **8.13**
Notation for a general *r*-by-*c* contingency table

Categories	B_1	B_2	B_3	$\cdots$	B_c	Total
A_1	f_{11}	f_{12}	f_{13}	$\cdots$	f_{1c}	$f_1 \cdot$
A_2	f_{21}	f_{22}	f_{23}	$\cdots$	f_{2c}	$f_2 \cdot$
A_3	f_{31}	f_{32}	f_{33}	$\cdots$	f_{3c}	$f_3 \cdot$
.	.	.	.		.	.
.	.	.	.		.	.
.	.	.	.		.	.
A_r	f_{r1}	f_{r2}	f_{r3}	$\cdots$	f_{rc}	$f_r \cdot$
Total	$f \cdot_1$	$f \cdot_2$	$f \cdot_3$	$\cdots$	$f \cdot_c$	n

T A B L E **8.14**

Individuals classified by blood-pressure status and age

Blood Pressure

Age		B_1	B_2	B_3	Total
	A_1	23	6	10	39
	A_2	40	12	6	58
	A_3	65	22	4	91
Total		128	40	20	188

hypothesis of independence. Note that $(rc - 1) - s = (rc - 1) - [(r - 1) + (c - 1)] = rc - r - c + 1 = (r - 1)(c - 1)$. The previous 2-by-2 table is a special case—$(r - 1)(c - 1) = (2 - 1)(2 - 1) = 1$.

The following example illustrates a chi-square analysis of an r-by-c contingency table. The categorical variable A is age (A_1 = young = less than 45 years old, A_2 = middle age = between 45 and 65 years old, A_3 = old = greater than 65 years old) and variable B is blood-pressure status (B_1 = normal, B_2 = hypertensive, B_3 = under a physician's care). Classifying 188 sampled individuals yields the 3-by-3 table that is Table 8.14. The estimated frequencies (expression 8.30) assuming that age (A) and blood-pressure status (B) are independent are given in Table 8.15. The Pearson chi-square statistic, therefore, is

$$X^2 = \frac{(23 - 26.553)^2}{26.553} + \cdots + \frac{(4 - 9.681)^2}{9.681} = 13.227 .$$

The X^2-value has an approximate chi-square distribution with $(r - 1)(c - 1) = (3 - 1)(3 - 1) = 4$ degrees of freedom when age and blood-pressure status are unrelated. The probability of observing a more extreme result when A and B are unrelated

T A B L E **8.15**

Expected frequencies of individuals classified by blood-pressure status and age

Blood Pressure

Age		B_1	B_2	B_3	Total
	A_1	26.553	8.298	4.149	39
	A_2	39.489	12.340	6.170	58
	A_3	61.957	19.362	9.681	91
Total		128	40	20	188

is p-value $= 0.010$, indicating that age and blood-pressure status are likely associated in the population sampled.

The Likelihood Ratio Approach

An r-by-c table can be alternatively analyzed with a likelihood ratio approach. The test statistic measuring the fit of the data to the conjecture of independence is

$$Y^2 = 2\left[\sum_{all\ cells} f_{ij} \log\left(\frac{f_{ij}}{\hat{f}_{ij}}\right)\right].$$

(8.32)

Like X^2, Y^2 has an approximate chi-square distribution with $(r-1)(c-1)$ degrees of freedom when the row and column variables are independent, and again $\hat{f}_{ij}$ represents the estimate $f_{i.}\,f_{.j}/n$. Continuing the blood-pressure example, the likelihood ratio chi-square statistic

$$Y^2 = 2\left[23 \log\left(\frac{23}{26.553}\right) + \cdots + 4 \log\left(\frac{4}{9.681}\right)\right] = 11.899$$

has an approximate chi-square distribution with 4 degrees of freedom, producing a p-value of 0.018.

The same rule of thumb mentioned in connection with applying a chi-square distribution to a 2-by-2 table, applies to an r-by-c table. An accurate chi-square value requires all cells to have expected values that are not too small (conservatively; expected frequencies greater than 5). The "5 or more" rule is overly cautious and as long as most cells have expected values greater than 5 and only a few cells have expectations greater than 1 (no more than 20% of the cells), a chi-square approximation will not be misleading [8]. However, tables of sparse data where cell frequency estimates fall below 1 are not good candidates for a chi-square analysis.

Relationships within a Contingency Table

Once a significant association is found between the categorical variables that define a contingency table, pertinent questions usually remain unanswered. It is indeed wasteful to state that a nonrandom association exists and look no further into the relationships under study. When an overall association is established, describing the specific relationships within a contingency table is generally the next step.

One approach to exploring several hypotheses concerning the same table depends on an application of Bonferroni's inequality. The investigator a priori selects the number and makeup of a series of subtables to be analyzed. Suppose m, not necessarily independent, subtables are of interest. A chi-square test performed on each of these m subtables is declared "significant" if the chi-square value exceeds the critical point associated with a significance level of α/m. The overall error rate—that is, the probability of wrongly rejecting one or more null hypotheses among the m tests—will be less than α (Bonferroni's inequality).

T A B L E **8.16**

Individuals tested for papilloma virus classified by three racial categories

	White	Black	Asian	Total
Present	47	25	25	97
Absent	50	15	27	92
Total	97	40	52	189

A series of tests conducted to contrast the frequency of human papilloma virus among three racial groups (Table 8.16) serves to illustrate the application of Bonferroni's inequality to test several relationships within a single table. Three comparisons ($m = 3$) each conducted at $\alpha = 0.05/3 = 0.017$ require the chi-square percentile of 5.731 (i.e., $P(X^2 > 5.731) = 0.05/3 = 0.017$ for one degree of freedom) and

white versus black

$$X^2 = 137[47(15) - 25(50)]^2/(72 \cdot 65 \cdot 97 \cdot 40) = 2.241 < 5.731; \text{implies no difference}$$

white versus Asian

$$X^2 = 149[47(27) - 25(50)]^2/(72 \cdot 77 \cdot 97 \cdot 52) = 0.002 < 5.731; \text{implies no difference}$$

black versus Asian

$$X^2 = 92[25(27) - 25(15)]^2/(50 \cdot 42 \cdot 40 \cdot 52) = 1.896 < 5.731; \text{implies no difference}.$$

None of the three tests indicates evidence of a difference in the prevalence of the virus among the racial groups and the overall error rate associated with the three tests is less than $\alpha = 0.05$.

Computing Adjusted Residual Values

Another effective way to describe specific differences between estimated and observed values within an r-by-c table is to compute a series of adjusted residual values. Adjusted residual values help locate the categories responsible for a large observed chi-square statistic. The standardized residual value for each cell is defined as

$$r_{ij} = \frac{f_{ij} - f_{i.}f_{.j}/n}{\sqrt{f_{i.}f_{.j}/n}}$$

(8.33)

T A B L E **8.17**

Standardized residual values constructed from individuals classified by blood-pressure status and age

Blood Pressure

R_{ij}		B_1	B_2	B_3
	< 45	−0.690	−0.798	2.873
Age	45–65	0.081	−0.097	−0.069
	> 65	0.387	0.600	−1.826

with approximate variance

$$variance(r_{ij}) = (1 - f_i/n)(1 - f_{\cdot j}/n),$$ (8.34)

yielding an adjusted residual value of

$$R_{ij} = \frac{r_{ij}}{\sqrt{variance(r_{ij})}}.$$ (8.35)

The adjusted residual values R_{ij} have approximate standard normal distributions ($\mu = 0$ and $\sigma^2 = 1$) but are not independent. Nevertheless, they offer an opportunity to locate important deviations (such as $|R_{ij}| > 2$ from the hypothesis that generated the estimated values and to identify patterns within a contingency table.

The residual values (r_{ij}) for the blood-pressure data (Table 8.14) are given in Table 8.17 to illustrate. The adjusted residual values (R_{ij}) are given in Table 8.18. The R_{ij} values indicate fewer than expected young people and more than expected older individuals appear in the sample among normal and hypertensive individuals (B_1 and B_2), whereas an opposite and stronger association exists among the individuals classified as "under a physician's care" (B_3).

T A B L E **8.18**

Adjusted residual values constructed from individuals classified by blood-pressure status and age

Blood Pressure

R_{ij}		B_1	B_2	B_3
	< 45	−1.371	−1.010	3.413
Age	45–65	0.173	−0.131	−0.087
	> 65	0.953	0.941	−2.689

Measures of Association in *r*-by-*c* Tables

Parallel to the measures of association described for the 2-by-2 table, it is useful to compute a number between two bounds (usually –1 and 1 or 0 and 1) designed to indicate the degree of association found in an *r*-by-*c* contingency table.

Cramer's *V* and Tschuprow's *T*

Two simple measures of association formed from transformations of the chi-square statistic are Cramer's *V*-statistic

$$\hat{V} = \sqrt{\frac{X^2}{n \times \min(r - 1, c - 1)}} \tag{8.36}$$

and Tschuprow's *T*-statistic

$$\hat{T} = \sqrt{\frac{X^2}{n\sqrt{(r - 1)(c - 1)}}} . \tag{8.37}$$

For the hypertension data (Table 8.14),

$$\hat{V} = \sqrt{\frac{13.227}{188(2)}} = 0.188$$

and

$$\hat{T} = \sqrt{\frac{13.227}{188\sqrt{(2)}(2)}} = 0.188 , \tag{8.38}$$

indicating that $\hat{V} = \hat{T}$ for square tables ($r = c$); otherwise $\hat{V} > \hat{T}$. Also, if $r = c = 2$, then $\hat{V} = \hat{T} = \hat{\phi}$, as discussed for 2-by-2 tables (expression 8.26). These two measures do not have probabilistic interpretations. Furthermore, for complete association (all off-diagonal values equal to zero), $\hat{V}$ and $\hat{T}$ may not produce a value of 1.0. These two traditional measures of association frequently appear in computer output from contingency table analysis programs, but more useful ways to describe associations between categorical variables exist.

Lambda Measures of Association

A lambda measure of association contrasts two estimated probabilities:

1 The maximum probability of predicting that an observation belongs to a specific row

with

2 the maximum probability of predicting that an observation belongs to a specific row given that the observation belongs to specific column.

The first probability (1) is estimated by $f_{m.}/n$, and the second probability (2) is estimated by $\Sigma_j f_{mj}/n$ where m indicates the row containing the largest value in column J. The lambda measure contrasts these two probabilities in terms of the probabilities of not predicting that an observation belongs to the right row, or

$$\hat{\lambda}_c = \frac{probability\ of\ error\ (1) - probability\ of\ error\ (2)}{probability\ of\ error\ (1)}$$

$$= \frac{(1 - f_{m.}/n) - (1 - \Sigma_j f_{mj}/n)}{1 - f_{m.}/n}$$

$$= \frac{\Sigma_j f_{mj} - f_{m.}}{n - f_{m.}}. \tag{8.39}$$

When categorical variables A and B are exactly independent, and the largest values within each column correspond to the row with the largest total, then $\Sigma_j f_{mj}$ equals $f_{m.}$ and $\hat{\lambda}_c = 0$. Therefore, values of $\hat{\lambda}_c$ near zero indicate that the row variable cannot be predicted from knowledge of the column variable, and a value of $\hat{\lambda}_c$ different from zero reflects the degree of association.

Hypothetical data (Table 8.19) illustrate a situation where variable A (rows) can be predicted from knowledge of variable B (columns). The λ_c-measure is then

$$\hat{\lambda}_c = \frac{(30 + 25 + 20) - 52}{150 - 52} = 0.235 .$$

A more intuitive argument justifying $\hat{\lambda}_c = 0.235$ follows:

If no information from the columns (B_1, B_2, B_3) is used, then the most likely event is A_1 and $\hat{P}(A_1) = 52/150$ based on the row marginal frequencies. So, if one predicts category A_1, the probability of being wrong is

$$\hat{P}(error_1) = 1 - \frac{52}{150} = 0.653 .$$

T A B L E **8.19**

Hypothetical data to illustrate the λ-measure of association

		Variable B			
		B_1	B_2	B_3	Total
Variable A	A_1	7	25	20	52
	A_2	12	20	16	48
	A_3	30	10	10	50
	Total	49	55	46	150

If information from the columns (B_1, B_2, B_3) is used, then the probability of correctly predicting A is

$$\hat{P}(A) = \hat{P}(A \mid B_1)\hat{P}(B_1) + \hat{P}(A \mid B_2)\hat{P}(B_2) + \hat{P}(A \mid B_3)\hat{P}(B_3) = \frac{30}{150} + \frac{25}{150} + \frac{20}{150} = \frac{75}{150}$$

and the probability of being wrong is

$$\hat{P}(error_2) = 1 - \frac{75}{150} = 0.500 .$$

The gain in accuracy of prediction from knowledge of the column variable relative to no information from the column variable is again $\hat{\lambda}_c = (0.653 - 0.500)/0.653 = 0.235$. The lambda measure of association is simply the proportional reduction in error.

A similar lambda measure of association is developed when interest is focused on the rows of a contingency table, or

$$\hat{\lambda}_r = \frac{\sum_i f_{il} - f_{.l}}{n - f_{.l}} \tag{8.40}$$

where l indicates the column containing the largest value in row i_x. Using the hypothetical data (Table 8.19) gives

$$\hat{\lambda}_r = \frac{(30 + 20 + 25) - 55}{150 - 55} = 0.211 .$$

Often no reason exists to measure association in terms of rows or columns. A summary lambda measure is then created by a weighted average of $\hat{\lambda}_c$ and $\hat{\lambda}_r$, or

$$\hat{\lambda} = \frac{(n - f_{.l})\,\hat{\lambda}_r + (n - f_{m.})\hat{\lambda}_c}{(n - f_{.l}) + (n - f_{m.})}$$

$$= \frac{\sum_i f_{il} + \sum_j f_{mj} - f_{.l} - f_{m.}}{2n - f_{.l} - f_{m.}} . \tag{8.41}$$

Using the hypothetical data again gives

$$\hat{\lambda} = \frac{(30 + 20 + 25) + (30 + 25 + 20) - 55 - 52}{300 - 55 - 52} = \frac{43}{193} = 0.223 .$$

Measures of Association in *r*-by-*c* Tables with Ordinal Categories

Not surprisingly, a bit of manipulation produces measures of association for tables with ordinal categories. These measures do not require specific numerical values be associated with each category but nevertheless take account of the ordered nature of ordinal categorical variables.

Gamma-Coefficient

A common measure of association for a table with ordinal categories with a probabilistic interpretation is traditionally called the gamma-coefficient. A gamma-coefficient is a special case of a general class of measures of association called rank correlation coefficients. As the name implies, these correlations are based on the ranks or orderings of the observations rather than on the values of the observations themselves. Before describing a gamma-coefficient applied to contingency table data, it is worthwhile to review briefly a particular rank correlation applied to nontabular data.

A rank correlation, like most correlations, applies to a sample of n bivariate observations (x_i, y_i). However, the computation and evaluation of a rank correlation does not require knowledge of the sampled distribution. Define a like pair as a pair of observations where each member is larger than the corresponding members of another pair (e.g., (10, 8), (6, 7)). Define an unlike pair as a pair where the first value of the pair is larger and the second value smaller than the corresponding values of another pair (e.g., (12, 3), (10, 8)). If there are no identical values, every comparison of two pairs of observations produces either a like pair or an unlike pair. Let P represent the total number of like pairs and Q the total number of unlike pairs among n pairs of data values; then the gamma-coefficient is the difference between the proportions of the like and unlike pairs, indicating the degree of association among n bivariate observations, or

$$\hat{\gamma} = \frac{P}{P+Q} - \frac{Q}{P+Q} = \frac{P-Q}{P+Q}.$$

(8.42)

Consider Table 8.20, where $n = 12$ pairs of observations are recorded—maternal prepregnancy weight and the birth weight of her infant. A systematic way to count like pairs and unlike pairs is suggested in Table 8.21, where the pairs are ordered by prepregnancy weight. When pairs are ordered by maternal prepregnancy weight, simply comparing each infant birth weight with the values appearing below it gives a count of the like pairs and unlike pairs. For example, consider the values below the fourth ranked pair (108, 7.06). There are six values greater than 7.06 (like pairs) and two values less than 7.06 (unlike pairs). The totals of all such counts are $P = 43$ and $Q = 23$, giving $\hat{\gamma} = (43 - 23)/(43 + 23) = 0.303$. Incidentally, $P + Q = n(n-1)/2$.

The counting process used to calculate P and Q is commonly employed in rank procedures. In this particular case, $\hat{\gamma}$ is the rank correlation called Kendall's τ-corre-

T A B L E   **8.20**

Twelve mother-infant pairs: Mother prepregnancy weight and infant birth weight

	1	2	3	4	5	6	7	8	9	10	11	12
Mother's weight	127	110	115	97	147	122	133	100	104	108	123	137
Infant's weight	5.06	7.25	7.50	6.19	9.50	8.25	9.13	8.13	7.56	7.06	5.63	9.00

T A B L E **8.21**

Computing the number of like and unlike pairs

Mother	Infant	Greater	Less
97	6.19	9	2
100	8.13	4	6
104	7.56	4	5
108	7.06	6	2
110	7.25	5	2
115	7.50	4	2
122	8.25	3	2
123	5.63	3	1
127	5.06	3	0
133	9.13	1	1
137	9.00	1	0
147	9.50	0	0
Total	—	43	23

lation (i.e., $\tau = (P - Q)/(P + Q)$ where $P + Q = n(n - 1)/2$ when no "ties" occur). The computation of the gamma-coefficient for tabular data is not very different.

Calculating *P* from Tabular Data

For tabular data, the value P is the total number of observations where both categorical values exceed the corresponding categorical values of a given pair of observations (like pairs). Mechanically, the value P is the sum over the table of all cells above and to the left of a given cell multiplied by the number of observations in that cell. It is assumed that the categories are ordered from "low" to "high." Table 8.22 displays illustrative hypothetical data. The "data" give a count P of the like pairs where the specific categorical variables A_i and B_j are larger than both the other categorical variables in the table (above and to the left), then

$$P = 10(20 + 12 + 25 + 7) + 10(12 + 7) + 16(25 + 7) + 20(7) = 1482 .$$

All pairs where the both categorical variables are smaller (below and to the right) could be counted but give the same total ($P = 1482$).

Calculating *Q* from Tabular Data

The value Q is similarly calculated and represents the total number of observations where the first categorical value (A_i) is larger and the other (B_j) is smaller for a given

T A B L E **8.22**

Hypothetical data (repeated from Table 8.19) to illustrate the gamma-coefficient

Variable *B*

		B_1	B_2	B_3	Total
	A_1	7	25	20	52
Variable *A*	A_2	12	20	16	48
	A_3	30	10	10	50
	Total	49	55	46	150

observation (above and to the right—unlike pairs). The value Q from the hypothetical data is

$$Q = 30(16 + 20 + 20 + 25) + 10(16 + 20) + 12(20 + 25) + 20(20) = 3730 \ .$$

Therefore, $\hat{\gamma}$ is

$$\hat{\gamma} = \frac{1482 - 3730}{1482 + 3730} = -0.431 \ .$$

The computation of P and Q involves comparing levels of contingency table variables but is essentially the same process used to calculate a rank correlation from a sample of n pairs of observations. A probabilistic interpretation of $\hat{\gamma}$ comes from the fact that $P/(P + Q)$ estimates the probability that pairs of observations chosen at random are a like-pair and $Q/(P + Q)$ estimates the probability that pairs of observations chosen at random are an unlike pair. The value $\hat{\gamma}$ is the difference between these quantities and produces a measure of association between –1 and 1. The gamma-coefficient is 1 when all data are on the main diagonal of the table (elements f_{ii}) and –1 when all data are on the other diagonal. Values of $\hat{\gamma}$ near zero indicate that the categorical variables are not highly associated (i.e., the number of like pairs is roughly equal to the number of unlike pairs). A probabilistic interpretation depends on the ability to order both categorical variables. Confusion arises over this point because $\hat{\gamma}$ can be calculated for any table of data but only reflects association when the categorical variables are ordered. Such calculations result from computer package programs that produce a number of measures of association regardless of the type of data contained in the table. One last point: When $r = c = 2$, $\hat{\gamma} = \hat{Y}$, Yule's measure of association in a 2-by-2 table (expression 8.24).

Correlation Coefficient

A Pearson's correlation coefficient (r_{XY}, expression 1.7) can be used to express the degree of association between two categorical variables when the categorical variables are numerical or are assigned meaningful numerical values. For example, a product-moment correlation coefficient can be directly computed by giving the categorical variables numerical values (e.g., $x_i = i$: $i = 1, 2, 3, \cdots, r$ and $y_j = j$: $j = 1, 2, 3, \cdots, c$). The quantities from an r-by-c table necessary to compute a correlation coefficient are

$$S_{YY} = \Sigma f_{.j} (y_j - \bar{y})^2 \quad \text{where} \quad \bar{y} = \Sigma f_{.j} \, y_j / n$$

$$S_{XX} = \Sigma f_{i.} (x_i - \bar{x})^2 \quad \text{where} \quad \bar{x} = \Sigma f_{i.} \, x_i / n, \text{ and}$$

$$S_{XY} = \Sigma\Sigma f_{ij} (x_i - \bar{x})(y_j - \bar{y}), \quad \text{giving} \quad r_{XY} = \frac{S_{XY}}{S_X \, S_Y}.$$

Applying the usual Pearson product-moment correlation coefficient gives a measure of association between the two categorical variables bounded by -1 and 1. For the case $r = c = 2$, $r_{XY} = \hat{\phi}$ (expression 8.26).

The previous hypothetical data (Table 8.19) serve to illustrate the calculation of the correlation coefficient applied to contingency table data where $x_i = i$ (rows) and $y_j = j$ (columns); then

$$S_{YY} = 94.940 \quad \text{where} \quad \bar{y} = 1.980,$$

$$S_{XX} = 101.973 \quad \text{where} \quad \bar{x} = 1.987, \text{ and}$$

$$S_{XY} = -33.040, \quad \text{giving} \quad r_{XY} = -0.336.$$

In general, correlation coefficients are descriptive quantities and tests of significance are relatively unimportant. Tests of significance, however, are available for evaluating measures of association derived from contingency table data [9].

Restricted Chi-Square Test

A restricted chi-square test is analogous to the F-to-remove test, producing an effective statistical tool for analyzing nested models generated from contingency table data. The first step is to compute a likelihood ratio chi-square goodness-of-fit statistic based on estimated frequencies generated under a specific hypothesis (labeled H_1). That is, comparing the one-way tabular frequencies $f_1, f_2, \cdots, f_k$ to the estimated frequencies $\hat{f}_1^{(1)}, \hat{f}_2^{(1)}, \cdots, \hat{f}_k^{(1)}$ gives

$$Y_1^2 = 2 \sum_{i=1}^{k} f_i \log\left(\frac{f_i}{\hat{f}_i^{(1)}}\right),$$

(8.43)

which has an approximate chi-square distribution with $(k - 1) - s_1$ degrees of freedom when H_1 is true (expression 8.8). The value s_1 is, as before, the number of independent estimates made from the data to specify the estimated values $\hat{f}_i^{(1)}$ for each cell in a table and k is the total number of cells.

Parallel to the analysis of variance strategy, a sub-hypothesis (nested within H_1—labeled H_0) is postulated generating a second set of estimated frequencies, $\hat{f}_1^{(0)}, \hat{f}_2^{(0)}, \cdots, \hat{f}_k^{(0)}$. A chi-square statistic measuring the fit of the more restricted hypothesis is

$$Y_0^2 = 2 \sum_{i=1}^{k} f_i \log\left(\frac{f_i}{\hat{f}_i^{(0)}}\right),$$

(8.44)

which also has an approximate chi-square distribution but with $(k-1) - s_0$ degrees of freedom when H_0 is true. The value represented by s_0 is again the number of independent estimates necessary to estimate the cell frequencies ($s_0 < s_1$). The more restrictive hypothesis requires fewer parameters and, therefore, fewer estimates to establish the cell frequencies.

The difference between these two chi-square statistics reflects the impact of the restriction placed on H_1, and

$$Y^2 = Y_0^2 - Y_1^2$$

has an approximate chi-square distribution with $(k-1-s_0) - (k-1-s_1) = s_1 - s_0$ degrees of freedom when H_0 and H_1 differ only because of random variation. A more compact expression for the restricted chi-square statistic Y^2 is

$$Y^2 = 2 \sum_{i=1}^{k} f_i \log\left(\frac{\hat{f}_i^{(1)}}{\hat{f}_i^{(0)}}\right),$$

(8.45)

but this does not allow inspection of the individual Y_1^2 or Y_0^2 chi-square statistics. The inspection of Y_1^2 is important because it indicates the adequacy of the model labeled H_1, supplying a check on this general description of the data. If the likelihood that Y_1^2 arose by chance is small (under H_1), a broader or different statistical structure should be considered. If the model H_1 adequately fits the observed values, then further restrictions potentially help refine the model chosen to represent the data. In other words, $Y^2 = Y_0^2 - Y_1^2$ addresses the question, "Is H_0 a useful and simpler representation of the observed data?" or "Are all components of H_1 necessary to adequately represent the data?"

Applying the Restricted Chi-Square Procedure (An Example)

For specific Mendelian genetic traits, offspring will be one of two types (represented by *AA* or *Aa* = nonaffected and *aa* = affected). If the characteristic under study follows Mendelian laws of inheritance, then the distribution of the numbers of affected offspring is binomially distributed; furthermore, under certain conditions the probability that an offspring is affected = 0.250. The counts of affected individuals within families of size four in Table 8.23 are part of a data set collected to investigate this mode of inheritance.

The conjecture of Mendelian inheritance can be explored using a restricted chi-square approach. The broader hypothesis is that the distribution of affected offspring is binomial (H_1). The probability estimated from the data that an individual is affected is $\hat{p} = 97/[4(90)] = 0.269$. Under H_1, the distribution of affected offspring among n families of size four is then given by

$$number\ affected = n \binom{4}{i}(0.269)^i (0.731)^{4-i} = \hat{f}_i^{(1)},$$

generating estimated frequencies

$$\hat{f}_0^{(1)} = 25.636, \ \hat{f}_1^{(1)} = 37.821, \ \hat{f}_2^{(1)} = 20.924, \ \hat{f}_3^{(1)} = 5.145 \text{ and } \hat{f}_4^{(1)} = 0.474$$

yielding

$$Y_1^2 = 2\left[20 \log\left(\frac{20}{25.636}\right) + \cdots + 1 \log\left(\frac{1}{0.474}\right)\right] = 6.400$$

with $(5-1)-1 = 3$ degrees of freedom.

A more restrictive hypothesis is that the distribution of families remains binomial but the probability of an affected individual is $p_0 = 0.250$. Under H_0, the distribution of affected individuals among n families is given by

$$number\ affected = n \binom{4}{i}(0.25)^i (0.75)^{4-i} = \hat{f}_i^{(0)},$$

generating estimated frequencies

$$\hat{f}_0^{(0)} = 28.477, \ \hat{f}_1^{(0)} = 37.969, \ \hat{f}_2^{(0)} = 18.984, \ \hat{f}_3^{(0)} = 4.219, \text{ and } \hat{f}_4^{(0)} = 0.352,$$

yielding

$$Y_0^2 = 2\left[20 \log\left(\frac{20}{28.477}\right) + \cdots + 1 \log\left(\frac{1}{0.352}\right)\right] = 7.114$$

with $(5-1)-0 = 4$ degrees of freedom. Therefore, the restricted chi-square statistic is

$$Y^2 = 7.114 - 6.400 = 0.714$$

 T A B L E **8.23**

Data collected to test the hypothesis that the trait under investigation follows a simple Mendelian mode of inheritance

Number Affected per Family	Number of Families	Total Number of Affected Offspring
0	20	0
1	47	47
2	20	40
3	2	6
4	1	4
Total	90	97

with $4 - 3 = 1$ degree of freedom. The chi-square value Y_1^2 indicates no strong evidence to reject the hypothesis that the affected offspring are binomially distributed (p-value = 0.094). The restricted chi-square statistic Y^2 further indicates that the data are consistent not only with the binomial distribution but also with the additional conjecture that $p_0 = 0.250$ (p-value = 0.398). The restricted chi-square procedure is a powerful tool to evaluate constraints placed on a model and is used in the next chapter, where a method to assess differences between nested log-linear models is essential.

Analysis of Trend in Tables with Quantitative Categories

A frequent configuration of tabular data is a 2-by-c contingency table where the question of trend or dose response is an issue. This pattern of data occurs when a binary variable is analyzed with respect to a categorical variable with a numerical value. Several analytic methods exist to account for the quantitative nature of the categories, improving the sensitivity of the chi-square procedure. A straightforward application of simple linear regression techniques usually works well.

The categories (columns) are described numerically by values x_j (e.g., $x_j = 1, 2, 3, \cdots, c$). These numeric variables serve as independent variables. The proportions $\hat{p}_j = f_{1j}/(f_{1j} + f_{2j}) = f_{1j}/f_{.j}$ become the dependent variables and a simple linear regression analysis is used to identify a trend (if any) among the categories. The notation for the data in a 2-by-c contingency table is displayed in Table 8.24. To evaluate the linear influence of variable B on the proportions $\hat{p}_j$ generated from each of the c categories, the following quantities are calculated:

$$SS_{XX} = \Sigma f_{.j}(x_j - \bar{x})^2 \quad \text{where} \quad \bar{x} = \Sigma x_j f_{.j}/n$$
$$SS_{YY} = n\hat{p}(1 - \hat{p}) \quad \text{where} \quad \hat{p} = f_1./n$$
$$SS_{XY} = \Sigma f_{.j}(x_j - \bar{x})(\hat{p}_j - \hat{p}) \quad \text{where} \quad \hat{p}_j = f_{1j}/f_{.j}$$

where $n = \Sigma_j f_{.j}$. Then, an estimate of the slope describing the influence of the categorical variable B on the proportions $\hat{p}_j$ ("the dose response") is

T A B L E 8.24

Notation for a 2-by-c contingency table

		Variable B					
		B_1	B_2	B_3	$\cdots$	B_c	Total
Variable A	A_1	f_{11}	f_{12}	f_{13}	$\cdots$	f_{1c}	$f_1.$
	A_2	f_{21}	f_{22}	f_{23}	$\cdots$	f_{2c}	$f_2.$
	Total	$f_{.1}$	$f_{.2}$	$f_{.3}$	$\cdots$	$f_{.c}$	n
	Proportion	$\hat{p}_1$	$\hat{p}_2$	$\hat{p}_3$	$\cdots$	$\hat{p}_c$	$\hat{p}$

$$\hat{b} = \frac{SS_{XY}}{SS_{XX}} \quad \text{with} \quad variance(\hat{b}) = \frac{SS_{YY}}{n\,SS_{XX}}\,.$$

(8.46)

A test for trend (H_0: $b = 0$) is

$$X_L^2 = \frac{\hat{b}^2}{variance(\hat{b})} \qquad (L \text{ for } linear)$$

(8.47)

where X_L^2 has an approximate chi-square distribution with one degree of freedom when the relative frequency of the binary variable (A) is not linearly related to the c-levels of the independent categorical variable B (i.e., $b = 0$).

The calculation and testing of the slope $\hat{b}$ does not differ from the process described in the simple linear regression case (Chapter 2). The same estimate $\hat{b}$ results from analyzing n pairs of observations in the form of (x_i, 0) and (x_i, 1) with least squares analysis, where again x_i represents the numeric categorical variable whereas $y_i = 1$ indicates that an observation belongs to category A_1 and $y_i = 0$ indicates that an observation belongs to category A_2. The least squares estimate (expression 2.6) is the identical value of $\hat{b}$ given by expression 8.46.

Partitioning the Chi-Square Statistic

The Pearson chi-square value X^2 (expression 8.31) calculated under the hypothesis of independence of row and column variables can be partitioned into two pieces: one measuring linear influence of the column variable and the other measuring the departure from linearity, represented in Table 8.25. To study the relationship between age and smoking exposure, 243 women were surveyed, and the data are presented in Table 8.26. Assigning age values, 20.5, 30.5, 40.5, 50.5, and 60.5 years, to the five age categories produces

$$SS_{XX} = 41896.3 \text{ where } \bar{x} = 40.623\,,$$
$$SS_{YY} = 49.407 \text{ where } \hat{p} = 0.284\,, \text{ and}$$
$$SS_{XY} = 371.482\,.$$

T A B L E **8.25**

The partitioning of the chi-square statistic into pieces reflecting linear and nonlinear influences

Source	Chi-Square	Degrees of Freedom
Due to linearity	$X_L^2 = \hat{b}^2 / variance(\hat{b})$	1
Departures from linearity	$X^2 - X_L^2$	$c - 2$
Total	X^2	$c - 1$

T A B L E   **8.26**

Women who smoke and do not smoke classified by age

	<25	25–35	36–45	46–55	>55	Total
Smokers	4	10	17	20	18	69
Nonsmokers	36	40	41	37	20	174
Total	40	50	58	57	38	243
$\hat{p}_i$	0.100	0.200	0.293	0.351	0.474	0.284

Then the estimated slope is $\hat{b} = 0.0089$ with $variance(\hat{b}) = 0.00000485$, yielding $X_L^2 = 16.200$ with a p-value < 0.001. The partitioned X^2-values are given in Table 8.27. These data show a remarkably linear relationship between the proportion of smokers and age (a plot of the data provides further evidence).

Three Features of X_L^2

Three features of applying simple linear regression techniques to proportions from a 2-by-c table should be noted. The value X_L^2 remains the same for all equally spaced choices of the numerical values for variable B. However, a certain amount of subjectivity enters the analysis when unequally spaced or nonlinear x_i-values are assigned to the categorical levels. Surprisingly, if variable A is treated as the independent variable and B becomes the dependent variable, the chi-square values do not change. Additionally, the X_L^2-value is simply nr^2 (i.e., $X_L^2 = nr^2$) where r is the Pearson product-moment correlation based on the n pairs (x_i, y_i) where x_i is a numeric categorical variable and y_i is a binary variable (0 or 1). For the data on smoking and age, $r = 0.258$ and $X_L^2 = nr^2 = 243 (0.258)^2 = 16.200$. Last, if the mean x-values for each of two levels of the binary categories are compared ($\bar{x}_1$ from A_1 and $\bar{x}_2$ from A_2) with a t-like statistic, the results are the same as the chi-square analysis. That is,

T A B L E   **8.27**

The partitioning of the chi-square statistic into pieces reflecting linear and nonlinear influences

Source	Chi-Square	Degrees of Freedom
Due to linearity	16.200	1
Departures from linearity	0.198	3
Total	16.398	4

$$z = \frac{\bar{x}_1 - \bar{x}_2}{\sqrt{variance(\bar{x}_1 - \bar{x}_2)}} = \frac{\bar{x}_1 - \bar{x}_2}{\sqrt{\frac{SS_{XX}}{n}\left(\frac{1}{f_{1.}} + \frac{1}{f_{2.}}\right)}}$$

(8.48)

and $z^2 = X_L^2$. Specifically, the mean age of smokers is $\bar{x}_1 = 46.007$ (A_1) and the mean age of nonsmokers is $\bar{x}_2 = 38.489$ (A_2) where age categories are again coded as 20.5, 30.5, 40.5, 50.5, and 60.5 years; however, any equally spaced five values would yield the same test statistic z. The t-like statistic $z = 4.025$ and $z^2 = 4.025^2 = 16.200$, which is identical to the previous chi-square value.

Applied Example

Measures of Association: *r*-by-*c* Table

Correlation between the amounts smoked by the mothers and fathers of infants in the CHDS set data is of interest because maternal exposure to father's smoking has been postulated as a cause of increased risk of low birth weight. The first step in a study of parental smoking patterns is to get some idea of the magnitude of the association between maternal and paternal smoking exposure. Parental smoking exposure classified into three categories (nonsmokers, smoke less than or equal to a pack/day, and smoke greater than a pack/day) produces a 3-by-3 contingency table. The 680 pairs of parents classified into these categories when analyzed for independence of smoking exposure pattern yield a chi-square value of 58.284 (p-value < 0.001). The large chi-square value gives assurance that quantitative measures of association between mother's and father's smoking exposure calculated from these data reflect more than chance association.

The value for Cramer's $\hat{V}$ is 0.207, which scales the chi-square value between 0 and 1, indicating a modest statistical association. However, in terms of predictability, the λ-statistics show that knowledge of one spouse's smoking exposure does not allow accurate prediction of the other's smoking exposure. Knowing the mother's smoking exposure (columns) is useless for predicting the father's smoking exposure ($\hat{\lambda}_c = 0$). Knowing the father's smoking exposure (rows) slightly predicts mother's smoking exposure ($\hat{\lambda}_r = 0.059$). The overall lambda correlation $\hat{\lambda}$ is 0.033.

The categorical variables are ordinal, giving the gamma-coefficient a probabilistic interpretation. The value $\hat{\gamma} = 0.344$ reflects the expected positive association between parental smoking exposure (more like pairs than unlike pairs). The additional assumption that the three categories are equally spaced means that a Pearson correlation coefficient reflects association (smoking levels coded 0, 1, and 2). The correlation $r = 0.242$ is consistent with the gamma value (modestly large and positive). When the actual numbers of reported cigarettes ($n = 680$) are used, the Pearson correlation between maternal and paternal amounts smoked is $r_{XY} = 0.262$ (example in Chapter 4). It is reassuring that the three approaches to measuring association, based on different ways of quantifying amounts smoked, give similar values. A large number of other methods exist to describe associations in contingency tables, and readers with energy and interest can certainly investigate the properties of these measures, which are described elsewhere (e.g., [10] or [11]).

Tabulation and Various Correlation Statistics

Using the CHDS data, a table is constructed for three levels of maternal and paternal smoking to calculate several measures of association.

chds.table

```
. use chds

. generate mcig=mnocig

. generate fcig =fnocig

. recode mcig 0=0 1/20=1 21/max=2
(299 changes made)

. recode fcig 0=0 1/20=1 21/max=2
(466 changes made)

. correlate mnocig fnocig
(obs=680)

          |   mnocig    fnocig
----------+------------------
   mnocig|   1.0000
   fnocig|   0.2617    1.0000


. tabulate fcig mcig, all

          | mcig
     fcig|         0          1          2 |     Total
----------+---------------------------------+----------
        0 |       159         34         21 |       214
        1 |        86         65         28 |       179
        2 |       136         70         81 |       287
----------+---------------------------------+----------
    Total|       381        169        130 |       680

          Pearson chi2(4) =   58.2838   Pr = 0.000
likelihood-ratio chi2(4) =   57.8315   Pr = 0.000
               Cramer's V =    0.2070
                    gamma =    0.3440   ASE = 0.051
```

```
Lambda = 0.0334   with fnocig (rows)  = 0.0585   with mnocig (columns)  = 0.000

Pearson correlation applied to table r = 0.242

Pearson correlation applied to the data r = 0.262

. exit,clear
```

Notation

Pearson chi-square statistic:

$$X^2 = \sum_{all\ cells} \frac{[x_i - E(X_i)]^2}{E(X_i)}$$

Likelihood ratio statistic:

$$Y^2 = 2 \sum_{all\ cells} x_i \log\left(\frac{x_i}{E(X_i)}\right)$$

Expression of the chi-square statistic:

the expected chi-square value $\approx$ *degrees of freedom* $+ n(lack\ of\ fit)^2$

Hypergeometric probabilities:

$$P(X = x) = \frac{\binom{m}{x}\binom{N-m}{n-x}}{\binom{N}{n}}$$

$$expected\ value = E(X) = n\left(\frac{m}{N}\right)$$

$$variance = \sigma_X^2 = n\left(\frac{m}{N}\right)\left(1 - \frac{m}{N}\right)\left(\frac{N-n}{N-1}\right)$$

Chi-square statistic to evaluate independence:

$$X^2 = \sum_i \sum_j \frac{(f_{ij} - \hat{f}_{ij})^2}{\hat{f}_{ij}} \quad where \quad \hat{f}_{ij} = \frac{f_{i.}f_{.j}}{n}$$

Chi-square shortcut formula for a 2-by-2 table:

$$X^2 = \frac{n(f_{11}f_{22} - f_{12}f_{21})^2}{f_{1.}f_{.1}f_{2.}f_{.2}}$$

McNemar's chi-square statistic:

$$X^2 = \frac{(b-c)^2}{b+c}$$

Standardized residual value:

$$r_{ij} = \frac{f_{ij} - f_{i.}f_{.j}/n}{\sqrt{f_{i.}f_{.j}/n}}$$

$$variance(r_{ij}) = (1 - f_{i.}/n)(1 - f_{.j}/n)$$

$$R_{ij} = \frac{r_{ij}}{\sqrt{variance(r_{ij})}}$$

Lambda measures of association (columns):

$$\hat{\lambda}_c = \frac{\text{probability of error (1)} - \text{probability of error (2)}}{\text{probability of error (1)}}$$

$$= \frac{(1 - f_{m.}/n) - (1 - \Sigma_j f_{mj}/n)}{1 - f_{m.}/n}$$

Restricted chi-square:

$$\text{Hypothesis 1:}\quad Y_1^2 = 2 \sum_{i=1}^{k} f_i \log\left(\frac{f_i}{\hat{f}_i^{(1)}}\right)$$

$$\text{Hypothesis 0:}\quad Y_0^2 = 2 \sum_{i=1}^{k} f_i \log\left(\frac{f_i}{\hat{f}_i^{(0)}}\right)$$

giving

$$Y^2 = Y_0^2 - Y_1^2.$$

Problems

1   Show that expression 8.6 is equivalent to expression 8.7.

2   The following table classifies individuals as to whether they died during the six months before or during the six months after their birthdays:

	Before	After	Total
Count	148	200	348

Compute X^2, Y^2, and z and their associated p-values to evaluate the hypothesis that date of death and date of birth are unrelated.

3   Consider the following 2-by-2 table from a study of coronary heart disease and behavior type (type A versus type B individuals), for men over 60 years of age:

	CHD	No CHD	
A	3	16	19
B	1	12	13
Total	4	28	32

Assess the association between coronary heart disease and behavior type with:

Fisher's exact test and the X^2 and Y^2 chi-square tests and report the associated p-values.

Compute the following measures of association:

Pearson's $\hat{P}$, Yule's $\hat{Y}$, the φ-correlation, and $\hat{OR}$

Calculate an approximate 95% confidence interval for the Y and OR.

4   Show $X^2 = nr_{XY}^2$ where X^2 is the chi-square statistic calculated from a 2-by-2 table (see expression 8.26).

5   The following table shows the data in problem 2 in terms of month of death relative to month of birth (e.g., a person who dies 3 months before his or her birthday = −3):

	−6	−5	−4	−3	−2	−1	1	2	3	4	5	6	Total
Count	25	23	30	17	26	27	42	40	36	30	28	24	348

Compute both X^2 and Y^2 and their associated p-values to evaluate the hypothesis that date of death and date of birth are unrelated.

6   If the categorical variables forming a 2-by-2 table are exactly independent, show that $\hat{Y} = 0$ (Yule's measure of association), $\hat{\varphi} = 0$, and $\hat{OR} = 1.0$. Also show that multiplying the rows or columns in a 2-by-2 table by a constant value changes the φ-measure of association but does not change the odds ratio $\hat{OR}$.

7   Consider the data where 272 white males (age > 54) are classified by current smoking habits and five economic levels. The data are:

	I (low)	II	III	IV	V (high)	Total
Never smoked	14	37	15	22	11	99
Past smoked	7	22	12	7	3	51
≤ 1 pack/day	3	25	6	9	3	46
> 1 pack/day	13	26	18	15	4	76
Total	37	110	51	53	21	272

Compute the chi-square statistics X^2 and Y^2 and report the corresponding p-values associated with a test of the hypothesis that smoking is unrelated to economic status. Also compute Cramer's measure of association $\hat{V}$, the lambda measure of association (columns), the gamma-coefficient, and the "correlation" coefficient based on equally spaced intervals for both smoking and economic levels.

8   The Oxford Survey of Childhood Cancer [12] provides data on malignancies in children under 10 years of age and information on the mother's exposure to x-rays. The data below (a small part of a large study containing data collected since 1953) show the numbers of prenatal x-rays received by mothers of children with a malignant disease and a series of controls (healthy children of the same age, sex, and similar areas of residence):

Films	0	1	2	3	4	5	Unk.	Total
Cases	7332	287	199	96	59	65	475	8513
Controls	7673	239	154	65	28	29	325	8513
Total	15005	526	353	161	87	94	800	17026
Proportion	0.489	0.546	0.564	0.596	0.678	0.691	—	—

Ignoring the unknown values, compute the Pearson chi-square statistic for these data based on the hypothesis that case/control status is unrelated to the number of x-rays. Estimate the slope describing the relationship between the number of x-rays (dependent variable) and the proportion of individuals classified as a case (independent variable). Partition the Pearson chi-square value into a piece associated with linearity and a piece describing departures from linearity. Use a t-like approach to analyze the same data (expression 8.48).

9 Consider the Poisson distribution (reviewed in the appendix to Chapter 12)

$$P(X = k) = \frac{e^{-\lambda}\lambda^k}{k!}$$

and the following data (frequency $= f_j$)

x	0	1	2	3	4	5	6
f_j	12	25	28	20	10	5	1

Estimate from the data the value λ (i.e., $\bar{x} = \hat{\lambda}$) and then estimate the seven frequencies $\hat{f}_j^{(1)}$ based on this estimated parameter. Calculate the likelihood chi-square statistic Y_1^2 for the fit of these estimates. Estimate the frequencies $\hat{f}_j^{(0)}$ based on a value of $\lambda = 2$ and produce a likelihood chi-square statistic Y_0^2. Use a restricted chi-square procedure to evaluate whether the data support the hypothesis that a Poisson distribution with $\lambda = 2$ describes the data.

References

1 Bishop, Y. M. M., Fienberg, S. E., and Holland, P. W. *Discrete Multivariate Analysis: Theory and Practice.* Cambridge: MIT Press, 1975.

2 Owen, D. B. *Handbook of Statistical Tables.* Reading, Mass.: Addison-Wesley, 1962.

3 Johnson, N. L., and Kotz, S. *Discrete Distributions.* New York: Houghton Mifflin, 1969.

4 Kendall, M. G., and Stuart, A. *The Advanced Theory of Statistics*, volume 1. London: Charles Griffin and Company, 1963.

5 Conover, W. J. *Practical Nonparametric Statistics*, 2nd ed. New York: John Wiley, 1980.

6 Garfinkel, J., Selvin, S., and Brown, S. B. Possible Increased Risk of Lung Cancer among Beauticians. *Journal of the National Cancer Institute* 58 (1977): 141–143.

7 Haldane, J. B. S. The Estimation and Significance of the Logarithm of a Ratio of Frequencies. *Annals of Human Genetics* 20 (1956): 309–311.

8 Cochran, W. G. Some Methods for Strengthening the Common χ^2 Test. *Biometrics* 10 (1954): 417–451.

9 Freeman, D. H. *Applied Categorical Data*. New York: Marcel Dekker, 1987.

10 Everitt, B. S. *The Analysis of Contingency Tables*. New York: John Wiley, 1977.

11 Upton, G. J. G. *The Analysis of Cross-Tabulated Data*. New York: John Wiley, 1978.

12 Bithell, J. F., and Steward, M. A. Prenatal Irradiation and Childhood Malignancy: A Review of British Data from the Oxford Study. *British Journal of Cancer* 31 (1975): 271–287.

9

Log-Linear Models

Background

A log-linear model, like the models employed in regression analysis, provides a relatively simple mathematical structure to represent the variables within a data set, making it possible to describe succinctly the relationships under investigation. This statistical structure leads to tractable and easily interpreted summary values and the analysis parallels the previously discussed pattern of evaluating a sequence of nested models, the "*F*-to-remove" strategy. Unlike regression analysis, however, the log-linear model is used to represent and explore tabular data. For example, a sample might consist of women classified by their number of children (none, one, two, three, or more), their age (35–44, 45–64, or ≥65), and their disease status (ovarian cancer—present or absent). A sequence of log-linear models provides a way to analyze the relationships among ovarian cancer, child-bearing status and influences of age where the count of the sampled women in each category is the dependent variable and the categorical variables used to classify the women are independent variables. Again, the goodness-of-fit of a series of nested models is the principal statistical measure employed to evaluate the influences on the dependent variable.

The first part of this chapter deals with the analysis of three-dimensional tables without a statistical model. The log-linear model is then introduced as a more general analytic tool effective for higher-dimensional tables. However, the major focus remains on log-linear analysis applied to the three-way table because the approaches and principles underlying a three-dimensional table are not very different from those of higher dimensions, making it unnecessary to develop a complex general notation. Part of one chapter is an introduction and clearly does not cover the entire topic of log-linear analysis, but a variety of books devoted to the subject are available (e.g., [1], [2], and [3]).

Introductory Notation

Before turning to the log-linear model, it is necessary to describe a natural extension of the "two-dimension" notation used in the previous chapter to the three-dimensional table. The cell frequency in a three-dimensional table with r rows, c columns, and l levels is represented by f_{ijk} ($i = 1, 2, \cdots, r$—r for rows; $j = 1, 2, \cdots, c$—c for columns; and $k = 1, 2, \cdots, l$—l for levels). Sums of these frequencies are represented by

$$f_{i..} = \sum_j \sum_k f_{ijk} = \sum_j f_{ij.} = \sum_k f_{i.k}$$

$$f_{.j.} = \sum_i \sum_k f_{ijk} = \sum_i f_{ij.} = \sum_k f_{.jk}$$

$$f_{..k} = \sum_i \sum_j f_{ijk} = \sum_i f_{i.k} = \sum_j f_{.jk}$$

and

$$n = \sum_i \sum_j \sum_k f_{ijk} = \sum_i f_{i..} = \sum_j f_{.j.} = \sum_k f_{..k} .$$

The "dot" in the subscript designates a marginal frequency. The position of the dot indicates which marginal frequency. For example, $f_{1.2}$ represents the marginal frequency where the values in the first row and the second level are summed over the columns. For $r = 2$, $c = 2$, and $l = 2$, Table 9.1 displays the notation for this simplest three-dimensional table, a 2-by-2-by-2 table. The total number of observations is $n = f_{1..} + f_{2..} = f_{.1.} + f_{.2.} = f_{..1} + f_{..2}$. The generalization of this notation to higher-dimensional tables follows the same pattern.

T A B L E **9.1**

Notation for a three-dimensional table based on binary variables A, B, and C—the 2-by-2-by-2 case

C_1	B_1	B_2	Total
A_1	f_{111}	f_{121}	$f_{1.1}$
A_2	f_{211}	f_{221}	$f_{2.1}$
Total	$f_{.11}$	$f_{.21}$	$f_{..1}$

C_2	B_1	B_2	Total
A_1	f_{112}	f_{122}	$f_{1.2}$
A_2	f_{212}	f_{222}	$f_{2.2}$
Total	$f_{.12}$	$f_{.22}$	$f_{..2}$

$C_1 + C_2$	B_1	B_2	Total
A_1	$f_{11.}$	$f_{12.}$	$f_{1..}$
A_2	$f_{21.}$	$f_{22.}$	$f_{2..}$
Total	$f_{.1.}$	$f_{.2.}$	n

T A B L E **9.2**

Artificial data that conform to exact complete independence of variables A, B, and C

Level 1

C_1	B_1	B_2	Total
A_1	16	64	80
A_2	64	256	320
Total	80	320	400

Level 2

C_2	B_1	B_2	Total
A_1	24	96	120
A_2	96	384	480
Total	120	480	600

Level 1 + Level 2

$C_1 + C_2$	B_1	B_2	Total
A_1	40	160	200
A_2	160	640	800
Total	200	800	1000

Complete Independence

Complete independence arises when all cell frequencies within a contingency table are strictly determined by the marginal probabilities. This means that all three categorical variables are unrelated. "Data" constructed with variables A, B, and C completely independent (no random variation) are displayed in Table 9.2. For these "data," $P(A_1) = 1 - P(A_2) = 0.2$; $P(B_1) = 1 - P(B_2) = 0.2$; $P(C_1) = 1 - P(C_2) = 0.4$; and, furthermore, $f_{ijk} = n\,P(A_i)P(B_j)P(C_k)$. The expression $\hat{f}_{ijk} = f_{i..}\,f_{.j.}\,f_{..k}/n^2$ estimates the cell frequencies when the three variables are completely independent. Notice that the estimated odds ratios (expression 8.27) for all three tables are 1.0, which directly follows from the definition of complete independence. For example, for level 1 of variable C, the odds ratio measuring the association between variables A and B is

$$\hat{OR}_{AB|C_1} = \frac{\hat{f}_{111}\hat{f}_{221}}{\hat{f}_{121}\hat{f}_{211}} = \frac{[(f_{1..}\,f_{.1.}\,f_{..1})/n^2][(f_{2..}\,f_{.2.}\,f_{..1})/n^2]}{[(f_{1..}\,f_{.2.}\,f_{..1})/n^2][(f_{2..}\,f_{.1.}\,f_{..1})/n^2]} = 1.0$$

and similarly $\hat{OR}_{AB|C_2}$ and $\hat{OR}_{AB}$ are exactly 1.0. Other measures of association would equally express perfect independence of variables A, B, and C. The odds ratio is used because it is a natural measure of association for log-linear analysis of a contingency table.

The analysis of a multidimensional table often starts with rejecting the hypothesis of complete independence. If no evidence exists to reject the hypothesis of complete independence, it rarely makes sense to search further for more complicated relationships among the categorical variables.

Example

Data from a survey of smoking patterns serve to illustrate a chi-square statistic applied to assess complete independence. The categorical variables are:

$$
\begin{array}{ll}
\text{Age:} & A_1 = \text{less than 30 years old} \\
& A_2 = \text{equal to or greater than 30 years old}
\end{array}
$$

$$
\begin{array}{ll}
\text{work status:} & B_1 = \text{working mother} \\
& B_2 = \text{nonworking mother}
\end{array}
$$

$$
\begin{array}{ll}
\text{smoking status:} & C_1 = \text{never smoked} \\
& C_2 = \text{past smoker} \\
& C_3 = \text{less than or equal to one pack/day} \\
& C_4 = \text{greater than one pack/day}
\end{array}
$$

The data are given in Table 9.3. Note that marginal frequencies are: for variable A: $f_{1..} = 216$ and $f_{2..} = 220$; for variable B: $f_{.1.} = 44$ and $f_{.2.} = 392$; and for variable C: $f_{..1} = 70, f_{..2} = 138, f_{..3} = 72,$ and $f_{..4} = 156$, with $n = 436$.

When the three categorical variables are completely independent, then $P(A_i\,B_j\,C_k) = P(A_i)\,P(B_j)\,P(C_k)$. The estimated cell frequencies (in parentheses in Table 9.3) are then generated from the expression $\hat{f}_{ijk} = f_{i..}\,f_{.j.}\,f_{..k}/n^2$. For example,

$$
\hat{f}_{111} = (216)\,(44)\,(70)/(436)^2 = 3.50 ,
$$

$$
\hat{f}_{112} = (216)\,(44)\,(138)/(436)^2 = 6.90 , \cdots , \text{etc.}
$$

The likelihood ratio chi-square measure comparing the fit of these estimated values to the observed data is $Y^2 = 16.448$ (expression 8.32). The degrees of freedom associated with the chi-square statistic are $(rcl - 1) - [(r-1) + (c-1) + (l-1)] = (2)\,(2)\,(4) - 1 - [1 + 1 + 3] = 10$. The degrees of freedom are calculated in the same way as a two-dimensional table. Specifically, the degrees of freedom equal $(rcl - 1) - s$ where rcl is the total number of cells and s is the number of independent estimates necessary to define the estimated cell frequencies (i.e., for complete independence, $s = (r-1) + (c-1) + (l-1)$). The probability of observing a Y^2-value as extreme as or more extreme than 16.448 when variables A, B, and C are completely independent is 0.088. Because the hypothesis of complete independence does not seem to produce estimated values that fit the observed data particularly well, it may be useful to look further for more complex relationships among the variables age, work status, and smoking exposure. The test for complete independence does not differ in principle from the test of independence in an r-by-c table, and the application to higher-dimensional tables is also analogous.

T A B L E **9.3**

Data on 436 individuals classified by age, work status, and smoking exposure—estimated values are in parentheses

C_1	B_1	B_2	Total
A_1	2 (3.50)	29 (31.18)	31
A_2	3 (3.56)	36 (31.76)	39
Total	5	65	70

C_2	B_1	B_2	Total
A_1	4 (6.90)	57 (61.47)	61
A_2	5 (7.03)	72 (62.61)	77
Total	9	129	138

C_3	B_1	B_2	Total
A_1	10 (3.60)	30 (32.07)	40
A_2	2 (3.67)	30 (32.66)	32
Total	12	60	72

C_4	B_1	B_2	Total
A_1	8 (7.80)	76 (69.49)	84
A_2	10 (7.94)	62 (70.77)	72
Total	18	138	156

Partial Independence

When data are classified by three categorical variables A, B, and C, three possible pairwise associations can exist: A with B, A with C, and B with C. Partial independence occurs when one pair of categorical variables is associated while the other two pairs are independent. This means one categorical variable is independent of the other two. Algebraically, partial independence generates one of the following sets of estimated values:

If A is independent of B and C whereas B and C are associated, then

$$\hat{f}_{ijk} = f_{\cdot jk}[f_{i\cdot\cdot}/n] ;$$

if B is independent of A and C whereas A and C are associated, then

$$\hat{f}_{ijk} = f_{i \cdot k} \, [f_{\cdot j \cdot} / n] \, ;$$

or if C is independent of A and B whereas A and B are associated, then

$$\hat{f}_{ijk} = f_{ij \cdot} \, [f_{\cdot \cdot k} / n] \, . \tag{9.1}$$

The artificial data in Table 9.4 show perfect partial independence of the variable C. Note that

$$\hat{OR}_{AB \mid C_1} = \frac{\hat{f}_{111} \hat{f}_{221}}{\hat{f}_{121} \hat{f}_{211}} = \frac{(f_{11 \cdot} f_{\cdot \cdot 1}/n)(f_{22 \cdot} f_{\cdot \cdot 1}/n)}{(f_{12 \cdot} f_{\cdot \cdot 1}/n)(f_{21 \cdot} f_{\cdot \cdot 1}/n)}$$

$$= \frac{f_{11 \cdot} f_{22 \cdot}}{f_{12 \cdot} f_{21 \cdot}} = \hat{OR}_{AB} = 13.5$$

where $\hat{OR}_{AB}$ comes from the table created by adding level 1 and level 2 values (Table 9.4). Similarly, $\hat{OR}_{AB \mid C_2} = \hat{OR}_{AB} = 13.5$, showing again that variables A and B are associated. Furthermore, the odds ratio measuring the association between A and C is 1.0 at both levels of B, or

$$\hat{OR}_{AC \mid B_1} = \frac{(f_{11 \cdot} f_{\cdot \cdot 1}/n)(f_{21 \cdot} f_{\cdot \cdot 2}/n)}{(f_{21 \cdot} f_{\cdot \cdot 2}/n)(f_{11 \cdot} f_{\cdot \cdot 1}/n)} = 1.0 \, .$$

T A B L E   **9.4**

Artificial data that conform to exact partial independence of variable C

Level 1					Level 2			
C_1	B_1	B_2	Total		C_2	B_1	B_2	Total
A_1	90	60	150		A_1	30	20	50
A_2	60	540	600		A_2	20	180	200
Total	150	600	750		Total	50	200	250

Level 1 + Level 2			
$C_1 + C_2$	B_1	B_2	Total
A_1	120	80	200
A_2	80	720	800
Total	200	800	1000

Similarly, $\hat{OR}_{AC|B_2} = 1.0$. Variables B and C are also exactly independent in this partially independent "data" at both levels of A ($\hat{OR}_{BC|A_1} = \hat{OR}_{BC|A_2} = 1.0$).

Partial independence implies that one of the three categorical variables does not influence the other two and further implies that a three-dimensional table can be reduced to a two-dimensional table without changing the relationship between the two associated variables by adding over the partially independent variable. For example, the "data" in Table 9.4 show that the odds ratio in the $(C_1 + C_2)$-table (summed over variable C) is identical to the odds ratio in each of the individual tables, namely 13.5.

Partial independence occurs when the total table is proportionally distributed among the levels of the partially independent categorical variables. For partial independence of the variable C, the cell frequencies found in the total table f_{ij} are distributed by the factor $f_{..k}/n$ among the two levels of the variable C to produce estimated cell frequencies. The artificial numbers in Table 9.4 are generated by $f_{ij}. \times [3/4]$ assigned to the cells in the C_1-table and $f_{ij}. \times [1/4]$ assigned to the cells in the C_2-table. This proportionality guarantees that variable C is independent of both variables A and B.

Example

Questionnaire data in Table 9.5 illustrate a chi-square test for partial independence (the estimated values are in parentheses). The categorical variables are:

Variable age:	A_1 = less than 45 years old
	A_2 = 45 to 65 years old
	A_3 = greater than 65 years old;
Question:	Do you believe exercise increases a person's lifetime?
Variable answer:	B_1 = yes and B_2 = no;
Variable gender:	C_1 = male and C_2 = female.

To test the hypothesis that gender (C) is independent of both age (A) and response (B), the estimated cell frequencies are given by $\hat{f}_{ijk} = f_{ij}. [f_{..k}/n]$. These estimated values are simply 113/204 of the $(C_1 + C_2)$-table assigned to the C_1-table and 91/204 of the $(C_1 + C_2)$-table assigned to the C_2-table. For example,

$$\hat{f}_{111} = 24[113/204] = 13.29 ,$$
$$\hat{f}_{112} = 24[91/204] = 10.71 , \cdots, \text{etc.}$$

The Y^2-measure of fit, 6.706, produces a significance probability of 0.243 with degrees of freedom = $(rcl - 1) - s = (3)(2)(2) - 1 - 6 = 5$ where $s = (l - 1) + (rc - 1) = 1 + 5 = 6$. Because no strong evidence exists that gender (C) is associated with either age (A) or response (B), the tables C_1 and C_2 can be combined without substantially changing the relationship between age and response.

The association between variables A and B can then be assessed in the resulting 3-by-2 table ($C_1 + C_2$ shown as part of Table 9.5) with estimated cell frequencies

T A B L E   **9.5**

Data classifying 204 individuals by age (A), response to a specific question (B) and gender (C)—estimated values are in parentheses

Response (male)

C_1	B_1	B_2	Total
A_1	10 (13.29)	2 (1.11)	12
A_2	27 (27.70)	3 (2.22)	30
A_3	46 (47.64)	25 (21.05)	71
Total	83	30	113

(Age)

Response (female)

C_2	B_1	B_1	Total
A_1	14 (10.71)	0 (0.89)	14
A_2	23 (22.30)	1 (1.78)	24
A_3	40 (38.36)	13 (16.95)	53
Total	77	14	91

(Age)

Response (male + female)

$C_1 + C_2$	B_1	B_2	Total
A_1	24 (20.39)	2 (5.61)	26
A_2	50 (42.35)	4 (11.65)	54
A_3	86 (97.25)	38 (26.75)	124
Total	160	44	204

(Age)

generated under the hypothesis of complete independence (i.e., $H_0 : \hat{f}_{ij.} = f_{i..} f_{.j.}/n$). The comparison of these estimated values (in parentheses) with the observed values produces a likelihood ratio chi-square statistic of $Y^2 = 17.283$ with $(r-1)(c-1) = (3-1)(2-1) = 2$ degrees of freedom yielding a corresponding p-value of less than 0.001. Therefore, the more easily understood two-way table shows that age and response are likely associated.

Conditional Independence

Conditional independence occurs when one pair of the categorical variables A, B, and C (AB, AC, or BC) is independent and the other two are associated. This means that at each level of a specific variable there is no association between the other two variables. If variables A and B are conditionally independent, then no association exists between A and B at each level of variable C. However, variable C is associated with both variables A and B. Conditional independence produces one of three sets of estimated cell frequencies. They are:

If A and B are conditionally independent, then

$$\hat{f}_{ijk} = f_{i.k} f_{.jk}/f_{..k};$$

If A and C are conditionally independent, then

$$\hat{f}_{ijk} = f_{.jk} f_{ij} / f_{.j.} \, ;$$

Or, if B and C are conditionally independent, then

$$\hat{f}_{ijk} = f_{ij.} f_{i.k} / f_{i..} \, . \tag{9.2}$$

Once again it is instructive to present artificial numbers (Table 9.6) that show exact conditional independence (no random variation) between two variables (A and B). Note that for table C_1, the odds ratio between A and B is

$$\hat{OR}_{AB \mid C_1} = \frac{\hat{f}_{111} \hat{f}_{221}}{\hat{f}_{121} \hat{f}_{211}} = \frac{(f_{1.1} f_{.11}/f_{..1})\,(f_{2.1} f_{.21}/f_{..1})}{(f_{1.1} f_{.21}/f_{..1})\,(f_{2.1} f_{.11}/f_{..1})} = 1.0 \, .$$

Similarly, $\hat{OR}_{AB \mid C_2} = 1.0$ but $\hat{OR}_{AB} \neq 1.0$, showing that variables A and B are independent at each level of C but not independent in the combined $(C_1 + C_2)$-table. This demonstrates that A and B are not independent of C.

For conditionally independent variables A and B, the relationship between variables B and C is unchanged in a table where A_1 and A_2 are combined. For example, the odds ratio in the subtables A_1 and A_2 is the same as the odds ratio in the combined table $A_1 + A_2$. This means that $\hat{OR}_{BC \mid A_1} = \hat{OR}_{BC \mid A_2} = OR_{BC} = 1.758$, as illustrated in Table 9.7. Table 9.7 is the same "data" displayed in Table 9.6 but rearranged to better show that variables B and C have identical relationships in all three tables. The rela-

T A B L E **9.6**

Artificial data that conform to exact conditional independence of variables A and B

	Level 1					Level 2		
C_1	B_1	B_2	Total		C_2	B_1	B_2	Total
A_1	32	48	80		A_1	33	87	120
A_2	48	72	120		A_2	187	493	680
Total	80	120	200		Total	220	580	800

	Level 1 + Level 2		
$C_1 + C_2$	B_1	B_2	Total
A_1	65	135	200
A_2	235	565	800
Total	300	700	1000

T A B L E **9.7**

Artificial data that conform to exact conditional independence of variables *A* and *B* (rearranged from Table 9.6)

Table 1

A_1	C_1	C_2	Total
B_1	32	33	65
B_2	48	87	135
Total	80	120	200

Table 2

A_2	C_1	C_2	Total
B_1	48	187	235
B_2	72	493	565
Total	120	680	800

Table 1 + Table 2

$A_1 + A_2$	C_1	C_2	Total
B_1	80	220	300
B_2	120	580	700
Total	200	800	1000

tionship between *A* and *C* also remains unchanged in a table formed by adding over variables B_1 and B_2. For example, the odds ratios for the illustrative "data" are the same in all three tables, B_1-table, B_2-table, and $(B_1 + B_2)$-table, or $\hat{OR}_{AC|B_1} = \hat{OR}_{AC|B_2} = \hat{OR}_{AC} = 3.778$. Conditional independence of a variable allows certain categories to be combined without changing the other associations.

Example

To illustrate conditional independence, consider data from a study of air quality restricted to 1061 U.S. counties with populations of more than 10,000 individuals, displayed in Table 9.8. The categorical variables are:

urban/rural status: A_1 = urban
                              A_2 = rural

average country income: B_1 = less than or equal to \$22,000
                              B_2 = greater than \$22,000

total suspended particulates (TSP): C_1 = low levels of TSP
                                        C_2 = intermediate levels of TSP
                                        C_3 = high levels of TSP.

T A B L E   **9.8**

Data on 1061 counties classified by urban/rural status (A), average income (B), and air quality (C)—estimated values are given in parentheses

C_1	B_1	B_2	Total
A_1	34 (17.97)	104 (120.03)	138
A_2	16 (32.03)	230 (213.97)	246
Total	50	334	384

C_2	B_1	B_2	Total
A_1	47 (30.37)	137 (153.63)	184
A_2	21 (37.63)	207 (190.37)	228
Total	68	344	412

C_3	B_1	B_2	Total
A_1	74 (49.66)	49 (73.34)	123
A_2	33 (57.34)	109 (84.66)	142
Total	107	158	265

The conjecture that urban/rural status (A) is conditionally independent of income status (B) generates the estimated cell frequencies $\hat{f}_{ijk} = f_{i \cdot k} f_{\cdot jk} / f_{\cdot \cdot k}$, or, specifically,

$$\hat{f}_{111} = (138)\,(50)/384 = 17.97 ,$$

$$\hat{f}_{112} = (184)\,(68)/412 = 30.37 , \cdots , \text{etc.}$$

The likelihood ratio chi-square test statistic is $Y^2 = 82.526$ with degrees of freedom = $(2)\,(2)\,(3) - 1 - (3 + 3 + 2) = 3$, yielding a p-value < 0.001 (expressions for the degrees of freedom are given later in the chapter in Table 9.29). This significance probability strongly indicates that urban/rural status and average county income are likely to be associated (variables A and B are associated at one or more levels of C).

An investigation of the conditional independence of TSP levels and the urban/rural status of a county is also possible. Shown in Table 9.9 are the observed and the estimated cell frequencies, $\hat{f}_{ijk} = f_{ij \cdot} f_{\cdot jk} / f_{\cdot j \cdot}$ (in parentheses). The likelihood ratio chi-square statistic of $Y^2 = 6.799$ with degrees of freedom $(2)\,(2)\,(3) - 2 - 2 - 4 = 4$ (p-value = 0.147) indicates no strong evidence of an association between urban/rural status and levels of TSP (variables A and C) within levels of income (variable B).

A third possible pattern of conditional independence occurs when income (B) and TSP (C) are unrelated but both variables are associated with urban/rural status (A). A test of this conjecture shows that this possibility is highly unlikely. Assuming conditional independence between income and TSP pollution levels yields a chi-square statistic of $Y^2 = 70.362$ with four degrees of freedom, producing a p-value of less than 0.001 (estimated values not shown).

T A B L E **9.9**

Data on TSP levels and the urban/rural status (*A*) of 1061 counties—estimated values are given in the parentheses

B_1	C_1 (TSP low)	C_2	C_3 (TSP high)	Total
A_1	34 (34.44)	47 (46.84)	74 (73.71)	155
A_2	16 (15.56)	21 (21.16)	33 (33.29)	70
Total	50	68	107	225

B_2	C_1 (TSP low)	C_2	C_3 (TSP high)	Total
A_1	104 (115.86)	137 (119.33)	49 (54.81)	290
A_2	230 (218.14)	207 (224.67)	109 (103.19)	546
Total	334	344	158	836

To summarize,

AB related?	*AC* related?	*BC* related?	Then
No	No	No	Complete independence
No	Yes	Yes	Conditional independence
Yes	No	Yes	Conditional independence
Yes	Yes	No	Conditional independence
No	No	Yes	Partial independence
No	Yes	No	Partial independence
Yes	No	No	Partial independence
Yes	Yes	Yes	No independence

Simpson's Paradox

Special conditions must exist, or forming lower-dimensional tables from higher-dimensional tables distorts the relationships under investigation. The rules for "combinability" of a table are discussed later. At this point, note that strange results can occur when tables are indiscriminately combined by adding over variables. A classic demonstration of this phenomenon is called Simpson's paradox [3]. Cancer "data" given in Table 9.10 illustrate the problems that arise from indiscriminately combining tables. The categorical variables are:

A_1 = died B_1 = usual treatment C_1 = stage IV cancer (severe)

A_2 = survived B_2 = new treatment C_2 = stage I cancer (not severe)

T A B L E **9.10**

Cancer "data" classified by survival (A), treatment (B), and stage (C) to illustrate Simpson's paradox

C_1	B_1	B_2	Total
A_1	950	9000	9950
A_2	50	1000	1050
Total	1000	10,000	11,000

C_2	B_1	B_2	Total
A_1	5000	5	5005
A_2	5000	95	5095
Total	10,000	100	10,100

$P(\text{survived} \mid \text{usual}) = P(A_2 \mid B_1) = 0.05$
$P(\text{survived} \mid \text{new}) = P(A_2 \mid B_2) = 0.10$

$P(\text{survived} \mid \text{usual}) = P(A_2 \mid B_1) = 0.50$
$P(\text{survived} \mid \text{new}) = P(A_2 \mid B_2) = 0.95$

The new treatment is roughly two times more effective for both stages C_1 and C_2. However, combining stages $(C_1 + C_2)$ gives Table 9.11. The combined table reverses the association. The usual treatment now appears about four times more effective. The reason for this paradox is that treatment (usual versus new) is highly associated with stage (I versus IV). Practically all stage IV patients get the new treatment and almost all stage I patients get the usual treatment. Therefore, when the tables are combined, contradictory results arise because

$$P(\text{survived} \mid \text{new}) = 0.11 \approx 0.10 = P(\text{survived} \mid \text{new and stage IV})$$
$$\text{from table } C_1 \text{ (stage IV)},$$
$$P(\text{survived} \mid \text{usual}) = 0.46 \approx 0.50 = P(\text{survived} \mid \text{usual and stage I})$$
$$\text{from table } C_2 \text{ (stage I)}.$$

The total table $(C_1 + C_2)$ is dominated by the relationship between treatment and stage, obscuring the association between survival and treatment. A log-linear model approach provides rules for combining tables, and Simpson's paradox ceases to be a paradox.

T A B L E **9.11**

Cancer "data" classified by survival (A) and treatment (B) ignoring stage $(C_1 + C_2)$ to illustrate Simpson's paradox

$C_1 + C_2$	B_1	B_2	Total
A_1	5950	9005	14,955
A_2	5050	1095	6145
Total	11,000	10,100	21,100

$P(\text{survived} \mid \text{usual}) = P(A_2 \mid B_1) = 0.46$
$P(\text{survived} \mid \text{new}) = P(A_2 \mid B_2) = 0.11$

T A B L E **9.12**

Artificial data illustrating the dangers of combining tables

C_1	B_1	B_2
A_1	6	8
A_2	6	9

C_2	B_1	B_2
A_1	6	22
A_2	3	12

C_3	B_1	B_2
A_1	4	17
A_2	6	27

C_4	B_1	B_2
A_1	4	43
A_2	3	33

Another Example

When tables are indiscriminately combined the situation can become even more paradoxical; consider Table 9.12. All four subtables show a positive association between variables A and B, which is seen by computing the odds ratios for each table ($\hat{OR}_1 = 1.13$, $\hat{OR}_2 = 1.09$, $\hat{OR}_3 = 1.06$, and $\hat{OR}_4 = 1.02$). If two tables are formed by adding tables C_1 and C_2 and tables C_3 and C_4, then

$C_1 + C_2$	B_1	B_2
A_1	12	30
A_2	9	21

$C_3 + C_4$	B_1	B_2
A_1	8	60
A_2	9	60

Now the association between variables A and B is negative (i.e., $\hat{OR}_{1+2} = 0.93$ and $\hat{OR}_{3+4} = 0.89$). Finally, if all four tables are combined, there is no association between variables A and B, and

$C_1 + C_2 + C_3 + C_4$	B_1	B_2
A_1	20	90
A_2	18	81

producing an odds ratio of $\hat{OR}_{1+2+3+4} = 1.0$.

The lesson of both Simpson's paradox and the artificial numbers in Table 9.12 is that the type and magnitude of an association between two variables may depend on the relationship to a third variable. Combining tables made up of such variables changes (sometimes dramatically) the relationships under investigation. However, complete, partial, and conditional independence allow certain variables to be combined while maintaining the relationships among the other variables. A log-linear model, one way to describe and identify relationships in a contingency table, provides a powerful tool to explore the possibility of combining tabular data to simplify analysis without distorting the relationships among the categorical variables.

Two-Way Tables

The representation of the relationships within a data set by a parametric model requires a certain amount of mathematical "overhead," but the effort pays off from a variety of perspectives. Statistical models (not only log-linear models) smooth the data, are effective descriptive tools, and focus the entire data set on estimating a few summary parameters. This last property makes a model approach particularly appealing when the data are sparsely distributed among the cells of a contingency table. The parameters of most statistical models have meaningful interpretations and serve as effective quantitative assessments of specific aspects of the relationships under investigation. A modeling approach also prescribes specific organizational strategies. For example, a log-linear analysis produces measures that reflect on the combinability of multidimensional data to simpler forms. Log-linear analysis does not automatically produce unique insights but is often a useful analytic tool. Like most tools, the quality of the results depends mostly on the skills of the user.

Two events A and B are independent when the probability of the joint occurrence is the product of the probabilities of each event, or $P(AB) = P(A)P(B)$. The essence of a log-linear model comes from the simple fact that when two events are independent then

$$\log[P(AB)] = \log[P(A)P(B)] = \log[P(A)] + \log[P(B)] .$$

Furthermore, the quantity

$$\Delta = \log[P(AB)] - (\log[P(A)] + \log[P(B)]) \tag{9.3}$$

is a linear measure of the magnitude of any association between events A and B. Stated simply, when two variables are independent, then the logarithm of the joint probability is a sum of logarithms of the individual probabilities; otherwise some degree of association exists.

A 2-by-2 contingency table provides the simplest possible introduction to the log-linear model. The notation for the 2-by-2 table is presented in Tables 9.13 (repeated from Table 8.8) and 9.14.

T A B L E **9.13**

Notation for the frequencies in a 2-by-2 contingency table

	B_1	B_2	Total
A_1	f_{11}	f_{12}	$f_{1.}$
A_2	f_{21}	f_{22}	$f_{2.}$
Total	$f_{.1}$	$f_{.2}$	n

T A B L E **9.14**

Notation for the logarithms of the frequencies in a 2-by-2 contingency table

	B_1	B_2	Mean
A_1	l_{11}	l_{12}	$\bar{l}_{1.}$
A_2	l_{21}	l_{22}	$\bar{l}_{2.}$
Mean	$\bar{l}_{.1}$	$\bar{l}_{.2}$	$\bar{l}$

A log-linear analysis of a 2-by-2 table, as well as more general tables, is conducted in terms of the logarithms of the row and column cell frequencies, or $l_{ij} = \log(f_{ij})$, yielding the 2-by-2 contingency table displayed in Table 9.14. The quantities $\bar{l}_{i.}$ and $\bar{l}_{.j}$ represent mean values of the logarithms of the row and column cell frequencies (e.g., $(l_{11} + l_{12})/2 = \bar{l}_{1.}$ or $(l_{12} + l_{22})/2 = \bar{l}_{.2}$). The l_{ij}-values and the corresponding mean values provide the opportunity to take advantage of the properties of a linear measure of association.

Estimation of the Model Parameters

A linear model of the logarithm of the frequencies used to represent the expected values of l_{ij} for a 2-by-2 table is

$$\textit{log-frequency} = u + u_1(i) + u_2(j) + u_{12}(i, j) \tag{9.4}$$

and, specifically, yields the expressions for the estimated value of the logarithm of the frequency for each cell in a 2-by-2 table (Table 9.15). The subscript on the u-terms refers to the categorical variables (variable A = variable 1 and variable B = variable 2) whereas the values inside the parentheses denote the row and column of the contingency table. For example, $\textit{log-frequency} = u + u_1(1) + u_2(2) + u_{12}(1, 2)$ represents the influence of variables A and B (subscript 1 and 2) in cell (1, 2). The parameter u denotes the overall mean level of the logarithms of the frequency, $u_1(1)$ denotes the influence of variable A, $u_2(2)$ denotes the influence of variable B, and

T A B L E **9.15**

Log-linear model for data contained in a 2-by-2 contingency table

	B_1	B_2	Mean Response
A_1	$u + u_1(1) + u_2(1) + u_{12}(1, 1)$	$u + u_1(1) + u_2(2) + u_{12}(1, 2)$	$u + u_1(1)$
A_2	$u + u_1(2) + u_2(1) + u_{12}(2, 1)$	$u + u_1(2) + u_2(2) + u_{12}(2, 2)$	$u + u_1(2)$
Mean response	$u + u_2(1)$	$u + u_2(2)$	u

$u_{12}(1, 2)$ represents the joint influence of both variables A and B. Also note that the model requires that:

$$u_1(2) = -u_1(1), \qquad u_2(2) = -u_2(1)$$

and

$$u_{12}(1, 1) = -u_{12}(1, 2) = -u_{12}(2, 1) = u_{12}(2, 2).$$

The linear nature of a log-linear representation of the data in a contingency table suggests that estimates of the u-terms are found using mean values. For the 2-by-2 table, four estimates specify all the u-terms necessary to represent the logarithms of the cell frequencies. They are:

overall mean $= u$:

$$\hat{u} = \bar{l} = \frac{l_{11} + l_{12} + l_{21} + l_{22}}{4} \tag{9.5}$$

the influence of the variable A_i, or $u_1(i)$:

$$\hat{u}_1(i) = \bar{l}_{i.} - \bar{l} = \frac{l_{i1} + l_{i2}}{2} - \bar{l} = \pm \frac{(l_{11} - l_{21}) + (l_{12} - l_{22})}{4} \tag{9.6}$$

the influence of variable B_j, or $u_2(j)$:

$$\hat{u}_2(j) = \bar{l}_{.j} - \bar{l} = \frac{l_{1j} + l_{2j}}{2} - \bar{l} = \pm \frac{(l_{11} - l_{12}) + (l_{21} - l_{22})}{4} \tag{9.7}$$

the joint influence of variables A_i and B_j, or $u_{12}(i, j)$:

$$\begin{aligned}
\hat{u}_{12}(i, j) &= (l_{ij} - \bar{l}) - (\bar{l}_{i.} - \bar{l}) - (\bar{l}_{.j} - \bar{l}) \\
&= l_{ij} - \bar{l}_{i.} - \bar{l}_{.j} + \bar{l} \\
&= \pm \frac{(l_{11} - l_{12}) - (l_{21} - l_{22})}{4}
\end{aligned} \tag{9.8}$$

These estimates have the properties that $\Sigma_i \hat{u}_1(i) = \Sigma_j \hat{u}_2(j) = \Sigma_i \hat{u}_{12}(i, j) = \Sigma_j \hat{u}_{12}(i, j) = 0$, which is consistent with the postulated log-linear model. The u_{12}-term is particular important because it is designed to reflect the amount of non-additivity of the u_1- and u_2-terms. In the context of a log-linear model, non-additivity of the logarithms of the frequencies indicates the degree of nonindependence between the frequencies themselves. In other words, $\hat{u}_{12}(i, j)$ is a linear measure of association.

As an example, consider two variables:

| age: | $A_1 =$ less than 30 years old |
| | $A_2 =$ equal to or greater than 30 years old; |

| work status: | $B_1 =$ working mother |
| | $B_2 =$ nonworking mother. |

T A B L E **9.16**

Data on 208 mothers classified by age ($< 30 = A_1$ or $\geq 30 = A_2$) and work status (working $= B_1$ or not working $= B_2$)

f_{ij}	B_1	B_2	Total
A_1	6	86	92
A_2	8	108	116
Total	14	194	208

yielding l_{ij}-values of:

l_{ij}	B_1	B_2	Mean
A_1	1.792	4.454	3.123
A_2	2.079	4.682	3.381
Mean	1.936	4.568	3.252

The data are displayed in Table 9.16. Estimates from these data (Table 9.16) of the parameters of a log-linear model are:

$$\hat{u} = \bar{l} = 3.252,$$
$$\hat{u}_1(1) = 3.123 - 3.252 = -0.129 \quad \text{and} \quad \hat{u}_1(2) = 0.129,$$
$$\hat{u}_2(1) = 1.936 - 3.252 = -1.316 \quad \text{and} \quad \hat{u}_2(2) = 1.316,$$
$$\hat{u}_{12}(1,1) = 1.792 + 0.129 + 1.316 - 3.252 = -0.015, \quad \text{and}$$
$$\hat{u}_{12}(1,2) = 0.015, \hat{u}_{12}(2,1) = 0.015, \hat{u}_{12}(2,2) = -0.015.$$

The four estimates $(\hat{u}, \hat{u}_1, \hat{u}_2,$ and $\hat{u}_{12})$ exactly reproduce the observed values for the four cells. For example,

$$\hat{l}_{11} = \hat{u} + \hat{u}_1(1) + \hat{u}_2(1) + \hat{u}_{12}(1,1)$$
$$= 3.252 - 0.129 - 1.316 - 0.015 = 1.792 = l_{11}$$

and

$$\hat{f}_{11} = f_{11} = e^{\hat{l}_{11}} = e^{1.792} = 6.000.$$

The Saturated Model

When the number of independent estimated values used to specify the model equals the number of cells in the table, the model is called saturated and the estimated values are identical to the observed values $(\hat{l}_{ij} = l_{ij}$ and $\hat{f}_{ij} = f_{ij})$. Because the model perfectly reproduces the data, a reasonable question is: What use is a saturated model? A saturated model has two important properties. First, it serves as a starting point for comparisons of models that do not fit the data perfectly. Second, the parameters of the saturated model separate the observed l_{ij}-values into components where each component has an interpretation. The u-term estimates the average level. The u_i-terms reflect the frequency of the ith variable, which is rarely of interest. The u_{ij}-terms indicate the joint influence of the categorical variables. An estimate of u_{ij} measures the degree of nonadditive influence on the l_{ij}-values from the categorical variables, producing, as noted, a linear measure of the magnitude of statistical association.

An alternative form of the estimate of the u_{12}-term from the 2-by-2 table provides insight into its interpretation. A few algebraic steps produce

$$\hat{u}_{12}(i,j) = \pm \frac{l_{11} - l_{12} - l_{21} + l_{22}}{4} = \pm \frac{1}{4} \log\left[\frac{f_{11}f_{22}}{f_{12}f_{21}}\right], \tag{9.9}$$

showing that the parameter u_{12} is directly related to the odds ratio (expression 8.27). Specifically,

$$\hat{u}_{12} = \pm \frac{1}{4} \log[\hat{OR}] \quad \text{because} \quad \hat{OR} = \frac{f_{11}f_{22}}{f_{12}f_{21}}. \tag{9.10}$$

A simpler form of notation for the u-terms (no column or row designation) is used when no ambiguity arises. When $u_{12} = 0$, the categorical variables are perfectly independent ($OR = 1.0$). Nonzero values of u_{12} indicate values of the odds ratio different from 1.0 and reflect the degree of association between variables 1 and 2. The basic difference between u_{12} and OR is that u_{12} is a linear combination of model parameters, which is the fundamental feature of a log-linear model in general.

Inference from a Log-Linear Model, 2-by-2 Case

A useful function of $\hat{u}_{12}$ is

$$z = \frac{\hat{u}_{12}}{\sqrt{variance(\hat{u}_{12})}}. \tag{9.11}$$

This expression has an approximate standard normal distribution when the categorical variables are independent, providing a test statistic for the hypothesis of independence (i.e., $H_0: u_{12} = 0$). The estimated variance of $\hat{u}_{12}$ (similar to expression 8.28) is

$$variance(\hat{u}_{12}) = \frac{1}{16}\left[\frac{1}{f_{11}} + \frac{1}{f_{12}} + \frac{1}{f_{21}} + \frac{1}{f_{22}}\right]. \tag{9.12}$$

Another approach to evaluating the association between two categorical variables A and B comes from postulating $u_{12} = 0$ and fitting the reduced log-linear model $u + u_1 + u_2$ to the observed data. This form of a log-linear model is referred to as an additive model. If the additive model closely fits the observed data, then no statistical evidence exists to indicate that the variables *A* and *B* are associated. In other words, the parameter u_{12} is not needed to describe the data because no important joint influence of *A* and *B* affects the frequencies observed in the 2-by-2 table. When the model without u_{12} fits the data poorly, the term u_{12} is then a valuable part of the description of the data implying that *A* and *B* are likely associated.

More technically, the log-linear model when *A* and *B* are independent (additive log-linear model) is

$$log\text{-}frequency = u + u_1(i) + u_2(j) \tag{9.13}$$

where the estimates of these parameters for the 2-by-2 table are given by

$$\hat{u} = \tfrac{1}{2}[\log(f_1.) + \log(f_{.1}) + \log(f_2.) + \log(f_{.2})] - \log(n) ,$$

$$\hat{u}_1(i) = \pm \tfrac{1}{2}[\log(f_1.) - \log(f_2.)] , \tag{9.14}$$

and

$$\hat{u}_2(j) = \pm \tfrac{1}{2}[\log(f_{.1}) - \log(f_{.2})] . \tag{9.15}$$

Continuing the previous example (Table 9.16), the estimated parameters are:

$$\hat{u} = 3.254 ,$$

$$\hat{u}_1(1) = -0.116, \quad \hat{u}_1(2) = 0.116, \text{ and}$$

$$\hat{u}_2(1) = -1.314, \quad \hat{u}_2(2) = 1.314 ,$$

yielding estimated l_{ij}-values and $\hat{f}_{ij} = e^{\hat{l}_{ij}}$ are the estimated cell frequencies when variables *A* and *B* are independent ($u_{12} = 0$). The estimated values ($\hat{l}_{ij}$ and $\hat{f}_{ij}$) given in Table 9.17 correspond to the observed l_{ij}-values and f_{ij}-values from the original data found in Table 9.16.

Estimated values from the additive log-linear model are the same as those produced by the usual approach to calculating estimated values under the hypothesis of independence, or,

$$\hat{l}_{ij} = \hat{u} + \hat{u}_1(i) + \hat{u}_2(j)$$

$$= \log(f_i. f_{.j}/n)$$

and then

$$\hat{f}_{ij} = e^{\hat{l}_{ij}} = f_i. f_{.j}/n .$$

The likelihood ratio chi-square statistic (expression 8.8) comparing the observed (Table 9.16, top) and estimated values (Table 9.17, bottom) is $Y^2 = 2[(6)\log(6/6.192) + \cdots + (108)\log(108/108.192)] = 0.012$ with one degree of freedom, yielding a *p*-value of 0.915. This is not very different from testing the H_0: $u_{12} = 0$ where $z = -0.015/0.140 = -0.107$ (expression 9.11) with an associated *p*-value of

T A B L E   **9.17**

Estimated values for the data in Table 9.16 based on independence of age and work status

$\hat{l}_{ij}$	B_1	B_2	Mean
A_1	1.823	4.452	3.138
A_2	2.055	4.684	3.370
Mean	1.939	4.568	3.253

$\hat{f}_{ij}$	B_1	B_2	Total
A_1	6.192	85.808	92
A_2	7.808	108.192	116
Total	14	194	208

0.915 ($Y^2 \approx z^2 = 0.011$). These two approaches are usually similar, particularly for tables with large numbers of observations in all cells.

Geometric Interpretation

Geometrically, independence means that the log-frequencies from a 2-by-2 contingency table can be displayed as two parallel lines. Conversely, nonindependence is measured by the degree to which two lines connecting the l_{ij}-values are not parallel. The estimated log-frequencies $\hat{l}_{11}$, $\hat{l}_{12}$, $\hat{l}_{21}$, and $\hat{l}_{22}$ are displayed in Figure 9.1 for the age and work-status data where the two variables are postulated as independent (Table 9.17, top). The geometry of the additive log-linear model shows $\hat{l}_{11} - \hat{l}_{12} = \hat{l}_{21} - \hat{l}_{22}$ and $\hat{l}_{11} - \hat{l}_{21} = \hat{l}_{12} - \hat{l}_{22}$, which are just other ways of expressing that $\hat{u}_{12} = (\hat{l}_{11} - \hat{l}_{12} - \hat{l}_{21} + \hat{l}_{22})/4 = 0$. Specifically for the estimates in Table 9.17 (top), $1.823 - 2.055 = 4.452 - 4.684$ (influence of A) and $1.823 - 4.452 = 2.005 - 4.684$ (influence of B); therefore, $(1.823 - 4.452 - 2.055 + 4.684)/4 = 0$. Using the logarithm of the cell frequencies produces an elegant geometric interpretation of independence in a 2-by-2 table.

One further property: If the rows contain the same values in both columns, making $u_2 = 0$, then necessarily $u_{12} = 0$ and, similarly, $u_1 = 0$ implies $u_{12} = 0$. This property indicates a hierarchical nature of the log-linear model, which also is found in the higher-dimensional table.

To summarize, the log-linear model applied to the 2-by-2 contingency table produces the same estimated values, the same chi-square statistic, and the same results as the previously described nonmodel approach (Chapter 8). However, a log-linear analysis of a 2-by-2 table provides an introduction to the log-linear approach that generalizes to the analysis of higher-dimensional contingency tables.

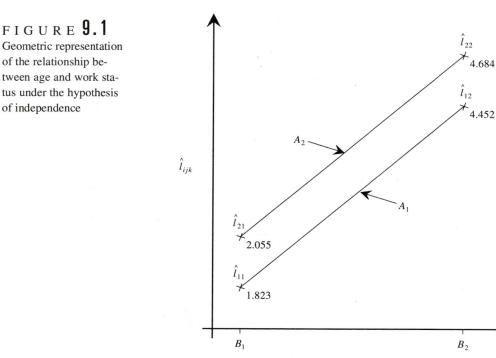

F I G U R E **9.1**
Geometric representation of the relationship between age and work status under the hypothesis of independence

The *r*-by-*c* Table

The log-linear analysis of a two-dimensional *r*-by-*c* contingency table is a direct extension of the 2-by-2 table. The saturated model is again given by the expression *log-frequency* = $u + u_1 + u_2 + u_{12}$. As before, the subscripts 1 and 2 denote the row and column categorical variables of the *r*-by-*c* table. The estimates of the *u*-terms are also analogous to the 2-by-2 case, and

$$\hat{u} = \bar{l} = \frac{\Sigma\Sigma l_{ij}}{rc},$$

$$\hat{u}_1(i) = \frac{1}{c}\sum_j l_{ij} - \bar{l} = \bar{l}_{i.} - \bar{l},$$

$$\hat{u}_2(j) = \frac{1}{r}\sum_i l_{ij} - \bar{l} = \bar{l}_{.j} - \bar{l}, \text{ and}$$

$$\hat{u}_{12}(i,j) = l_{ij} - [\hat{u} + \hat{u}_1(i) + \hat{u}_2(j)]$$
$$= (l_{ij} - \bar{l}) - (\bar{l}_{i.} - \bar{l}) - (\bar{l}_{.j} - \bar{l})$$
$$= l_{ij} - \bar{l}_{i.} - \bar{l}_{.j} + \bar{l}. \tag{9.16}$$

For the additive model (complete independence; $u_{12} = 0$), the estimated cell frequencies are $\hat{l}_{ij} = \hat{u} + \hat{u}_1 + \hat{u}_2$ where $\hat{l}_{ij} = \log(f_{i.}\,f_{.j}/n)$; this is again a direct extension of the 2-by-2 table.

The data concerning response to a specific question (yes or no) and age (Table 9.5; combined table—$C_1 + C_2$) produce the following parameter estimates for the log-linear model:

$$log\text{-}frequency = u + u_1 + u_2 + u_{12} \text{ (saturated model):}$$

mean: $\hat{u} = 2.877$

age: $\hat{u}_1(1) = -0.941$ response: $\hat{u}_2(1) = 0.971$

$\hat{u}_1(2) = -0.228$ $\hat{u}_2(1) = -0.971$

$\hat{u}_1(3) = 1.169$

age-by-response association: $\hat{u}_{12}(1,1) = 0.271$ $\hat{u}_{12}(1,2) = -0.271$

$\hat{u}_{12}(2,1) = 0.292$ $\hat{u}_{12}(2,2) = -0.292$

$\hat{u}_{12}(3,1) = -0.563$ $\hat{u}_{12}(3,2) = 0.563.$

It is easily verified that the parameters exactly reproduce the data. For example,

$$\hat{l}_{12} = 2.877 - 0.941 - 0.971 - 0.271 = 0.693$$

$$\hat{f}_{12} = e^{\hat{l}_{12}} = e^{0.693} = 2.000 = f_{12}\,.$$

The additive model $log\text{-}frequency = u + u_1 + u_2$ ($u_{12} = 0$ or response is independent of age) produces the estimated values observed previously (Table 9.5). For example,

$$\hat{l}_{12} = \hat{u} + \hat{u}_1(1) + \hat{u}_2(2)$$

$$= \log[(26)\,(44)/204] = 1.724$$

and

$$\hat{f}_{12} = e^{\hat{l}_{12}} = e^{1.724} = 5.608\,.$$

Like the 2-by-2 table, the log-linear model reproduces the previous estimated values (expression 8.14) and is presented as an introduction to the three-dimensional table, where several more complex issues arise.

Three-Way Tables

Extension of the log-linear model approach to a three-dimensional table begins with the analysis of a 2-by-2-by-2 contingency table, followed by the more general and parallel approach to analyzing three-way tables of any size. The notation for a table of logarithms of the cell frequencies ($l_{ijk} = \log[f_{ijk}]$) is shown in Table 9.18.

T A B L E   **9.18**

Notation for the logarithms of the frequencies in a 2-by-2-by-2 contingency table

C_1	B_1	B_2	Mean		C_2	B_1	B_2	Mean
A_1	l_{111}	l_{121}	$\bar{l}_{1 \cdot 1}$		A_1	l_{112}	l_{122}	$\bar{l}_{1 \cdot 2}$
A_2	l_{211}	l_{221}	$\bar{l}_{2 \cdot 1}$		A_2	l_{212}	l_{222}	$\bar{l}_{2 \cdot 2}$
Mean	$\bar{l}_{\cdot 11}$	$\bar{l}_{\cdot 21}$	$\bar{l}_{\cdot \cdot 1}$		Mean	$\bar{l}_{\cdot 12}$	$\bar{l}_{\cdot 22}$	$\bar{l}_{\cdot \cdot 2}$

Estimation

For the saturated log-linear model, the logarithm of each cell frequency in a three-way table is represented as a sum of eight u-terms, or

$$\text{log–frequency} = u + u_1(i) + u_2(j) + u_3(k) \tag{9.17}$$
$$+ u_{12}(i, j) + u_{13}(i, k) + u_{23}(j, k) + u_{123}(i, j, k) ,$$

where i designates the row, j the column, and k the level of any observation classified into a three-dimensional table. Like all saturated log-linear models, the number of terms in the model is equal to the number of observed values.

Single subscript terms are again a function of the frequencies of the categorical variables, the double subscript terms quantify the pairwise associations, and u_{123}-terms indicate differences in the pairwise associations at different levels of a third variable. The estimates of these components are analogous to the estimates used in the analysis of variance and parallel the estimates made from a 2-by-2 contingency table. They are again linear combinations of mean values, and

overall mean: $\hat{u} = \bar{l}$

main effects: $\hat{u}_1(i) = \bar{l}_{i \cdot \cdot} - \bar{l}$, $\hat{u}_2(j) = \bar{l}_{\cdot j \cdot} - \bar{l}$, $\hat{u}_3(k) = \bar{l}_{\cdot \cdot k} - \bar{l}$

2-way interaction: $\hat{u}_{12}(i, j) = \bar{l}_{ij \cdot} - \bar{l}_{i \cdot \cdot} - \bar{l}_{\cdot j \cdot} + \bar{l}$,
$\hat{u}_{13}(i, k) = \bar{l}_{i \cdot k} - \bar{l}_{i \cdot \cdot} - \bar{l}_{\cdot \cdot k} + \bar{l}$,
$\hat{u}_{23}(j, k) = \bar{l}_{\cdot jk} - \bar{l}_{\cdot j \cdot} - \bar{l}_{\cdot \cdot k} + \bar{l}$

3-way interaction: $\hat{u}_{123}(i, j, k) = \hat{l}_{ijk} - [\hat{u} + \hat{u}_{12}(i, j) + \hat{u}_{13}(i, k) + \hat{u}_{23}(j, k) + \hat{u}_1(i) + \hat{u}_2(j) + \hat{u}_3(k)]$

The estimate of the logarithm of the cell frequency is then the sum of eight estimated components, or

$$\hat{l}_{ijk} = \hat{u} + \hat{u}_1(i) + \hat{u}_2(j) + \hat{u}_3(k)$$
$$+ \hat{u}_{12}(i, j) + \hat{u}_{13}(i, k) + \hat{u}_{23}(j, k) + \hat{u}_{123}(i, j, k) . \tag{9.18}$$

For the 2-by-2-by-2 table, the estimates of the u-terms are:

overall mean: $\quad \hat{u} = \bar{l} = (l_{111} + l_{121} + l_{211} + l_{221} + l_{112} + l_{122} + l_{212} + l_{222})/8$

main effects: $\quad \hat{u}_1(i) = \pm (l_{111} + l_{121} - l_{211} - l_{221} + l_{112} + l_{122} - l_{212} - l_{222})/8$

$\qquad\qquad\quad \hat{u}_2(j) = \pm (l_{111} - l_{121} + l_{211} - l_{221} + l_{112} - l_{122} + l_{212} - l_{222})/8$

$\qquad\qquad\quad \hat{u}_3(k) = \pm (l_{111} + l_{121} + l_{211} + l_{221} - l_{112} - l_{122} - l_{212} - l_{222})/8$

2-way interaction: $\quad \hat{u}_{12}(i,j) = \pm (l_{111} - l_{121} - l_{211} + l_{221} + l_{112} - l_{122} - l_{212} + l_{222})/8$

$\qquad\qquad\qquad\quad \hat{u}_{13}(i,k) = \pm (l_{111} + l_{121} - l_{211} - l_{221} - l_{112} - l_{122} + l_{212} + l_{222})/8$

$\qquad\qquad\qquad\quad \hat{u}_{23}(j,k) = \pm (l_{111} - l_{121} + l_{211} - l_{221} - l_{112} + l_{122} - l_{212} + l_{222})/8$

3-way interaction: $\quad \hat{u}_{123}(i,j,k) = \pm (l_{111} - l_{121} - l_{211} + l_{221} - l_{112} + l_{122} + l_{212} - l_{222})/8$.

As with the saturated model for the 2-by-2 table, the estimated values exactly reproduce the data using the pattern of plus and minus signs shown in Table 9.19 (i.e., $l_{ijk} = \hat{l}_{ijk}$). For example, $\hat{l}_{122} = \hat{u} + \hat{u}_1 - \hat{u}_2 - \hat{u}_3 - \hat{u}_{12} - \hat{u}_{13} + \hat{u}_{23} + \hat{u}_{123} = l_{122}$, and in general $e^{\hat{l}_{ijk}} = \hat{f}_{ijk} = f_{ijk}$.

Average of Two Measures of Association

The u_{ij}-terms in the model for the 2-by-2-by-2 table have alternative expressions as functions of odds ratios. The pairwise u_{ij}-terms are the logarithms of the product (or sums of logarithms) of two odds ratios. For example,

$$\hat{u}_{12} = \pm ([l_{111} - l_{121} - l_{211} + l_{221}]/4 + [l_{112} - l_{122} - l_{212} + l_{222}]/4)/2$$

$$\hat{u}_{12} = \pm \frac{1}{8} \log([\hat{OR}_1][\hat{OR}_2]) = \pm \frac{1}{8} (\log[\hat{OR}_1] + \log[\hat{OR}_2]) \tag{9.19}$$

where $\hat{OR}_1$ and $\hat{OR}_2$ are the odds ratios reflecting the association between variables 1 and 2 calculated from the 2-by-2 subtables of variable 3. The estimate $\hat{u}_{ij}$ is an aver-

T A B L E **9.19**

Estimates for the saturated model from the estimated components

	$\hat{u}$	$\hat{u}_1$	$\hat{u}_2$	$\hat{u}_3$	$\hat{u}_{12}$	$\hat{u}_{13}$	$\hat{u}_{23}$	$\hat{u}_{123}$
$\hat{l}_{111}$	+	+	+	+	+	+	+	+
$\hat{l}_{211}$	+	−	+	+	−	−	+	−
$\hat{l}_{121}$	+	+	−	+	−	+	−	−
$\hat{l}_{221}$	+	−	−	+	+	−	−	+
$\hat{l}_{112}$	+	+	+	−	+	−	−	−
$\hat{l}_{212}$	+	−	+	−	−	+	−	+
$\hat{l}_{122}$	+	+	−	−	−	−	+	+
$\hat{l}_{222}$	+	−	−	−	+	+	+	−

age of two measures of the association between variables i and j, one from each sub-table or, for example,

$$estimated\ association = \frac{\hat{u}_{12}[from\ table\ 1] + \hat{u}_{12}[from\ table\ 2]}{2}.$$

Like most averages, an average of two measures of association only makes sense if the two components estimate the same quantity. The u_{123}-term is a function of the difference in the two components that make up the u_{ij}-term and indicates differences in the pairwise associations. The interpretation of the pairwise u_{ij}-terms depends on the magnitude of u_{123}.

The estimate of u_{123} from a 2-by-2-by-2 table also has a useful expression in terms of odds ratios, which is

$$\hat{u}_{123} = \pm\ ([l_{111} - l_{121} - l_{211} + l_{221}]/4 - [l_{112} - l_{122} - l_{212} + l_{222}]/4)/2 \quad or$$

$$\hat{u}_{123} = \pm\frac{1}{8} \log\left[\frac{f_{111}f_{221}}{f_{121}f_{211}} \Big/ \frac{f_{112}f_{222}}{f_{122}f_{212}}\right]$$

and

$$\hat{u}_{123} = \pm\frac{1}{8} \log[\hat{OR}_1/\hat{OR}_2] = \pm\frac{1}{8} (\log[\hat{OR}_1] - \log[\hat{OR}_2])$$

(9.20)

where $\hat{OR}_1$ and $\hat{OR}_2$ are again the odds ratios from the two subtables formed from a third variable. The term $\hat{u}_{123}$, for example, is a function of the difference in the strength of association between variables 1 and 2 measured at each level of variable 3 (expression 9.20). In fact, u_{123} reflects the difference in odds ratios between any pair of variables compared at each level of any third variable. In general, $u_{123} = 0$ implies that the degree of association is the same between any two variables at all levels of a third variable. For example, $u_{123} = 0$ implies that the average of measures of association between variables 1 and 2 over the levels of variable 3 (u_{12}) is a meaningful combination of the two pairwise associations. To the extent that u_{123} is not zero, the average of the pairwise measures of association becomes a less accurate assessment of statistical independence because u_{ij} then represents an average of differing quantities.

An artificial but concrete example of the interplay between u_{ij} and u_{ijk} is given in Table 9.20. For Table 9.20, a summary of the joint influence of variables A and B is

$$\hat{u}_{12} = \frac{1}{8} \log\left\{\left[\frac{(20)\,(20)}{(5)\,(5)}\right]\left[\frac{(5)\,(5)}{(20)\,(20)}\right]\right\} = 0\ ,$$

which is misleading because variables A and B are obviously related. The odds ratio from table C_1 is 16 and from table C_2 is 1/16. However,

$$\hat{u}_{123} = \frac{1}{8} \log\left[\frac{(20)\,(20)}{(5)\,(5)}\right]\left[\frac{(5)\,(5)}{(20)\,(20)}\right] = 0.693$$

indicating that $\hat{u}_{12}$ is the average of two differing (in fact, opposite) pairwise associations between variables A and B. To repeat, the estimate $\hat{u}_{ij}$ accurately measures the association between variables i and j only when $\hat{u}_{123} = 0$, or, at least, is small.

T A B L E   **9.20**

Artificial data illustrating the relationship between u_{ij} and u_{ijk}

C_1	B_1	B_2	Total		C_2	B_1	B_2	Total
A_1	20	5	25		A_1	5	20	25
A_2	5	20	25		A_2	20	5	25
Total	25	25	50		Total	25	25	50

A particularly noteworthy case arises when two variables are conditionally independent. Conditional independence of variables 1 and 2 means that the odds ratio at both levels of variable 3 is 1.0, making $u_{12} = 0$; their ratio (OR_1/OR_2) is therefore 1.0, also making $u_{123} = 0$.

Example

Data collected on male (A_1) and female (A_2) hypertension patients to investigate the difference between younger patients (age ≤ 45 years old—B_1) and older patients (age > 45 years old—B_2) are shown in Table 9.21, classified as not often depressed (C_1) and often depressed (C_2). Estimates of the components of the saturated log-linear model are:

$$\hat{u} = 0.985$$
$$\hat{u}_1(1) = 0.063 = -\hat{u}_1(2)$$
$$\hat{u}_2(1) = 0.308 = -\hat{u}_2(2)$$
$$\hat{u}_3(1) = 0.039 = -\hat{u}_3(2)$$
$$\hat{u}_{12}(1,1) = \frac{1}{8} \log \left\{ \left[\frac{(2)(5)}{(2)(3)} \right] \left[\frac{(11)(1)}{(4)(1)} \right] \right\} = 0.190$$
$$= -\hat{u}_{12}(1,2) = -\hat{u}_{12}(2,1) = \hat{u}_{12}(2,2).$$

Similarly,

$$\hat{u}_{13}(i,k) = \pm 0.190$$
$$\hat{u}_{23}(j,k) = \pm 0.638$$

and the $\hat{u}_{123}$ term is

$$\hat{u}_{123}(i,j,k) = \pm \frac{1}{8} \log \left[\frac{(2)(5)}{(2)(3)} \Big/ \frac{(11)(1)}{(4)(1)} \right]$$
$$= \pm \frac{1}{8} \log[1.667/2.750)] = \pm 0.063.$$

T A B L E **9.21**

Patients classified by gender (*A*), age (*B*), and mental depression status (*C*)

C_1 = Status	B_1 = Young	B_2 = Old	Total	C_2 = Status	B_1 = Young	B_2 = Old	Total
A_1 = Male	2	3	5	A_1 = Male	11	1	12
A_2 = Female	2	5	7	A_2 = Female	4	1	5
Total	4	8	12	Total	15	2	17

The estimates exactly reproduce the data (saturated model); for example,

$$\hat{l}_{111} = \hat{u} + \hat{u}_1(1) + \hat{u}_2(1) + \cdots + \hat{u}_{123}(1,1,1)$$
$$= 0.985 + 0.063 + 0.308 + 0.039 + 0.190 - 0.190 - 0.638 - 0.063$$
$$= 0.693 \quad \text{and} \quad e^{\hat{l}_{111}} = 2.000 = f_{111}.$$

Inference

The parameter estimates from a saturated model, like all estimates, are subject to sampling variation and can be statistically evaluated if an estimate of this variation is available. An approximate variance of the logarithm of a cell frequency f_{ijk} is $variance(\hat{l}_{ijk}) = 1/f_{ijk}$ [4]. Using this property, a linear combination of independent $\hat{l}_{ijk}$-values

$$L = \Sigma\Sigma\Sigma \, a_{ijk}\hat{l}_{ijk} \tag{9.21}$$

has an approximate variance given by

$$variance(L) = \Sigma\Sigma\Sigma \, a_{ijk}^2 \, variance(\hat{l}_{ijk}) = \Sigma\Sigma\Sigma \, \frac{a_{ijk}^2}{f_{ijk}}. \tag{9.22}$$

To evaluate $\hat{u}_{123}$, the variance is then estimated by

$$variance(\hat{u}_{123}) = \frac{1}{64} \, \Sigma\Sigma\Sigma \, \frac{1}{f_{ijk}}$$

and the standardized value

$$z = \hat{u}_{123}/\sqrt{variance(\hat{u}_{123})}$$

has an approximate standard normal distribution when $u_{123} = 0$. For the example data (Table 9.21),

$$variance(\hat{u}_{123}) = \frac{1}{64}\left[\frac{1}{2} + \frac{1}{3} + \cdots + \frac{1}{1}\right] = 0.061$$

and

$$z = -\frac{0.063}{\sqrt{0.061}} = -0.254,$$

indicating that $\hat{u}_{123}$ is not likely different from zero (p-value = 0.799). The previous estimate $\hat{u}_{12}$ (expression 9.8) is a linear combination of the l_{ij}-values, and the variance (expression 9.12) is also a special case of expression 9.22.

A small value of $\hat{u}_{123}$ is a prerequisite for assessing the summary pairwise associations $\hat{u}_{ij}$. Table 9.22 contains the parameter estimates and summarizes the testing of the important components of the log-linear model applied to the data in Table 9.21. The u_{123}-term (Table 9.22) is small (likely by chance), implying that the inferences drawn from the u_{ij}-terms will not be misleading. For the 2-by-2-by-2 table, the estimated *variance*($\hat{u}_{ij}$) is the same as *variance*($\hat{u}_{123}$) because all estimates of the u-terms are made up of the same eight l_{ijk}-values and only differ in the configuration of the signs, which has no effect on the estimated variance (i.e., $a_{ijk}^2 = 1/64$ for all terms). The estimated variance (0.061) and the three z-statistics

$$z = \hat{u}_{ij} / \sqrt{variance(\hat{u}_{ij})}$$

allow an evaluation of the magnitude of each pairwise term in the log-linear model. The results indicate that variables A and B (gender and age, p-value = 0.439) and variables A and C (gender and depression, p-value = 0.439) are not likely to be associated whereas variables B and C (age and depression, p-value = 0.009) are likely to be associated. The data are consistent with inferring that the variable A is independent of both variables B and C—partial independence.

Analyzing a contingency table by evaluating the estimates made from a saturated model has limitations. When data are classified into categories with more than two levels, the u-terms have more than one absolute value and computing standardized values for all estimates becomes awkward to interpret. Also, like regression coefficients, the estimates of the u-terms depend on the model employed. If the model is inappropriate, the estimates are inappropriate. A more general method to assess the role of each categorical variable in a contingency table is discussed in the next section.

T A B L E **9.22**

Summary of four components of the log-linear model applied to age/gender/depression data (Table 9.21)

Variables	Component	Estimate	Variance	z	p-value
A and B and C	$\hat{u}_{123}$	± 0.063	0.061	± 0.254	0.798
A and B	$\hat{u}_{12}$	± 0.190	0.061	± 0.773	0.439
A and C	$\hat{u}_{13}$	± 0.190	0.061	± 0.773	0.439
B and C	$\hat{u}_{23}$	± 0.638	0.061	± 2.594	0.009

Contrasting Log-Linear Models

All Pairwise Associations Present—No Independence

Components of a log-linear model are readily evaluated in much the same manner as the components of a regression model—namely, by comparing nested models. A model containing the components to be assessed is constructed and compared to another constructed without these specific components. To evaluate the influence of the u_{123}-term, the fit of the saturated log-linear model (u_{123} included) is compared to the fit of a constrained log-linear model with u_{123} set equal to zero where

$$log\text{-}frequency = u + u_1 + u_2 + u_3 + u_{12} + u_{13} + u_{23} \ (u_{123} = 0) . \qquad (9.23)$$

To estimate the parameters based on the constrained model ($u_{123} = 0$) requires iterative (computer) techniques. For the hypertension data (Table 9.21), the estimated cell frequencies generated from the log-linear model with $u_{123} = 0$ are given in Table 9.23. The $\hat{l}_{ij}$-values are estimated under the constrained model and the estimated cell frequencies $\hat{f}_{ijk} = e^{\hat{l}_{ij}}$ are computed to evaluate the fit of the postulated model. The odds ratios in both tables are identical ($\hat{OR}_{AB|C_1} = \hat{OR}_{AB|C_2} = 1.998$), because u_{123} was set to 0 while the marginal frequencies of the estimated values are the same as the observed marginal frequencies (Table 9.21). Contrasting these estimated cell frequencies with the original data (Table 9.21) produces a likelihood ratio chi-square value for goodness-of-fit of $Y^2 = 0.065$, which has an approximate chi-square distribution with one degree of freedom (p-value = 0.799) when u_{123} is zero. Because the saturated model fits the data perfectly ($Y^2 = 0$), the observed chi-square value of $Y^2 = 0.065$ is totally attributable to the influence of removing u_{123} from the model. Therefore, a reasonable conclusion to draw is that the u_{123}-term is not an important part of the description of the data and the pairwise u-terms directly summarize statistical association. Expectedly, this result is similar to the previous test of $u_{123} = 0$ where the p-value was 0.798 ($z^2 = (-0.254)^2 = 0.065$). The details of the procedure for estimating the parameters of a log-linear model with $u_{123} = 0$ are not given ([3] or [4]). Statistical computer programs to produce estimates and calculate chi-square statistics for applications of log-linear models are widely available.

T A B L E **9.23**

Estimated cell frequencies for patients classified by gender (A), age (B), and mental depression status (C) under the condition $u_{123} = 0$

C_1	B_1	B_2	Total		C_2	B_1	B_2	Total
A_1	2.13	2.87	5.00		A_1	10.88	1.12	12.00
A_2	1.86	5.14	7.00		A_2	4.13	0.87	5.00
Total	4.00	8.00	12.00		Total	15.00	2.00	17.00

T A B L E   **9.24**

Estimated cell frequencies for patients classified by gender (*A*), age (*B*), and mental depression status (*C*) under the constraint of conditional independence

C_1	B_1	B_2	Total		C_2	B_1	B_2	Total
A_1	1.67	3.33	5.00		A_1	10.59	1.41	12.00
A_2	2.33	4.67	7.00		A_2	4.41	0.59	5.00
Total	4.00	8.00	12.00		Total	15.00	2.00	17.00

For evaluating a model with $u_{123} = 0$, the degrees of freedom for the chi-square statistic are a function of the number of terms in the log-linear model. The degrees of freedom are calculated by subtracting the number of independent parameter estimates from the total number of cells in the table. When u_{123} is set to zero, seven independent estimates are necessary ($\hat{u}$, $\hat{u}_1$, $\hat{u}_2$, $\hat{u}_3$, $\hat{u}_{12}$, $\hat{u}_{13}$, $\hat{u}_{23}$) to estimate the eight cell frequencies in a 2-by-2-by-2 table, giving $8 - 7 = 1$ degree of freedom.

Conditional Independence Model

To explore further the relationships among the variables in a three-way table, other models are postulated and comparisons made. An important model postulates conditional independence—one pairwise u_{ij}-term is zero and the other two u_{ij}-terms are not zero. To evaluate this conjecture, where the hypothesis $u_{12} = 0$ is assessed, a six-term log-linear model

$$\text{log-frequency} = u + u_1 + u_2 + u_3 + u_{13} + u_{23} \quad (u_{12} = u_{123} = 0) \qquad (9.24)$$

is the fit to the observed cell frequencies. Because this model is an alternative way of expressing conditional independence, the estimated cell frequencies have already been discussed (i.e., $\hat{l}_{ijk} = \log[f_{i \cdot k} f_{\cdot jk}/f_{\cdot \cdot k}]$).

The gender/age/depression data (Table 9.21) yield the estimated cell frequencies under the hypothesis that gender and age are conditionally independent ($u_{12} = u_{123} = 0$), given in Table 9.24. The resulting chi-square value $Y^2 = 0.598$ (Table 9.21 compared to Table 9.24) with degrees of freedom $= 8 - 6 = 2$ produces a *p*-value $= 0.742$, showing no evidence that variables *A* and *B* (gender and age) are associated in tables C_1 and C_2. Note that the estimated cell frequencies in Table 9.24 produce $\hat{OR}_{AB \mid C_1} = \hat{OR}_{AB \mid C_2} = 1$, as expected, for estimates derived under the constraint of conditional independence ($u_{123} = u_{12} = 0$).

Partial Independence Model

Another valuable model postulates partial independence—no association between a specific variable and the two other variables. If variable 1 is postulated to be partially independent of variables 2 and 3, the resulting five-term log-linear model is

T A B L E **9.25**

Estimated cell frequencies for patients classified by gender (*A*), age (*B*), and mental depression status (*C*) under the conditions of partial independence

C_1	B_1	B_2	Total
A_1	2.34	4.69	7.03
A_2	1.66	3.31	4.97
Total	4.00	8.00	12.00

C_2	B_1	B_2	Total
A_1	8.79	1.17	9.96
A_2	6.21	0.83	7.04
Total	15.00	2.00	17.00

$$\text{log-frequency} = u + u_1 + u_2 + u_3 + u_{23} \quad (u_{12} = u_{13} = u_{123} = 0) . \tag{9.25}$$

This partial independence model is an explicit expression of the previously discussed partial independence case (i.e., $\hat{l}_{ijk} = \log[f_{.jk}f_{i..}/n]$) and produces the estimated cell frequencies given in Table 9.25 from the data in Table 9.21 when gender (1) is postulated as unrelated to age (2) and depression status (3). These estimated values compared to the original data (Table 9.21) generate a chi-square statistic of $Y^2 = 3.036$ with degrees of freedom $= 8 - 5 = 3$ (*p*-value $= 0.386$), yielding no evidence that variable 1 (*A*, gender) is associated with either variable 2 (*B*, age) or variable 3 (*C*, depression status). The estimated values in Table 9.25 produce $OR_{AC|B_1} = OR_{AC|B_2} = OR_{AB|C_1} = OR_{AB|C_2} = 1.0$ whereas $OR_{BC|A_1} = OR_{BC|A_2} = OR_{BC} = 0.067$, which is no more than an alternative statement of the partial independence model (expression 9.25) for a 2-by-2-by-2 table.

Complete Independence Model

The most specific model postulates complete independence—no association between any variables. When all three variables are unrelated the model is

$$\text{log-frequency} = u + u_1 + u_2 + u_3 \quad (u_{12} = u_{13} = u_{23} = u_{123} = 0) . \tag{9.26}$$

This complete independence model is not different from the previously discussed complete independence case (i.e., $\hat{l}_{ijk} = \log[f_{i..}f_{.j.}f_{..k}/n^2]$) and produces the estimated cell frequencies given in Table 9.26 from the data in Table 9.21 when gender (1), age (2), and depression status (3) are postulated as unrelated. The estimates in Table 9.26 produce estimated cell frequencies that yield odds ratios of one for any pair of variables in any table (subtables and combined tables). However, these estimated cell frequencies compared to the original data (Table 9.21) generate a chi-square statistic of $Y^2 = 12.807$ with degrees of freedom $= 8 - 4 = 4$ (*p*-value $= 0.012$), yielding strong evidence that among the variables *A* (gender), *B* (age), or *C* (depression status) at least one pair is related.

T A B L E **9.26**

Estimated cell frequencies for patients classified by gender (*A*), age (*B*), and mental depression status (*C*) under the conditions of complete independence

C_1	B_1	B_2	Total		C_2	B_1	B_2	Total
A_1	4.61	2.43	7.04		A_1	6.53	3.44	9.97
A_2	3.25	1.71	4.96		A_2	4.61	2.43	7.04
Total	7.86	4.14	12.00		Total	11.14	5.87	17.00

Summary of All Four Models

A three-dimensional table with two levels for each variable provides the possibility of eight different log-linear models (*k*-dimensional tables produce 2^k possible log-linear models). All possible models and likelihood chi-square statistics for the data on the hypertension patients (Table 9.21) are presented in Table 9.27.

A more systematic pattern for examining the data on hypertension patients (Table 9.21) might be:

1 Fit the log-linear model with $u_{123} = 0$ and note that the resulting fit is extremely good ($Y_1^2 = 0.065$), which warrants the assumption that the two-factor *u*-terms accurately measure statistical association.

2 Following from 1, fit the model with $u_{12} = u_{123} = 0$. This model also closely fits the data $Y_2^2 = 0.598$. Furthermore, the restricted chi-square difference $Y_2^2 - Y_1^2 = 0.533$, solely attributable to the association between variables 1 and 2, is small (degrees of freedom = 2 − 1 = 1; *p*-value = 0.465), producing the inference that variables 1 and 2 are not likely associated.

3 Following from 1 and 2, fit the model with $u_{12} = u_{23} = u_{123} = 0$, producing $Y_3^2 = 10.369$. The portion attributable strictly to the association between variables 2 and 3 is $Y_3^2 - Y_2^2 = 9.771$ (degrees of freedom = 3 − 2 = 1). A restricted chi-square value of 9.771 is not likely to occur by chance (*p*-value = 0.002), indicating that age (2) is likely associated with depression status (3).

4 Finally, fit the additive model where $u_{12} = u_{13} = u_{23} = u_{123} = 0$, producing a chi-square statistic $Y_4^2 = 12.807$. The restricted chi-square $Y_4^2 - Y_3^2 = 2.438$ (degrees of freedom = 1) measures exclusively the association between variables 1 and 3 and has a *p*-value of 0.118. The variables gender (1) and depression (3) show no strong evidence of an association.

Steps 1–4 are an examination of a sequence of hierarchical models. Specifically, the sequence of hierarchical models is:

T A B L E **9.27**

All possible models for patients classified by gender (A), age (B), and mental depression status (C)

Model	Y^2	Degrees of Freedom	p-Value
All Pairwise Associations			
1. $u_{123} = 0$	0.065	1	0.799
Conditional Independence			
2. $u_{12} = u_{123} = 0$	0.598	2	0.742
3. $u_{13} = u_{123} = 0$	0.859	2	0.651
4. $u_{23} = u_{123} = 0$	8.192	2	0.017
Partial Independence			
5. $u_{12} = u_{13} = u_{123} = 0$	3.036	3	0.386
6. $u_{12} = u_{23} = u_{123} = 0$	10.369	3	0.016
7. $u_{13} = u_{23} = u_{123} = 0$	10.630	3	0.014
Complete Independence			
8. $u_{12} = u_{13} = u_{23} = u_{123} = 0$	12.807	4	0.012

1 *log-frequency* $= u + u_1 + u_2 + u_3 + u_{12} + u_{13} + u_{23} + u_{123}$ ($Y^2 = 0.0$)

2 *log-frequency* $= u + u_1 + u_2 + u_3 + u_{12} + u_{13} + u_{23}$ ($Y^2 = 0.065$)

3 *log-frequency* $= u + u_1 + u_2 + u_3 + u_{13} + u_{23}$ ($Y^2 = 0.598$)

4 *log-frequency* $= u + u_1 + u_2 + u_3 + u_{13}$ ($Y^2 = 10.369$)

5 *log-frequency* $= u + u_1 + u_2 + u_3$ ($Y^2 = 12.807$)

For a sequence of such nested models, the Y^2-values will always increase because each model is less complicated (contains fewer parameters). The simpler models have less flexibility and necessarily are less accurate representations of data. Other patterns of hierarchal models could be similarly investigated but the results are not substantially different. For the gender/age/depression data the logarithms of the estimated cell frequencies for the four models are displayed in Figure 9.2, giving a geometric summary of the modeled relationship among the three variables (Tables 9.23, 9.24, 9.25, and 9.26).

A log-linear model approach produces a series of restricted chi-square values ($Y^2 = Y_m^2 - Y_n^2$) with degrees of freedom equal to the difference between the degrees of freedom associated with each model. However, the nature of a nested model approach is such that any "procedure-wise" significance probability is not exact. Examining a series of models in a more or less ad hoc fashion defies accurate calcu-

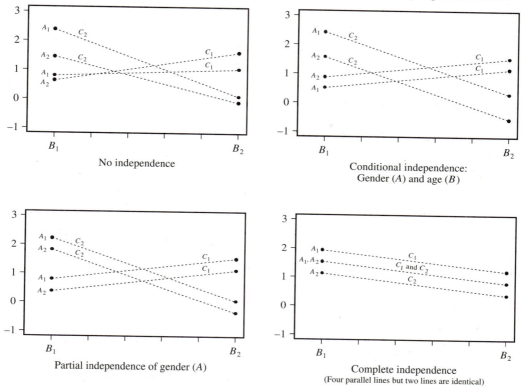

FIGURE **9.2**

The estimated logarithms of the cell frequencies for the depression data (Table 9.21) calculated from four log-linear models

lation of a procedure-wise significance probability. Only when a single a priori sequence of models is postulated and tested without regard to the observed results can approximate probabilities be calculated (e.g., Bonferroni's inequality—Chapter 8). The lack of an accurate overall probability statement should not inhibit a full exploration of the data.

r-by-*c*-by-*l* Table

An extension of the log-linear model approach to a general *r*-by-*c*-by-*l* table follows the pattern of the 2-by-2-by-2 table. A major difference is that the *u*-terms are no longer simple functions of odds ratios. Nevertheless, the interpretation of the fit of a sequence of hierarchical log-linear models is essentially the same. The three-factor term u_{123} indicates whether the association between pairs of variables (u_{ij}-terms) is the same at all levels of a third variable. Large u_{123}-values imply that different types of associations are likely at some or all levels of a third variable. Small u_{123}-values mean an average measure of pairwise association (u_{ij}) accurately reflects the associa-

T A B L E   **9.28**

A set of 874 mothers classified by body mass index (A), their infant's weight (B),
and smoking status (C)—a 3-by-2-by-4 table

C_1	B_1	B_2	Total
A_1	10	50	60
A_2	12	97	109
A_3	5	82	87
Total	27	229	256

C_2	B_1	B_2	Total
A_1	3	54	57
A_2	6	70	76
A_3	2	85	87
Total	11	209	220

C_3	B_1	B_2	Total
A_1	4	56	60
A_2	8	98	106
A_3	1	89	90
Total	13	243	256

C_4	B_1	B_2	Total
A_1	1	34	35
A_2	4	58	62
A_3	1	44	45
Total	6	136	142

tions found at all levels of a third variable. Interpretation of two-factor u-terms also presents nothing new—$u_{ij} = 0$ implies exact statistical independence between variables i and j whereas $u_{ij} \neq 0$ measures the degree of association between these two variables when u_{123} is zero or, at least, small.

The data in Table 9.28 (a 3-by-2-by-4 table) are a classification of mothers by a body mass index (A_1 = slight, A_2 = normal, and A_3 = heavy), weight of their newborn infant ($B_1 \leq 6$ pounds, and $B_2 > 6$ pounds), and smoking status (C_1 = nonsmoker, C_2 = past smoker, C_3 = smoke $\leq$ pack/day, and C_4 = smoke > pack/day).

One approach to identifying the important and unimportant relationships among these three variables takes the following form:

1   To begin the search for a simple but useful model to describe the relationships within the data, a seven-term log-linear model ($u + u_1 + u_2 + u_3 + u_{12} + u_{13} + u_{23}$) with only $u_{123} = 0$ is postulated as a first step. The small chi-square value ($Y^2 = 3.100$ with 6 degrees of freedom with p-value = 0.796) implies that using a reduced log-linear model with u_{123} set equal to zero will not severely distort further analyses.

2   Postulating $u_{13} = u_{123} = 0$ yields a further reduced model with a chi-square value of $Y^2 = 7.788$ (degrees of freedom = 12; p-value = 0.801), producing a restricted chi-square = $Y^2 = 7.788 - 3.100 = 4.688$ with degrees of freedom = 12 – 6 = 6 and a p-value of 0.584. This indicates that the body mass index (A) is not strongly associated with smoking (C).

3   However, further postulating $u_{12} = u_{13} = u_{123} = 0$ yields a chi-square value of $Y^2 = 19.324$ (degrees of freedom = 14; p-value = 0.153) and a restricted chi-square value of $Y^2 = 19.324 - 7.788 = 11.536$ (degrees of freedom = 2; p-value = 0.003). This indicates that the data provide strong evidence of an association between body mass index (A) and the birth weight of a newborn (B).

4 Finally, postulating the further restriction $u_{12} = u_{13} = u_{23} = u_{123} = 0$ ($Y^2 = 28.363$ with degrees of freedom = 17; p-value = 0.041) and making a comparison to the model with $u_{12} = u_{13} = u_{23} = 0$ ($Y^2 = 19.324$ with degrees of freedom = 14) yields a restricted chi-square value of $Y^2 = 28.363 - 19.324 = 9.039$ (degrees of freedom = 3; p-value = 0.029). This shows that it is likely that maternal smoking (C) is also related to an infant's birth weight (B).

Again, a log-linear analysis can be viewed as examining a sequence of hierarchical models. Specifically, a sequence of hierarchical models for the data in Table 9.28 is:

1 $\quad log\text{-}frequency = u + u_1 + u_2 + u_3 + u_{12} + u_{13} + u_{23} + u_{123}$ $\quad (Y^2 = 0.0)$
2 $\quad log\text{-}frequency = u + u_1 + u_2 + u_3 + u_{12} + u_{13} + u_{23}$ $\quad (Y^2 = 3.100)$
3 $\quad log\text{-}frequency = u + u_1 + u_2 + u_3 + u_{12} + u_{23}$ $\quad (Y^2 = 7.788)$
4 $\quad log\text{-}frequency = u + u_1 + u_2 + u_3 + u_{23}$ $\quad (Y^2 = 19.324)$
5 $\quad log\text{-}frequency = u + u_1 + u_2 + u_3$ $\quad (Y^2 = 28.363)$

The sequence of chi-square values 1–5 indicates that the body mass index (1) and maternal smoking (3) are not strongly associated, whereas both body weight (1) and smoking (3) appear related to birth weight (2)—conditional independence.

Although it is generally impractical for higher-dimensional tables, the analysis of all eight possible models applied to the body mass index data are given in a summary table (Table 9.29); also included are the Pearson and likelihood chi-square values (X^2 and Y^2).

An Example of Estimation

Many times a table is formed to reduce extraneous influences from a specific variable or variables. For example, it is well known that the frequency of most diseases increases with age. To investigate the relationship between a disease and a risk factor, data are often stratified to remove an obscuring influence of a variable such as age. Interest is focused on measuring the degree of association rather than statistical testing. The data in Table 9.30 are a series of males classified by the occurrence of a coronary event (CHD), behavior type (type A: aggressive personality with a stressful lifestyle and a sense of time urgency, and type B: nonaggressive person with a much more relaxed lifestyle), and age. These data are a small piece of a study of behavior type and coronary disease [5]. The data consist of five 2-by-2 tables, one for each of five age strata. The odds ratios measuring the association between behavior type and coronary heart disease frequency within these five age strata are $\hat{OR}_1 = 2.045$, $\hat{OR}_2 = 2.012$, $\hat{OR}_3 = 2.375$, $\hat{OR}_4 = 1.968$, and $\hat{OR}_5 = 2.893$, respectively. It is generally assumed that CHD, behavior, and age are related but whether this association is con-

TABLE 9.29

Summary of all eight log-linear models for three-way tables applied to the body mass index data set

Terms set to zero	$u_{123} = 0$	$u_{12} = u_{123} = 0$	$u_{13} = u_{123} = 0$	$u_{23} = u_{123} = 0$	$u_{12} = u_{13} = u_{123} = 0$	$u_{12} = u_{23} = u_{123} = 0$	$u_{13} = u_{23} = u_{123} = 0$	$u_{12} = u_{13} = u_{23} = u_{123} = 0$
Model	$u + u_1 + u_2 + u_3 + u_{12} + u_{13} + u_{23}$	$u + u_1 + u_2 + u_3 + u_{13} + u_{23}$	$u + u_1 + u_2 + u_3 + u_{12} + u_{23}$	$u + u_1 + u_2 + u_3 + u_{12} + u_{13}$	$u + u_1 + u_2 + u_3 + u_{23}$	$u + u_1 + u_2 + u_3 + u_{13}$	$u + u_1 + u_2 + u_3 + u_{12}$	$u + u_1 + u_2 + u_3$
Estimated values	iterative	$f_{i \cdot k} f_{\cdot jk}/f_{\cdot \cdot k}$	$f_{ij \cdot} f_{\cdot jk}/f_{\cdot j \cdot}$	$f_{i \cdot k} f_{ij \cdot}/f_{i \cdot \cdot}$	$f_{\cdot jk} f_{i \cdot \cdot}/n$	$f_{i \cdot k} f_{\cdot j \cdot}/n$	$f_{ij \cdot} f_{\cdot \cdot k}/n$	$f_{i \cdot \cdot} f_{\cdot j \cdot} f_{\cdot \cdot k}/n^2$
Degrees of freedom	$(r-1)(c-1)(l-1)$	$(r-1)(c-1)(l)$	$(r-1)c(l-1)$	$r(c-1)(l-1)$	$(r-1)(cl-1)$	$(rl-1)(c-1)$	$(rc-1)(l-1)$	$rcl-r-c-l+2$
Degrees of freedom	6	8	12	9	14	11	15	17
X^2	2.908	13.177	7.515	12.686	17.780	23.386	17.331	28.140
p-values	0.820	0.106	0.822	0.177	0.217	0.016	0.299	0.043
Y^2	3.100	14.491	7.788	11.994	19.324	23.529	16.827	28.363
p-values	0.796	0.070	0.801	0.214	0.153	0.015	0.329	0.041

T A B L E **9.30**

Behavior/coronary risk: Age by behavior type by CHD—data

| | Type A | | Type B | | |
	CHD	No CHD	CHD	No CHD	Total
Age	f_{11i}	f_{12i}	f_{21i}	f_{22i}	$f_{..i}$
< 40	20	241	11	271	543
40–44	34	462	21	574	1091
45–49	49	337	21	343	750
50–54	38	209	17	184	448
>54	37	162	9	114	322
Total	178	1411	79	1486	3154

sistent for all age categories and measuring the common association between behavior type and CHD are questions of primary interest.

The estimated cell frequencies from a log-linear model calculated under the hypothesis that the relationship between CHD and behavior is identical within each age stratum ($u_{123} = 0$) are given in Table 9.31. These estimated values show little difference when compared to the observed values. That is, the model with u_{123} set equal to zero fits the data well—$Y^2 = 0.854$ with degrees of freedom = 4, yielding a *p*-value of 0.931. Also, a comparison of Table 9.30 with Table 9.31 shows informally the high degree of accuracy of the model as a representation of the data. The pairwise

T A B L E **9.31**

Behavior/coronary risk: Age by behavior type by CHD—estimated values ($u_{123} = 0$)

| | Type A | | Type B | | |
	CHD	No CHD	CHD	No CHD	Total
Age	f_{11i}	f_{12i}	f_{21i}	f_{22i}	$f_{..i}$
< 40	20.52	240.48	10.48	271.52	543
40–44	35.15	460.85	19.85	575.15	1091
45–49	48.01	337.99	21.99	342.01	750
50–54	39.19	207.81	15.81	185.19	448
>54	35.12	163.88	10.88	112.12	322
Total	178	1411	79	1486	3154

associations between these three variables are expectedly strong and not likely to have occurred by chance:

Behavior/CHD, $u_{12} = u_{123} = 0$, $Y^2 = 34.4$, p-value < 0.001

Behavior/age, $u_{13} = u_{123} = 0$, $Y^2 = 26.6$, p-value $= 0.001$

CHD/age, $u_{23} = u_{123} = 0$, $Y^2 = 38.2$, p-value < 0.001

The important issue, however, is a summary measure of association for the relationship between behavior type and a coronary event taking into account the influence of age. A summary measure is sensible because the data support (low chi-square value) the contention that the degree of association is the same in each table ($u_{123} = 0$). An excellent estimate is found by computing the odds ratio from the estimated cell frequencies for any age stratum (Table 9.31). The odds ratios are the same for each stratum because the model forces the behavior/CHD relationship to be identical in each age group. A summary odds ratio is, then,

$$odds\ ratio = \hat{OR} = \frac{(20.52)\ (271.52)}{(240.48)\ (10.48)} = 2.211 \ .$$

The odds of coronary disease are about 2.2 times higher in type A individuals than in type B individuals accounting for the influence of age. The reduction in bias from stratifying the data into five age categories emerges from comparing the odds ratio ignoring age (i.e., $\hat{OR}' = (178)\ (1486)/(1411)\ (79) = 2.373$; last line in Table 9.30) to the odds ratio taking age into account ($\hat{OR} = 2.211$). This comparison measures confounder bias.

Degrees of Freedom

Probably the least intuitive aspect of the analysis of contingency table data is the determination of the degrees of freedom. The calculation is relatively mechanical for a chi-square statistic calculated from a log-linear model. The degrees of freedom are the number of cells in the table minus the number of independent estimates necessary to specify the parameters of the log-linear model. The enumeration of the number of independent estimates to establish a log-linear model is not complicated, as it follows a simple pattern [4]. The constant term (u) requires one estimate and, therefore, uses one degree of freedom. The number of independent estimates of the u-terms with a single subscript is one less than the number of levels associated with the categorical variable. For example, when u_1 has r values, one for each level of variable 1, then $r - 1$ independent estimates are required to define all u_1-values because these u_1-values add to zero ($\Sigma\, u_1(i) = 0$). The number of u-terms with two subscripts is equal to the number of pairwise combinations of the two associated variables. For example, r rows of variable 1 and c columns of variable 2 produce rc pairs, but only $(r - 1)\ (c - 1)$ independent estimates are necessary to specify all rc values of $u_{12}(i, j)$ because $\Sigma_i\, u_{12}(i, j) = \Sigma_j\, u_{12}(i, j) = 0$. Similarly, the u-terms associated with u_{123} have rcl values but require $(r - 1)\ (c - 1)\ (l - 1)$ independent estimates. The calculation of the number of independent estimates associated with u-terms with more than three levels follows a similar pattern.

The number of independent u-terms necessary to specify a saturated log-linear model is exactly equal to the total number of cells in the contingency table. Specifically, for a three-way table,

Terms	Independent Estimates
u	1
u_1	$r - 1$
u_2	$c - 1$
u_3	$l - 1$
u_{12}	$(r - 1)(c - 1)$
u_{13}	$(r - 1)(l - 1)$
u_{23}	$(c - 1)(l - 1)$
u_{123}	$(r - 1)(c - 1)(l - 1)$
Total	rcl

Because the sum of the number of independent estimates equals the number of cells in the table, the degrees of freedom for log-linear models can be calculated two ways—directly or by subtraction. Again the three-dimensional table illustrates. The model $u + u_1 + u_2 + u_3 + u_{23}$ requires $[1 + (r - 1) + (c - 1) + (l - 1) + (c - 1)(l - 1)]$ independent estimates to specify the postulated model, yielding $rcl - [1 + (r - 1) + (c - 1) + (l - 1) + (c - 1)(l - 1)] = (r - 1)(cl - 1)$ degrees of freedom. Alternatively, because $u_{12} = u_{13} = u_{123} = 0$, then $(r - 1)(c - 1) + (r - 1)(l - 1) + (r - 1)(c - 1)(l - 1)$ independent u-terms are not estimated (set equal to zero), making this sum necessarily the degrees of freedom (i.e., $(r - 1)(cl - 1)$). That is,

$$[1 + (r - 1) + (c - 1) + (l - 1) + (c - 1)(l - 1)] + [(r - 1)(cl - 1)] = rcl$$

or

[number of independent estimates to specify the model]
+ [number of independent parameters not in the model]
= [total number of cells in the table] .

A final example: For the model of complete independence $(u + u_1 + u_2 + u_3)$,

number of independent estimates to specify the model
$$= 1 + (r - 1) + (c - 1) + (l - 1)$$

number of independent parameters not in the model
$$= (r - 1)(c - 1) + (r - 1)(l - 1) + (c - 1)(l - 1) + (r - 1)(c - 1)(l - 1)$$

and the total number of cells in the table is rcl, giving the degrees of freedom = $rcl - r - c - l + 2$ calculated either directly or by subtraction. The degrees of freedom for all eight possible log-linear models for a three-dimensional table are given in Table 9.29.

Combinability

An excellent method for simplifying complicated multivariate tabular data is to reduce the complexity of a table by combining subtables whenever possible, decreasing the dimensionality. A more parsimonious description of the relationships under investigation is always a primary analytic goal. However, the reduction of dimensionality must be done with care, as already noted. The following rule is a basis for combinability:

> In a three-dimensional table, the association between two variables can be assessed from a table created by summing over a third variable if and only if the third variable is independent of at least one of the other two variables.

This rule applies only to three-dimensional contingency tables. A general theorem exists for higher-dimensional tables but a formal statement is not given (e.g., [4]). However, in less formal terms, when two variables are independent, such as 1 and 2 ($u_{12} = 0$), then the relationship of either 1 or 2 to other variables can be investigated in tables that result from adding over and eliminating the other variable regardless of the dimensionality. A special case arises when a specific variable is independent of all other variables; then a contingency table can be created by completely eliminating that variable ("removability") and not affect the relationships among the remaining variables. For partially independent variables, if $u_{12} = u_{13} = u_{123} = 0$ in a three-dimensional table, then variable 1 can be eliminated completely by combining (adding) all levels of variable 1. The resulting two-way table, consisting of variables 2 and 3, exactly reflects the structural features of the data. For conditionally independent variables, if $u_{12} = u_{123} = 0$, then variables 2 and 3 can be investigated in a two-way table summed over variable 1 and variables 1 and 3 can be investigated in another two-way table summed over variable 2. However, if no pairwise associations are zero, then summing a table over any variable obscures the underlying structure and, potentially, produces misleading results (e.g., Simpson's paradox).

An analogy to the combinability property [4] involves partial correlation for continuous variables written as (expression 3.23)

$$r_{23 \cdot 1} = \frac{r_{23} - r_{12}r_{13}}{\sqrt{1 - r_{12}^2}\sqrt{1 - r_{13}^2}} .$$

If $r_{12} = r_{13} = 0$, then $r_{23 \cdot 1} = r_{23}$, showing that considering variable 1 is unnecessary in the study of variables 2 and 3, which is analogous to "removability" of variable 1. Furthermore, if only $r_{12} = 0$, then

$$r_{23 \cdot 1} = r_{23}/\sqrt{1 - r_{13}^2}$$

T A B L E **9.32**

Data formed by adding categories C_1 and C_2 from Table 9.4

$C_1 + C_2$	B_1	B_2	Total
A_1	120	80	200
A_2	80	720	800
Total	200	800	1000

and $r_{23.1}$ is a simple multiple of r_{23}. If $r_{23} = 0$, then $r_{23.1} = 0$. The association between variables 2 and 3 can be studied while ignoring variable 1, much like the case of conditional independence.

To illustrate, variable C is partially independent ($u_{13} = u_{23} = u_{123} = 0$) in the previous "data" shown in Table 9.4, and when the two subtables are summed over the categories of variable C, Table 9.32 emerges. Table 9.32 perfectly reflects the relationships observed between variables 1 and 2 in each subtable. The odds ratios, for example, are $\hat{OR}_{AB|C_1} = \hat{OR}_{AB|C_2} = 13.5$ (Table 9.4) and $\hat{OR}_{AB} = 13.5$ (Table 9.32). Other measures of association will be similarly unaffected by the combining of tables C_1 and C_2 into one table because C is partially independent of variables A and B.

Exact conditionally independent categorical variables A and B ($u_{12} = u_{123} = 0$), given previously (Table 9.6), summed over the categories of variable B are shown in Table 9.33. Table 9.33 illustrates that the identical relationship between variables A and C in each subtable ($\hat{OR}_{AC|B_1} = \hat{OR}_{AC|B_2} = 3.778$; Table 9.6) is observed in the combined table ($\hat{OR}_{AC} = 3.778$; Table 9.33).

The "data" presented to demonstrate Simpson's paradox (Table 9.10) illustrate the converse of the combinability rule. If no pair of variables is independent, then combining tables will obscure the underlying structure. In the case of Simpson's paradox, the association is completely reversed, which is an extreme example; nevertheless, when the conditions for combinability are violated, nonsensical results can arise.

T A B L E **9.33**

Data formed by adding categories B_1 and B_2 from Table 9.6

$B_1 + B_2$	C_1	C_2	Total
A_1	80	120	200
A_2	120	680	800
Total	200	800	1000

T A B L E   **9.34**

Hypertensive individuals classified by age (B) and depression (C)
for both males and females ($A_1 + A_2$)

$A_1 + A_2$	C_1	C_2	Total
B_1	4	15	19
B_2	8	2	10
Total	12	17	29

Data Examples of Combining Tables

The analysis of the possibility that gender is partially independent of age and depression in hypertensive individuals (Table 9.21) yields a chi-square statistic $Y^2 = 3.036$ with three degrees of freedom (p-value = 0.386), indicating that the hypothesis $u_{12} = u_{13} = u_{123} = 0$ adequately describes the data. This moderate chi-square value implies that the relationship between age and depression is not substantially changed when the variable A (gender) is eliminated by combining male and female patients into one table, which produces Table 9.34. The reduced table more simply shows that younger hypertensive patients are more often depressed ($15/19 = 0.789$ versus $2/10 = 0.200$), a result consistent with the previous analysis of the three-way table.

The air pollution data show no evidence that urban/rural status (variable A) and TSP levels (variable C) are associated (Table 9.9). Therefore, assuming conditional independence between variables A and C ($u_{13} = u_{123} = 0$), the relationship between variable B (average county income) and variable C (TSP) can be studied using a table formed by adding urban and rural data ($A_1 + A_2$; Table 9.35) without substantially changing the relationship between county income and air quality. The reduced table clearly shows an increasing proportion of low income counties associated with increasing TSP levels. Under the hypothesis that income and air quality are unrelated, the likelihood chi-square value is $Y^2 = 72.743$ with two degrees of freedom (p-value < 0.001). This result, like the analysis of the three-dimensional table, indicates

T A B L E   **9.35**

Counties classified by income (B) and air quality (C) for both rural and urban counties

	$A_1 + A_2$	C_1 (TSP low)	C_2	C_3 (TSP high)	Total
≤ $22,000	B_1	50 81.43)	68 (87.37)	107 (56.20)	225
> $22,000	B_2	334 (302.57)	344 (324.63)	158 (208.80)	836
	Total	384	412	265	1061
	Proportion B_1	0.130	0.165	0.404	0.212

a systematic association between income and TSP levels. To repeat, the validity of the inference drawn from the combined table depends on the assumption that $u_{13} = u_{123} = 0$, which is supported by the observed data.

One last type of combinability is not related to the structure of the data but rather relates to the definition of the categorical variables. Opportunities occur to combine categories by broadening the definition of a variable and, therefore, reducing the dimensionality of a table. Instead of two variables with r levels of one variable and c levels of another variable, a combined single variable with rc categories can be created.

A study of lung cancer and survival employs the following four dimensions:

survival: A_1 = survived more than 5 years
         A_2 = died during the first 5 years;

histology: B_1 = adenocarcinoma
         B_2 = other;

stage: C_1 = advanced
         C_2 = not advanced; and

socioeconomic status: D_1 = upper class

         D_2 = upper-middle class
         D_3 = lower-middle class
         D_4 = lower class.

These variables produce a 2-by-2-by-2-by-4 table. However, histology and stage could be combined, forming four categories:

histology/stage: B'_1 = adenocarcinoma advanced
         B'_2 = adenocarcinoma not advanced
         B'_3 = other histology advanced
         B'_4 = other histology not advanced

The new three-dimensional table (2-by-4-by-4) is more easily interpreted. Combinability of categories does not depend on any statistical relationship but rather on the utility of the newly created composite variable.

Fixed Marginal Frequencies

Data are occasionally collected in such a way that some of the marginal frequencies are fixed, which means they are not subject to random variation. To guarantee that the values generated by a log-linear model have the same marginal frequencies as those fixed by the data collection design, u-terms associated with the fixed frequencies are kept in the model. If variable 2 has the same marginal frequencies at all levels of variable 3, then maintaining u_{23} in the log-linear model assures that these marginal values are exactly reproduced in tables of estimated cell frequencies.

Example

Blood from eight medical students was sampled and analyzed (80 cells each) for a chromosomal abnormality called sister chromatid exchanges (SCEs). After a series of intense exposures to formaldehyde (anatomy class), the eight students were retested (again, 80 cells each). The categorical variables are:

$$\text{SCEs:} \quad A_1 = \text{less than 11 SCEs observed}$$
$$A_2 = \text{equal to or greater than 11 SCEs observed;}$$

$$\text{exposure status:} \quad B_1 = \text{before class (unexposed)}$$
$$B_2 = \text{after class (exposed);}$$

$$\text{individuals:} \quad C_1, C_2, \cdots, C_8$$

The complete data set is given in Table 9.36.

These data were collected in an experimental setting and, as is often the case with experimental data, some of the marginal frequencies are fixed. Specifically, the exposure categories (B_1 and B_2) sum to 80 by design because 80 cells from each student were examined before and after exposure. To account for these fixed values, u_{23}

T A B L E 9.36

Data on sister chromatid exchanges ($A_1 < 11$ and $A_2 \geq 11$) for eight students (C_i) exposed to formaldehyde (B_1 = unexposed and B_2 = exposed)

C_1	B_1	B_2	Total
A_1	75	60	135
A_2	5	20	25
Total	80	80	160

C_2	B_1	B_2	Total
A_1	71	59	130
A_2	9	21	30
Total	80	80	160

C_3	B_1	B_2	Total
A_1	71	70	141
A_2	9	10	19
Total	80	80	160

C_4	B_1	B_2	Total
A_1	75	74	149
A_2	5	6	11
Total	80	80	160

C_5	B_1	B_2	Total
A_1	75	78	153
A_2	5	2	7
Total	80	80	160

C_6	B_1	B_2	Total
A_1	72	60	132
A_2	8	20	28
Total	80	80	160

C_7	B_1	B_2	Total
A_1	72	68	140
A_2	8	12	20
Total	80	80	160

C_8	B_1	B_2	Total
A_1	74	71	145
A_2	6	9	15
Total	80	80	160

$C_1 + \cdots + C_8$	B_1	B_2	Total
A_1	585	540	1125
A_2	55	100	155
Total	640	640	1280

is kept in the log-linear model so that the estimated values generated by various log-linear models also sum to the fixed totals (i.e., 80 for the *B*-marginal frequencies and, therefore, 160 for the *C*-marginal frequencies).

Model I

The simplest hypothesis concerning the SCE data is that no difference exists in SCE rates among the eight students ($u_{13} = 0$) and exposure to formaldehyde also has no influence on the SCE rate ($u_{12} = 0$). The log-linear model that reflects this hypothesis and maintains the fixed marginal frequencies is *log-frequency* = $u + u_1 + u_2 + u_3 + u_{23}$ ($u_{12} = u_{13} = u_{123} = 0$). This model generates the same table of estimated cell frequencies for each of the eight students (Table 9.37) with an odds ratio reflecting the risk from formaldehyde exposure equal to 1.0. The fit of this model to the data is poor ($Y_1^2 = 55.908$ with 15 degrees of freedom; *p*-value < 0.001).

Model II

A slightly more sophisticated hypothesis postulates that individual SCE rates vary among students ($u_{13} \neq 0$) but still no association exists between SCE rate and exposure to formaldehyde ($u_{12} = 0$). This hypothesis, along with the desire to have estimated values sum to the marginal frequencies fixed by the data collection design, generates the log-linear model *log-frequency* = $u + u_1 + u_2 + u_3 + u_{23} + u_{13}$ ($u_{12} = u_{123} = 0$). The estimated cell frequencies are given in the summary Table 9.37. Again, the odds ratios measuring the association between SCEs and exposure are 1.0 for all eight individuals whereas the level of SCEs differs among the students tested.

The fit of this model to the data is improved over model I: $Y_2^2 = 26.890$ with degrees of freedom = 8 (*p*-value < 0.001). The restricted chi-square value is $Y^2 = Y_1^2 - Y_2^2 = 29.018$ with degrees of freedom = 7, yielding *p*-value < 0.001. This indicates that the u_{13}-terms are valuable in the description of the data.

Model III

An extension of the previous hypothesis is to postulate an elevated but constant risk from formaldehyde exposure ($u_{12} \neq 0$). This hypothesis generates the log-linear model *log-frequency* = $u + u_1 + u_2 + u_3 + u_{23} + u_{13} + u_{12}$ (only $u_{123} = 0$). The estimated cell frequencies generated by this model are also given in the summary Table 9.37. Model III begins to adequately represent the data ($Y_3^2 = 11.510$ with degrees of freedom = 7, yielding *p*-value = 0.118). Including u_{12} certainly represents a substantial improvement over model II; the restricted chi-square statistic is $Y^2 = Y_2^2 - Y_3^2 = 26.890 - 11.510 = 15.380$ with degrees of freedom = $8 - 7 = 1$, giving *p*-value < 0.001. A feature of these estimated values is that each table provides the same measure of risk from formaldehyde exposure, a summary odds ratio of $\hat{OR} = 1.999$.

One last perspective on the SCE data comes from calculating the odds ratio and its standardized value ($z = \log(\hat{OR})/\sqrt{variance[\log(\hat{OR})]}$) for each individual, shown in Table 9.38. These values are equivalent to estimating the parameters from a saturated model (i.e., $Y^2 = 0$). Large standardized values (such as $|z_k| > 2$) identify those

T A B L E   **9.37**
Summary table of the analyses of the SCE data

A_1	A_2	B_1	B_2	C_i	Data	Estimate, Model I	Estimate, Model II	Estimate, Model III
1	0	1	0	1	75	70.31	67.50	71.07
1	0	0	1	1	60	70.31	67.50	63.93
0	1	1	0	1	5	9.69	12.50	8.93
0	1	0	1	1	20	9.69	12.50	16.07
1	0	1	0	2	71	70.31	65.00	69.13
1	0	0	1	2	59	70.31	65.00	60.87
0	1	1	0	2	9	9.69	15.00	10.87
0	1	0	1	2	21	9.69	15.00	19.13
1	0	1	0	3	71	70.31	70.50	73.32
1	0	0	1	3	70	70.31	70.50	67.68
0	1	1	0	3	9	9.69	9.50	6.68
0	1	0	1	3	10	9.69	9.50	12.32
1	0	1	0	4	75	70.31	74.50	76.22
1	0	0	1	4	74	70.31	74.50	72.78
0	1	1	0	4	5	9.69	5.50	3.78
0	1	0	1	4	6	9.69	5.50	7.22
1	0	1	0	5	75	70.31	76.50	77.62
1	0	0	1	5	78	70.31	76.50	75.38
0	1	1	0	5	5	9.69	3.50	2.38
0	1	0	1	5	2	9.69	3.50	4.62
1	0	1	0	6	72	70.31	66.00	69.91
1	0	0	1	6	60	70.31	66.00	62.09
0	1	1	0	6	8	9.69	14.00	10.09
0	1	0	1	6	20	9.69	14.00	17.91
1	0	1	0	7	72	70.31	70.00	72.95
1	0	0	1	7	68	70.31	70.00	67.05
0	1	1	0	7	8	9.69	10.00	7.05
0	1	0	1	7	12	9.69	10.00	12.95
1	0	1	0	8	74	70.31	72.50	74.79
1	0	0	1	8	71	70.31	72.50	70.21
0	1	1	0	8	6	9.69	7.50	5.21
0	1	0	1	8	9	9.69	7.50	9.79

T A B L E **9.38**

The odds ratios for each student measuring the association between level of sister chromatid exchanges and exposure status

Individual	$\hat{OR}_k$	$variance[\log(\hat{OR}_k)]$	z_k
1.	5.000	0.280	3.042
2.	2.808	0.190	2.370
3.	1.127	0.239	0.244
4.	1.216	0.394	0.312
5.	0.385	0.726	−1.121
6.	3.000	0.206	2.423
7.	1.588	0.237	0.950
8.	1.563	0.305	0.809

individuals who appear to have a strong association between SCE rates and exposure. Three of the eight students (1, 2, and 6) show likely associations whereas the other five individuals seem relatively unaffected by exposure.

Overall, evidence exists of an association between 11 or more SCEs (*A*) and exposure to formaldehyde (*B*). However, this inference should be tempered by the possibility that the exposure to formaldehyde may differ among the eight students tested (e.g., perhaps some of the students didn't go to all of the classes). Eliminating u_{123} from the saturated model increases the lack of fit from $Y^2 = 0$ to $Y_3^2 = 11.510$ ($u_{123} = 0$), producing a *p*-value of 0.118 and showing, perhaps, some indication of heterogeneity of the relationship between SCEs and exposure among the eight tested individuals.

Zero Cell Frequencies

Two types of zeros appear in contingency tables: structural zeros and random zeros. Random zeros occur because of sampling variation and, at least in theory, disappear if the sample size is sufficiently large. Random zeros often do not cause concern. Contingency table data exist where estimates are unaffected by cell frequencies of zero. To illustrate, the estimated values for the "data" in Table 9.39 are routinely calculated (e.g., $\hat{f}_{12} = (6)(2)/12 = 1$). Random zeros also occur in ways so that estimated values cannot be calculated. For example, the estimated cell frequencies for the "data" in Table 9.40 cannot be calculated. The easiest way to avoid random zeros is to either increase the sample size or combine categories to eliminate all cells with zero frequencies. Yet another strategy (used automatically in some computer package programs) is to add 0.5 to all cells, eliminating empty cells and making it possible to produce estimated values. None of these solutions is ideal and the fact remains that statistical analysis does not make up for sparse data.

Structural zeros occur because specific combinations of the independent variables are impossible. A table including the variables of gender and cancer site will

T A B L E **9.39**

A 2-by-2 table where the analysis is unaffected by the presence of an empty cell

	B_1	B_2	Total
A_1	6	0	6
A_2	4	2	6
Total	10	2	12

T A B L E **9.40**

A 2-by-2 table where the analysis is affected by the presence of empty cells

	B_1	B_2	Total
A_1	0	0	0
A_2	6	4	10
Total	6	4	10

necessarily have structural zeros for female prostatic cancer or male ovarian cancer. In some cases, these cells can be avoided by the design of the data collection. Alternatively, special analytic methods exist to deal with incomplete data [4]. In the next section, a specific type of data with a structural zero is considered. In the next section, a specific type of data with a structural zero is considered.

Capture-Recapture Model

The capture-recapture model is presented not only as a useful approach to analyzing a specific type of data but also for two other important reasons. First, the model provides an application of the log-linear notation and concepts. Second, the capture-recapture model illustrates some of the issues involved in statistical estimation—a topic not yet emphasized.

A capture-recapture model is typically proposed to estimate the size of a closed population. Closed in this context means the population is unchanged by such things as migration, births, deaths, and so forth. The capture-recapture process (model) has been used effectively to estimate population sizes of free-living wild animals (such as whales, wild dogs, fish species, and cougars). A sample of animals is captured and tagged. These tagged animals are then released and allowed to mix freely with the population under study. Another capture phase is then conducted and the number of tagged animals is recorded. Under specific conditions, the proportion of recaptured (tagged) animals is related to the population size. The estimation of the size of a closed population has received attention from both applied and theoretical researchers [6].

T A B L E **9.41**
Structure and notation for the capture-recapture model

Source 1	Source 2	Source 3	Frequency
Included	Included	Included	f_{111}
Included	Included	Not Included	f_{112}
Included	Not Included	Included	f_{121}
Included	Not Included	Not Included	f_{122}
Not Included	Included	Included	f_{211}
Not Included	Included	Not Included	f_{212}
Not Included	Not Included	Included	f_{221}
Not Included	Not Included	Not Included	*

An estimate of population size is improved when more than one recapture phase is conducted. The data collected from multiple recaptures consist of counts of the number of times each member of the population is caught $(1, 2, 3, \cdots)$. Such "capture-recapture" data arise from a series of lists. For example, a number of sources report information on children with congenital malformations: hospital records, birth certificates, records from rehabilitation centers, and welfare records. A data set can be constructed from counts of how many times each child is identified ("recaptured") by these sources and, like the animal population estimation, these data produce an estimate of the total number of congenitally malformed children in a population under specific conditions. Notice that children not identified by any source surveyed will necessarily not be included in the data set. The sample is said to be truncated, causing a structural zero.

The capture-recapture model is discussed under four sets of conditions (models) for the case of three sources of information. That is, every individual in the sample is identified by either one, two, or three specific sources. Table 9.41 defines the notation. The value f_{222}^* is the number of individuals in the population not included on any of the lists used to construct the data set. The symbol "*" is a reminder that the frequency f_{222}^* is not observed. The total number of individuals observed plus the estimated number of individuals not observed $(\hat{f}_{222}^*)$ is an estimate of the size of the closed population. Capture-recapture data are also equivalently displayed in a 2-by-2-by-2 contingency table with one cell missing (structural zero), shown in Table 9.42. The term $n' = f_{111} + f_{121} + f_{211} + f_{221} + f_{112} + f_{122} + f_{212}$ represents the total number of individuals counted. The four models are described below.

Independence

A traditional capture-recapture model involves the assumption that all sources of information are independent. Data from three independent sources of information can be represented by the additive log-linear model

T A B L E   **9.42**

An alternative form of the structure and notation for the capture-recapture model

Source 3	Source 2	Not Source 2		Not Source 3	Source 2	Not Source 2
Source 1	f_{111}	f_{121}		Source 1	f_{112}	f_{122}
Not source 1	f_{211}	f_{221}		Not source 1	f_{212}	*

$$log\text{-}frequency = u + u_1 + u_2 + u_3 \quad \text{and} \quad u_{12} = u_{13} = u_{23} = u_{123} = 0 \, .$$

Complete independence implies that the number of nonincluded individuals (the missing cell frequency) f_{222}^* could be expressed by the relationship

$$f_{222}^* = \frac{f_{2..}\,f_{.2.}\,f_{..2}}{n^2} \, ,$$

but the totals $f_{2..}, f_{.2.}, f_{..2}$, and n are too low, biased by the absence of f_{222}. However, "correcting" for the missing data gives

$$f_{222}^* = \frac{(x + f_{222}^*)\,(y + f_{222}^*)\,(z + f_{222}^*)}{(n' + f_{222}^*)^2} \, , \tag{9.27}$$

with $x = f_{211} + f_{221} + f_{212}$, $y = f_{121} + f_{221} + f_{122}$, and $z = f_{112} + f_{122} + f_{212}$. Expression 9.27 simplifies to a quadratic equation where

$$(x + y + z - 2n')\,[f_{222}^*]^2 + [xy + xz + yz - (n')^2]f_{222}^* + xyz = 0$$

and can be solved for an estimate of $\hat{f}_{222}^*$. The estimate of the population size is then $\hat{n} = n' + \hat{f}_{222}^*$. The estimate $\hat{n}$ has the estimated variance [4].

$$variance(\hat{n}) = \frac{\hat{n}\,\hat{f}_{222}^*}{\hat{f}_{111} + \hat{f}_{121} + \hat{f}_{211} + \hat{f}_{112}} \, . \tag{9.28}$$

Hidden in the estimation process is the assumption that n' is fixed, which is not a very realistic assumption for some applications.

The data in Table 9.43 illustrate the estimation of population size $\hat{n}$. The frequency f_{ijk} is the count of infants with Downs syndrome listed by at least one of three sources of information (source 1 = birth certificates, source 2 = hospital records, and source 3 = county social service agency). The symbol S_i represents infants identified by source i and $\bar{S}_i$ represents those not identified by source i. It follows directly that

$$x = 114, \quad y = 68, \quad z = 186, \quad \text{and} \quad n' = 214 \, ,$$

T A B L E **9.43**

Data from three sources (birth certificates (1), hospital records (2), and county agency (3)) of information on infants with Downs syndrome

Sources (1)	(2)	(3)	f_{ijk}
1	1	1	4
1	1	0	38
1	0	1	6
1	0	0	52
0	1	1	8
0	1	0	96
0	0	1	10

or

	S_3	S_2	$\bar{S}_2$	Total
S_1	4	6	10
$\bar{S}_1$	8	10	18
Total	12	16	28

	$\bar{S}_3$	S_2	$\bar{S}_2$	Total
S_1	38	52	90
$\bar{S}_1$	96	*	—
Total	134	—	—

giving

$$(-60)\,[f_{222}^*]^2 - 4192\,f_{222}^* + 1{,}441{,}872 = 0\,,$$

which yields

$$\hat{f}_{222}^* = 123.974 \quad \text{and} \quad \hat{n} = 214 + 123.974 = 337.974\,.$$

Based on the model of independence and the estimated value $\hat{f}_{222}^*$, the estimated frequencies are given in Table 9.44 when the three sources recording malformations are independent. The assumption that the sources of information are independent can be evaluated. The estimated values (Table 9.44) compared to the data (Table 9.43) yield a chi-square statistic of $Y^2 = 0.617$ with three degrees of freedom (p-value = 0.888), showing that the assumption of independence of the three sources of information is consistent with the data. An estimate of the variance of $\hat{n}$ [4] based on these data is

$$variance(\hat{n}) = (337.974)\,(123.974)/56.43 = 742.513\,.$$

T A B L E **9.44**

Estimated frequencies when all sources of identification are independent

	S_3	S_2	$\bar{S}_2$		$\bar{S}_3$	S_2	$\bar{S}_2$
S_1		3.58	4.71	S_1		39.62	52.10
$\bar{S}_1$		8.52	11.20	$\bar{S}_1$		94.28	123.974

Partial Independence

The assumption of complete independence relaxed slightly produces a model with two independent and one nonindependent association. If sources 1 and 2 are associated whereas sources 1 and 3 and 2 and 3 are independent, then the log-linear model describing the data is

$$\textit{log-frequency} = u + u_1 + u_2 + u_3 + u_{12} \quad \text{or} \quad u_{13} = u_{23} = u_{123} = 0 \,.$$

Other combinations of partially independent categorical variables are also possible. The assumption of partial independence dictates that

$$f_{222}^* = \frac{f_{22} \cdot f_{\cdot \cdot 2}}{n} \,.$$

But, again, unbiased values of the totals f_{22}, $f_{\cdot \cdot 2}$, and n are not observable because f_{222}^* is not included in the data collection process. However, "adding" the missing frequency gives

$$f_{222}^* = \frac{(f_{221} + f_{222}^*)\,(f_{112} + f_{122} + f_{212} + f_{222}^*)}{n' + f_{222}^*} \,.$$

A bit of manipulation yields

$$\hat{f}_{222}^* = \frac{f_{221}\,(f_{112} + f_{122} + f_{212})}{f_{111} + f_{121} + f_{211}} \tag{9.29}$$

which estimates the number of individuals not included in the sample data. Applying this estimate to the illustrative data produces

$$\hat{f}_{222}^* = 10(186)/18 = 103.333$$

and $\hat{n} = 214 + 103.333 = 317.333$. The assumption of partial independence and $\hat{f}_{222}^* = 103.33$ produce the estimated values displayed in Table 9.45. Comparison of these estimated values with those observed yields a chi-square summary of the goodness-

T A B L E   9.45

Estimated frequencies when source 3 is independent of sources 1 and 2

S_3	S_2	$\overline{S}_2$	$\overline{S}_3$	S_2	$\overline{S}_2$
S_1	3.71	5.12	S_1	38.29	52.88
$\overline{S}_1$	9.18	10.00	$\overline{S}_1$	94.82	103.33

of-fit of $Y^2 = 0.356$ with two degrees of freedom (*p*-value = 0.837). The estimated value $\hat{n}$ has an estimated variance [4] of

$$variance(\hat{n}) = (\hat{f}_{222}^*)^2 \left\{ \frac{1}{f_{111} + f_{121} + f_{211}} + \frac{1}{f_{112} + f_{122} + f_{212}} + \frac{1}{f_{221}} + \frac{1}{\hat{f}_{222}^*} \right\} \quad (9.30)$$

$$= (103.33)^2 \left\{ \frac{1}{18} + \frac{1}{186} + \frac{1}{10} + \frac{1}{103.33} \right\} = 1821.73 \; .$$

Conditional Independence

Yet another approach (model) postulates that among the three sources of information, two pairs are associated and the other pair is not. For example, conditional independence implies that

$$log\text{-}frequency = u + u_1 + u_2 + u_3 + u_{12} + u_{23} \quad \text{or} \quad u_{13} = u_{123} = 0$$

when only sources 1 and 3 are assumed independent. Then

$$\frac{f_{221} f_{122}}{f_{222}^* f_{121}} = 1.0 \; ,$$

yielding an estimate for f_{222}^* of

$$\hat{f}_{222}^* = \frac{f_{221} f_{122}}{f_{121}} \; . \quad (9.31)$$

The example data give

$$\hat{f}_{222}^* = (10)\,(52)/6 = 86.667 \; ,$$

producing $\hat{n} = 214 + 86.667 = 300.667$. The estimated values based on conditional independence are shown in Table 9.46. The estimated values in Table 9.46 yield a

T A B L E **9.46**

Estimated frequencies when only sources 1 and 3 are independent

S_3	S_2	$\bar{S}_2$		$\bar{S}_3$	S_2	$\bar{S}_2$
S_1	3.45	6.00		S_1	38.55	52.00
$\bar{S}_1$	8.55	10.00		$\bar{S}_1$	95.45	86.67

likelihood ratio chi-square statistic of $Y^2 = 0.130$ with one degree of freedom (p-value $= 0.718$). The estimated variance of $\hat{n}$ [4] is

$$variance(\hat{n}) = (\hat{f}_{222}^*)^2 \left(\frac{1}{f_{121}} + \frac{1}{f_{221}} + \frac{1}{f_{122}} + \frac{1}{\hat{f}_{222}^*} \right)$$

(9.32)

$$= (86.67)^2 \left(\frac{1}{6} + \frac{1}{10} + \frac{1}{96} + \frac{1}{86.667} \right)$$

$$= 2234.07 .$$

All Pairwise Associations Exist

The most general model contains all pairwise associations and generates the log-linear model

$$log\text{-}frequency = u + u_1 + u_2 + u_3 + u_{12} + u_{13} + u_{23} \quad \text{or} \quad u_{123} = 0 .$$

An estimate of the frequency of nonincluded individuals f_{222}^* is again readily found. When $u_{123} = 0$, then

$$\frac{f_{111}f_{221}/f_{121}f_{211}}{f_{112}f_{222}^*/f_{122}f_{212}} = 1.0 ,$$

giving

$$\hat{f}_{222}^* = \frac{f_{111}f_{221}f_{122}f_{212}}{f_{121}f_{211}f_{112}}$$

(9.33)

and, as with the other models, $\hat{n} = n' + \hat{f}_{222}^*$. The illustrative data produce the estimate

$$\hat{f}_{222}^* = \frac{(4)\,(10)\,(52)\,(96)}{(6)\,(8)\,(38)} = 109.474$$

and the estimated population size is $\hat{n} = 214 + 109.474 = 323.474$. The fit of the model with $u_{123} = 0$ to the data is perfect ($Y^2 = 0$). The number of parameters in the model (degrees of freedom) equals the number of observed values in the table (seven), producing a saturated model. The estimated variance of $\hat{n}$ [4] is

$$variance(\hat{n}) = (\hat{f}_{222}^*)^2 \left[\frac{1}{f_{111}} + \frac{1}{f_{121}} + \cdots + \frac{1}{\hat{f}_{222}^*} \right]$$

(9.34)

$$= (109.474)^2 \left[\frac{1}{4} + \frac{1}{6} + \cdots + \frac{1}{109.474} \right]$$

$$= 8470.21 .$$

T A B L E **9.47**

Summary of the four estimates of the population size n

Model	$\hat{f}_{222}^{*}$	$\hat{n}$	$\sqrt{variance(\hat{n})}$	Y^2	Degrees of freedom
Independence	123.97	337.97	27.25	0.62	3
$u_{13}=u_{23}=u_{123}=0$	103.33	317.33	42.68	0.36	2
$u_{13}=u_{123}=0$	86.67	300.67	47.27	0.13	1
$u_{123}=0$	109.47	323.47	92.03	0.00	0

Comparison of the Four Models

Contrasting the results from the application of these four models (Table 9.47) demonstrates two issues that are true in general. The more complex models fit the data better. Also, the opposite is often true with regard to the variability. The more complicated models produce estimates with the higher variances. The example data present no problems in selecting an optimum model (estimate) because the simplest model (independence) fits the data extremely well ($Y^2 = 0.62$) and produces an estimate of the population size with the lowest estimated variance ($variance(\hat{n}) = 742.513$). It is easy to imagine, however, cases where the choice is more difficult, such as simple models giving estimates with relatively low variability but that fit the data poorly or complicated models that fit the data well but give highly variable estimates.

Test Case

Computer Implementation

Not much is learned from applying a log-linear model to a 2-by-2 table because the estimated values and the chi-square statistic are identical to the nonmodel approach. Nevertheless, data analyzed earlier are repeated here (from Table 9.17) to illustrate the computer implementation. The estimates for the log-linear models $u + u_1 + u_2 + u_{12}$ (saturated) and $u + u_1 + u_2$ (independence) are summarized in Table 9.48.

Using these estimated parameters produces the estimated cell frequencies (independence) in Table 9.49. As before, the difference between these estimated and the observed cell frequencies yields $Y^2 = 0.012$ with one degree of freedom (p-value = 0.914).

The estimates of the model parameters from a particular computer program can differ depending on how the model is constructed to represent the data, particularly the way the constant term is defined. The STATA system employs a representation of the model parameters that produces estimates of the u-terms that do not add to zero. The parameters from the STATA representation are denoted u_i'. The logarithms of the

T A B L E **9.48**

Work status by age (data and parameter estimates)

Nonsmokers	Working	Not Working	Total
Age ≤ 30	6	86	92
Age > 30	8	108	116
Total	14	194	208

Estimated parameters, saturated and independence models

Estimates	Saturated	Independence
$\hat{u}$	3.252	3.254
$\hat{u}_1(1)$	−0.129	−0.116
$\hat{u}_1(2)$	0.129	0.116
$\hat{u}_2(1)$	−1.316	−1.314
$\hat{u}_2(2)$	1.316	1.314
$\hat{u}_{12}$	±0.015	0.0

cell frequencies (denoted, as before, l_{ij}) for the case of independence in a 2-by-2 table are estimated by:

$$\hat{l}_{11} = \hat{u}' + \hat{u}_1' + \hat{u}_2'$$
$$\hat{l}_{12} = \hat{u}' + \hat{u}_1'$$
$$\hat{l}_{21} = \hat{u}' + \hat{u}_2'$$
$$\hat{l}_{22} = \hat{u}'.$$

Although the estimated parameters are different, the estimated values $\hat{l}_{ijk}$ do not change. Furthermore, the estimated cell frequencies and, therefore, the assessments

T A B L E **9.49**

Estimated values for the work status by age (independence)

Nonsmokers	Working	Not Working	Total
Age ≤ 30	6.19	85.81	92
Age > 30	7.81	108.19	116
Total	14	194	208

of goodness-of-fit are also unaffected by the parameterization of the model, producing statistical tests based on nested models that are identical to those discussed.

example.log-linear

```
. infile count age work using twoby.dat
(4 observations read)

. loglin count age work, fit(age,work) resid
Variable age = A
Variable work = B
Margins fit: age,work
Note: Regression-like constraints are assumed.  The first level of each
variable (and all iteractions with it) will be dropped from estimation.

Iteration 0: Log Likelihood = -104.47705
Iteration 1: Log Likelihood = -30.745178
Iteration 2: Log Likelihood = -12.063293
Iteration 3: Log Likelihood = -10.273682
Iteration 4: Log Likelihood = -10.211548
Iteration 5: Log Likelihood = -10.211426
```

Y^2 p-value   Number of cells

```
Poisson regression                            Number of obs    =        4
Goodness-of-fit chi2(1)    =      0.012       Model chi2(2)    = 188.531
Prob  chi2                 =      0.9142      Prob  chi2       =  0.0000
Log Likelihood             =    -10.211       Pseudo R2        =  0.9023
```

```
---------------------------------------------------------------------------
   count |      Coef.   Std. Err.       t    P|t|    [95% Conf. Interval]
---------+-----------------------------------------------------------------
      A2 |  -.2318014   .1396075    -1.660   0.098   -.5070284    .0434257
      B2 |  -2.628801   .2767367    -9.499   0.000   -3.174369   -2.083232
   _cons |    4.68391   .0946976    49.462   0.000     4.49722     4.8706
---------------------------------------------------------------------------
```

```
count age work  cellhat   resid   stdres
108    0    0   108.192   -0.192  -0.018
  8    0    1     7.808    0.192   0.069
 86    1    0    85.808    0.192   0.021
  6    1    1     6.192   -0.192  -0.077
```

f_{ij} $\hat{f}_{ij}$ $f_{ij} - \hat{f}_{ij}$ $\dfrac{f_{ij} - \hat{f}_{ij}}{\sqrt{\hat{f}_{ij}}}$

```
. loglin count age work, fit(age,work,age work) resid
Variable age = A
Variable work = B
Margins fit: age,work,age work
Note: Regression-like constraints are assumed.  The first level of each
variable (and all iteractions with it) will be dropped from estimation.

Iteration 0: Log Likelihood = -104.47705
Iteration 1: Log Likelihood = -32.44104
Iteration 2: Log Likelihood = -12.160828
```

```
Iteration 3: Log Likelihood = -10.275085
Iteration 4: Log Likelihood = -10.205811
Iteration 5: Log Likelihood = -10.205627


Poisson regression                              Number of obs   =        4
Goodness-of-fit chi2(0)     =     -0.000        Model chi2(3)   = 188.543
Prob chi2                   =       .           Prob  chi2      =   0.0000
Log Likelihood              =    -10.206        Pseudo R2       =   0.9023


---------------------------------------------------------------------------
   count |     Coef.    Std. Err.       t     P|t|    [95% Conf. Interval]
---------+-----------------------------------------------------------------
      A2 |  -.2277842    .1445239    -1.576   0.117    -.5127117    .0571434
    AB22 |  -.0598979    .5590651    -0.107   0.915    -1.162089   1.042294
      B2 |   -2.60269    .3664141    -7.103   0.000    -3.325072  -1.880308
   _cons |   4.682131    .096225     48.658   0.000     4.492425   4.871838
---------------------------------------------------------------------------


count  age  work  cellhat   resid   stdres
108    0    0     108.000  -0.000  -0.000
  8    0    1       8.000   0.000   0.000
 86    1    0      86.000   0.000   0.000
  6    1    1       6.000   0.000   0.000


. exit
```

Applied Example **Log-Linear Model**

Log-Linear Analysis of a 2-by-2-by-3 Table

Three variables tabulated from the CHDS data set are (1) birth weight (BWT) divided into two categories (less than or equal to 6 pounds and greater than 6 pounds), (2) maternal age also divided into two categories (mother's age less than or equal to 30 years and mother's age greater than 30 years), and (3) smoking divided into three categories (nonsmokers, 1 to 20 cigarettes, and more than 20 cigarettes per day). The data form the 2-by-2-by-3 contingency table given in Table 9.50.

The analysis using log-linear techniques applied to these data yields the following summary results for all eight possible models where 1 = birth weight = *bwt*, 2 = *age*, and 3 = amount smoked = *smk*:

1   design = *bwt*, *age*, *smk* and *bwt* by *smk*, *bwt* by *age*, *age* by *smk*

$$\text{model: } \textit{log-frequency} = u + u_1 + u_2 + u_3 + u_{12} + u_{13} + u_{23}$$

Likelihood ratio chi-square = Y^2 = 2.999, degrees of freedom = 2; p-value = 0.223

T A B L E   **9.50**

Infants classified by birth weight, mother's age, and mother's smoking exposure

Nonsmoker

	Age ≤ 30	Age > 30	Total
BWT ≤ 6	12	5	17
BWT > 6	293	71	364
Total	305	76	381

Smoker: 1 to 20 cigarettes per day

	Age ≤ 30	Age > 30	Total
BWT ≤ 6	14	7	21
BWT > 6	123	25	148
Total	137	32	169

Smoker: 20+ cigarettes per day

	Age ≤ 30	Age > 30	Total
BWT ≤ 6	16	2	18
BWT > 6	91	21	112
Total	107	23	130

2 design = *bwt*, *age*, *smk* and *bwt* by *smk*, *age* by *smk*

$$model: log\text{-}frequency = u + u_1 + u_2 + u_3 + u_{13} + u_{23}$$

Likelihood ratio chi-square = $Y^2 = 4.450$; degrees of freedom = 3; p-value = 0.217

3 design = *bwt*, *age*, *smk* and *bwt* by *age*, *age* by *smk*

$$model: log\text{-}frequency = u + u_1 + u_2 + u_3 + u_{12} + u_{23}$$

Likelihood ratio chi-square = $Y^2 = 19.749$; degrees of freedom = 4; p-value = 0.001

4 design = *bwt*, *age*, *smk* and *bwt* by *age*, *bwt* by *smk*

$$model: log\text{-}frequency = u + u_1 + u_2 + u_3 + u_{12} + u_{13}$$

Likelihood ratio chi-square = Y^2 = 3.572; degrees of freedom = 4; p-value = 0.467

5 design = *bwt*, *age*, *smk* and *age* by *smk*

$$\text{model: log-frequency} = u + u_1 + u_2 + u_3 + u_{23}$$

Likelihood ratio chi-square = Y^2 = 20.963; degrees of freedom = 5; p-value = 0.001

6 design = *bwt*, *age*, *smk* and *bwt* by *smk*

$$\text{model: log-frequency} = u + u_1 + u_2 + u_3 + u_{13}$$

Likelihood ratio chi-square = Y^2 = 4.786; degrees of freedom = 5; p-value = 0.443

7 design = *bwt*, *age*, *smk* and *bwt* by *age*

$$\text{model: log-frequency} = u + u_1 + u_2 + u_3 + u_{12}$$

Likelihood ratio chi-square = Y^2 = 20.084; degrees of freedom = 6; p-value = 0.003

8 design = *bwt*, *age*, *smk*

$$\text{model: log-frequency} = u + u_1 + u_2 + u_3$$

Likelihood ratio chi-square = Y^2 = 21.299; degrees of freedom = 7; p-value = 0.003

The log-linear model without a three-way interaction term (design = *bwt*, *age*, *smk* and *bwt* by *age*, *bwt* by *smk*, *age* by *smk*) gives no strong evidence that the term u_{123} is necessary to describe the data (p-value = 0.223). Therefore, the two-way interaction terms are likely to be useful measures of the associations among age, smoking, and birth weight. The model with all nonadditive terms set to zero (design = *bwt*, *age*, *smk*) indicates evidence of some associations among the categorical variables (p-value = 0.003). The three models that eliminate only one pairwise association show that the terms representing the "age–birth weight" (p-value = 0.217) and the "age-smoking" (p-value = 0.467) associations are not extremely important. The model taking into account only the "smoking–birth weight" association (design = *bwt*, *age*, *smk* and *bwt* by *smk*) is a reasonable fit to the data (p-value = 0.443). In addition, all models that do not account for an association between smoking and birth weight (design = *bwt*, *age*, *smk* and *age* by *smk* and design = *bwt*, *age*, *smk* and *bwt* by *age*) have large chi-square statistics (p-values of 0.001 and 0.003, respectively). Therefore, age appears to have little or no association with smoking and birth weight whereas smoking seems to be related to infant birth weight—partial independence. A table formed by adding over maternal age provides a simpler and realistic description of these data, giving Table 9.51. It is more easily seen in this combined

T A B L E **9.51**

Smoking status by birth weight (added over maternal age)

Cigarettes	None	1 to 20	20+	Total
BWT ≤ 6	17	21	18	56
BWT > 6	364	148	112	624
Total	381	169	130	680
$\hat{p}_i$	0.045	0.124	0.138	0.082

table that the proportion of low birth-weight infants increases as the amount smoked by the mother increases. The chi-square for linear trend is $X_L^2 = 14.854$ (expression 8.47), yielding a p-value < 0.001.

```
chds.log-linear

. infile count bwt age smk using bwt.dat
(12 observations read)

. loglin count bwt age smk, fit(bwt,age,smk,age smk,bwt smk,bwt age) resid
Variable bwt = A
Variable age = B
Variable smk = C
Margins fit: bwt,age,smk,age smk,bwt smk,bwt age
Note: Regression-like constraints are assumed.  The first level of each
variable (and all iterations with it) will be dropped from estimation.

Poisson regression                              Number of obs   =        12
Goodness-of-fit chi2(2)    =       2.999        Model chi2(9)   =1002.261
Prob  chi2                 =       0.2233        Prob  chi2      =   0.0000
Log Likelihood             =     -31.462        Pseudo R2       =   0.9409

count  bwt  age  smk   cellhat    resid   stdres
  71    0    0    1    71.431   -0.431   -0.051
  25    0    0    2    26.762   -1.762   -0.341
  21    0    0    3    18.806    2.194    0.506
 293    0    1    1   292.569    0.431    0.025
 123    0    1    2   121.238    1.762    0.160
  91    0    1    3    93.194   -2.194   -0.227
   5    1    0    1     4.569    0.431    0.202
   7    1    0    2     5.238    1.762    0.770
   2    1    0    3     4.194   -2.194   -1.071
  12    1    1    1    12.431   -0.431   -0.122
  14    1    1    2    15.762   -1.762   -0.444
  16    1    1    3    13.806    2.194    0.590

. loglin count bwt age smk, fit(bwt,age,smk,age smk,bwt smk) resid
Variable bwt = A
Variable age = B
```

Variable smk = C
Margins fit: bwt,age,smk,age smk,bwt smk
Note: Regression-like constraints are assumed. The first level of each
variable (and all iteractions with it) will be dropped from estimation.

Poisson regression   Number of obs   =   12
Goodness-of-fit chi2(3)   =   4.450   Model chi2(8)   =1000.810
Prob chi2   =   0.2168   Prob chi2   = 0.0000
Log Likelihood   = -32.188   Pseudo R2   = 0.9396

count	bwt	age	smk	cellhat	resid	stdres
71	0	0	1	72.609	-1.609	-0.189
25	0	0	2	28.024	-3.024	-0.571
21	0	0	3	19.815	1.185	0.266
293	0	1	1	291.391	1.609	0.094
123	0	1	2	119.976	3.024	0.276
91	0	1	3	92.185	-1.185	-0.123
5	1	0	1	3.391	1.609	0.874
7	1	0	2	3.976	3.024	1.516
2	1	0	3	3.185	-1.185	-0.664
12	1	1	1	13.609	-1.609	-0.436
14	1	1	2	17.024	-3.024	-0.733
16	1	1	3	14.815	1.185	0.308

. loglin count bwt age smk, fit(bwt,age,smk,bwt age,age smk) resid
Variable bwt = A
Variable age = B
Variable smk = C
Margins fit: bwt,age,smk,bwt age,age smk
Note: Regression-like constraints are assumed. The first level of each
variable (and all iteractions with it) will be dropped from estimation.

Poisson regression   Number of obs   =   12
Goodness-of-fit chi2(4)   =   19.749   Model chi2(7)   = 985.511
Prob chi2   =   0.0006   Prob chi2   = 0.0000
Log Likelihood   = -39.837   Pseudo R2   = 0.9252

count	bwt	age	smk	cellhat	resid	stdres
71	0	0	1	67.878	3.122	0.379
25	0	0	2	28.580	-3.580	-0.670
21	0	0	3	20.542	0.458	0.101
293	0	1	1	281.667	11.333	0.675
123	0	1	2	126.519	-3.519	-0.313
91	0	1	3	98.814	-7.814	-0.786
5	1	0	1	8.122	-3.122	-1.096
7	1	0	2	3.420	3.580	1.936
2	1	0	3	2.458	-0.458	-0.292
12	1	1	1	23.333	-11.333	-2.346
14	1	1	2	10.481	3.519	1.087
16	1	1	3	8.186	7.814	2.731

```
. loglin count bwt age smk, fit(bwt,age,smk,bwt age,bwt smk) resid
Variable bwt = A
Variable age = B
Variable smk = C
Margins fit: bwt,age,smk,bwt age,bwt smk
Note: Regression-like constraints are assumed.  The first level of each
variable (and all iteractions with it) will be dropped from estimation.
```

Poisson regression				Number of obs	=	12
Goodness-of-fit chi2(4)	=	3.572		Model chi2(7)	=1001.688
Prob chi2	=	0.4670		Prob chi2	=	0.0000
Log Likelihood	=	-31.749		Pseudo R2	=	0.9404

count	bwt	age	smk	cellhat	resid	stdres
71	0	0	1	68.250	2.750	0.333
25	0	0	2	27.750	-2.750	-0.522
21	0	0	3	21.000	-0.000	-0.000
293	0	1	1	295.750	-2.750	-0.160
123	0	1	2	120.250	2.750	0.251
91	0	1	3	91.000	0.000	0.000
5	1	0	1	4.250	0.750	0.364
7	1	0	2	5.250	1.750	0.764
2	1	0	3	4.500	-2.500	-1.179
12	1	1	1	12.750	-0.750	-0.210
14	1	1	2	15.750	-1.750	-0.441
16	1	1	3	13.500	2.500	0.680

```
. loglin count bwt age smk, fit(bwt,age,smk,age smk) resid
Variable bwt = A
Variable age = B
Variable smk = C
Margins fit: bwt,age,smk,age smk
Note: Regression-like constraints are assumed.  The first level of each
variable (and all iteractions with it) will be dropped from estimation.
```

Poisson regression				Number of obs	=	12
Goodness-of-fit chi2(5)	=	20.962		Model chi2(6)	=	984.297
Prob chi2	=	0.0008		Prob chi2	=	0.0000
Log Likelihood	=	-40.444		Pseudo R2	=	0.9241

count	bwt	age	smk	cellhat	resid	stdres
71	0	0	1	69.741	1.259	0.151
25	0	0	2	29.365	-4.365	-0.805
21	0	0	3	21.106	-0.106	-0.023
293	0	1	1	279.882	13.118	0.784
123	0	1	2	125.718	-2.718	-0.242
91	0	1	3	98.188	-7.188	-0.725
5	1	0	1	6.259	-1.259	-0.503
7	1	0	2	2.635	4.365	2.689
2	1	0	3	1.894	0.106	0.077
12	1	1	1	25.118	-13.118	-2.617
14	1	1	2	11.282	2.718	0.809
16	1	1	3	8.812	7.188	2.422

```
. loglin count bwt age smk, fit(bwt,age,smk,bwt smk) resid
Variable bwt = A
Variable age = B
Variable smk = C
Margins fit: bwt,age,smk,bwt smk
Note: Regression-like constraints are assumed.  The first level of each
variable (and all iteractions with it) will be dropped from estimation.
```

Poisson regression				Number of obs	=	12
Goodness-of-fit chi2(5)		=	4.786	Model chi2(6)	=1000.474
Prob chi2		=	0.4425	Prob chi2	=	0.0000
Log Likelihood		=	-32.356	Pseudo R2	=	0.9392

count	bwt	age	smk	cellhat	resid	stdres
71	0	0	1	70.124	0.876	0.105
25	0	0	2	28.512	-3.512	-0.658
21	0	0	3	21.576	-0.576	-0.124
293	0	1	1	293.877	-0.877	-0.051
123	0	1	2	119.488	3.512	0.321
91	0	1	3	90.424	0.576	0.061
5	1	0	1	3.275	1.725	0.953
7	1	0	2	4.046	2.954	1.469
2	1	0	3	3.468	-1.468	-0.788
12	1	1	1	13.725	-1.725	-0.466
14	1	1	2	16.954	-2.954	-0.718
16	1	1	3	14.532	1.468	0.385

```
. loglin count bwt age smk, fit(bwt,age,smk,bwt age) resid
Variable bwt = A
Variable age = B
Variable smk = C
Margins fit: bwt,age,smk,bwt age
Note: Regression-like constraints are assumed.  The first level of each
variable (and all iteractions with it) will be dropped from estimation.
```

Poisson regression				Number of obs	=	12
Goodness-of-fit chi2(6)		=	20.084	Model chi2(5)	=	985.175
Prob chi2		=	0.0027	Prob chi2	=	0.0000
Log Likelihood		=	-40.005	Pseudo R2	=	0.9249

count	bwt	age	smk	cellhat	resid	stdres
71	0	0	1	65.554	5.446	0.673
25	0	0	2	29.078	-4.078	-0.756
21	0	0	3	22.368	-1.368	-0.289
293	0	1	1	284.069	8.931	0.530
123	0	1	2	126.004	-3.004	-0.268
91	0	1	3	96.926	-5.926	-0.602
5	1	0	1	7.844	-2.844	-1.015
7	1	0	2	3.479	3.521	1.887
2	1	0	3	2.676	-0.676	-0.413
12	1	1	1	23.532	-11.532	-2.377
14	1	1	2	10.438	3.562	1.102
16	1	1	3	8.029	7.971	2.813

```
. loglin count bwt age smk, fit(age,age,smk) resid
Variable bwt = A
Variable age = B
Variable smk = C
Margins fit: age,age,smk
Note: Regression-like constraints are assumed.  The first level of each
variable (and all iteractions with it) will be dropped from estimation.
```

Poisson regression				Number of obs	=	12
Goodness-of-fit chi2(7)	=	21.299		Model chi2(4)	=	983.961
Prob chi2	=	0.0034		Prob chi2	=	0.0000
Log Likelihood	=	-40.612		Pseudo R2	=	0.9237

count	bwt	age	smk	cellhat	resid	stdres
71	0	0	1	67.354	3.646	0.444
25	0	0	2	29.876	-4.876	-0.892
21	0	0	3	22.982	-1.982	-0.413
293	0	1	1	282.270	10.730	0.639
123	0	1	2	125.206	-2.206	-0.197
91	0	1	3	96.312	-5.312	-0.541
5	1	0	1	6.045	-1.045	-0.425
7	1	0	2	2.681	4.319	2.638
2	1	0	3	2.062	-0.062	-0.043
12	1	1	1	25.332	-13.332	-2.649
14	1	1	2	11.236	2.764	0.824
16	1	1	3	8.643	7.357	2.502

```
. loglin count bwt age smk, fit(bwt,age,smk,bwt age, bwt smk, age smk, bwt age
  smk) resid
Variable bwt = A
Variable age = B
Variable smk = C
Margins fit: bwt,age,smk,bwt age, bwt smk, age smk, bwt age smk
Note: Regression-like constraints are assumed.  The first level of each
variable (and all iteractions with it) will be dropped from estimation.
```

Poisson regression				Number of obs	=	12
Goodness-of-fit chi2(0)	=	-0.000		Model chi2(11)	=	1005.260
Prob chi2	=	.		Prob chi2	=	0.0000
Log Likelihood	=	-29.963		Pseudo R2	=	0.9437

count	bwt	age	smk	cellhat	resid	stdres
71	0	0	1	71.000	-0.000	-0.000
25	0	0	2	25.000	-0.000	-0.000
21	0	0	3	21.000	-0.000	-0.000
293	0	1	1	293.000	0.000	0.000
123	0	1	2	123.000	0.000	0.000
91	0	1	3	91.000	0.000	0.000
5	1	0	1	5.000	0.000	0.000
7	1	0	2	7.000	-0.000	-0.000
2	1	0	3	2.000	0.000	0.000

```
12   1   1   1    12.000    0.000    0.000
14   1   1   2    14.000   -0.000   -0.000
16   1   1   3    16.000    0.000    0.000


. exit
```

Notation

"Dot" notation for an *r*-by-*c*-by-*l* table:

$$f_{i\cdot\cdot} = \sum_j \sum_k f_{ijk} = \sum_j f_{ij\cdot} = \sum_k f_{i\cdot k}$$

$$f_{\cdot j\cdot} = \sum_i \sum_k f_{ijk} = \sum_i f_{ij\cdot} = \sum_k f_{\cdot jk}$$

$$f_{\cdot\cdot k} = \sum_i \sum_j f_{ijk} = \sum_i f_{i\cdot k} = \sum_j f_{\cdot jk}$$

and

$$n = \sum_i \sum_j \sum_k f_{ijk} = \sum_i f_{i\cdot\cdot} = \sum_j f_{\cdot j\cdot} = \sum_k f_{\cdot\cdot k}$$

Estimated log-linear model for a 2-way table:

$$log\text{-}frequency = u + u_1(i) + u_2(j) + u_{12}(i, j)$$

or

	B_1	B_2	Mean Response
A_1	$u + u_1(1) + u_2(1) + u_{12}(1, 1)$	$u + u_1(1) + u_2(2) + u_{12}(1, 2)$	$u + u_1(1)$
A_2	$u + u_1(2) + u_2(1) + u_{12}(2, 1)$	$u + u_1(2) + u_2(2) + u_{12}(2, 2)$	$u + u_1(2)$
Mean response	$u + u_2(1)$	$u + u_2(2)$	u

Log-linear for a 3-way table:

$$\hat{l}_{ijk} = \hat{u} + \hat{u}_1(i) + \hat{u}_2(j) + \hat{u}_3(k) + \hat{u}_{12}(i, j) + \hat{u}_{13}(i, k) + \hat{u}_{23}(j, k) + \hat{u}_{123}(i, j, k)$$

Interpretation of u_{12} in terms of odds ratios:

$$\hat{u}_{12} = \pm\frac{1}{8} \log([\hat{OR}_1][\hat{OR}_2])$$

Interpretation of u_{123} in terms of odds ratios:

$$\hat{u}_{123} = \pm\frac{1}{8} \log[\hat{OR}_1 / \hat{OR}_2]$$

Problems

T A B L E   **9.52**

Data on maternal age (<35, ≥35), twinning (single births = 0, twin births = 1), and race (white, black, and Hispanic)

Age	Twin	Race	Count
1	0	1	4993
1	0	3	1378
1	0	4	1111
1	0	5	204
1	1	1	107
1	1	2	30
1	1	3	33
1	1	4	6
1	1	5	3
0	0	1	929
0	0	2	100
0	0	3	172
0	0	4	166
0	0	5	50
0	1	1	23
0	1	2	1
0	1	3	4
0	1	4	1
0	1	5	2

Note: Table 9.52 contains births by maternal age (<35, ≥ 35), twin status (twin births = 1 versus single births = 0), and race (1 = white, 2 = black, 3 = Hispanic, 4 = Japanese, and 5 = Chinese).

1   Using the data in Table 9.52, employ a log-linear model to describe the relationships among the three variables, maternal age, twin birth, and race. Compute Y^2 and its associated *p*-value for all eight possible log-linear models and describe the role of each of the three variables.

2   For the 2-by-2 table, show that $\hat{u}_{12} = \log[\hat{OR}]/4$ (see expression 9.10).

3   For the 2-by-2 table, show that if $\hat{l}_{21} = \hat{u} + \hat{u}_2 + \hat{u}_1$, then

$$\hat{f}_{21} = e^{\hat{l}_{21}} = \frac{f_{2.}f_{.1}}{n}.$$

4   Given the data:

	≤ 45 years	>45 years	Total
Male	2	3	5
Female	2	5	7
Total	4	8	12

Find estimated values for the parameters $\hat{u}$, $\hat{u}_1$, $\hat{u}_2$, and $\hat{u}_{12}$ from the saturated model for these data and calculate the associated values. Demonstrate that these parameters produce a saturated model (i.e., reproduce the data). Find estimated values of the parameters $\hat{u}$, $\hat{u}_1$, and $\hat{u}_2$ from the additive model for these data and calculate the associated estimated values.

5   Consider the following data:

C_1	B_1	B_2	Total
A_1	14	6	20
A_2	6	4	10
Total	20	10	30

C_2	B_1	B_2	Total
A_1	28	22	50
A_2	12	8	20
Total	40	30	70

Find the estimates for u, u_1, u_2, u_3, u_{12}, u_{13}, u_{23}, and u_{123}. Find estimates of the variance associated with the estimates $\hat{u}_{123}$, $\hat{u}_{12}$, $\hat{u}_{13}$, and $\hat{u}_{23}$. Test the following hypotheses:

$$H_1: u_{123} = 0, \quad H_2: u_{12} = 0, \quad H_3: u_{13} = 0, \quad \text{and} \quad H_4: u_{23} = 0.$$

6   If the rows of a 2-by-2 table contain the same values in both columns, then show $u_1 = 0$ and $u_{12} = 0$.

7   For the following table, find an estimate of f_{22}^* under the requirement that variables A and B are independent:

		Variable B		
		$B = 1$	$B = 0$	Total
Variable A	$A = 1$	f_{11}	f_{12}	$f._1$
	$A = 0$	f_{21}	*	?
	Total	$f._1$	?	?

That is, $f_{22}^* = ?$

8   Data from three sources (police records (1), drug rehabilitation center records (2), and hospital emergency rooms (3)) provide information on serious drug addition:

(1)	(2)	(3)	f_{ijk}
1	1	1	40
1	1	0	382
1	0	1	45
1	0	0	621
0	1	1	77
0	1	0	888
0	0	1	123
0	0	0	*

Use four models (independence, partial independence, conditional independence, and all pairwise associations exist) to estimate the total number of drug-addicted individuals in this population.

9   Consider data collected on individuals convicted of felony crimes where the type of crime (violent = 0 and nonviolent = 1), offender status (first offender = 0 and repeat offender = 1), and race (white = 0 and black = 1) are classified into a three-dimensional table:

Violent	Repeat	Race	Count
1	0	0	300
1	0	1	400
1	1	0	378
1	1	1	525
0	0	0	17
0	0	1	20
0	1	0	5
0	1	1	8

Conduct an analysis using a log-linear model to find a simpler way to classify these data (i.e., a combined table that does not distort the relationships within the data set). Analyze the relationship in this simpler two-dimensional table.

References

1   Fienberg, S. E. *The Analysis of Cross-Classification Categorical Data*. Cambridge: MIT Press, 1977.
2   Fingleton, B. *Model of Category Counts*. Cambridge, U.K.: Cambridge University Press, 1984.
3   Upton, G. G. J. *The Analysis of Cross-Tabulated Data*. New York: John Wiley, 1978.
4   Bishop, Y. M. M., Fienberg, S. E., and Holland, P. W. *Discrete Multivariate Analysis: Theory and Practice*. Cambridge: MIT Press, 1975.

5 Roseman R. H., Brand, R. J., and Jenkins, C. C. Coronary Heart Disease in the Western Collaborative Group Study. *Journal of the American Medical Association* 223 (1975): 872–877.

6 Seber, G. A. F. *The Estimation of Animal Abundance and Related Parameters*. London: Charles Griffin and Company, 1982.

10

Logistic Regression Analysis

Background

Logistic regression analysis is a particularly important analytic tool in epidemiological and biomedical research. The logistic approach overlaps several techniques already discussed. Linear regression, log-linear, and discriminant analysis resemble certain aspects of a linear logistic model. However, differences exist and the logistic approach is usually considered separately from these related techniques. For a logistic regression, the independent variables can be continuous or discrete but the dependent variable is typically binary. The importance of the logistic model comes from the fact that many data sets naturally occur in a pattern of a binary outcome thought to be related to a series of independent variables. For example, the occurrence of coronary heart disease is thought to be related to factors such as blood pressure, cholesterol level, body weight, age, and behavior pattern. The role of such variables can be investigated with a logistic regression analysis when the outcome is one of two possible events: the occurrence of a coronary event or the absence a coronary event.

The presence or absence of a disease, the success or failure of a new drug, or almost any variable characterized by two outcomes (coded 0 and 1) can be the subject of a logistic regression analysis. Underlying this binary response variable is a probability (e.g., P(disease) or P(success)), and a logistic regression analysis explores the relationship of a set of independent variables to this probability (dependent variable). At the center of the logistic regression strategy is a transformation equating the probabilities associated with a binary dependent variable to the independent variables. A rich and mathematically simple model emerges when the logarithm of the odds is represented by a linear combination of independent variables. As will be seen, describing the logarithm of the odds with a linear combination of independent variables is the same as describing the probability associated with a binary outcome with a logistic function. Although the logarithm of the odds is not an intuitive measure of association, experience indicates that this transformation produces an effective description of many binary variables in terms of a linear combination of independent variables. In addition, the application of the logistic model yields a set of estimated odds ratios as a summary statistic that are useful and readily interpretable measures of association.

The 2-by-2 Table

The simplest linear logistic model arises from the application to data from a 2-by-2 table. The notation for the elements of a 2-by-2 table is the same as before (Table 8.7) and is repeated in Table 10.1. In the context of a logistic regression analysis, a natural measure of association is the odds ratio. An odds ratio, defined previously (expression 8.27), is a ratio of two odds, where the odds are calculated under different conditions. Algebraically, the odds are $p/(1-p)$ where p is a probability (odds are a single quantity but are usually referred to as a plural noun). If the odds under one set of conditions are $p_0/(1-p_0)$ and the odds under another set of conditions are $p_1/(1-p_1)$, then the odds ratio summarizing the magnitude of any difference is

$$odds\ ratio = \frac{p_1/(1-p_1)}{p_0/(1-p_0)} = \frac{p_1(1-p_0)}{p_0(1-p_1)}.$$

(10.1)

It is convenient to investigate association between variables in terms of the logarithm of the odds, called the log-odds. Specifically for a 2-by-2 table, when $G = 1$ (first row), then $log\text{-}odds = log[P(D \mid G = 1)/P(\overline{D} \mid G = 1)]$ and is estimated by $y_1 = log(f_{11}/f_{12})$. For $G = 0$ (second row), then $log\text{-}odds = log[P(D \mid G = 0)/P(\overline{D} \mid G = 0)]$ and is estimated by $y_0 = log(f_{21}/f_{22})$. Therefore, the odds ratio reflecting the association between group status (labeled $G = 1$ and $G = 0$) and the binary outcome categories (labeled D and $\overline{D}$) in a 2-by-2 table is

$$OR = \frac{odds(G=1)}{odds(G=0)} = \frac{P(D \mid G = 1)/P(\overline{D} \mid G = 1)}{P(D \mid G = 0)/P(\overline{D} \mid G = 0)}$$

(10.2)

and the odds ratio OR is estimated by

$$\hat{OR} = \frac{f_{11}/f_{12}}{f_{21}/f_{22}} = \frac{f_{11}f_{22}}{f_{12}f_{21}} \quad \text{or} \quad log(\hat{OR}) = y_1 - y_0 \quad \text{making} \quad \hat{OR} = e^{y_1 - y_0}.$$

(10.3)

T A B L E **10.1**

The notation for a 2-by-2 table (repeated from Table 8.8)

		Outcome		
		D	$\overline{D}$	Total
Group	$G = 1$	f_{11}	f_{12}	$f_{1\cdot}$
	$G = 0$	f_{21}	f_{22}	$f_{2\cdot}$
	Total	$f_{\cdot 1}$	$f_{\cdot 2}$	n

The basic difference between assessing an association in terms of either an odds ratio or the difference in log-odds lies in their statistical properties.

The same odds ratio occurs if instead of rows (G), the columns (D) are considered. The odds ratio based on the columns of a 2-by-2 table is

$$OR = \frac{odds(D)}{odds(\overline{D})} = \frac{P(G=1 \mid D)/P(G=0 \mid D)}{P(G=1 \mid \overline{D})/P(G=0 \mid \overline{D})}.$$
(10.4)

The odds ratio given in expression 10.4 is made up of different odds, but a bit of manipulation shows that it is identical to the previous odds ratio based on the rows (expression 10.2). The estimated odds ratio $\hat{OR}$ is also unchanged:

$$\hat{OR} = \frac{f_{11}/f_{21}}{f_{12}/f_{22}} = \frac{f_{11}f_{22}}{f_{12}f_{21}}.$$
(10.5)

Like all estimates, to evaluate the estimated odds ratio it is necessary to estimate its variability. The variance of the logarithm of $\hat{OR}$ is estimated by (repeated from expression 8.28)

$$variance[\log(\hat{OR})] = S^2_{\log(\hat{OR})} = \frac{1}{f_{11}} + \frac{1}{f_{12}} + \frac{1}{f_{21}} + \frac{1}{f_{22}}.$$
(10.6)

The estimated odds ratio $\hat{OR}$ can be compared to a postulated value OR_0 (typically, $OR_0 = 1$). This comparison is usually in terms of $\log(\hat{OR})$ because the distribution of the logarithms of the odds ratio has a more symmetric distribution than the distribution of the odds ratios themselves, making it possible to use more accurately a normal distribution as an approximation. Using the relatively simple normal distribution as an approximation to a more or less symmetric but complex distribution is a typical analytic strategy. In symbols,

$$z = \frac{\log(\hat{OR}) - \log(OR_0)}{S_{\log(\hat{OR})}}$$
(10.7)

where z has an approximate standard normal distribution when the estimated odds ratio $\hat{OR}$ differs from OR_0 only because of random variation.

Along the same lines, an approximate confidence interval can be constructed based on $\log(\hat{OR})$ also using a normal distribution approximation. The confidence interval is

$$lower\ bound = \log(\hat{OR}) - z_{1-\alpha/2}S_{\log(\hat{OR})} \quad \text{and}$$

$$upper\ bound = \log(\hat{OR}) + z_{1-\alpha/2}S_{\log(\hat{OR})}$$
(10.8)

where $z_{1-\alpha/2}$ is the $(1 - \alpha/2)$-percentile of the standard normal distribution. In the case of logarithms, an approximate confidence interval for the logarithm of a quantity can be used to construct a confidence interval for the anti-logarithms. Or, an approximate $(1 - \alpha)$% confidence interval for $OR = e^{\log(OR)}$ is (e^{lower}, e^{upper}), giving a confidence interval for an odds ratio of

$$lower\ bound = \hat{OR}\ e^{-z_{1-\alpha/2}S_{\log(\hat{OR})}} \quad \text{and}\quad upper\ bound = \hat{OR}\ e^{+z_{1-\alpha/2}S_{\log(\hat{OR})}}.$$
(10.9)

Logistic Model

A linear logistic model applied to a 2-by-2 table serves as a starting point for a discussion of more general and complicated situations. The linear model is

$$log\text{-}odds = a + bg_i \qquad (10.10)$$

where $g_i = 1$ makes the $log\text{-}odds = a + b$ for observations belonging to one category and $g_i = 0$ makes the $log\text{-}odds = a$ for observations belonging to the other category (rows of Table 10.1).

To estimate the model parameters a and b, the observed log-odds are used where

$$\hat{a} = y_0 = \log(f_{21}/f_{22}) \text{ and } \hat{b} = y_1 - y_0 = \log(f_{11}/f_{12}) - \hat{a} . \qquad (10.11)$$

The quantity of fundamental importance is the parameter b. This parameter reflects the change in log-odds associated with comparing groups $G = 1$ to $G = 0$ ("baseline"). If the estimate of b is zero or near zero, then group status G is, at best, weakly associated with outcome status D and $\overline{D}$. The degree to which $\hat{b}$ differs from zero estimates the magnitude of the association between binary variables G and D. Alternatively, $e^{\hat{b}} = e^{y_1 - y_0} = \hat{OR}$, which also estimates the association between two binary variables, as noted earlier (expression 8.27). The parameter b directly reflects the magnitude of an association between group status and outcome in terms of a component of a linear model. This pattern is repeated in more complicated logistic models.

To illustrate the application of the linear logistic model (expression 10.10), a 2-by-2 table relating low birth-weight infants (D = infant < 6 pounds and $\overline{D}$ = infant ≥ 6 pounds) and maternal smoking ($G = 0$ = nonsmoker and $G = 1$ = smoker) is used (Table 10.2). These data are from the CHDS data described in Chapter 2. The observed log-odds are $y_0 = \log(13/368) = -3.343$ and $y_1 = \log(28/271) = -2.270$, giving estimates $\hat{a} = -3.343$ and $\hat{b} = -2.270 - (-3.343) = 1.073$. Therefore, the estimated linear model is

$$\hat{y}_i = \hat{a} + \hat{b}g_i = -3.343 + 1.073g_i$$

and

$$\hat{OR} = e^{\hat{b}} = e^{1.073} = 2.925 .$$

T A B L E **10.2**

Birth weight of infants by maternal smoking exposure

		Weight		
		< 6 Pounds (*D*)	≥ 6 Pounds ($\overline{D}$)	Total
Group	Smokers (*G* = 1)	28	271	299
	Nonsmokers (*G* = 0)	13	368	381
	Total	41	639	680

FIGURE **10.1**

Linear logistic model for a 2-by-2 table using smoking and low birth-weight data

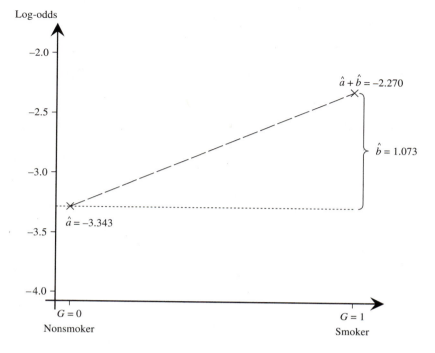

The odds of a low birth-weight infant are slightly more than 2.9 times higher in smoking mothers than in nonsmoking mothers. This simple example shows a property that is characteristic of all logistic models. The estimated coefficient $\hat{b}$ measures association as a linear change in the log-odds and the estimate $e^{\hat{b}}$ reflects the same association but as a multiplicative change in the odds themselves. Figure 10.1 depicts the association between smoking and the risk of a low birth-weight infant on a log-odds scale.

The estimated variance of $\hat{b}$ is 0.119 (i.e., $S_{\hat{b}}^2 = S_{\log(\hat{O}R)}^2 = 0.119$; expression 10.6). To test the conjecture that $\hat{O}R = 2.925$ arose by chance alone and no association exits between smoking and low birth-weight status ($OR_0 = 1.0$), then

$$z = \frac{\log(\hat{O}R) - \log(1.0)}{S_{\log(\hat{O}R)}} = \frac{1.073}{0.345} = 3.110 \qquad \text{(expression 10.7)},$$

yielding a p-value $= 0.002$. The approximate 95% confidence interval for $b = \log(OR)$ based on $\hat{b} = \log(\hat{O}R) = 1.073$ is (0.397, 1.749) and the corresponding confidence interval for the odds ratio $e^{\hat{b}} = 2.925$ is ($e^{0.397}, e^{1.749}$), or (1.487, 5.751). Table 10.3 summarizes the two parameters involved in estimating a logistic model to describe the smoking and birth-weight data classified into a 2-by-2 table.

The identical odds ratio can be calculated directly from Table 10.2, because

$$\hat{O}R = \frac{28/271}{13/368} = 2.925 .$$

T A B L E   **10.3**

Analysis of low birth weight by maternal smoking—simplest logistic model

Variable	Coefficient	Estimate	Standard Error	p-Value	$\hat{OR}$
Constant	a	−3.343	0.282	—	—
Smoking	b	1.073	0.345	0.002	2.925

$$-2\log(L) = 299.289 \text{ (to be discussed)}$$

Note: The standard errors are calculated as part of the computer algorithm and a discussion of these calculations is beyond the scope of this text.

When the number of parameters equals the number of observations (log-odds in this case), the values calculated directly from the data are identical to those produced by a model because the model is saturated.

Logistic Function

A logistic function is defined by

$$f(x) = \frac{1}{1 + e^{-x}}.$$ (10.12)

If x is small (a large negative value), then $f(x) \approx 0.0$; if $x = 0$, then $f(0) = 0.5$; and if x is large, then $f(x) \approx 1.0$. The logistic function expresses the values of x as a "dose-response" relationship (Figure 10.2) that is always between 0 and 1, making the logistic function ideal for modeling probabilities.

The linear model of the log-odds (expression 10.10) is directly related to the logistic function. If the probabilities of D and $\overline{D}$ associated with G are modeled as a logistic function, then, for a 2-by-2 table,

$$P(D \mid G = 0) = \frac{1}{1 + e^{-a}} \quad \text{and} \quad P(\overline{D} \mid G = 0) = \frac{e^{-a}}{1 + e^{-a}}; \quad \text{also}$$

$$P(D \mid G = 1) = \frac{1}{1 + e^{-(a+b)}} \quad \text{and} \quad P(\overline{D} \mid G = 1) = \frac{e^{-(a+b)}}{1 + e^{-(a+b)}}.$$ (10.13)

These probabilities are equivalent to expressing the log-odds as a linear combination of a and b, or

$$log\text{-}odds = \log\left[\frac{P(D \mid G = 0)}{P(\overline{D} \mid G = 0)}\right] = a, \quad \text{and}$$ (10.14)

$$log\text{-}odds = \log\left[\frac{P(D \mid G = 1)}{P(\overline{D} \mid G = 1)}\right] = a + b.$$ (10.15)

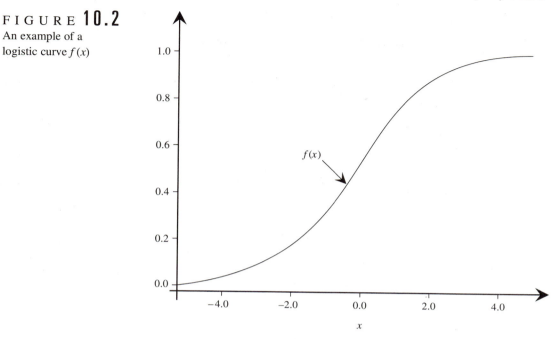

FIGURE **10.2**
An example of a
logistic curve $f(x)$

In short, if the log-odds values from a 2-by-2 table can be represented as a linear combination of two parameters, then the corresponding probabilities are described by a logistic function and vice versa.

The 2-by-2-by-2 Table

The notation used for the 2-by-2-by-2 table is the same as before (Table 9.1). An alternative display of data collected to analyze a binary group variable G, a binary strata variable H, and a binary outcome variable D is given in Table 10.4. Representing each cell of a contingency table with a set of indicator variables (Table 10.4) generalizes to multidimensional tables of any size and is a common form of input for logistic regression computer programs.

A 2-by-2-by-2 table is two 2-by-2 tables, and the considerations of a single 2-by-2 table often apply to each subtable. For example, there are two estimated odds ratios relating G and D, one for each level of H, namely

$$\hat{OR}_{GD\,|\,H=1}=\frac{f_{111}/f_{121}}{f_{211}/f_{221}} \quad \text{for } H=1 \quad \text{and} \quad \hat{OR}_{GD\,|\,H=0}=\frac{f_{112}/f_{122}}{f_{212}/f_{222}} \quad \text{for } H=0 \,.$$

$$(10.16)$$

T A B L E   **10.4**

Alternate representation of a 2-by-2-by-2 table

G	D	H	Frequency
1	1	1	f_{111}
1	0	1	f_{121}
0	1	1	f_{211}
0	0	1	f_{221}
1	1	0	f_{112}
1	0	0	f_{122}
0	1	0	f_{212}
0	0	0	f_{222}

Logistic Model

A logistic model describing a linear relationship among three binary variables (a 2-by-2-by-2 table) in terms of the log-odds is

$$log\text{-}odds = a + b_1 g_i + b_2 h_j + b_3 g_j h_j \tag{10.17}$$

where g_i is 0 or 1, indicating group membership, and h_j is 0 or 1, indicating stratum membership. This model defines the values of the log-odds associated with each of the four combinations of the variables G and H as a linear combination of four logistic model parameters. Specifically, the logistic model (expression 10.17) gives

when $g_i = 1$ and $h_j = 1$, then $log\text{-}odds = a + b_1 + b_2 + b_3$,

when $g_i = 1$ and $h_j = 0$, then $log\text{-}odds = a + b_1$,

when $g_i = 0$ and $h_j = 1$, then $log\text{-}odds = a + b_2$, and

when $g_i = 0$ and $h_j = 0$, then $log\text{-}odds = a$. $\tag{10.18}$

Figure 10.3 shows the geometry of this four-parameter linear representation of the log-odds values associated with binary variables represented by H, G, and D (solid lines).

The estimates of the four model parameters are straightforward manipulations of the observed log-odds values given by

$$\hat{a} = \log \left[\frac{f_{212}}{f_{222}} \right],$$

$$\hat{b}_1 = \log \left[\frac{f_{112}}{f_{122}} \right] - \hat{a} ,$$

$$\hat{b}_2 = \log\left[\frac{f_{211}}{f_{221}}\right] - \hat{a}, \text{ and}$$

$$\hat{b}_3 = \log\left[\frac{f_{111}}{f_{121}}\right] - \log\left[\frac{f_{211}}{f_{221}}\right] - \log\left[\frac{f_{112}}{f_{122}}\right] + \log\left[\frac{f_{212}}{f_{222}}\right].$$

(10.19)

Of particular importance is the parameter b_3. If b_3 is zero, then the effects of the variables represented by G and H are additive; sometimes the effects are then said to be independent. Geometrically, $b_3 = 0$ means that the lines in Figure 10.3 are parallel (dotted line). When b_3 is not zero, the magnitude of b_3 measures the degree of non-additivity. Another term for non-additivity is interaction. In Chapter 5, the analysis of covariance introduced the term interaction where a central issue is also whether the data can be represented by parallel lines (Model II) or nonparallel lines (Model I). The concept of non-additivity (interaction) is discussed further in several contexts.

Example

To illustrate the four-parameter logistic model (expression 10.17), mothers are classified as smokers and nonsmokers and, additionally, by whether their newborn infant is their first child or not (Table 10.5). The binary outcome variable is again low birth weight (< 6 pounds and ≥ 6 pounds).

The log-odds values associated with the four combinations of parity and smoking calculated directly for the tabled data are given in Table 10.6. The logistic model (expression 10.17) has the same number of parameters as values of the log-odds, making it saturated. The four log-odds values directly yield estimates of the model

F I G U R E 10.3

Saturated and additive
logistic models for the
2-by-2-by-2 table

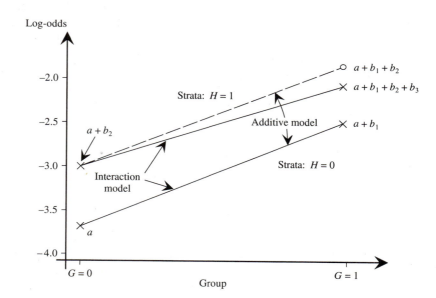

T A B L E **10.5**

Data on low birth-weight infants by smoking and parity

Firstborn Child

	< 6 Pounds	≥ 6 Pounds	Total
Smokers	17	132	149
Nonsmokers	9	194	203
Total	26	326	352

Not Firstborn Child

	< 6 Pounds	≥ 6 Pounds	Total
Smokers	11	139	150
Nonsmokers	4	174	178
Total	15	313	328

parameters (expression 10.19), namely $\hat{a} = -3.773$, $\hat{b}_1 = 1.236$, $\hat{b}_2 = 0.702$, and $\hat{b}_3 = -0.215$. The estimated logistic model is, then,

$$\hat{y}_{ij} = -3.773 + 1.236g_i + 0.702h_j - 0.215g_ih_j .$$

A summary of this saturated model is given in Table 10.7.

The odds ratios calculated directly or from the model give the same results, as expected for a saturated model. For example, among births with parity > 1, the odds ratio reflecting the association between smoking and low birth weight is

$$\hat{OR}_{GD \mid H=0} = \frac{(11)(174)}{(4)(139)} = 3.442 \quad \text{or} \quad \hat{OR}_{GD \mid H=0} = e^{\hat{b}_1} = e^{1.236} = 3.442 .$$

T A B L E **10.6**

Log-odds for the saturated model

G	H	Log–Odds	y_{ij}
Smoker	Firstborn	log(17/132)	$y_{11} = -2.050$
Nonsmoker	Firstborn	log(9/194)	$y_{01} = -3.071$
Smoker	Not firstborn	log(11/139)	$y_{10} = -2.537$
Nonsmoker	Not firstborn	log(4/174)	$y_{00} = -3.773$

T A B L E **10.7**

Analysis of low birth-weight infants by smoking and parity—saturated model

Variable	Coefficient	Estimate	Standard Error	p-Value	$\hat{OR}$
Constant	a	−3.773	0.497	—	—
Smoking	b_1	1.236	0.587	0.035	3.442
Parity	b_2	0.702	0.602	0.244	2.018
Smoking × parity	b_3	−0.215	0.726	0.768	0.806

$$-2\log(L) = 296.396$$

A saturated model is little more than a compact way of expressing the relationship among the variables, but it provides a starting point to develop models that summarize the data: models with fewer parameters than observations.

Additive Logistic Model for the 2-by-2-by-2 Table

The additive logistic model is expressed in the same form as the saturated model but, as mentioned, b_3 is set to zero, or

$$log\text{-}odds = a + b_1 g_i + b_2 h_j , \tag{10.20}$$

defining the specific log-odds values as

when $g_i = 1$ and $h_j = 1$, then $log\text{-}odds = a + b_1 + b_2$,

when $g_i = 1$ and $h_j = 0$, then $log\text{-}odds = a + b_1$,

when $g_i = 0$ and $h_j = 1$, then $log\text{-}odds = a + b_2$,   and

when $g_i = 0$ and $h_j = 0$, then $log\text{-}odds = a$. \hfill (10.21)

The necessarily parallel lines of the additive model are displayed in Figure 10.3 (dotted line).

The relationships among the log-odds suggest a process to estimate the parameters of the additive model. To produce perfectly additive log-odds values it is necessary to find a value δ to increase or decrease the eight observed cell frequencies f_{ijk} so that $b_3 = 0$, or

$$b_3 = \log\left[\frac{f_{111} + \delta}{f_{121} - \delta}\right] - \log\left[\frac{f_{211} - \delta}{f_{221} + \delta}\right] - \log\left[\frac{f_{112} - \delta}{f_{122} + \delta}\right] + \log\left[\frac{f_{212} + \delta}{f_{222} - \delta}\right] = 0 . \tag{10.22}$$

The value δ creates perfectly additive "data" (i.e., $b_3 = 0$), which maintains the observed marginal frequencies. These estimated values are then used to estimate the parameters of the additive model. Although a value of δ is difficult to calculate directly, a trial-and-error iteration process produces δ' so that b_3 is zero (expression

10.22). The estimate δ' allows estimates of the three parameters of the additive model as

$$\hat{a} = \log\left[\frac{f_{212} + \delta'}{f_{222} - \delta'}\right],$$

$$\hat{b}_1 = \log\left[\frac{f_{112} - \delta'}{f_{122} + \delta'}\right] - \hat{a}, \text{ and}$$

$$\hat{b}_2 = \log\left[\frac{f_{211} - \delta'}{f_{221} + \delta'}\right] - \hat{a}, \tag{10.23}$$

giving the estimated log-odds as

$$\hat{y}_{ij} = \hat{a} + \hat{b}_1 g_i + \hat{b}_2 h_j.$$

The data on smoking and parity (Table 10.5) yield a value of $\delta' = 0.407$. Using the estimated value of δ and the additive model produces the exactly additive "data" given in Tables 10.8 and 10.9. Table 10.9 is derived from the model-generated (expression 10.20) additive "data" but expressed in terms of log-odds values. The log-odds values in Table 10.9 correspond to the values in Table 10.6 calculated directly from the observed data. These estimated log-odds values give estimates of the additive model parameters of $\hat{a} = -3.674$, $\hat{b}_1 = 1.096$, and $\hat{b}_2 = 0.554$ (expression 10.21), and necessarily $\hat{b}_3 = \hat{y}_{11} - \hat{y}_{10} - \hat{y}_{01} + \hat{y}_{00} = -2.023 + 3.119 + 2.577 - 3.674 = 0$. The additive model is summarized in Table 10.10.

T A B L E   **10.8**

"Data" (additive model) on low birth-weight infants by smoking and parity

Firstborn Child

	< 6 Pounds	≥ 6 Pounds	Total
Smokers	17.407	131.593	149
Nonsmokers	8.593	194.407	203
Total	26	326	352

Not Firstborn Child

	< 6 Pounds	≥ 6 Pounds	Total
Smokers	10.593	139.407	150
Nonsmokers	4.407	173.593	178
Total	15	313	328

T A B L E **10.9**

Log-odds from the additive model

G	H	Log-Odds	$\hat{y}_{ij}$
Smoker	Firstborn	log(17.407/131.593)	$\hat{y}_{11} = -2.023$
Nonsmoker	Firstborn	log(8.593/194.407)	$\hat{y}_{01} = -3.119$
Smoker	Not firstborn	log(10.593/139.407)	$\hat{y}_{10} = -2.577$
Nonsmoker	Not firstborn	log(4.407/173.593)	$\hat{y}_{00} = -3.674$

The estimated parameter $\hat{b}_1$ produces an estimated odds ratio measuring the association between smoking and a low birth-weight infant accounting for the influence of parity, or $\hat{OR} = 2.993$; that is,

$$\hat{OR}_{GD \mid H=1} = \hat{OR}_{GD \mid H=0} = e^{\hat{b}_1} = 2.993 .$$

The phrase "accounting for the influence of parity" means that the change in the logarithm of odds (b_1) associated with smoking categories is taken to be the same for both levels of parity. Because the log-odds for one group ($G = 0$) is *log-odds* $= a + b_2 h_j$ and the log-odds for the other group ($G = 1$) is *log-odds*$_1 = a + b_1 + b_2 h_j$, then

$$difference = (log\text{-}odds_1) - (log\text{-}odds_0) = b_1 .$$

This shows that b_1 measures the association between the outcome variable D and the group variable G for any constant value of the stratifying variable H ($H = 0$ or $H = 1$). This interpretation of the parameter b_1 is analogous to the interpretation of the regression coefficient b_1 in a bivariate linear regression model (Chapter 3). The additive model guarantees that the odds ratios are identical at the two levels of the other variable and that $e^{\hat{b}_1}$ is an estimate of this common odds ratio. The properties of the estimate $\hat{b}_2$ are identical, producing a parallel interpretation.

Inference

Any 2-by-2-by-2 table can be represented by a three-parameter additive (no interaction) model. However, the question arises: Does an additive model accurately repre-

T A B L E **10.10**

Analysis of low birth weight by smoking and parity—additive model

Variable	Coefficient	Estimate	Standard Error	p-Value	$\hat{OR}$
Constant	a	−3.674	0.357	—	—
Smoking	b_1	1.096	0.345	0.001	2.993
Parity	b_2	0.555	0.336	0.099	1.741

$-2\log(L) = 296.483$

sent the relationships within the data? The goodness-of-fit of an additive model can be evaluated in three ways: by assessing the coefficient $\hat{b}_3$, by computing a Pearson chi-square goodness-of-fit statistic, or by comparing likelihood measures.

The magnitude of the estimated coefficient $\hat{b}_3$ (from the saturated model) is one measure of non-additivity. A statistical test consists of the usual approach of postulating that an additive model describes the data, or H_0: $b_3 = 0$, and computing

$$z = \frac{\hat{b}_3}{S_{\hat{b}_3}}$$

where z has an approximate standard normal distribution when $b_3 = 0$. Using the smoking/parity data, $\hat{b}_3 = -0.215$ with a standard error of 0.726 (Table 10.7), giving $z = -0.296$. The associated p-value is 0.768, providing little evidence that b_3 differs from zero. Again, it should be noted that the estimated standard error of $\hat{b}_3$ is a complicated computation but is generally found in the output of logistic regression analysis programs.

The additive model produces eight estimated cell frequencies $\hat{f}_{ijk}$ that can be contrasted to the eight observed cell frequencies f_{ijk} to evaluate the fit of the additive model. The Pearson chi-square statistic

$$X^2 = \sum_{all\ cells} (f_{ijk} - \hat{f}_{ijk})^2 / \hat{f}_{ijk}$$

has an approximate chi-square distribution with one degree of freedom if no systematic effect occurs when b_3 is deleted from the model. The illustrative data (f_{ijk}-values from Table 10.5 and $\hat{f}_{ijk}$-values from Table 10.8) give

$$X^2 = (17 - 17.407)^2 / 17.407 + \cdots + (174 - 173.593)^2 / 173.593 = 0.086\ ,$$

which is essentially equal to the previous result ($z^2 = (-0.296)^2 = 0.088$).

Likelihood Approach

A more general evaluation of additivity (i.e., choosing between expressions 10.17 and 10.20) involves the comparison of likelihood values. Likelihood values are a relative measure of the goodness-of-fit of a model. A rigorous definition of the likelihood value L is beyond the scope of this text but, roughly, it is the probability of the occurrence of the observed data for a specific set of parameter values. For most data sets, L is a small probability. A related quantity is $-2\log(L)$ where L is the likelihood associated with the data and a specific model. Parallel to the Pearson chi-square statistic X^2 and the likelihood chi-square statistic Y^2, this transformed likelihood value is a measure of goodness-of-fit. In addition, the transformed likelihood $-2\log(L)$ also has an approximate chi-square distribution but, more importantly, the difference between two transformed likelihood statistics associated with nested models has an approximate chi-square distribution. This difference in goodness-of-fit is a basic tool for comparing nested models, like the comparison of the sums of squares with an F-statistic (F-to-remove) or, more specifically, like the restricted chi-square test.

Typically a model is proposed and the likelihood L_1 is found by computer techniques; then a restriction of the model is made, producing another likelihood value L_0. The restricted model is usually created by setting specific parameters to zero, giving a simpler representation of the data. Each model has an associated degrees of freedom, denoted by df_1 and df_0. Parallel to the log-linear models applied to tabular data, the degrees of freedom are equal to the number log-odds values in the table minus the number of independent estimated parameters necessary to establish the logistic model. Therefore, the difference $df_0 - df_1$ is the number of parameters set to zero to form the restricted model. The difference between two transformed likelihood measures, called a deviance, reflects the difference in the fit of the two models, producing a test statistic with an approximate chi-square distribution that makes it possible to assess the impact of the imposed restrictions. Directly comparing the likelihood values L is not easy, but the deviance provides the straightforward comparison

$$Y^2 = [-2\log(L_0)] - [-2\log(L_1)] . \tag{10.24}$$

The value Y^2 has an approximate chi-square distribution when the restrictions that generate L_0 have no systematic influence on the fit of the model. The degrees of freedom equal the number of parameters set to zero to establish the restricted model (i.e., $df_0 - df_1$). The deviance Y^2-statistic is a likelihood ratio test statistic and is related to the likelihood ratio tests discussed in Chapter 8.

The deviance calculated from the smoking/parity data compares $-2\log(L_1) = 296.396$ (Table 10.7) from the saturated model to $-2\log(L_0) = 296.483$ (Table 10.10) from the additive model. That is, $Y^2 = 296.483 - 296.396 = 0.087$ and has an approximate chi-square distribution with one degree of freedom ($df_1 = 0$ and $df_0 = 1$ gives $df_0 - df_1 = 1$) when no systematic difference exists between the saturated and the additive models. The associated p-value is 0.768. This third approach to evaluating additivity ($b_3 = 0$) produces basically the same results as the previous two methods.

All three approaches indicate that the additive model accurately represents the smoking/parity data and that considering the effects from smoking and parity as independent influences on the probability of a low birth-weight infant is supported by the data. The three approaches to evaluating the additive model address the same issue but generally differ somewhat because they are all approximate. Typically these differences are small, especially when the sample size is large. The evaluation of additivity with a z-statistic only applies to the assessment of a single parameter. A Pearson chi-square statistic can be calculated only when the data are classified into a series of categories. However, comparing likelihood values (deviance) remains a valid approach for multiparameter statistical comparisons of both continuous and discrete independent variables.

Continuous Data

Data analyzed using a logistic model do not have to consist of discrete values classified into a table such as the smoking/parity data. Independent variables that take on a large number of values or continuous independent variables are readily analyzed with a logistic model. For a single continuous independent variable x, the probability of an event again modeled by the logistic function using parameters a and b is

$$P(D \mid x) = \frac{1}{1 + e^{-(a + bx)}}$$

(10.25)

or, equivalently, the log-odds is a linear combination of the same two parameters, giving

$$log\text{-}odds = a + bx .$$

(10.26)

The interpretation of the parameter b is not very different from a simple linear regression model (Chapter 2). The value b is the change in the log-odds for a one-unit increase in the independent variable x. Furthermore, e^b is the multiplicative change in the odds associated with a one-unit increase in x. That is, if x is increased to $x + 1$, then the log-odds changes by b and the odds change by a factor of e^b.

Consider the influence of a mother's prepregnancy weight on the probability of a low birth-weight infant estimated from the same 680 observations used in the previous illustrations (Tables 10.2 and 10.5). The estimates of the logistic parameters a and b require a computer algorithm and are $\hat{a} = -2.022$ and $\hat{b} = -0.006$. The estimated logistic probabilities are then given by

$$\hat{P}(D \mid x) = \frac{1}{1 + e^{-(-2.022 - 0.006x)}}$$

or, equivalently,

$$estimated\ log\text{-}odds = \hat{y}_x = -2.022 - 0.006\ x .$$

Also, the estimate of the odds ratio associated with maternal prepregnancy weight is $\hat{OR} = e^{\hat{b}} = e^{-0.006} = 0.994$, which, at first glance, might seem essentially equal to 1.0 (no relationship between maternal weight and low birth-weight). However, the magnitude of the odds ratio depends on the units used to measure the independent variable, in this case pounds. The odds ratio 0.994 shows a decrease in the probability of a low birth-weight infant for each pound of increase in maternal weight. For example, if two mothers are compared where one weighs 20 pounds more than the other, then the risk of a low birth-weight infant for the heavier mother, measured by the odds relative to the lighter mother, decreases by a factor of $e^{20b} = e^{20(-0.006)} = 0.887$.

Another property of the linear logistic model is clear from this example. The decrease in probability depends only on the difference in weight of the mothers compared. The difference between a 100-pound and a 120-pound woman or between a 200-pound and a 220-pound woman influences the log-odds and the odds ratio measures of risk of a low birth-weight infant by identical amounts. The property that logistic derived summaries of association are not related to the level of the variable being studied results from the assumption that the log-odds is a linear function of the independent variable and, in many cases, may not be biologically plausible. Nonlin-

T A B L E **10.11**

Analysis of low birth-weight infants by maternal prepregnancy weight—continuous data

Variable	Coefficient	Estimate	Standard Error	p-Value	$\hat{OR}$
Constant	a	−2.022	1.206	—	—
Weight	b	−0.006	0.010	0.547	0.994

$$-2\log(L) = 309.400$$

ear relationships are, perhaps, more realistic in some situations and easily incorporated into a logistic analysis. This special type of non-additivity is discussed.

The statistical evaluation of the relationship between a continuous variable and a binary outcome follows the usual pattern. For the continuous variable maternal weight, the test statistic $z = \hat{b}/S_{\hat{b}} = -0.006/0.010 = -0.603$ has an approximate standard normal distribution when $b = 0$ and yields a p-value of 0.547. The estimated standard error of $S_{\hat{b}} = 0.010$ is produced as part of the same computer estimation process that produces the estimate $\hat{b} = -0.006$. A likelihood comparison yields almost the same result where $-2\log(L_0) = 309.774$ when b is set to zero ($b = 0$) and $-2\log(L_1) = 309.400$ (Table 10.11) when b is not equal to zero ($b \neq 0$). The difference $Y^2 = 309.774 - 309.400 = 0.374$ has a chi-square distribution when $b = 0$ with one degree of freedom (the number of parameters set to zero to establish the restricted model), producing a p-value of 0.541. The result is expectedly similar to $z^2 = (-0.603)^2 = 0.364$. The results of the logistic analysis applied to prepregnancy weight (a single continuous variable) are summarized in Table 10.11.

Interaction

The term "interaction" is often carelessly used and many times the concept is not well understood. An interaction refers to a relationship between two variables with reference to a third variable. If no interaction exists, then the relationship between two variables is the same regardless of the value of the third variable. For example, if a person's weight, cholesterol level, and the probability of a coronary event are being studied, then no interaction exists when the relationship between the probability of a heart attack and the level of cholesterol is the same regardless of the weight of the individual. On the other hand, an interaction exists when the relationship between two variables depends on the value of a third variable. Cholesterol and weight might interact with respect to a coronary event. This interaction could be such that the probability of a coronary event for a man who weighs 100 pounds and increases his cholesterol by 50 mg/100 ml is much greater than for a man who weighs 200 pounds and increases his cholesterol the same amount. The relationship of cholesterol level to the probability of a heart attack depends on the weight of the individual being considered.

From the perspective of a logistic regression model relating two independent variables x_1 and x_2 to the probability of a binary event, an interaction involves the inclusion of a second-order term, or

$$log\text{-}odds = a + b_1x_1 + b_2x_2 + b_3x_1x_2 \,. \tag{10.27}$$

The logistic model with an interaction term is similar to the previously discussed linear regression model (expression 4.26). Rewriting this model shows how the influence of x_1 depends on the level of x_2 when b_3 is not zero, or

$$log\text{-}odds = a + (b_1 + b_3x_2)x_1 + b_2x_2 \,. \tag{10.28}$$

The coefficient $(b_1 + b_3x_2)$ associated with x_1 depends on x_2 when $b_3 \neq 0$; the variables x_1 and x_2 interact. Therefore, the relationship between the log-odds and variable x_1 depends on the value of x_2. This interaction translates to the odds ratio because the change in odds is $e^{(b_1 + b_3x_2)}$ for a one-unit increase in x_1 and, like the log-odds, the amount of change depends on the level of x_2.

It is important to emphasize that an interaction between x_1 and x_2 means the influences measured by the coefficients $\hat{b}_1$ and $\hat{b}_2$ are biased and not simply interpreted. These estimated coefficients cleanly reflect the separate impacts of the independent variables only when the influences are additive. As noted earlier, if b_3 is zero, then no interaction exists, producing an additive model where the influence on the dependent variable from any single independent variable is isolated from the other variables in the regression equation. In a nonadditive model, influences of a single variable are not directly measured by a single coefficient.

An analysis of maternal height (x_1) and maternal age (x_2) from the CHDS data shows two variables likely to interact with respect to the probability of a low birth-weight infant (Table 10.12). Figure 10.4 displays three lines representing the relationship between the log-odds and maternal height for three specific maternal ages (20 years, 25 years, and 30 years) on log-odds and probability scales based on the estimates given in Table 10.12. Mothers who are young have decreasing probabilities of a low birth-weight infant with increasing height. The amount of reduction, however, depends on the age of the mother—an interaction. At about age 30, mothers begin to have an increasing probability (rather than decreasing probability) of a low birth-weight infant with increasing height. For example, the $log\text{-}odds = 10.960 - 0.226x_1$ for age 20 years, $log\text{-}odds = 3.235 - 0.106x_1$ for age 25 years, and $log\text{-}odds =$

T A B L E **10.12**

Analysis of low birth-weight infants by maternal age and maternal height—interaction

Variable	Coefficient	Estimate	Standard Error	p-Value	$\hat{OR}$
Constant	a	41.860	19.919	—	—
Height	b_1	−0.706	0.311	0.023	0.494
Age	b_2	−1.545	0.738	0.036	0.213
Age × height	b_3	0.024	0.011	0.033	1.025

$$-2\log(L) = 303.842$$

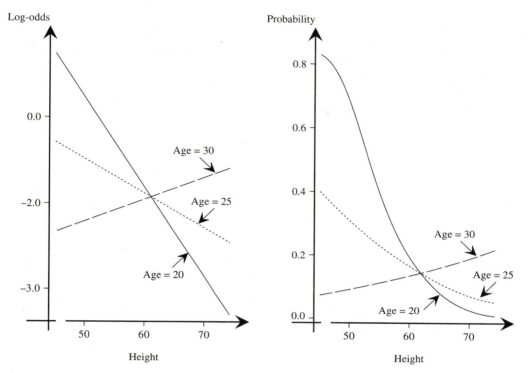

The relationship between height and risk of a low birth-weight infant by maternal age (log-odds and probability scales)

$-4.490 + 0.014x_1$ for age 30 years where x_1 represents a mother's height (Figure 10.4). In addition, the estimated odds ratio associated with a 1-inch increase in height is $e^{-0.226} = 0.798$ for 20-year-olds where an increase of 1 inch produces an odds ratio, of $e^{0.014} = 1.014$ for mothers 30 years old. Whether measured by the log-odds, an odds ratio, or a probability, the estimated relationship between low birth weight and a mother's height depends on her age.

Continuous and Discrete Data

A logistic analysis is not restricted to one type of data. A model combining both discrete and continuous independent variables is

$$log\text{-}odds = a + b_1 x + b_2 g_i + b_3 g_i x \qquad (10.29)$$

where, as before, x represents a continuous measurement and g_i represents a binary categorical variable. Expression 10.29 allows the sampled data to be described as two estimated straight lines with different slopes and intercepts on the log-odds scale. That is, when $g_i = 0$, then $\hat{y}_{x0} = \hat{a} + \hat{b}_1 x$, and

when $g_i = 1$, then $\hat{y}_{x1} = (\hat{a} + \hat{b}_2) + (\hat{b}_1 + \hat{b}_3)\, x = \hat{A} + \hat{B}x$. (10.30)

If b_3 is not zero, the model incorporates an interaction between the continuous variable x and the group variable G—the relationship between log-odds and x is not the same within the two groups $G = 1$ and $G = 0$, producing two different straight lines. Although the outcome variable is the log-odds, this model is similar to the analysis of covariance model described in Chapter 5 (expression 5.1).

Dividing the data on maternal prepregnancy weight (x) into two groups based on maternal age (g_i) illustrates an application of a logistic model with both continuous and discrete independent variables. One group consists of mothers under 35 years of age and the other consists of mothers whose age is greater than or equal to 35 years, where low birth-weight once again is the binary dependent variable. The computer estimates of the logistic model parameters are $\hat{a} = -1.678$, $\hat{b}_1 = -0.009$, $\hat{b}_2 = -0.599$, and $\hat{b}_3 = 0.0114$. The two estimated straight lines become

when $g_i = 0$, then $\hat{y}_{x0} = -1.678 - 0.009x$, and

when $g_i = 1$, then $\hat{y}_{x1} = (-1.678 - 0.599) + (-0.009 + 0.0114)x$

$$= -2.277 + 0.0024x.$$

A summary of the interaction model (expression 10.29) is contained in Table 10.13 and depicted in Figure 10.5 (solid lines).

To test whether maternal age (two groups) and maternal prepregnancy weight interact, an additive model is proposed ($b_3 = 0$), or

$$log\text{-}odds = a + b_1 x + b_2 g_i.$$ (10.31)

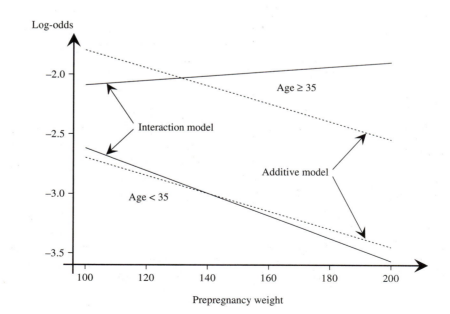

FIGURE 10.5

Analysis of low birth weight by maternal age and prepregnancy weight—interaction and additive models

T A B L E **10.13**

Analysis of low birth-weight infants by maternal age and prepregnancy weight—interaction model

Variable	Coefficient	Estimate	Standard Error	p-Value	$\hat{OR}$
Constant	a	−1.678	1.346	—	—
Weight	b_1	−0.009	0.011	0.385	0.991
Age	b_2	−0.599	3.538	0.866	0.549
Age × weight	b_3	0.011	0.027	0.671	1.011

$$-2\log(L) = 305.813$$

The additive model describes two estimated straight lines on a log-odds scale with different intercepts but with the same slope, or

when $g_i = 0$, then $\hat{y}_{x0} = \hat{a} + \hat{b}_1 x$, and

when $g_i = 1$, then $\hat{y}_{x1} = (\hat{a} + \hat{b}_2) + \hat{b}_1 x = \hat{A} + \hat{b}_1 x$.

The contrast of the three-parameter additive model to the four-parameter interaction model indicates the adequacy of describing the data by two lines with equal slopes ($b_3 = 0$). The likelihood ratio statistic is the difference between $-2\log(L_1) = 305.813$ (interaction model; Table 10.13) and $-2\log(L_0) = 305.990$ (additive model; Table 10.14), yielding a deviance of $Y^2 = 305.990 - 305.813 = 0.177$ with one degree of freedom (one parameter set to zero) and producing an associated p-value of 0.674. The deletion of the interaction term ($b_3 = 0$) has only a minor impact on the fit of the model. The estimated additive model is

when $g_i = 0$, then $\hat{y}_{x0} = -1.895 - 0.008x$, and

when $g_i = 1$, then $\hat{y}_{x1} = (-1.895 + 0.885) - 0.008x = -1.010 - 0.008x$.

This model is shown in Figure 10.5 (dotted lines) where two estimated straight lines with equal slopes represent the relationship between the log-odds and maternal weight within the two age groups (age < 35 and age ≥ 35). A summary of the additive model is given in Table 10.14.

A test of the usefulness of the binary maternal age classification follows the same pattern of comparing likelihood values. A linear model based only on prepregnancy weight (x) is postulated ($b_3 = b_2 = 0$), or

$$log\text{-}odds = a + b_1 x \tag{10.32}$$

and the likelihood value from this model is compared to the one generated by the additive model ($b_2 \neq 0$). The transformed likelihood measure for the additive model is 305.990 (Table 10.14) and for the model with maternal age removed is 309.400 (Table 10.11), giving a deviance $Y^2 = 309.400 - 305.990 = 3.410$. The p-value of

T A B L E   **10.14**

Analysis of low birth-weight infants by maternal age and prepregnancy weight—additive model

Variable	Coefficient	Estimate	Standard Error	p-Value	$\hat{OR}$
Constant	a	−1.895	1.232	—	—
Weight	b_1	−0.008	0.010	0.438	0.992
Age	b_2	0.885	0.443	0.046	2.423

$$-2\log(L) = 305.990$$

0.065 (degrees of freedom = 1) implies that the binary classification of maternal age is likely a useful variable in the description of the relationship between prepregnancy weight and the probability of a low birth-weight infant.

To summarize, the analysis of maternal weight and age (<35 and ≥ 35) as factors in the risk of a low birth-weight infant can be viewed as a sequential contrasting three hierarchical models or,

1 $log\text{-}odds = a + b_1 x + b_2 g_i + b_3 g_i x$, $-2\log(L) = 305.813$;
2 $log\text{-}odds = a + b_1 x + b_2 g_i$, $-2\log(L) = 305.990$; and
3 $log\text{-}odds = a + b_1 x$, $-2\log(L) = 309.400$.

An additive model allows the impact of the independent variables to be isolated and summarized separately. The comparison of women of the same weight who belong to different age groups is summarized by an odds ratio of $e^{\hat{b}_2} = e^{0.885} = 2.423$. Similarly, mothers in the same age category reduce their odds of a low birth-weight child by a factor of $e^{-0.008} = 0.992$ for each additional pound of prepregnancy weight.

The parameters associated with maternal age and prepregnancy weight yield the estimated additive logistic regression equation $\hat{y}_{xi} = -1.895 - 0.008x + 0.885g_i$, and for women older than 35 ($g_i = 1$)

$$\text{difference in estimated log-odds} = \hat{y}_{x_1 1} - \hat{y}_{x_0 1} = \hat{b}_1(x_1 - x_0) + \hat{b}_2 .$$

Specifically,

$$\text{difference in estimated log-odds} = -0.008(x_1 - x_0) + 0.885$$

and the associated odds ratio is

$$\hat{OR} = e^{0.885}[e^{-0.008}]^{(x_1 - x_0)} = 2.423(0.992)^{(x_1 - x_0)}$$

for specific prepregnancy weights x_0 and x_1. For example, if the risk of a low birth-weight infant for two women who differ by 20 pounds in prepregnancy weight and who are older than 35 is of concern, then the overall odds ratio is estimated by the product of two components, $\hat{OR} = 2.423 \times (0.992)^{20} = 2.423(0.851) = 2.063$, one component reflecting a constant maternal age category and the other reflecting the influence of prepregnancy weight. Similarly, for a difference of 40 pounds, $\hat{OR} = 2.423(0.725) = 1.757$. An additive logistic model induces a multiplicative relationship between two separate odds ratios; one for maternal age classification (increasing risk) and one for maternal weight (decreasing risk).

The General *k*-Variate Logistic Model

In general, coefficients of an additive logistic model translate into a series of multiplicative factors each associated with one of the independent variables in the equation, or for the *k*-variate additive model

$$log\text{-}odds(i^{th}\ observation) = a + b_1 x_{1i} + b_2 x_{2i} + b_3 x_{3i} + \cdots + b_k x_{ki} \quad \text{and}$$

$$log\text{-}odds(j^{th}\ observation) = a + b_1 x_{1j} + b_2 x_{2j} + b_3 x_{3j} + \cdots + b_k x_{kj}, \tag{10.33}$$

giving

$$OR_{ij} = [e^{b_1(x_{1i} - x_{1j})}]\ [e^{b_2(x_{2i} - x_{2j})}]\ [e^{b_3(x_{3i} - x_{3j})}] \cdots [e^{b_k(x_{ki} - x_{kj})}]. \tag{10.34}$$

Two properties of this *k*-variable linear logistic equation result from additivity. First, if all differences $x_{li} - x_{lj}$ are zero (independent variables held constant) except one, such as variable 3, then $OR_{ij} = e^{b_3(x_{3i} - x_{3j})}$, showing that b_3 reflects the influence of variable 3 on the dependent variable while accounting for the influences of the other $k - 1$ variables in the regression equation. The coefficients of an additive model are a way of comparing two groups or two individuals as though the other independent variables are the same. Often it is said that the individual coefficients b_i reflect the influence of the ith variable adjusted for the influences of the other $k - 1$ variables. This interpretation of the coefficients applies in the *k*-variable linear regression model (Chapters 3 and 4).

Second, as already noted, an additive model on the log-odds scale means the overall odds ratio *OR* is the product of a series of *k* individual odds ratios, each contributing to *OR* adjusted for the other variables in the equation. If the adjusted odds ratio from an estimated logistic regression equation for one variable is OR_1 and for another OR_2, then the odds ratio describing the impact of both variables, for an additive model, is $OR = OR_1\ OR_2$. If a third variable is involved, then $OR = OR_1\ OR_2\ OR_3$ and so forth. It is this multiplicative relationship among the odds ratios that leads to referring to the *k* variables as having independent influences on the outcome variable, which occurs only when the data are accurately described by an additive model.

Categorical Variables

Non-numeric variables often play a role in the occurrence of binary outcomes. For example, when considering a disease (present versus absent), the race of an individual is frequently relevant to the analysis. A non-numeric variable such as race consists of a number of unordered categories (e.g., white, black, Japanese, Chinese, and

T A B L E **10.15**

Example of the four design variables representing five categories of race

Race	z_{2j}	z_{3j}	z_{4j}	z_{5j}
White	0	0	0	0
Black	1	0	0	0
Japanese	0	1	0	0
Chinese	0	0	1	0
Other	0	0	0	1

other). Such variables are included in a logistic regression analysis using dummy variables or design variables (introduced in Chapter 5). The simplest design variable is a binary variable coded with a one or a zero indicating the presence or absence of a specific property. It takes one design variable made up of zeros and ones to characterize a non-numeric variable with two classes. Such a design variable was previously illustrated by mothers less than 35 years old who were coded 0 and mothers age 35 years or older who were coded 1. The concept of design variables generalizes to any number of categories. For a non-numeric independent variable with m categories, it is necessary to construct $m - 1$ design variables to characterize the categorical variable to be analyzed. Consider the variable race with five categories (white, black, Japanese, Chinese, and other; $m = 5$); a set of design variables characterizing race consists of four variables each made up of zeros and ones. An illustration of how these design variables, denoted by z, represent race is given in Table 10.15. That is, if the jth individual is black, then $z_{2j} = 1$ and $z_{3j} = z_{4j} = z_{5j} = 0$; if the jth individual is Japanese, then $z_{3j} = 1$ and $z_{2j} = z_{4j} = z_{5j} = 0$; if the jth individual is Chinese, then $z_{4j} = 1$ and $z_{2j} = z_{3j} = z_{5j} = 0$; and if the jth individual is not white, black, Japanese, or Chinese, then $z_{5j} = 1$ and $z_{2j} = z_{3j} = z_{4j} = 0$. For the variable race with five classes, only four design variables are necessary and, in this example, a variable directly indicating white individuals is left out ($z_{1j} = 0$). These design variables are then treated like any other variable entered in a logistic regression equation. The choice of which category is not included does not influence the fit of the model (the likelihood is unaffected). However, the interpretation of the coefficients associated with each of the design variables is affected. The coefficient and, therefore, the adjusted odds ratios are relative to the category not included. For example, an odds ratio of 2.0 associated with z_2 (Table 10.15) implies that the odds estimated for a black individual are twice the odds relative to a white individual.

An obvious use of design variables is to assess the influence of a non-numeric independent variable on a specific outcome. Evaluation takes the form of computing the likelihood with the $m - 1$ design variables included in the model and comparing this likelihood to one calculated with all $m - 1$ design variables excluded from the model. The difference, summarized by a deviance likelihood statistic, is interpreted in the same way as before. To evaluate the influence of race, the four design variables would be included and then excluded from the logistic model. The difference in likelihoods, measured by the deviance Y^2, reflects the impact of race on the outcome variable.

A less obvious role for a set of design variables is to explore the relationship of a continuous independent variable to a binary dependent variable. If a variable such as reported maternal prepregnancy weight is directly entered into a logistic regression equation, the implicit assumption is that weight is linearly related to the log-odds (expression 10.26). Clearly, this is not necessarily the case. Using design variables provides an opportunity to identify more complicated relationships between an independent variable and a binary outcome. Because design variables do not impose a specific relationship between an independent variable and the dependent variable, the estimated coefficients from the $m - 1$ categories can suggest nonlinear ways a continuous x-variable relates to the binary outcome. The continuous independent variable is classified into m categories and represented by $m - 1$ design variables. The resulting $m - 1$ coefficients, estimated from the logistic model and the value 0.0 from the reference category ($z_{1j} = 0$), produce an unconstrained picture of the relationship of a continuous independent variable to the binary outcome variable (illustrated in the next section). The pattern of these estimated coefficients can lead to identifying specific nonlinear relationships, raising the possibility of creating more informative independent variables.

Two Applications of Design Variables

Example: Summary Odds Ratio for k Individual 2-by-2 Tables

A popular approach to estimating a common odds ratio from a series of 2-by-2 tables is called the Mantel-Haenszel procedure. Each 2-by-2 table is viewed as a stratum of a third variable and the Mantel-Haenszel estimate provides an assessment of association between the two binary variables by combining information from each stratum (table), producing an estimate of a summary odds ratio (for details see [1]). Using the notation from Chapter 9, particularly from Table 9.30, the Mantel-Haenszel summary odds ratio estimated from k individual 2-by-2 tables is

$$\hat{OR}_{MH} = \frac{\sum_{i=1}^{k} \frac{f_{11i} f_{22i}}{f_{..i}}}{\sum_{i=1}^{k} \frac{f_{12i} f_{21i}}{f_{..i}}} \, .$$

(10.35)

The statistical summary $\hat{OR}_{MH}$ is an estimate of the odds ratio common to the k tables when the odds ratios from each table differ only because of random variation (no interaction). Using the data on behavior type and coronary disease stratified by age presented in Table 9.30, the estimated summary odds ratio is

$$\hat{OR}_{MH} = \frac{(20)\,(271)/543 + (34)\,(574)/1091 + \cdots + (37)\,(114)/322}{(241)\,(11)/543 + (462)\,(21)/1091 + \cdots + (162)\,(9)/322} = 2.214 \, .$$

The estimate 2.214 provides a summary measure of the association between behavior type and coronary disease adjusted for the influence of age, providing no interaction exits among the $k = 5$ age strata.

Employing design variables and a logistic model also provides an estimate of the common odds ratio among a series of 2-by-2 tables. The additive logistic model that allows an estimate of the common odds ratio from a series of k strata is

$$log\text{-}odds = a + bf + c_2 z_{2j} + c_3 z_{3j} + \cdots + c_k z_{kj} \tag{10.36}$$

where f represents a binary independent variable (within each stratum, $f = 1 =$ the binary independent variable is present and $f = 0 =$ the binary independent variable is absent) and the values z_{ij} are dummy variables representing the k levels of the strata variable. The coefficients $c_2, c_3, \cdots, c_k$ and the variables $z_2, z_3, \cdots, z_k$ account for the influence of the k-level strata variable. The coefficient b measures the common association between the outcome and the binary independent variable using the data from all k strata. For the ith strata, $log\text{-}odds = a + bf + c_i = A_i + bf$; that is, the log-odds differs among the strata (measured by A_i) but the association between the binary outcome and the binary independent variable (measured by b) remains the same for each of the k strata. This model incorporates the same basic assumption underlying the Mantel-Haenszel estimate—additivity or no interaction.

The assumption of no interaction is addressed by comparing a model that adds the possibility of interaction to the additive model. A fully saturated model that allows each log-odds to differ at each level of the strata variable contains $k - 1$ additional interaction terms and is

$$log\text{-}odds = a + bf + c_2 z_{2j} + c_3 z_{3j} + \cdots + c_k z_{kj} + d_2 f z_{2j} + d_3 f z_{3j} + \cdots + d_k f z_{kj}. \tag{10.37}$$

For the ith strata, $log\text{-}odds = a + bf + c_i + d_i f = (a + c_i) + (b + d_i)f = A_i + B_i f$; that is, not only does the log-odds differ among the strata (A_i) but also the association between the binary independent variable and the outcome (measured by B_i) differs among the k strata. The tractability of an additive model to represent the data contained in k individual 2-by-2 tables is evaluated by comparing the likelihood value from the saturated model to the likelihood from the additive model. If the interaction terms add little to the representation of the data, then the additive model produces a meaningful estimate of the common odds ratio, namely $\hat{OR} = e^{\hat{b}}$.

Applying these two models to the coronary/behavior data given in Table 9.30 yields a likelihood value for the saturated model of $-2\log(L_1) = 1702.156$ (10 parameters) and for the additive model of $-2\log(L_0) = 1703.010$ (6 parameters). Deleting the interaction terms hardly changes the likelihood (deviance $= Y^2 = 1703.010 - 1702.156 = 0.854$ with four degrees of freedom and the p-value $= 0.931$). Using the additive model produces the estimates in Table 10.16.

The key summary coefficient is $\hat{b} = 0.793$, producing an estimated common odds ratio combining information from all five age strata of $\hat{OR} = e^{0.793} = 2.210$. That is, the odds of a coronary event associated with type A individuals are 2.210 times the odds associated with type B individuals accounting for the influence of age. The value of the summary odds ratio from the additive logistic model is similar to the

T A B L E **10.16**

Logistic analysis (additive model) relating the risk of coronary heart disease to A/B behavior type for five age categories (data from Table 9.29)

Variable	Coefficient	Estimate	Standard Error	p-Value	$\hat{OR}$
Constant	a	−3.254	0.207	—	—
A/B	b	0.793	0.141	< 0.001	2.210
Age 40–44	c_2	−0.112	0.232	0.629	0.894
Age 45–49	c_3	0.510	0.225	0.023	1.665
Age 50–54	c_4	0.793	0.236	0.001	2.211
Age > 54	c_5	0.921	0.246	<0.001	2.512

$$-2\log(L) = 1703.010$$

Mantel-Haenszel estimate, 2.214 versus 2.210. These two estimates are usually similar. The odds ratio estimated from the logistic model is slightly more efficient (smaller variance) but practically, no important difference exists between the two estimates of a common odds ratio from k individual 2-by-2 tables. This logistic analysis is similar to the log-linear model used in Chapter 9 applied to the same coronary heart disease data (Table 9.30) where the estimated common odds ratio is $\hat{OR} = 2.211$ ($u_{123} = 0$). The two estimates will usually be similar especially when all cells of k individual 2-by-2 tables contain substantial numbers of observations. However, the ability to assess the degree of homogeneity among the k odds ratios (assess the amount of interaction; expression 10.37) is a definite advantage to a logistic approach.

The influence of age classified into five age intervals can also be assessed by contrasting two likelihood statistics: the likelihood from the additive model ($-2\log(L_1) = 1703.010$) with the likelihood from the model with age ignored ($-2\log(L_0) = 1740.344$). The formal likelihood ratio statistic is $Y^2 = 1740.344 - 1703.010 = 37.334$ with one degree of freedom (b set to zero), producing a p-value < 0.001 showing that age, as suspected, is an important variable in the description of coronary heart disease.

Example: Nonlinear Logistic Regression

Consider the variable maternal age (x) and its influence on the probability of a low birth-weight infant from the same $n = 680$ observations used in the previous examples. The model *log-odds* $= a + b_1 x$ imposes a linear relationship. A summary of this simple two-parameter logistic model directly measuring the linear influence of reported age is given in Table 10.17.

A set of design variables constructed for a series of age categories produces estimated coefficients that potentially suggest alternative ways to describe the relationship between reported age and the log-odds associated with low birth-weight. Such design variables are given in Table 10.18 where the reference category is mothers

T A B L E **10.17**

Analysis of low birth-weight infants by maternal age

Variable	Coefficient	Estimate	Standard Error	p-Value	$\hat{OR}$
Constant	a	−3.423	0.772	—	—
Age	b_1	0.026	0.028	0.363	1.026

$$-2\log(L) = 308.968$$

less than 25 years old (i.e., $z_{1j} = 0$). Categorizing the continuous variable maternal age into four intervals (< 25, 25–29, 30–35, and >35) and the logistic model

$$log\text{-}odds = c_0 + c_2 z_{2j} + c_3 z_{3j} + c_4 z_{4j} \tag{10.38}$$

allows an unconstrained description of the relationship between maternal age and the probability of a low birth-weight infant (Table 10.19). By "an unconstrained description" it is meant that the log-odds can take on any pattern where a constrained model requires the log-odds to follow a specific pattern (e.g., linear requires an increasing or decreasing pattern).

The estimated coefficients show a possible nonlinear relationship between maternal age and the log-odds associated with a low birth-weight infant (note: for

T A B L E **10.18**

Design variables for four age categories

Age	z_{2j}	z_{3j}	z_{4j}
Age < 25	0	0	0
Age 25–29	1	0	0
Age 30–35	0	1	0
Age > 35	0	0	1

T A B L E **10.19**

Analysis of low birth-weight infants by age—categorical model

Variable	Coefficient	Estimate	Standard Error	p-Value	$\hat{OR}$
Constant	c_0	−2.698	0.231	—	—
Age 25–29	c_2	−0.608	0.448	0.174	0.544
Age 30–35	c_3	0.049	0.454	0.915	1.050
Age > 35	c_4	0.693	0.464	0.136	2.000

$$-2\log(L) = 304.330$$

T A B L E **10.20**

Analysis of low birth-weight infants by maternal age—quadratic model

Variable	Coefficient	Estimate	Standard Error	p-Value	$\hat{OR}$
Constant	a	2.966	—	—	—
Age	b_1	−0.449	0.222	0.043	0.638
Age × age	b_2	0.008	0.004	0.030	1.008

$$-2\log(L) = 304.610$$

age < 25, $c_1 = 0$ by design). The odds ratios for increasing age categories are 1.00 (age < 25 = "baseline"), 0.54 (age 25–29), 1.05 (age 30–35), and 2.00 (age > 35), indicating a relationship where young and older mothers are at the greatest risk of having a low birth-weight infant. The minimum odds ratio occurs for category 25–30 years of age (0.54). One way to use this information is to treat reported maternal age as a nonlinear but continuous variable in a logistic regression model.

To incorporate a "u-shaped" response from a variable such as maternal age, the relationship between the log-odds can be represented by a quadratic equation, or

$$log\text{-}odds = a + b_1 x + b_2 x^2 \tag{10.39}$$

where x represents a continuous independent variable. Other than adding a new variable (x^2) to the equation, the logistic regression analysis is no different in principle from other logistic models. The results of including the second-order term x^2 in the equation are given in Table 10.20 where x = reported maternal age.

A statistical test of the coefficient $\hat{b}_2$ yields $z = \hat{b}_2/S_{\hat{b}_2} = 2.166$ with a p-value of 0.030, showing that adding a quadratic term to the model improves the description of the relationship between reported age and the probability of a low birth-weight infant. Comparing likelihood values produces almost the same result (*deviance* = $Y^2 = 308.968 - 304.610 = 4.358$ from Tables 10.17 and 10.20; degrees of freedom = 1 yielding a p-value = 0.037).

The quadratic term in the regression model is a special type of interaction. That is, the change in log-odds or odds ratio for a one-year increase in maternal age differs depending on the age of the mother. If a mother age 30 is compared to a mother age 35, the increase in the log-odds is 0.355 ($\hat{OR} = 1.426$). For a mother age 35 compared to a mother age 40, however, the log-odds increases 0.755 ($\hat{OR} = 2.128$). A quadratic term in a logistic model allows a variable to "interact" with itself. When the log-odds is an additive function of the independent variables, the change in risk is the same for all values of the independent variables, as noted previously.

There is a cost to including a quadratic term in the analysis. Generally the variables x and x^2 are strongly correlated and, like the linear regression situation (Chapters 3 and 4), using correlated variables in a regression equation decreases the precision of the estimated coefficients. For example, the standard error of $\hat{b}_1$ associated with maternal age and low birth-weight is 0.028 when x^2 is not included in the analysis (Table 10.17) and increases to 0.222 when x^2 is included in the model (Table 10.20). Again like the linear regression model, correlation between x and x^2 is often reduced by replacing the independent variable x with the transformed variable $x - \bar{x}$.

Comparison of the Regression Coefficients

The general k-variate additive logistic model is expressed as

$$P(D \mid x_{1j}, x_{2j}, x_{3j}, \cdots, x_{kj}) = \frac{1}{1 + e^{-(a + \sum\limits_{i=1}^{k} b_i x_{ij})}}$$ (10.40)

or

$$log\text{-}odds = a + \sum_{i=1}^{k} b_i x_{ij} .$$ (10.41)

The relationship between the logistic model and the log-odds for the k-variable case is described in Box 10.1.

B O X **10.1** RELATIONSHIP BETWEEN LOG-ODDS AND A LOGISTIC MODEL

The general k-variable additive logistic model is

$$P(D \mid x_1, x_2, \cdots, x_k) = \frac{1}{1 + e^{-(a + \Sigma b_i x_{ij})}}$$

making

$$P(\bar{D} \mid x_1, x_2, \ldots, x_k) = \frac{e^{-(a + \Sigma b_i x_{ij})}}{1 + e^{-(a + \Sigma b_i x_{ij})}} .$$

Then,

$$odds = \frac{P(D \mid x_1, x_2, \cdots, x_k)}{P(\bar{D} \mid x_1, x_2, \cdots, x_k)} = e^{a + \Sigma b_i x_{ij}}$$

$$log\text{-}odds = \log(odds) = \log \left[\frac{P(D \mid x_1, x_2, \cdots, x_k)}{P(\bar{D} \mid x_1, x_2, \cdots, x_k)} \right] = \log[e^{a + \Sigma b_i x_{ij}}] ;$$

therefore,

$$log\text{-}odds = a + \Sigma b_i x_{ij} .$$

T A B L E **10.21**

Analysis of low birth-weight infants by maternal age, height, weight, and smoking exposure—four-variable additive model

Variable	Coefficient	Estimate	Standard Error	p-Value	$\hat{OR}$
Constant	a	0.027	—	—	—
Age	b_1	0.022	0.029	0.453	1.022
Height	b_2	−0.052	0.076	0.498	0.949
Weight	b_3	−0.003	0.010	0.813	0.997
Cigarettes	b_4	0.033	0.012	0.006	1.033

$$-2\log(L) = 301.319$$

Once a model to represent the data is chosen, interest usually shifts to interpreting the estimated regression coefficients. Direct comparisons among the estimated coefficients are not generally useful because the magnitude of the coefficients depends on the units used to measure the independent variables. Of course, if variables are measured in the same units, then the relative impact of these variables is directly reflected by the coefficient values; otherwise, the coefficients are not commensurate. A number of ways exist to create commensurate coefficients.

To illustrate, a four-variable logistic regression analysis is used with the independent variables maternal age (x_1), height (x_2), prepregnancy weight (x_3), and reported number of cigarettes smoked per day (x_4). Low birth-weight is again the dependent variable. Applying the additive logistic regression model

$$log\text{-}odds = a + b_1 x_{1j} + b_2 x_{2j} + b_3 x_{3j} + b_4 x_{4j} \tag{10.42}$$

to the 680 observations used in the previous examples yields the estimated coefficients given in Table 10.21.

The relative influence of these four variables on the probability of a low birth-weight infant is not easily judged without standardizing the coefficients. Three possible ways to standardize estimated regression coefficients are:

1 The ith estimated coefficient multiplied by the standard deviation of its associated independent variable produces a measure of response for an increase of one standard deviation of the ith variable (analogous to path coefficients—Chapter 4).

2 The ith coefficient divided by its standard error produces a unitless quantity that is related to the unique contribution of the ith variable to the regression equation.

3 The difference between transformed likelihood with the ith variable included in the regression equation ($-2\log[L]$) and the transformed likelihood with the ith variable excluded ($-2\log[L_{(i)}]$) produces another unitless measure of the unique contribution of the ith variable to the regression equation.

These suggestions are not very different from those made in Chapter 4 for standardizing coefficients from a linear regression analysis.

T A B L E   **10.22**
An illustration of three types of standardized coefficients

Variable	$\hat{b}_i$	S_i	$\hat{b}_i S_i$ (1)	$\hat{b}_i / S_{\hat{b}_i}$ (2)	Likelihood (3)
Age	0.022	5.463	0.120	0.750	0.552
Height	−0.052	2.483	−0.129	−0.678	0.456
Weight	−0.003	17.878	−0.046	−0.237	0.058
Cigarettes	0.033	11.272	0.366	2.748	6.826

The standardized coefficients in Table 10.22 are examples of these three options using the coefficients from Table 10.21. Cigarette exposure, by far, has the most important influence; maternal age and height have similar but weak contributions; and maternal weight is the least effective variable in describing the probability of a low birth-weight infant. Standardization by S_{b_i} and the likelihood method are similar (the squared values from method 2 are about equal to the values from method 3). The likelihood approach (method 3) can be extended to assess the simultaneous impact of more than one variable on the probability of a specific outcome (see applied example at the end of the chapter).

Goodness-of-Fit

When a logistic regression analysis is applied to a sample of observations contained in a contingency table, a goodness-of-fit evaluation is based on estimating a value for each cell in the table under specific conditions. These estimated values are generated using the logistic probabilities calculated from the estimated model. The statistical assessment of the differences between the estimated frequencies and the observed frequencies is usually accomplished with a chi-square statistic. The situation becomes more complicated when independent variables consist of a large number of values or are continuous.

One strategy for dealing with continuous data involves creating a somewhat arbitrary table based on the estimated logistic probabilities for each observation. Once the table is constructed, the Pearson chi-square statistic is applied in much the same manner as the evaluation of any contingency table. Typically, 10 groups are formed so the first group contains the 10% of the observations with the lowest estimated logistic probabilities, the second group contains the 10% with the next lowest estimated probabilities, and so forth until the tenth group contains the 10% of the observations with the highest estimated logistic probabilities. If the data do not divide evenly into 10 groups, groups are formed so each category contains approximately 10% of the observations. This process divides the data into percentiles based on logistic probabilities, called "deciles of risk." For each decile the mean logistic probability is estimated. If $\hat{p}_j$ represents the estimated logistic probability for the jth observation among a sample of n observations, then for a specific decile (denoted by k) the mean value is

$$\overline{p}_k = \frac{1}{n_k} \sum_{j=1}^{n_k} \hat{p}_j$$

(10.43)

and the summation is over the n_k estimated logistic probabilities of the kth decile where, for example,

$$\hat{p}_j = \frac{1}{1 + e^{-(\hat{a} + \Sigma \hat{b}_i x_{ij})}}$$

and n_k is as close as possible to $n/10$. An example of the calculation of $\hat{p}_j$ is given in Box 10.2.

Based on each of the 10 mean logistic probabilities, the number of binary outcomes in a specific category is estimated by $\hat{e}_k = \Sigma \hat{p}_j = n_k \overline{p}_k$ and $n_k - \hat{e}_k = n_k(1 - \overline{p}_k)$ for the kth decile of risk. This process applied to all 10 groups produces 20 estimated values, which are compared to the 20 observed values derived from classifying the observed binary outcomes into the same deciles of risk (represented as o_k and $n_k - o_k$). The Pearson chi-square statistic is then applied where

---

B O X **10.2**   LOGISTIC PROBABILITY

Consider a woman with the following specific values:

x_1 = age = 35 years, x_2 = height = 65 inches, x_3 = weight = 122 pounds, and x_4 = smokes one pack a day = 20 cigarettes per day

Using the coefficients estimated from the additive logistic regression analysis employing these four variables (Table 10.21) gives

$$estimated\ log\text{-}odds = \hat{y}_j = \hat{a} + \hat{b}_1 x_{1j} + \hat{b}_2 x_{2j} + \hat{b}_3 x_{3j} + \hat{b}_4 x_{4j}$$

and, specifically,

$$estimated\ log\text{-}odds = 0.027 + 0.022(35) - 0.052(65) - 0.003(122) + 0.033(20)$$
$$= -2.289 .$$

The estimated probability of a low birth-weight infant for this woman is then

$$\hat{p} = \hat{P}(D \mid x_1 = 35, x_2 = 65, x_3 = 122, x_4 = 20) = \frac{1}{1 + e^{-(-2.289)}} ;$$

therefore, $\hat{p}_j = 0.092$.

$$X^2 = \sum_{i=1}^{10} \frac{(o_i - \hat{e}_i)^2}{\hat{e}_i} + \sum_{i=1}^{10} \frac{[(n_i - o_i) - (n_i - \hat{e}_i)]^2}{n_i - \hat{e}_i} = \sum_{i=1}^{10} \frac{(o_i - \hat{e}_i)^2}{n_i \bar{p}_i (1 - \bar{p}_i)}.$$

(10.44)

The test statistic X^2 has an approximate chi-square distribution with eight degrees of freedom when the logistic regression model produces values that differ from the observed data only by chance variation (more on the goodness-of-fit of the logistic model is found in [2]). The degrees of freedom are eight, which is neither intuitive nor easily justified [1].

To illustrate, the previous logistic model based on four variables (maternal age, height, weight, and smoking) is used (Table 10.21). The average logistic probabilities ($\bar{p}_k$), the estimated values ($\hat{e}_k$), and the observed values (o_k) are given in Table 10.23. The estimated values in Table 10.23 are $\hat{e}_k = 68\bar{p}_k$ and $68 - \hat{e}_k$ because $n = 680$ observations produce 68 observations per decile. For example, $\hat{e}_5 = 68(0.0483) = 3.28$ and $n_5 - \hat{e}_5 = 68 - 3.28 = 64.72$, which are compared to the corresponding observed values $o_5 = 4$ and $n_5 - o_5 = 64$. The test statistic calculated from these 20 comparisons is $X^2 = 14.728$ and has an approximate chi-square distribution with eight degrees of freedom, generating a p-value of 0.065. The chi-square analysis indicates that the fit of the four-variable additive logistic model to the low birthweight data is not extremely good and consideration should be given to modifying the additive model to improve it as a description of the 680 observed values. Perhaps specific interaction terms are necessary, some of the independent variables would be better described by nonlinear relationships, or additional independent variables need to be added to the analysis.

T A B L E **10.23**

Illustration of the fit of a logistic regression equation

Decile	$\bar{p}_k$	$\hat{e}_k$	o_k	$n_k - \hat{e}_k$	$n_k - o_k$
1	0.0340	2.31	1	65.69	67
2	0.0388	2.64	0	65.36	68
3	0.0417	2.83	2	65.17	66
4	0.0450	3.06	2	64.94	66
5	0.0483	3.28	4	64.72	64
6	0.0525	3.57	7	64.43	61
7	0.0587	3.99	3	64.01	65
8	0.0704	4.79	9	63.21	59
9	0.0857	5.83	8	62.17	60
10	0.1279	8.69	5	59.31	63
Total	0.0603	41	41	639	639

Case/Control Data

The pattern of data collection most often associated with a logistic regression analysis involves observing a group of individuals and noting the occurrence or nonoccurrence of a binary outcome. Effective analysis of such cohort or cross-sectional data becomes difficult when the event of interest is rare. For example, if the frequency of a disease such as lung cancer is 1/1000, then many thousands of observations would have to be collected for the sample to include a sufficient number of cases for analysis. Usually collecting large numbers of observations to study a rare phenomenon is not practical. An alternative approach is to collect observations in such a way that guarantees a sufficient number of cases are included. A sample is collected so a fixed number of cases and corresponding controls make up the sample to be analyzed. In a study of lung cancer, information on a case could be obtained from cancer patient's hospital records and, at the same time, information on a corresponding control individual also could be collected. In this way, a specific number of binary dependent variables (cases and controls) are collected to study the influence of the independent variables. It is not obvious that a logistic regression analysis is the appropriate statistical tool when the data collection is based on the dependent binary variable.

The distinction between sampling based on the dependent variable and sampling based on cohort or cross-sectional approaches is seen in a 2-by-2 table. A cohort pattern of sampling corresponds to collecting a fixed number of individuals who belong to one group ($G = 1$) and a fixed number of individuals who belong to another group ($G = 0$) and comparing the frequency of occurrences of the outcome D in each group. Therefore, the rows of a 2-by-2 table contain fixed numbers of observations (Table 10.1). The proportion of occurrence of D or $\overline{D}$ in each row is an estimate of the proportion in the sampled population. A cross-sectional sample consists of a fixed number of individuals collected without regard to outcome or to group status, but again the proportion of occurrences of the dependent variable, D or $\overline{D}$, is an estimate of the proportion in the sampled population. On the other hand, a case/control sample requires collecting a specified number of D-individuals and a specified number of $\overline{D}$-individuals and comparing the frequency of individuals belonging to the two groups ($G = 0$ and $G = 1$). Obviously, the proportion of cases in the sample does not estimate the proportion in the sampled population because this proportion is determined by the sampling scheme. Case/control data can be viewed as a 2-by-2 table with specified numbers of observations in the columns (Table 10.1). As noted, regardless of the pattern of data collection, the estimated odds ratio is the same in both situations and has the same interpretation (expressions 10.3 and 10.5). Remarkably, logistic regression analysis has the same property. Whether the data are collected in a cohort pattern, a cross-sectional pattern, or a case/control pattern, the application of the logistic model is valid and produces the identical interpretation of the coefficients (shown for a simple case in Box 10.3). The only difference involves the interpretation of the constant term and this term is relatively unimportant in most logistic analyses.

For the logistic model *log-odds* = $a + bx$, the estimated coefficient $\hat{b}$ associated with the independent variable x is not influenced by case/control sampling. The estimated coefficient $\hat{b}$, as usual, estimates the change in log-odds for a 1-unit increase in x for cohort, cross-sectional, or case/control patterns of data collection. Only the constant term in the logistic model (a) is affected by the choice of the sampling pattern.

Similarly, regression coefficients from a k-variable logistic analysis are not influenced by case/control sampling of data. Again, only the constant term is affected by the sampling scheme. The value of the estimated constant term for case/control sampling is made up of an unknown mixture of two elements, namely the probability of being included in the sample as well as the frequency of the disease in the sampled population. However, the constant term in the logistic model does not play an important role and can be ignored. Therefore, most properties of a logistic regression analysis apply equally to case/control data, adding flexibility to this analytic technique and, more important, providing a powerful tool for investigating rare binary outcomes. Texts are available that are entirely devoted to the analysis of case/control data (e.g., [1] and [3]).

B O X   **10.3**   CASE/CONTROL SAMPLING

Consider the logistic equation where p_x represents the probability of a specific outcome, or as before

$$p_x = \frac{1}{1 + e^{-(a+bx)}} \quad \text{and, therefore,} \quad \log\left[\frac{p_x}{1 - p_x}\right] = a + bx.$$

Let $s_1 = P(sampled \mid case) = P(sampled \mid disease)$ and $s_0 = P(sampled \mid control) = P(sampled \mid No\ disease)$. Then, specifically

$$P(Disease \mid x \text{ and } sampled) = \frac{s_1 p_x}{s_1 p_x + s_0(1 - p_x)}.$$

The log-odds is

$$log\text{-}odds = \log\left[\frac{P(disease \mid x \text{ and } sampled)}{P(No\ disease \mid x \text{ and } sampled)}\right],$$

and it follows that

$$log\text{-}odds = \log\left[\frac{s_1 p_x/[s_1 p_x + s_0(1 - p_x)]}{s_0(1 - p_x)/[s_1 p_x + s_0(1 - p_x)]}\right] = \log\left[\frac{s_1}{s_0}\frac{p_x}{1 - p_x}\right] = A + \log\left[\frac{p_x}{1 - p_x}\right]$$

where $A = \log(s_1/s_0)$, giving

$$log\text{-}odds = A + a + bx = A' + bx \quad \text{where} \quad A' = A + a.$$

Applied Example	# Joint Analysis of Maternal and Paternal Data

Continuing the example of low birth-weight infants, eight maternal and paternal variables from the CHDS data set are used to illustrate the application of a logistic regression analysis. The variables are:

maternal: prepregnancy weight, age, height, cigarettes smoked per day

paternal: years of education, age, height, and cigarettes smoked per day

Each of these eight parental variables potentially influences the probability of a low birth-weight infant. A more specific question concerns the joint influence of the maternal and paternal variables: What is the relative impact of the maternal variables compared to the paternal variables? A similar question is addressed with a multivariable linear regression analysis (Chapter 4) where the dependent variable is the directly measured birth-weight rather than the probability of a low birth-weight infant. To provide a baseline analysis, the additive logistic regression model using all eight variables is estimated. The goodness-of-fit chi-square statistic $X^2 = 3.70$ (degrees of freedom = 8; p-value = 0.833) shows no reason to reject an eight-variable additive logistic model as a representation of the relationship of the parental variables to the probability of a low birth-weight infant.

To evaluate the influence of the four maternal variables, they are deleted from the model and a logistic regression equation is estimated containing only the four paternal variables. The likelihood measure associated with this model is $-2\log[L_0] = 306.518$. The transformed likelihood value from the eight-variable model serves as a relative measure of fit ($-2\log[L_1] = 296.798$). A contrast of these measures of goodness-of-fit gives a deviance of $Y^2 = 306.518 - 296.798 = 9.720$ with four degrees of freedom (four parameters set to zero) and yields a p-value of 0.045. The comparison of these two models shows that at least some of the maternal variables are likely an important influence on the probability of a low birth-weight infant.

The same strategy applies to evaluating the role of the paternal variables. The model with the four paternal variables removed yields a transformed likelihood value of $-2\log[L_0'] = 301.319$. The difference between the eight-variable additive model and the model with the four paternal variables excluded gives a deviance of $Y^2 = 301.319 - 296.798 = 4.521$, again with four degrees of freedom. The p-value associated with this contrast is 0.340, indicating a weak or nonexistent association between the four paternal variables and the probability of a low birth-weight infant.

The two likelihood goodness-of-fit measures also produce a commensurate comparison of maternal and paternal influences indicating the relative importance of these two sets of variables (i.e., 9.7 versus 4.5).

Linear Logistic Regression

The following STATA code contrasts three logistic regression analyses using the CHDS data. Eight parental variables are analyzed with the goal of better understanding the relative influence of the maternal and paternal variables on the probability of having a low birth-weight infant. A brief description of these nine variables is given in Chapter 2.

chds.logistic

```
. use chds

. generate lwt=bwt=6

. tabulate lwt
      lwt|     Freq.      Percent       Cum.
----------+-----------------------------------
        0 |        41        6.03        6.03
        1 |       639       93.97      100.00
----------+-----------------------------------
    Total |       680      100.00

. logistic lwt mage mnocig mheight mppwt fage fedyrs fnocig fheight
```

$\log(L)$

Logit Estimates

	Number of obs	=	680
	chi2(8)	=	12.98
	Prob chi2	=	0.1126
Log Likelihood = -148.39889	Pseudo R2	=	0.0419

$\hat{OR}_i$ $S_{\hat{OR}_i}$

lwt	Odds Ratio	Std. Err.	t	P|t|	[95% Conf. Interval]	
mage	.934341	.0455341	-1.394	0.164	.8490789	1.028165
mnocig	.9661056	.0122307	-2.724	0.007	.9423865	.9904216
mheight	1.014738	.0806415	0.184	0.854	.8681338	1.1861
mppwt	1.001886	.0108908	0.173	0.862	.9807289	1.0235
fage	1.054522	.0496255	1.128	0.260	.9614487	1.156606
fedyrs	1.05998	.0823599	0.750	0.454	.909998	1.234683
fnocig	1.005932	.0122849	0.484	0.628	.9820977	1.030345
fheight	1.110496	.0754319	1.543	0.123	.9718376	1.268938

```
. logit
```

Logit Estimates

	Number of obs	=	680
	chi2(8)	=	12.98
	Prob chi2	=	0.1126
	Pseudo R2	=	0.0419
Log Likelihood = -148.39889			

$\hat{b}_i$ $S_{\hat{b}_i}$ *p*-values

lwt	Coef.	Std. Err.	t	P|t|	[95% Conf. Interval]	
mage	-.0679138	.0487339	-1.394	0.164	-.1636031	.0277755
mnocig	-.0344822	.0126598	-2.724	0.007	-.0593397	-.0096246
mheight	.0146307	.0794702	0.184	0.854	-.1414094	.1706709
mppwt	.0018847	.0108703	0.173	0.862	-.0194592	.0232285
fage	.0530878	.0470597	1.128	0.260	-.0393141	.1454897
fedyrs	.0582505	.0776995	0.750	0.454	-.0943129	.2108139
fnocig	.0059148	.0122125	0.484	0.628	-.0180644	.029894
fheight	.104807	.0679263	1.543	0.123	-.0285666	.2381806
_cons	-6.085187	5.669659	-1.073	0.284	-17.21759	5.047221

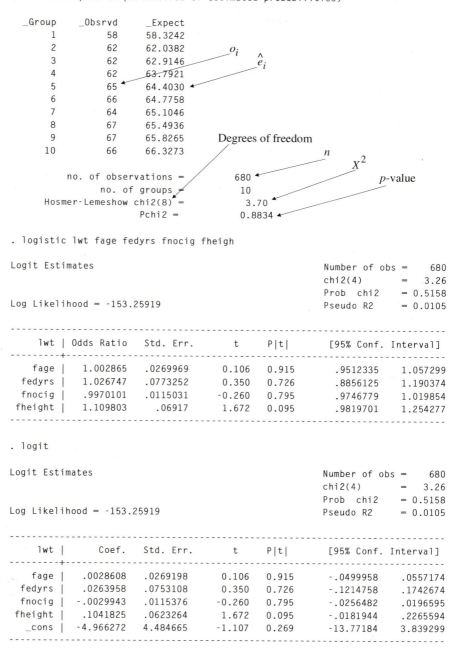

```
. lfit, group(10) table

Logistic estimates for lwt, goodness-of-fit test
(Table collapsed on percentiles of estimated probabilities)

     _Group    _Obsrvd    _Expect
          1         58    58.3242
          2         62    62.0382
          3         62    62.9146
          4         62    63.7921
          5         65    64.4030
          6         66    64.7758
          7         64    65.1046
          8         67    65.4936
          9         67    65.8265
         10         66    66.3273

              no. of observations =          680
                 no. of groups =             10
          Hosmer-Lemeshow chi2(8) =         3.70
                        Pchi2 =           0.8834
```

o_i $\hat{e}_i$

Degrees of freedom

n x^2 *p*-value

```
. logistic lwt fage fedyrs fnocig fheigh

Logit Estimates                            Number of obs =      680
                                           chi2(4)       =     3.26
                                           Prob  chi2    = 0.5158
Log Likelihood = -153.25919                Pseudo R2     = 0.0105

-----------------------------------------------------------------------------
    lwt | Odds Ratio   Std. Err.       t     P|t|     [95% Conf. Interval]
--------+--------------------------------------------------------------------
   fage |   1.002865    .0269969     0.106   0.915     .9512335    1.057299
 fedyrs |   1.026747    .0773252     0.350   0.726     .8856125    1.190374
 fnocig |    .9970101    .0115031    -0.260   0.795     .9746779    1.019854
fheight |   1.109803     .06917      1.672   0.095     .9819701    1.254277
-----------------------------------------------------------------------------

. logit

Logit Estimates                            Number of obs =      680
                                           chi2(4)       =     3.26
                                           Prob  chi2    = 0.5158
Log Likelihood = -153.25919                Pseudo R2     = 0.0105

-----------------------------------------------------------------------------
    lwt |     Coef.    Std. Err.       t     P|t|     [95% Conf. Interval]
--------+--------------------------------------------------------------------
   fage |   .0028608    .0269198     0.106   0.915    -.0499958    .0557174
 fedyrs |   .0263958    .0753108     0.350   0.726    -.1214758    .1742674
 fnocig |  -.0029943    .0115376    -0.260   0.795    -.0256482    .0196595
fheight |   .1041825    .0623264     1.672   0.095    -.0181944    .2265594
  _cons |  -4.966272    4.484665    -1.107   0.269   -13.77184    3.839299
-----------------------------------------------------------------------------
```

```
. logistic lwt mage mnocig mheight mppwt
```

Logit Estimates

```
                                              Number of obs =      680
                                              chi2(4)       =     8.46
                                              Prob  chi2    = 0.0762
Log Likelihood = -150.65941                   Pseudo R2     = 0.0273
```

```
-----------------------------------------------------------------------
    lwt | Odds Ratio   Std. Err.       t     P|t|    [95% Conf. Interval]
--------+--------------------------------------------------------------
   mage |   .9782811   .0286566    -0.750   0.454    .9236018   1.036198
 mnocig |    .968017   .0114555    -2.747   0.006    .9457837    .990773
mheight |   1.053223   .0805877     0.678   0.498    .9063033   1.223961
  mppwt |   1.002567   .0108349     0.237   0.813    .9815166   1.024068
-----------------------------------------------------------------------
```

```
. logit
```

Logit Estimates

```
                                              Number of obs =      680
                                              chi2(4)       =     8.46
                                              Prob  chi2    = 0.0762
Log Likelihood = -150.65941                   Pseudo R2     = 0.0273
```

```
-----------------------------------------------------------------------
    lwt |     Coef.   Std. Err.       t     P|t|    [95% Conf. Interval]
--------+--------------------------------------------------------------
   mage | -.0219582   .0292929    -0.750   0.454    -.0794743    .0355579
 mnocig | -.0325056   .0118339    -2.747   0.006    -.0557414   -.0092698
mheight |  .0518554   .0765153     0.678   0.498    -.0983813    .202092
  mppwt |  .0025635   .0108072     0.237   0.813    -.0186563    .0237833
  _cons | -.0268581   4.47587     -0.006   0.995    -8.815161    8.761444
-----------------------------------------------------------------------
```

```
. exit,clear
```

Notation

Odds ratio:

$$odds\text{-}ratio = \frac{p_1/(1-p_1)}{p_0/(1-p_0)} = \frac{p_1(1-p_0)}{p_0(1-p_1)}$$

Odds ratio applied to a 2-by-2 table:

$$OR = \frac{P(D\mid G=1)/P(\overline{D}\mid G=1)}{P(D\mid G=0)/P(\overline{D}\mid G=0)} = \frac{P(G=1\mid D)/P(G=0\mid D)}{P(G=1\mid\overline{D})/P(G=0\mid\overline{D})} \quad \text{and} \quad \hat{OR} = \frac{f_{11}/f_{12}}{f_{21}/f_{22}}.$$

Variance of the logarithm of the estimated odds ratio:

$$variance(\log(\hat{OR})) = S^2_{\log(\hat{OR})} = \frac{1}{f_{11}} + \frac{1}{f_{12}} + \frac{1}{f_{21}} + \frac{1}{f_{22}}$$

Confidence interval for log-odds and odds ratio:

$$lower\ bound = \log(\hat{OR}) - z_{1-\alpha/2}S_{\log(\hat{OR})}$$
$$upper\ bound = \log(\hat{OR}) + z_{1-\alpha/2}S_{\log(\hat{OR})}$$

and

$$lower\ bound = \hat{OR}\ e^{-z_{1-\alpha/2}S_{\log(\hat{OR})}}$$
$$upper\ bound = \hat{OR}\ e^{+z_{1-\alpha/2}S_{\log(\hat{OR})}}$$

Logistic model for the 2-by-2 table:

$$log\text{-}odds = a + bg_i$$

Logistic function:

$$f(x) = \frac{1}{1 + e^{-x}}$$

Saturated logistic model for the 2-by-2-by-2 table:

$$log\text{-}odds = a + b_1g_i + b_2h_j + b_3g_jh_j$$

Additive logistic model for the 2-by-2-by-2 table:

$$log\text{-}odds = a + b_1g_i + b_2h_j$$

Deviance:

$$Y^2 = [-2\log(L_0)] - [-2\log(L_1)]$$

Logistic model for a continuous variable:

$$P(D \mid x) = \frac{1}{1 + e^{-(a+bx)}}$$

or equivalently,

$$log\text{-}odds = a + bx$$

General additive k-variate logistic model:

$$log\text{-}odds = a + b_1x_{1j} + b_2x_{2j} + b_3x_{3j} + \cdots + b_kx_{kj}, \quad \text{or}$$
$$log\text{-}odds = a + \sum_{i=1}^{k} b_ix_{ij}$$

Problems

1   Using the data in Table 10.24, employ a logistic regression analysis to explore the relationship between the variables parity, age, and race (independent variables) and the probability of a twin birth (dependent variable). Which variable, age or parity, has the greatest impact on the probability of a twin? Assess the variable age using a standard normal statistic z and contrast this result with the same evaluation using a

T A B L E **10.24**

Data on parity, maternal age, twinning, and race

Parity	Age	Twin	Race	Count
0	1	0	1	3186
0	1	0	2	800
0	1	0	3	728
0	1	0	4	742
0	1	0	5	145
0	1	1	1	70
0	1	1	2	11
0	1	1	3	13
0	1	1	4	6
0	1	1	5	3
0	0	0	1	474
0	0	0	2	26
0	0	0	3	50
0	0	0	4	65
0	0	0	5	20
0	0	1	1	10
0	0	1	2	0
0	0	1	3	1
0	0	1	4	1
0	0	1	5	2
1	1	0	1	1807
1	1	0	2	600
1	1	0	3	650
1	1	0	4	369
1	1	0	5	50
1	1	1	1	37
1	1	1	2	19
1	1	1	3	20
1	1	1	4	0
1	1	1	5	0
1	0	0	1	455
1	0	0	2	74
1	0	0	3	122
1	0	0	4	121
1	0	0	5	30
1	0	1	1	13
1	0	1	2	1
1	0	1	3	3
1	0	1	4	0
1	0	1	5	0

Note: Table 10.24 contains births by parity ($0 = 1$, $1 = 2^+$), maternal age (0 = less than 35, 1 = greater than or equal to 35), twin status (1 = twin births, 0 = single births), and race (1 = white, 2 = black, 3 = Hispanic, 4 = Japanese, and 5 = Chinese).

likelihood approach. Formally assess the impact of race on the probability of a twin birth.

2 Using the painted turtle data (Table 6.7), let the carapace measurements be independent variables (i.e., length, width, and height) and the dependent variable be sex; use a logistic regression to investigate the relationship between size and sex. Compare this approach to the discriminant analysis used in Chapter 6, Problem 2.

3 Show that:

$$\frac{odds(G=1)}{odds(G=0)} = \frac{odds(D=1)}{odds(D=0)} \quad \text{(expression 10.2 equals expression 10.4)}$$

Hint: $P(A \mid B) = P(B \mid A)P(A)/P(B)$

4 Show that, when the model is saturated,

$$e^{\hat{b}_3} = \frac{\hat{OR}_1}{\hat{OR}_2}$$

for a 2-by-2-by-2 table.

5 Derive the estimated cell frequencies given in Table 10.8.

6 Show that $\hat{OR}_{GD \mid H=0} = \hat{OR}_{GD \mid H=1}$ for an additive model.
Also show that if $\hat{OR}_{GD \mid H=1} = \hat{OR}_{GD \mid H=0}$, then $\hat{OR}_{HD \mid G=1} = \hat{OR}_{HD \mid G=0}$.

7 Sketch the saturated logistic model representing k individual 2-by-2 tables with the presence of interaction; that is,

$$log\text{-}odds = a + bf + c_2 z_2 + c_3 z_3 + \cdots + c_k z_k + d_2 f z_2 + d_3 f z_3 + \cdots + d_k f z_k .$$

Also sketch the same model with no interaction; that is, sketch the model with the $k - 1$ interaction terms removed, or

$$log\text{-}odds = a + bf + c_2 z_2 + c_3 z_3 + \cdots + c_k z_k .$$

8 Again consider the set of crime data where the type of crime (violent = 0 and non-violent = 1), offender status (first offender = 0 and repeat offender = 1), and race (white = 0 and black = 1) are classified into a three-dimensional table:

Violent	Repeat	Race	Count
1	0	0	300
1	0	1	400
1	1	0	378
1	1	1	525
0	0	0	17
0	0	1	20
0	1	0	5
0	1	1	8

Conduct a logistic analysis of this variable where the dependent variable is the nature of the crime (violent $= 0$ and nonviolent $= 1$). Determine whether the variables offender status and race have an impact on the type of crime. Contrast these results with the analysis that was done using a log-linear approach in Chapter 9 (Problem 9).

References

1 Breslow, N. E., and Day, N. E. *Statistical Methods in Cancer Research,* Volume 1. New York: Oxford University Press, 1980.

2 Hosmer, D. W., and Lemeshow, S. *Applied Logistic Regression.* New York: John Wiley, 1989.

3 Schlesselman, J. J. *Case-Control Studies.* New York: Oxford University Press, 1982.

11

Survival Data Analysis

Background

In human populations, typically survival data consist of observing a sample of individuals over a period of time and recording changes in status. For example, a well person could become sick, a person who is HIV-positive could be diagnosed as having AIDS, a diseased person could die, or a cancer patient in remission could once again show symptoms of the disease. In more general terms, subjects are observed for a period of time until failure occurs, the study ends, or for some reason the subject is lost to further observation. The variable of primary interest is the time to the occurrence of these events. For example, individuals in remission from acute leukemia are given a new treatment and the time is recorded until one of two things happens: They fail (leukemia recurs) or the study ends. The statistical analysis of such failure-time data constitutes the topic of survival analysis. Survival analysis also explores the relationship of failure time to other relevant variates, in the hope of better understanding the reasons for failure. In this chapter, the word "failure" is used in a general sense but the techniques and issues discussed focus on failures in human populations, which usually means the occurrence of disease or death. This chapter is a brief introduction to survival analysis and is only a small part of the extensive literature on the topic (e.g., [1], [2], and [3]).

The first half of the chapter is devoted to defining and using a survival curve. The related concept of a hazard rate is also introduced and explored as a basis for comparing survival times. Then, to further compare survival experiences, a brief introduction to the proportional hazards model is presented. This regression model allows the assessment of the influences of a series of variates on the survival experience of a group or an individual in much the same way as the linear and logistic regression models.

Survival Probabilities

Survival Function

Two formal descriptions of failure times are a survival probability and a hazard rate. These two ways to describe the occurrence of failures are fundamental to the study of disease and mortality in human populations. A survival probability is defined as the probability of surviving over a given period (the description of a hazard rate is presented later in the chapter). In symbols,

$$survival\ probability = S(t) = P(surviving\ time\ 0\ to\ t) .$$

$$(11.1)$$

Equivalently, a survival probability is defined as the probability at *time* = 0 of living beyond a specific time, or

$$survival\ probability = S(t) = P(surviving\ beyond\ t) .$$

$$(11.2)$$

Similar to probabilities in general, a survival probability refers to a specific sample or population. For example, an observed probability of living beyond age 65 for a male resident of California born in 1980 is $S(65) = 0.707$ and for a female is $S(65) = 0.831$; in addition, surviving beyond age 90 for males is $S(90) = 0.083$ and for females is $S(90) = 0.195$. A series of survival probabilities calculated over a period can be formed into a survival function. Two typical survival functions are shown in Figure 11.1.

The value of a survival function at the beginning of observation (*time* = 0) is 1.0 (i.e., $S(0) = 1.0$); as time increases, the function generally decreases. In Figure 11.1, the survival function at the top is derived from theoretical considerations (exponential model) and the survival function at the bottom is constructed from observed mortality data for 1980 California male residents. Both plots show the probability of surviving, $S(t)$ on the vertical axis, from time 0 to any time t on the horizontal axis. The two examples illustrate that survival functions are developed from theoretical models and also estimated directly from data.

Exponential Model

A simple and often effective approach to the study of survival data is based on properties of the exponential model where the survival probabilities are given by an exponential function. Specifically,

$$survival\ probability = S(t) = P(surviving\ beyond\ t) = e^{-\lambda t} \quad with\ \lambda > 0 . \quad (11.3)$$

An exponential survival function starts at $S(0) = 1.0$ and steadily decreases as t increases; the rate of decrease is governed by the size of the parameter λ (illustrated in Figure 11.1, top). Comparing the two survival functions in Figure 11.1 shows that an exponential survival model (top) is not going to be a good model for the survival in human populations over a life span of 100 years (bottom). However, for specific

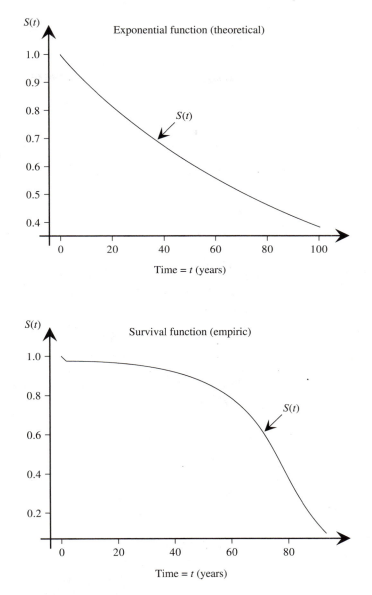

situations or over short periods the exponential model often usefully describes survival experience.

Exponential survival has a special property: The population does not "age." This means the likelihood of future failure does not depend on the amount of time that has passed. In a human population this property would mean that older individuals have no greater chance of dying than younger individuals, which begins to explain the lack of similarity between the two survival functions in Figure 11.1. Because

$$S(t_1) = P(\text{surviving beyond time } t_1) = e^{-\lambda t_1}$$

and

$$S(t_2) = P(\text{surviving beyond time } t_2) = e^{-\lambda t_2} ,$$

then, for $t_2 > t_1$

$$P(\text{surviving beyond time } t_2) = P(\text{surviving beyond } t_1) \, P(\text{surviving from } t_1 \text{ to } t_2) ,$$

or

$$e^{-\lambda t_2} = e^{-\lambda t_1} P(\text{surviving from } t_1 \text{ to } t_2) ,$$

giving

$$P(\text{surviving from time } t_1 \text{ to } t_2) = e^{-\lambda(t_2 - t_1)} . \tag{11.4}$$

The probability of surviving from time t_1 to t_2 only depends on the difference $t_2 - t_1$ and not on the values of t themselves. If $t_2 - t_1 = 10$ months, then the probability of surviving an additional 10 months is the same for individuals observed for 6 months or for 60 months; all that affects the survival probability is the parameter λ and the difference $t_2 - t_1 = 10$. Important consequences follow from the property that time to failure is unrelated to the amount of time already observed.

The parameter λ is a rate and is constant over time. To understand fully the properties of the rate λ, it is necessary to digress from the discussion of exponential survival to describe an average rate. For disease or mortality, an average rate is the number of cases of disease or death divided by the total amount of time the population was at risk. That is,

$$\text{average rate} = \frac{\text{number of events}}{\text{total time exposed}}$$

or, for example,

$$\text{average mortality rate} = \frac{\text{number of deaths}}{\text{total person--time--at--risk}} . \tag{11.5}$$

For a period t to $t + \delta$, the expected number of deaths among N individuals is $N[S(t) - S(t + \delta)]$ and is estimated by the observed number of deaths, denoted d_t. If the times of death are known exactly, then the total time-at-risk is the total time lived by those who died plus the total time lived by those who survived. However, data are reported where the exact time of failure is unknown. In this case an approximate value is calculated to estimate the total time-at-risk. To calculate an approximate total person-time-at-risk, two types of individuals need to be considered: those who survived the entire interval and those who died during the interval. For those who survived the entire interval, each lived δ-time units, producing a total of $N\delta S(t + \delta)$ person-time-at-risk when N is the total number of individuals under study at *time* $= 0$ and δ is the length of the time interval considered. For those who died during the

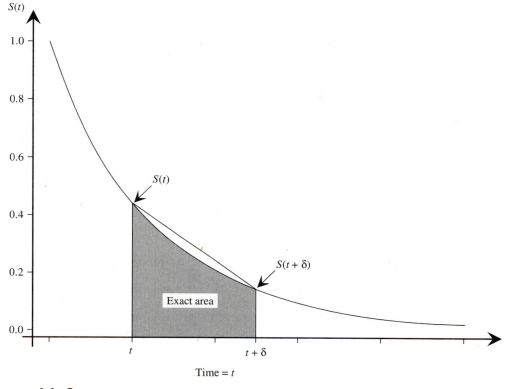

FIGURE **11.2**

A linear approximation to a survival function

interval, it is assumed that they lived, on the average, approximately $(\delta/2)$-time units each. If the survival function $S(t)$ is linear, then this approximation is the exact time each individual, on the average, would be expected to contribute to the total time-at-risk. The approximate total time accumulated by those who failed is then $N(\delta/2)[S(t) - S(t + \delta)]$.

Using a linear approximation to the survival curve gives the approximate total time accumulated by N individuals between times t and $t + \delta$ as

$$total\ time\text{-}at\text{-}risk \approx N\,\delta\,S(t + \delta) + N\frac{\delta}{2}\,[S(t) - S(t + \delta)]\,,$$

(11.6)

or δ times the number of individuals who completed the interval plus $\delta/2$ times the number of individuals who became diseased or died within the interval approximates the total time-at-risk. Figure 11.2 shows a linear approximation to a nonlinear survival function, which is most accurate over short periods.

If $NS(t)$ is estimated by the number of persons free of disease or alive at time t, denoted l_t, then an approximate average rate is

$$r_t = \frac{d_t}{\delta(l_{t+\delta} + \frac{1}{2} d_t)} \cdot$$

(11.7)

Such estimated rates are usually slightly biased because the total person-time-at-risk is calculated as though the survival function is linear. When the time unit is one year ($\delta = 1$), which is often the case, then

$$total\ person\text{-}years \approx l_{t+1} + \frac{1}{2} d_t = l_t - \frac{1}{2} d_t = \frac{1}{2} (l_{t+1} + l_t)$$

(11.8)

and the midyear population (m_t) is typically used as an estimate of total person-years accumulated over a single year, making the estimated average rate

$$r_t = \frac{d_t}{m_t} \cdot$$

(11.9)

Although m_t is the number of individuals alive at midyear, it is an approximation of the total person-years for individuals at risk at the beginning of the year (expression 11.8).

The number of deaths from lung cancer among males living in California in 1988 is $d_{1988} = 9468$ and the midyear 1988 population is $m_{1988} = 13,966,886$, producing an estimate of the average mortality rate of $r_{1988} = 0.000678$. This rate is usually multiplied by an arbitrary base such as 100,000 to yield a value greater than 1.0 (e.g., 67.8 deaths per 100,000 person-years). For females from the same population, the lung cancer mortality rate is 45.8 per 100,000 person-years. The rate ratio 67.8/45.8 = 1.48 shows almost a 50% excess of deaths in the male population.

To statistically assess the impact of sampling variation on an estimated average rate, an estimate of the variance is needed and given by

$$variance(r_t) = S_{r_t}^2 = \frac{r_t^2}{d_t} \left(1 - \frac{\delta^2 r_t^2}{4} \right).$$

(11.10)

When a rate is expressed relative to a base (b), the variance is increased by b^2. That is, if $rate = b \times r_t$, then $variance(br_t) = b^2\ variance(r_t)$. In many situations an estimate of a rate is based on a large number of observations such as the 1988 California population and the need for an estimate of the variance is minimal. But situations arise in which the number of individuals sampled is small and an estimate of the variance allows the construction of approximate confidence intervals or other assessments of the impact of sampling variation.

Geometrically, the total person-time-at-risk is the area under the survival function (Figure 11.2) for the time interval being considered multiplied by the number of individuals in the relevant population. Therefore, the area under the survival curve is the mean survival time. When a survival function is linear, simple geometry produces the area under the survival function and leads to a simple expression for the mean person-time-at-risk (the area of a rectangle plus a triangle).

When the survival function is not linear but has a known form, then the area under the survival function (mean person-time-at-risk) is calculated exactly with calculus techniques. In general, for the interval from t to $t + \delta$, the total time-at-risk for N individuals during the interval t to $t + \delta$ is given by

$$total\ person\text{-}time\text{-}at\text{-}risk = N \int_{t}^{t+\delta} S(u)du\ .$$

(11.11)

The integral

$$\int_{t}^{t+\delta} S(u)du$$

is the mean time-at-risk for the interval t to $t + \delta$ and gives the total person-time-at-risk accumulated by N individuals when multiplied by N. An average rate calculated from a known survival function $S(t)$ is, therefore,

$$average\ rate = \frac{S(t) - S(t + \delta)}{\int_{t}^{t+\delta} S(u)du}\ .$$

(11.12)

For the exponential survival function, the expected number of deaths among N individuals is

$$number\ of\ deaths = N\ (e^{-\lambda t} - e^{-\lambda(t+\delta)})$$

(11.13)

and the total time-at-risk for these N individuals is

$$total\ person\text{-}time\text{-}at\text{-}risk = N \int_{t}^{t+\delta} e^{-\lambda u}du = \frac{N}{\lambda}\ (e^{-\lambda t} - e^{-\lambda(t+\delta)})\ ,$$

(11.14)

producing

$$average\ mortality\ rate = \frac{number\ of\ deaths}{total\ person\text{-}time\text{-}at\text{-}risk} = \lambda\ .$$

(11.15)

An important property of exponentially distributed survival data is that the average rate is constant and equal to the parameter λ. This means that regardless of the time interval, the mortality rate is constant.

A constant rate arises when failure occurs at random. Probably the best examples of random failures occur in certain types of accidents: struck by falling objects, injured by high winds, or attacked by wild animals. Other more common examples of approximately constant rates are deaths among very old people or deaths among extremely ill people. In general, random failure often adequately describes human mortality experience over a short time interval.

To summarize, for the exponential survival function $S(t) = e^{-\lambda t}$ the failure rate is constant and vice versa; for a constant failure rate, the survival times follow an exponential distribution.

Application

The exponential model provides one way to compare disease or mortality rates among groups with different age distributions. If the change in U.S. cancer mortality from 1940 to 1960 is considered, then it is necessary to take into account that the population in 1960 has a higher percentage of older individuals (Table 11.1). Data on cancer mortality for all sites are given in Table 11.1 for the years 1940 and 1960.

A comparison of the overall mortality rate (crude rate = {total deaths}/{total person-years}) of 271.19 (1940) to 359.34 (1960) per 100,000 person-years is not satisfactory because a proportion of the difference is caused by the larger numbers of older individuals in the 1960 population (e.g., ages 75–84 is 4.2% in 1940 and 5.7% in 1960). To compensate for the difference in mortality caused by the older, high-risk individuals, it is postulated that an exponential model adequately describes the cancer mortality experience during a 10-year age interval, or

$$P(\text{surviving the interval } t \text{ to } t+10) = \hat{p}_i = e^{-\hat{\lambda}_i(t_i - t_{i-1})} = e^{-\hat{\lambda}_i(10)}$$

where $\hat{\lambda}_i$ is the estimated average mortality rate in the ith interval. For example, for 1960 individuals 65 to 74 years old,

$$P(\text{surviving from } 65 \text{ to } 74 \mid 1960) = \hat{p}_4 = e^{-\hat{\lambda}_4(10)} = e^{-0.00887(10)} = e^{-0.0887} = 0.915$$

where the value $\hat{\lambda}_4 = 0.00887$ is the estimated age-specific average rate for the age interval 65–74 from the 1960 population. An estimate of the probability of not dying from cancer between the ages 35 to 84 is then

$$\hat{P} = P(\text{surviving } 35\text{–}84) = \hat{p}_1 \hat{p}_2 \hat{p}_3 \hat{p}_4 \hat{p}_5 = \prod_{i=1}^{5} \hat{p}_i. \tag{11.16}$$

These age-specific probabilities for both years and each age category, based on the exponential distribution, are given in Table 11.2.

T A B L E **11.1**

Comparison of U.S. cancer mortality for the years 1940 and 1960 (white males)

	1940			1960		
Age	Deaths	Population	*Rate*[*]	Deaths	Population	*Rate*[*]
35–44	3160	8,249,558	38.3	4891	10,563,872	46.3
45–54	9723	7,294,330	133.3	14,956	9,114,202	164.1
55–64	17,935	5,022,499	357.1	30,888	6,850,263	450.9
65–74	22,179	2,920,220	759.5	41,725	4,702,482	887.3
75–84	13,461	1,019,504	1320.3	26,501	1,874,619	1413.7
Total	66,458	24,506,111	271.19	118,961	33,105,438	359.34

* Rate = age-specific rate per 100,000 person-years = $r_i \times 100,000$

T A B L E **11.2**

Probability of not dying from cancer for each 10-year age interval for data from 1940 and 1960

Age	1940	1960
35–44	0.996	0.995
45–54	0.987	0.984
55–64	0.965	0.956
65–74	0.927	0.915
75–84	0.876	0.868
$\prod p_i$	0.770	0.744

The product of the five age-specific probabilities gives an estimated survival probability (probability of not dying from cancer) between the ages of 35 to 84 of $\hat{P}_{1940} = 0.770$ for 1940 and $\hat{P}_{1960} = 0.744$ for 1960. The difference in these two probabilities is not due to the difference in age distributions between the two groups. The estimated probabilities based on age-specific survival probabilities produce a comparison "free" of the influence from differences in age distributions. The probabilities of death from cancer for the years 1940 and 1960 between age 35 and 84 are $\hat{Q}_{1940} = 1 - 0.770 = 0.230$ and $\hat{Q}_{1960} = 1 - 0.744 = 0.256$, giving a ratio of $0.256/0.230 = 1.11$. The 11% increase in the 1960 cancer mortality is likely due primarily to the increase in smoking-related cancer mortality.

A simple approximate probability of death is achieved by adding the estimated age-specific rates multiplied by the length of the associated time interval where the rates are considered constant within each interval (exponential survival). When the disease or mortality rates are constant and small, then the probability of failure is approximately

$$\hat{Q} \approx \Sigma \hat{\lambda}_i (t_i - t_{i-1}) \tag{11.17}$$

and from Table 11.1, $\hat{Q}_{1940} \approx 10\Sigma\hat{\lambda}_i = 0.261$ and $\hat{Q}_{1960} \approx 10\Sigma\hat{\lambda}_i = 0.296$. Other methods exist to "adjust" rates so differing age distributions are not an issue in making comparisons. These methods involve a standard population and are discussed in the next chapter or in other texts ([4] or [5]). Multiplying a series of interval-specific survival probabilities to produce an overall summary survival probability is a useful technique employed in several contexts in the following discussion.

Estimates from Survival Data

Ideally a study of survival times starts with a defined number of individuals (*time* = $t = 0$) and these study participants are followed until each reaches the endpoint of the study (e.g., occurrence of disease or death). For example, a series of patients with severe cancer given a new treatment could be observed from the time of treatment until they all died, and then the total survival time could be computed. Such survival data are called complete because every participant has a completed survival time. In most situations collecting complete data is not practical, particularly in human populations. The amount of observation time necessary until the entire study cohort reaches the endpoint is usually too long for efficient study. Most data are analyzed after a period of time has passed, leaving a number of the participants with unknown survival times. In these situations, individuals who have not reached the endpoint (e.g., disease or death) would add more survival time if the study is continued. Survival data of this sort are not complete and are said to be censored. A censored observation means that a certain amount of survival time is recorded but the time the endpoint occurred is unknown because it did not occur during the study period. More technically, a measurement is censored when its exact failure time is unknown but it is known that the observation time exceeds a specific value. Two views of the same survival data with three censored failure times are illustrated in Figure 11.3.

If mortality data are collected and the mean survival time is calculated in the usual manner, the estimated mean is unbiased only when all the data are complete (no censored observations). The estimated mean survival time derived from complete data is the total amount of time lived divided by the number of study participants ($\bar{t} = \Sigma t_i / n$ where t_i is the observed survival time of each of n individuals). A bias from incomplete data is incurred because the total time Σt_i is too small due to the fact that censored individuals are still alive at the end of the study period. As mentioned, these individuals if observed for more time would contribute additional survival time.

Mean Survival Time: Parametric Estimation

An unbiased estimate of the mean survival time from incomplete data requires special techniques. One approach uses the exponential model (expression 11.3). If survival times are exponentially distributed, each censored observation is "missing," on the average, the same amount of time. Because past survival time has no influence on subsequent survival for exponentially distributed data, then each censored individual has the same expected additional time to the endpoint.

A survival study typically consists of two types of survival times represented by t_i where the endpoint is observed (complete) and represented by t_i^+ where the endpoint is not observed (censored). The "+" sign denotes censored observations. A censored observation can be made "complete" under the exponential model by adding to each censored time the expected amount of additional survival time to the time already observed: $t_i + \mu$ where μ is the expected mean survival time from the sampled population of exponentially distributed survival times. Using these "complete" data, the estimated mean survival time is

$$\hat{\mu} = \frac{total\ time}{number\ of\ individuals} = \frac{\sum\limits_{i=1}^{d} t_i + \sum\limits_{i=1}^{n-d} (t_i^+ + \hat{\mu})}{n}.$$

FIGURE **11.3**

Two views of censored survival data: Eight individuals observed for a maximum of 60 months with five complete and three censored observations

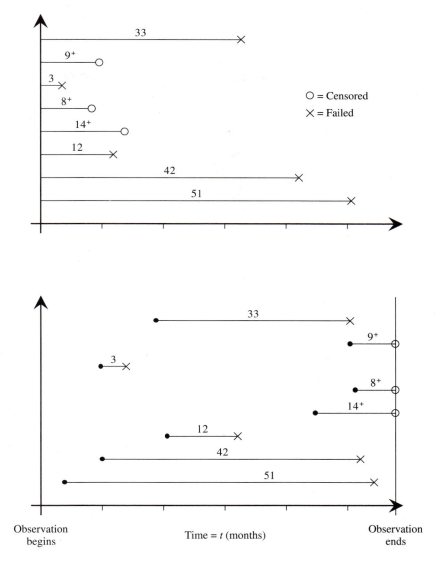

The value n represents the number of observations sampled (number of t_i plus t_i^+ observations) with d complete survival times (t_i) and $n - d$ incomplete observations (t_i^+). Elementary manipulations yield an estimate of the mean survival time μ,

$$\hat{\mu} = \frac{\sum_{i=1}^{n} t_i}{d} .$$

(11.18)

T A B L E   **11.3**

A hypothetical group of eight complete survival times (no censored values) in months

Observation	1	2	3	4	5	6	7	8
Time	3	8	9	12	14	33	42	51

The estimated mean $\hat{\mu}$ is unbiased (unaffected by censored survival times) as long as an exponential distribution describes the sample of survival times.

Again based on the exponential distribution, an estimated variance for the estimated mean survival time is given by

$$variance(\hat{\mu}) = S_{\hat{\mu}}^2 = \frac{\hat{\mu}^2}{d}.$$

(11.19)

Table 11.3 displays a small set of hypothetical complete survival times (in months) where the endpoint is death. For complete data, estimation of the mean survival time and its variance are made in the usual way. From Table 11.3, the estimated mean survival time is $\bar{t} = 172/8 = 21.5$ months and the estimated variance of $t = S_t^2 = 321.429$. The estimated standard deviation of $\bar{t}$ follows as $S_{\bar{t}} = \sqrt{321.429/8} = 6.339$.

If three of these eight values are incomplete (Table 11.4 and Figure 11.3; $n = 8$ and $d = 5$), then a direct calculation of the mean is biased. It is generally too small because the censored individuals would add more time to the numerator of the calculated mean if the study period was extended. The mean value treating the data as complete ($\bar{t} = 21.5$ months) is likely an understatement of the expected survival time. Making these data "complete" yields Table 11.5 when it is assumed that the data are a sample from a population of exponential survival times. Adding a constant expected amount of additional survival time to each censored observation yields an estimated mean of $\hat{u} = 172/5 = 34.4$ months (expression 11.18). The estimate is not

T A B L E   **11.4**

A hypothetical group of eight survival times (three censored values) in months

Observation	1	2	3	4	5	6	7	8
Time	3	8^+	9^+	12	14^+	33	42	51

T A B L E   **11.5**

A hypothetical group of eight survival times (three "completed" values) in months

Observation	1	2	3	4	5	6	7	8	Total
Time	3	$8 + \mu$	$9 + \mu$	12	$14 + \mu$	33	42	51	$172 + 3\mu$

biased by the presence of the three censored observations. The estimated standard deviation associated with the estimate $\hat{\mu} = 34.4$ is $S_{\hat{\mu}} = 15.384$ months.

An estimate of the average mortality rate (expression 11.15) calculated from the same hypothetical data (Table 11.4) is

$$average\ mortality\ rate = \hat{\lambda} = \frac{number\ of\ deaths}{total\ person-time-at-risk} = \frac{d}{\Sigma t_i} = \frac{5}{172} = 0.029 \ . \tag{11.20}$$

In addition, the estimated variance of this estimate is

$$S_{\hat{\lambda}}^2 = \frac{\hat{\lambda}^2}{d} = 0.000169 \ .$$

An exponential survival function is estimated in a straightforward way by using the estimate $\hat{\lambda}$ and

$$estimated\ survival\ probability = \hat{S}(t) = e^{-\hat{\lambda}t} \ . \tag{11.21}$$

Based on the eight observations given in Table 11.4, the estimated survival function is $\hat{S}(t) = e^{-0.029t}$.

The rate of failure is the reciprocal of the mean survival time ($\hat{\mu} = 1/\hat{\lambda}$) for exponential survival data. This relationship shows specifically that changes in survival time are inversely related to changes in risk, measured by a rate; this relationship holds regardless of the distribution of survival times.

Median Survival Time

An alternative summary of the survival experience of a sample of individuals is the median survival time. The median survival time is defined as the point t_m where the probability is 0.5 that a random individual from the sampled population will survive beyond t_m. In terms of mortality data, it is the point beyond which half the sample are expected to survive. If again the survival data are sampled from an exponential distribution, then for the median time t_m

$$S(t_m) = e^{-\lambda t_m} = 0.5 \quad \text{and the estimated median is} \quad \hat{t}_m = \frac{-\log(0.50)}{\hat{\lambda}} \ . \tag{11.22}$$

For the example data (Table 11.4), the estimated median is $\hat{t}_m = -\log(0.5)/\hat{\lambda} = 0.693/0.029 = 23.844$ months. For survival data from an exponential distribution, the estimated median is the estimated mean multiplied by a constant value (i.e., $\hat{t}_m = \hat{\mu}\log(2)$).

To illustrate, times between the diagnosis of HIV and the onset of AIDS are given in Table 11.6. The term "survival" means the time from diagnosis to the symptoms of AIDS. Using these data, the estimated mean "survival" time to AIDS is $\hat{\mu} = 2766/26 = 106.385$ months (expression 11.18) with an estimated standard error $S_{\hat{\mu}} = 20.864$ months (expression 11.19) based on assuming that the time to AIDS is adequately modeled by an exponential survival function. Using the estimated standard error, an approximate 95% confidence interval is $\hat{\mu} \pm 1.96 S_{\hat{\mu}} = 106.385 \pm$

T A B L E   **11.6**
Number of months between diagnosis of HIV infection and AIDS for 50 individuals

46	20	65	21	84	68	76^+	77^+	10^+	81^+	71^+	67	13
60	80^+	80^+	52	8	80^+	79^+	11	66	77^+	67	31	83^+
11	28	83^+	81^+	80^+	66^+	82^+	27	52	59	75^+	80^+	2
78^+	17^+	75^+	45	78^+	79^+	18	73^+	62	22	20		

1.96(20.864) or (65.5, 147.3). Similarly, the rate of HIV individuals converting to AIDS is estimated as $\hat{\lambda} = 1/106.385 = 0.0094$ with an estimated standard error of $S_{\hat{\lambda}} = 0.00184$. Of course, these estimates are useful only if the distribution of "survival" times has at least an approximate exponential distribution.

Product-Limit Estimation of a Survival Function

The estimate of a survival function is straightforward (expression 11.21) when the exponential distribution is a good description of the sampled population. However, it is certain that situations arise in which this survival function is not a reasonable statistical model. An alternative approach to estimating a survival function without a parametric model is the Kaplan-Meier product-limit estimate.

To understand the product-limit estimate, first consider a set of complete survival data (Table 11.3). The product-limit estimate results from classifying each observed survival time into a table based on the time of the endpoint. A series of categories is constructed so one failure occurs in each class (if there are identical survival times, then these values are placed in the same category). The complete data from Table 11.3 form eight classes ($n = d = 8$), one for each death, and yield Table 11.7. Each interval contains one death, producing the interval-specific estimated probability of death ($\hat{q}_k$) as one divided by the number of individuals who could have died in that interval. For example, in the fourth interval (9 to 12 months) one death occurred

T A B L E   **11.7**
Hypothetical survival data: No censoring

k	Interval	d_k	$\hat{q}_k$	$\hat{p}_k$	$\hat{P}_k = \prod \hat{p}_i$	$S_{\hat{P}_k}$
1	0–3	1	1/8	7/8	7/8	0.117
2	3–8	1	1/7	6/7	6/8	0.153
3	8–9	1	1/6	5/6	5/8	0.171
4	9–12	1	1/5	4/5	4/8	0.177
5	12–14	1	1/4	3/4	3/8	0.171
6	14–33	1	1/3	2/3	2/8	0.153
7	33–42	1	1/2	1/2	1/8	0.117
8	42–51	1	1/1	0/1	0/8	0.000

among five individuals, giving an estimated probability of death for that interval of $\hat{q}_4 = 1/5 = 0.200$. The estimated probability of surviving a specific interval for an individual who began the interval is $\hat{p}_i = 1 - \hat{q}_i$ (e.g., for interval $k = 4$, $\hat{p}_4 = 1 - 0.200 = 0.800$). The product-limit estimate of a survival probability is the product of these interval-specific probabilities, giving

$$P(surviving\ beyond\ the\ k^{th}\ interval) = \hat{P}_k = \hat{p}_1\,\hat{p}_2 \cdots \hat{p}_k = \prod_{i=1}^{k} \hat{p}_i\,.$$

(11.23)

For example, for the eight complete observations, the probability of surviving beyond 12 months (through the fourth interval) is

$$P(surviving\ beyond\ 12\ months) = \hat{P}_4 = \hat{p}_1\,\hat{p}_2\,\hat{p}_3\,\hat{p}_4$$
$$= (7/8)\,(6/7)\,(5/6)\,(4/5) = 4/8 = 1/2\,.$$

(11.24)

For complete data, the product-limit estimate is equal to the natural estimate given by the number of individuals who survive beyond the kth interval divided by the number of individuals sampled. Specifically, four individuals from the original eight survived beyond 12 months, so the probability of surviving beyond 12 months is estimated by $\hat{P}_4 = 4/8 = 1/2$. The product-limit estimate and this intuitive estimate are identical because

$$\hat{P}_k = \hat{p}_1\,\hat{p}_2 \cdots \hat{p}_k = \left(\frac{n-1}{n}\right)\left(\frac{n-2}{n-1}\right)\left(\frac{n-3}{n-2}\right) \cdots \left(\frac{n-k}{n-k+1}\right) = \frac{n-k}{n}$$

(11.25)

where $n - k$ is the number of individuals surviving beyond the end of the kth interval out of the n observed individuals. In addition, for complete data, the variance of the product-limit estimated survival probability is derived by noting that the number of deaths beyond a given time has an approximate binomial distribution. Therefore, the variance of $\hat{P}_k$ is estimated by

$$variance(\hat{P}_k) = S_{P_k}^2 = \frac{\hat{P}_k(1 - \hat{P}_k)}{n} \quad \text{(see appendix, Chapter 12)}\,.$$

(11.26)

For example, the estimated variance of $\hat{P}_4 = 1/2$ is $S_{\hat{P}_4}^2 = (1/2)\,(1/2)/8 = 0.0313$ and the estimated standard error follows as $S_{\hat{P}_4} = 0.177$.

The product-limit estimate from incomplete data is not much different. Again, a series of intervals is constructed based on the observed time of failure. Ideally, these intervals contain one failure. Each probability of failure is based on all individuals completing the interval or who died within the interval. Individuals with censored survival times within an interval are not used to calculate the interval-specific probability of failure (i.e., $\hat{q}_i$) and, furthermore, do not contribute to any remaining calculations. If no failure times are identical, then d intervals are created ($d \le n$), each with an associated estimated probability $\hat{q}_i$. The hypothetical data with three incomplete observations again serve to illustrate (Table 11.4) where $n = 8$ and $d = 5$. The estimated probabilities $\hat{q}_k$, $\hat{p}_k$, and $\hat{P}_k$ are shown in Table 11.8. For example, in the 3–12 months interval five individuals are at risk (not censored), making the estimated

T A B L E   **11.8**

Hypothetical survival data: Censoring

k	Interval	d_k	$\hat{q}_k$	$\hat{p}_k$	$\hat{P}_k = \prod \hat{p}_i$	$S_{\hat{p}_k}$
1	0–3	1	1/8	7/8	0.875	0.117
2	3–12	1	1/5	4/5	0.700	0.182
3	12–33	1	1/3	2/3	0.467	0.226
4	33–42	1	1/2	1/2	0.233	0.200
5	42–51	1	1/1	0/1	0.000	—

probability of failure $\hat{q}_2 = 1/5 = 0.2$. As before, an estimate of the probability of surviving the interval is $\hat{p}_2 = 1 - \hat{q}_2 = 0.8$ for the five individuals at risk. The probability of surviving beyond 12 months is estimated as $\hat{P}_2 = (7/8)(4/5) = 0.700$ ($k = 2$). In general, for both complete and censored data, the estimated survival probability is

$$\hat{P}_k = \prod \hat{p}_i \quad \text{(expression 11.23)}$$

where the probabilities $\hat{p}_i$ are based on interval-specific counts of deaths and persons at risk.

The estimate of the variance of the estimated survival probabilities with censoring is not as simple as the complete data situation. An estimate of the variance of $\hat{P}_k$ developed by an early contributor to biostatistics, Major Greenwood, is

$$variance(\hat{P}_k) = S_{\hat{P}_k}^2 = \hat{P}_k^2 \sum_{i=1}^{k} \frac{\hat{q}_i}{n_i \hat{p}_i}$$

where n_i individuals are at risk (not censored) for the ith interval (e.g., $n_2 = 5$). Applied to the estimate $\hat{P}_2 = 0.700$, Greenwood's variance formula gives an estimate of

$$S_{\hat{P}_2}^2 = (0.700)^2 \left[\left(\frac{1}{8}\right)\left(\frac{1}{7}\right) + \left(\frac{1}{5}\right)\left(\frac{1}{4}\right) \right] = 0.0333 \quad \text{and} \quad S_{\hat{P}_2} = 0.182 .$$

When no censoring ($n = d$) is present Greenwood's variance formula is identical to expression 11.26.

Using the estimates $\hat{P}_k$, a plot of the survival function (a step function) is readily constructed. Figure 11.4 shows the estimated survival function derived from the eight survival times in Table 11.8 (5 complete and 3 censored individuals). The $\hat{P}_k$-values come from the product-limit table (Table 11.8).

Median

An estimate of the median is easily extracted from a table of product-limit estimated survival probabilities. Using survival probabilities $\hat{P}_j$ and $\hat{P}_{j+1}$ where $\hat{P}_{j+1} < 0.50 < \hat{P}_j$, an estimate of the median, denoted $\tilde{t}_m$, is

$$\tilde{t}_m = t_{j+1} - \frac{(t_{j+1} - t_j)\,(0.5 - \hat{P}_{j+1})}{\hat{P}_j - \hat{P}_{j+1}} \; . \tag{11.27}$$

The estimate of the median is simply a linear interpolation of the values from a product-limit table of survival probabilities. Again, for the example data in Table 11.8, $\hat{P}_2 = 0.700$, $\hat{P}_3 = 0.467$, and $\tilde{t}_m = 33 - (33 - 12)(0.5 - 0.467)/(0.700 - 0.467) = 30.0$.

Mean Survival Time: Nonparametric Estimation

The estimation of the mean survival time from complete data, as noted, is routine. However, a more general approach exists and produces an estimate of the mean for both complete and incomplete data. The estimated mean can be expressed as

$$\hat{\mu} = \sum_{i=1}^{d} \hat{P}_{i-1}(t_i - t_{i-1}) \quad \text{with } \hat{P}_0 = 1.0 \text{ and } t_0 = 0 \tag{11.28}$$

where $\hat{P}_{i-1}$ is an estimated survival probability and t_{i-1} and t_i are the endpoints of the ith interval from each of a total of d intervals. To illustrate this approach to estimating

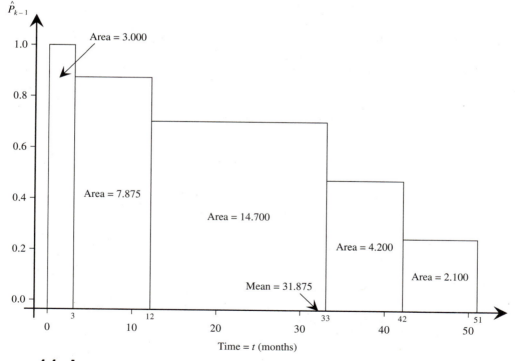

FIGURE **11.4**

A plot of the product-limit estimate of the survival function from the hypothetical data in Table 11.8

T A B L E **11.9**

Calculation of the mean from the hypothetical survival data: No censoring

k	Interval	$t_k - t_{k-1}$	$\hat{P}_{k-1}$	$\hat{P}_{k-1}(t_k - t_{k-1})$
1	0–3	3	8/8	24/8 = 3.000
2	3–8	5	7/8	35/8 = 4.375
3	8–9	1	6/8	6/8 = 0.750
4	9–12	3	5/8	15/8 = 1.875
5	12–14	2	4/8	8/8 = 1.000
6	14–33	19	3/8	57/8 = 7.125
7	33–42	9	2/8	18/8 = 2.250
8	42–51	9	1/8	9/8 = 1.125
Total	—	—	—	172/8 = 21.500

mean survival time, the complete data ($n = d = 8$) from Table 11.3 are used and Table 11.9 is constructed, again based on each observed time of death. The result is the same as the previous calculation. For complete data, $\bar{t}$ is identical to $\hat{\mu}$ (expression 11.28 when $n = d$).

When censored data are included, the process is the same but accounts for the censored observations because estimates of the survival probabilities, $\hat{P}_k$, are available from the product-limit table unbiased by incomplete data. To illustrate, the censored data in Table 11.4 are once again used and produce Table 11.10. The estimate $\hat{\mu} = 31.875$ is a straightforward application of expression 11.28 ($d = 5$) to the elements of the product-limit Table 11.8.

The estimate of the variance of the estimate $\hat{\mu}$ from data containing incomplete observations is a bit involved. The expression for the estimate of the variance associated with $\hat{\mu}$ is

$$S_{\hat{\mu}}^2 = \sum_{i=1}^{d-1} \frac{\hat{q}_i}{n_i \hat{p}_i} A_i^2 \quad \text{where} \quad A_i = \hat{\mu} - \sum_{j=1}^{i} \hat{P}_{j-1}(t_j - t_{j-1}).$$

(11.29)

T A B L E **11.10**

The calculation of the mean from the hypothetical survival data: Censoring (Table 11.4)

k	Interval	$t_k - t_{k-1}$	$\hat{P}_{k-1}$	$\hat{P}_{k-1}(t_k - t_{k-1})$
1	0–3	3	1.000	3.000
2	3–12	9	0.875	7.875
3	12–33	21	0.700	14.700
4	33–42	9	0.467	4.200
5	42–51	9	0.233	2.100
Total	—	—	—	31.875

T A B L E **11.11**

The calculation of the variance from the hypothetical survival data: Censoring (Table 11.4)

k	Interval	$\hat{q}_k$	$\hat{p}_k$	n_k	$\hat{q}_k/(n_k\hat{p}_k)$	A_k	$[\hat{q}_k/(n_k\hat{p}_k)]A_k^2$
1	0–3	0.125	0.875	8	0.0179	28.875	14.889
2	3–12	0.200	0.800	5	0.0500	21.000	22.050
3	12–33	0.333	0.667	3	0.1667	6.300	6.615
4	33–42	0.500	0.500	2	0.5000	2.001	2.199
5	42–51	1.000	0.000	1	—	—	—
Total	—	—	—	—	—	—	45.752

Continuing the previous example, the process of estimating the variance associated with $\hat{\mu} = 31.875$ is displayed in Table 11.11. The estimated variance of $\hat{\mu}$ is $S_{\hat{\mu}}^2 = 45.752$, making the estimated standard error $S_{\hat{\mu}} = 6.765$.

To summarize, the data from Table 11.4 produce

	Mean (months)	Standard Error	Median (months)
Exponential (parametric)	34.400	15.384	23.844
Product-limit (nonparametric)	31.875	6.765	30.000

Because the estimated mean value from complete data $\hat{\mu}$ (expression 11.28) is identical to the usual mean value $\bar{t}$ (expression 1.1), the estimated variance of $\hat{\mu}$ (expression 11.29) should be equal to $S_{\bar{t}}^2$. However, for complete data, $S_{\hat{\mu}}^2 = [(n-1)/n]S_{\bar{t}}^2$, showing a slight bias that has no important impact, particularly for moderate or large sample sizes.

When the longest survival period ends with a censored observation, the estimated mean value (expression 11.28) is biased. Because each interval included in the product-limit estimate must contain at least one failure, the last interval is undefined and the longest survival time does not fully contribute to the calculation. One conventional practice is to consider the longest survival time as always ending in a failure; a lesser bias is incurred when this artificial endpoint is necessary.

Geometrically, the estimated mean survival time $\hat{\mu}$ has a simple interpretation. The estimated mean is the total area enclosed by the product-limit survival function. As noted earlier, the mean survival time is equal to the area under the survival function so, not surprisingly, the area under the estimated survival curve is an estimate of the mean survival time. The product-limit estimate produces a series of rectangles (e.g., Figure 11.4) with height $\hat{P}_{i-1}$ and width $t_i - t_{i-1}$. The area of each rectangle is $\hat{P}_{i-1}(t_i - t_{i-1})$ and the estimated mean is the sum of the areas of these rectangles (expression 11.28). For example, Table 11.10 and Figure 11.4 show

$$area = (1.00)\,(3) + (0.875)\,(9) + (0.700)\,(21) + (0.467)\,(9) + (0.233)\,(9) = 31.875 \,.$$

T A B L E **11.12**

The "survival" probabilities for conversion to AIDS after a diagnosis of HIV infection (from the 50 observations given in Table 11.6)

k	Interval	d_k	$\hat{P}_k = \prod \hat{p}_i$	$S_{\hat{P}_k}$
1	0–2	1	0.980	0.020
2	2–8	1	0.960	0.028
3	8–11	2	0.919	0.039
4	11–13	1	0.899	0.043
5	13–18	1	0.878	0.047
6	18–20	2	0.836	0.053
7	20–21	1	0.815	0.056
8	21–22	1	0.794	0.058
9	22–27	1	0.773	0.060
10	27–28	1	0.752	0.062
11	28–31	1	0.732	0.064
12	31–45	1	0.711	0.065
13	45–46	1	0.690	0.067
14	46–52	2	0.648	0.069
15	52–59	1	0.627	0.070
16	59–60	1	0.606	0.070
17	60–62	1	0.585	0.071
18	62–65	1	0.564	0.072
19	65–66	1	0.543	0.072
20	66–67	2	0.500	0.072
21	67–68	1	0.478	0.072
22	68–84	1	0.000	0.000

Whether the survival function is estimated by a continuous function $\hat{S}(t)$ or estimated by a discrete "step function" $\hat{P}_k$, the area under the curve is an estimate of the mean survival time.

The time-to-AIDS data (Table 11.6) provide a realistic example of the product-limit estimation of a survival function. Table 11.12 shows the estimated survival function based on the data from the 50 HIV-infected individuals (Table 11.6). The plot of the HIV "survival" function (product-limit estimate) in Table 11.12 is displayed in Figure 11.5.

The HIV data produce an estimate of the mean time to AIDS by applying expression 11.28. The mean time from HIV diagnosis to AIDS is $\hat{\mu} = 59.983$ months and the estimated standard error associated with this estimate is $S_{\hat{\mu}} = 4.025$ months (expression 11.29). An approximate 95% confidence interval is $59.983 \pm 1.96(4.025)$ or (52.1, 67.9).

A Transformed Survival Function

The Weibull distribution, defined by

$$S(t) = e^{-(\lambda t)^a} , \tag{11.30}$$

is another distribution used to describe survival data. The exponential distribution is a special case of this distribution; when $a = 1$, then $S(t) = e^{-\lambda t}$. A natural transformation is suggested by the Weibull distribution because

$$\log(-\log[S(t)]) = a\log(\lambda) + a\log(t) . \tag{11.31}$$

This log-log transformation causes survival times from a Weibull distribution to have a linear relationship with the logarithm of time, $\log(t)$ (i.e., *intercept* $= a \log(\lambda)$ and *slope* $= a$). The transformation is useful for plotting survival data because data that fall approximately on a straight line can then be accurately represented by a Weibull distribution. Specifically, if the plot of a transformed survival function is approximately linear with a slope of 1, then the data can be accurately described by an exponential distribution ($a = 1$; dotted line in Figure 11.6). Figure 11.6 illustrates the

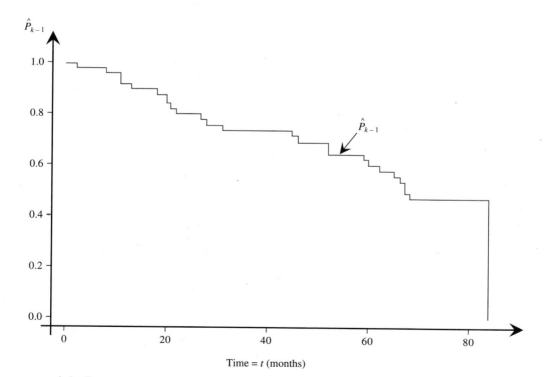

FIGURE **11.5**
The probability of AIDS after diagnosis of HIV infection

FIGURE **11.6**
Three Weibull distributions ($a = 1.5$, $a = 1.0$, and $a = 0.5$) and their log-log transformations

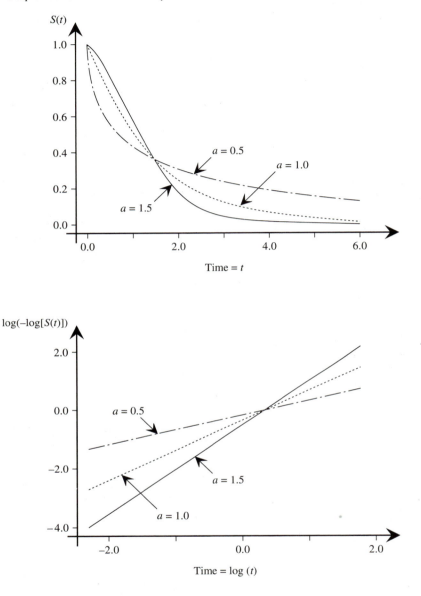

Weibull distribution for $a = 1.5$, $a = 1.0$, and $a = 0.5$ (top) and the linearizing transformation for these three distributions (bottom).

As noted, when $\log(-\log[S(t)]) = \log(\lambda) + \log(t)$, the data have an exponential distribution. Figure 11.7 displays a plot of the log-log transformed survival probabilities (Table 11.12) for the 50 HIV individuals against the $\log(t)$, showing a slope not equal to one but a fairly linear pattern. Figure 11.7 indicates that parametric analysis

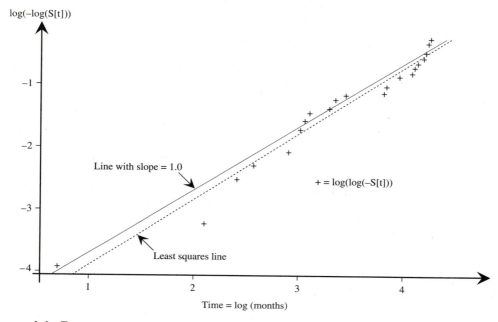

$\log(-\log(S[t]))$

-1

-2

Line with slope = 1.0

$+ = \log(\log(-S[t]))$

-3

Least squares line

-4

1   2   3   4

Time = log (months)

F I G U R E **11.7**

A log-log plot of $S(t)$ against the $\log(t)$ for the 50 HIV-positive individuals

based on an exponential distribution is a reasonable choice to represent these data, but perhaps another Weibull distribution would be a better description. A log-log plot is one way to begin to evaluate the goodness-of-fit of the model selected to represent survival data among a number of other possible choices [6].

Comparison of Survival Times from Two Samples

Survival data are often collected to make specific comparisons. For example, a new treatment might be compared to an old treatment, individuals from one category might be compared to individuals from another, or an experimental group might be compared to a control group. Such comparisons can be made by computing the mean survival times for each group and assessing the difference with a statistical test or by constructing confidence intervals. A better approach involves classifying data into a series of 2-by-2 tables. Similar to the product-limit estimate, one table is generated for each observed failure time. Survival data are then analyzed as a series of 2-by-2 tables, such as Table 11.13. Table 11.13 reflects the status of two groups compared at the failure time T_i. These tables are used to assess the association between group status and survival status. Essentially the same approach to assessing an association between two binary variables described in Chapter 8 is applied to each of a series of 2-by-2 tables.

T A B L E **11.13**

Survival status of n_i individuals at time $= T_i$

T_i = Time	Failed	Not Yet Failed	Total
Group 1	a_i	b_i	$a_i + b_i$
Group 2	c_i	d_i	$c_i + d_i$
Total	$a_i + c_i$	$b_i + d_i$	n_i

Consider the comparison of hypothetical mortality data between two groups given in Table 11.14. These data are displayed (Table 11.15) as 11 individual 2-by-2 tables (one for each complete observation) where the rows are counts of individuals in groups 1 and 2 and the columns are counts of those who failed and those who have not yet failed at time T_i. As with the product-limit estimate, individuals with incomplete survival times are included only in tables where they survive the complete interval.

The Mantel-Haenszel Chi-Square Statistic

Estimated values are generated under the hypothesis that row classification (group membership) is independent of column classification (survival status). The estimates from each table are identical to expression 8.12. The estimated number of individuals who belong to group 1 and who failed at time t_i is

$$\hat{A}_i = \frac{(a_i + b_i)(a_i + c_i)}{n_i} \tag{11.32}$$

where $n_i = a_i + b_i + c_i + d_i$ (notation defined in Table 11.13) when group status and survival status are unrelated. The estimated variance of a_i is given by (expression 8.13, repeated with different notation)

$$variance(a_i) = \frac{(a_i + b_i)(a_i + c_i)(c_i + d_i)(b_i + d_i)}{n_i^2(n_i - 1)} . \tag{11.33}$$

T A B L E **11.14**

Two hypothetical groups each with eight survival times (months)

Observation	1	2	3	4	5	6	7	8
Group 1	3	8^+	9^+	12	14^+	33	42	51
Group 2	7	10^+	24	28^+	45	54	62	64

T A B L E **11.15**

A display of the survival status at the 11 times of death in a series of 2-by-2 tables

$T_1 = 3$	Died	Alive	Total
Group 1	1	7	8
Group 2	0	8	8
Total	1	15	16

$T_2 = 7$	Died	Alive	Total
Group 1	0	7	7
Group 2	1	7	8
Total	1	14	15

$T_3 = 12$	Died	Alive	Total
Group 1	1	4	5
Group 2	0	6	6
Total	1	10	11

$T_4 = 24$	Died	Alive	Total
Group 1	0	3	3
Group 2	1	5	6
Total	1	8	9

$T_5 = 33$	Died	Alive	Total
Group 1	1	2	3
Group 2	0	4	4
Total	1	6	7

$T_6 = 42$	Died	Alive	Total
Group 1	1	1	2
Group 2	0	4	4
Total	1	5	6

$T_7 = 45$	Died	Alive	Total
Group 1	0	1	1
Group 2	1	3	4
Total	1	4	5

$T_8 = 51$	Died	Alive	Total
Group 1	1	0	1
Group 2	0	3	3
Total	1	3	4

$T_9 = 54$	Died	Alive	Total
Group 1	0	0	0
Group 2	1	2	3
Total	1	2	3

$T_{10} = 62$	Died	Alive	Total
Group 1	0	0	0
Group 2	1	1	2
Total	1	1	2

$T_{11} = 64$	Died	Alive	Total
Group 1	0	0	0
Group 2	1	0	1
Total	1	0	1

T A B L E **11.16**

Summary of the hypothetical data displayed in Table 11.15 (11 2-by-2 tables)

t_i	a_i	b_i	c_i	d_i	n_i	$\hat{A}_i$ (expected)	Variance(a_i)
$T_1 = 3$	1	7	0	8	16	0.500	0.250
$T_2 = 7$	0	7	1	7	15	0.467	0.249
$T_3 = 12$	1	4	0	6	11	0.455	0.248
$T_4 = 24$	0	3	1	5	9	0.333	0.222
$T_5 = 33$	1	2	0	4	7	0.429	0.245
$T_6 = 42$	1	1	0	4	6	0.333	0.222
$T_7 = 45$	0	1	1	3	5	0.200	0.160
$T_8 = 51$	1	0	0	3	4	0.250	0.188
$T_9 = 54$	0	0	1	2	3	0.000	—
$T_{10} = 62$	0	0	1	1	2	0.000	—
$T_{11} = 64$	0	0	1	0	1	0.000	—
Total	5	25	6	43	—	2.966	1.784

A test statistic to compare the observed a_i-values with the hypothesis-generated $\hat{A}_i$-values is

$$X^2_{MH} = \frac{\left[\displaystyle\sum_{i=1}^{d} a_i - \sum_{i=1}^{d} \hat{A}_i \right]^2}{\displaystyle\sum_{i=1}^{d} variance(a_i)}$$

(11.34)

where again d is the number of intervals (i.e., d = the number of unique complete survival times = number of 2-by-2 tables). The test statistic X^2_{MH} has an approximate chi-square distribution with one degree of freedom when row status (group) is independent of column status (survival). This test statistic is often called the Mantel-Haenszel chi-square statistic [6]. Continuing the example, the data in Table 11.15 are more compactly displayed in Table 11.16 along with expected values $\hat{A}_i$ and the estimated variances of a_i for each table. The total number of deaths in group 1 is estimated to be 2.966 when group status is unrelated to failure time. The corresponding observed value is five. The Mantel-Haenszel chi-square statistic is

$$X^2_{MH} = \frac{[5 - 2.966]^2}{1.784} = 2.318 \, ,$$

(11.35)

producing a p-value of 0.128. Sometimes the Mantel-Haenszel chi-square statistic is calculated with a correction factor given by

$$X_{MH_c}^2 = \frac{\left[\left|\sum_{i=1}^{d} a_i - \sum_{i=1}^{d} \hat{A}_i\right| - 0.5\right]^2}{\sum_{i=1}^{d} variance(a_i)}.$$

(11.36)

Using the corrected chi-square statistic, the hypothetical data produce $X_{MH_c}^2 = 1.319$ also with one degree of freedom, yielding a *p*-value of 0.251.

The Log-Rank Test

Another approach to testing the difference between expected and observed numbers of failures in a series of 2-by-2 tables involves another version of the chi-square statistic. The numbers of failures are estimated for both groups 1 and 2 based on the conjecture that group status is unrelated to survival. Specifically, the expected values for each group are, respectively, $\hat{D}_1 = \Sigma \hat{A}_i$ and $\hat{D}_2 = \Sigma \hat{C}_i = \Sigma(1 - \hat{A}_i) = D - \hat{D}_1$ where D is the total number of deaths (2-by-2 tables). These two estimated values are compared to the observed values $d_1 = \Sigma a_i$ and $d_2 = \Sigma c_i = d - d_1$ by

$$X^2 = \frac{[d_1 - \hat{D}_1]^2}{\hat{D}_1} + \frac{[d_2 - \hat{D}_2]^2}{\hat{D}_2}$$

where X^2 has an approximate chi-square distribution with one degree of freedom when group status and survival time are unrelated. This chi-square statistic is a version of a procedure called the log-rank test. For the illustrative data, $X^2 = (5 - 2.966)^2/2.966 + (6 - 8.034)^2/8.034 = 1.910$ and the associated *p*-value is 0.167.

The Mantel-Haenszel and the log-rank chi-square approaches usually yield similar results, particularly when the sample sizes are large. The log-rank approach has the advantage that it generalizes to allow comparison of survival data from more than two groups. The expected values are similarly estimated from each group under the hypothesis that group status and survival status are independent and then compared to the observed values by means of a chi-square statistic summed over all groups (further described in [1] or [6]).

A data set from a study of risk factors and coronary heart disease [6] provides an application of the comparison of survival times from two groups. The data (presented in Table 11.17) are a sample of white males with high levels of cholesterol (greater than 340mg/100ml) and are used in a number of contexts in the rest of the chapter [7].

For the risk factor data set, "survival" means free from a coronary event from the time of entry into the study until a coronary event or the end of the study period (no individuals were lost). The "survival" times are given in days and interest is focused on comparing type A individuals (A = 1) to type B individuals (B = 0) for differences in time to a coronary event. The observed number of coronary events among type A

T A B L E **11.17**

Data on five risk factors related to survival of 35 individuals with high cholesterol

	Weight	Sys. BP[*]	Chol.	Cigs.[**]	A = 1/B = 0	Censored[***]	Time (days)
1	156	146	349	35	1	1	1257
2	196	120	414	35	1	1	2059
3	156	114	343	0	0	0	2987
4	208	152	349	15	1	0	773
5	175	128	390	25	1	0	2959
6	205	124	346	40	1	0	2943
7	205	130	354	0	1	0	3044
8	165	120	386	30	1	1	2613
9	147	104	353	30	0	0	2864
10	225	144	344	0	0	0	2875
11	159	106	348	0	1	0	2840
12	198	140	364	30	0	0	2839
13	120	104	369	0	1	0	2862
14	152	110	390	0	0	0	2857
15	154	140	349	30	0	1	1076
16	180	146	358	0	1	0	3000
17	166	124	346	30	1	0	3009
18	155	120	383	20	0	0	1845
19	210	154	645	0	1	1	2296
20	186	124	353	30	1	1	3122
21	145	146	354	20	0	0	3141
22	160	130	400	20	0	0	3053
23	148	124	344	15	0	0	2926
24	170	130	346	25	0	0	2927
25	172	148	353	0	0	0	2925
26	157	124	365	25	0	0	2930
27	170	142	349	30	0	0	2890
28	162	144	385	20	1	0	2940
29	195	140	348	20	1	0	2424
30	160	124	351	0	1	0	2042
31	135	120	376	20	1	1	1825
32	150	116	357	20	0	0	3048
33	180	140	348	20	1	0	2881
34	175	146	368	0	0	0	1889
35	156	116	348	15	0	1	2361

A = aggressive individuals with perceived stressful lives and B = more relaxed and less aggressive individuals
* = systolic blood pressure
** = cigarettes smoked per day
*** 1 = coronary event, 0 = no coronary event during study period

individuals is 6 and assuming behavior type is independent of a coronary event, the estimated number of heart attacks among these $n = 35$ individuals is $\Sigma \hat{A}_i = 3.917$ with $\Sigma variance(a_i) = 1.996$. The Mantel-Haenszel chi-square statistic is $X^2_{MH} = (6 - 3.917)^2 / 1.996 = 2.174$ (p-value = 0.140) or the corrected chi-square statistic is $X^2_{MH_c} = 1.255$ (p-value = 0.262). Alternatively, the number of coronary events among type A individuals is $d_A = 6$ with an estimated number $\hat{D}_A = 3.917$, and the count of coronary events among type B individuals is $d_B = 2$ with an estimated number $\hat{D}_B = 4.083$. As before, the values $\hat{D}_A$ and $\hat{D}_B$ are estimated under the hypothesis that behavior type and a coronary event are unrelated. When these two variables (behavior and coronary disease) are independent, the value $X^2 = (6 - 3.917)^2 / 3.917 + (2 - 4.083)^2 / 4.083 = 2.170$ has an approximate chi-square distribution with one degree of freedom, giving a p-value of 0.141. Behavior type appears unrelated or, perhaps, weakly related to the risk of a coronary event among individuals with extremely high cholesterol levels (see STATA output at the end of the chapter).

Hazard Function

An average mortality rate is usually expressed as the number of deaths per person-years. This quantity represents an average over a specific period. The period could be years, months, weeks, or days. The hazard rate is a specialized rate where the period is instantaneous. Clearly, this is a theoretical quantity that cannot be calculated exactly from data but only approximated. The formal definition of a hazard rate at a specific time t is

$$hazard\ rate = \lambda(t) = -\frac{dS(t)/dt}{S(t)} \tag{11.37}$$

where $S(t)$ is a survival function. Geometrically, a hazard rate measures the rate of change in a survival function at time t relative to the probability of surviving beyond time t. Figure 11.8 shows a typical hazard function derived from the 1980 mortality data for residents of California and corresponds to the survival function in Figure 11.1 (bottom). Roughly, a hazard rate at a specific point in time t equals the probability of failure during a small interval of time about the point t. For example, the probability that an individual 60 years old dies before the age of 61 is approximately equal to the hazard function evaluated at age 60.5, or $\lambda(60.5)$. Sometimes called the force of mortality, a hazard rate is a sensitive reflection of the risk of failure at a given time and is a major element in the study of disease and mortality.

For an exponential survival function the hazard rate is constant, or $\lambda(t) = \lambda$. If the average rate does not depend on time (expression 11.15), then the hazard rate also will not involve time. Direct application of the definition of a hazard rate (expression 11.37) to the exponential survival function, $S(t) = e^{-\lambda t}$, verifies that the hazard rate is constant.

A hazard rate can be approximated by an average rate, particularly when the time interval considered is short. The hazard rate at the middle of a time interval, denoted t_c, is estimated by $\hat{\lambda}(t_c) \approx r_t$ where r_t is the estimated average rate discussed earlier

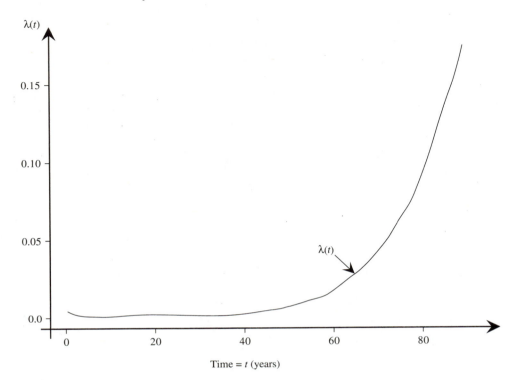

FIGURE 11.8
Hazard function of California, 1980

(expression 11.7). Therefore, an estimate of the variance of an estimated hazard rate $\hat{\lambda}(t_c)$ is given by expression 11.10. A hazard rate is related to a survival function (demonstrated in Box 11.1). Small hazard rates are associated with large survival probabilities and conversely. An important case arises when two hazard rates are proportional. That is, if two groups are compared with hazard rate $\lambda_1(t)$ and $\lambda_0(t)$, where

$$\lambda_1(t) = \lambda_0(t)c, \quad \text{then} \quad S_1(t) = [S_0(t)]^c. \tag{11.38}$$

When two hazard functions are proportional, the constant of proportionality c can be estimated (more detail follows in the next section). If the hazard function associated with the risk of a coronary event among type B individuals is proportional to the hazard function associated with type A individuals, then an estimate of the ratio of these two hazard functions is $\hat{c} = 3.133$, estimated from the data in Table 11.17 (columns 5 and 7). Specifically, $\lambda_A(t) = 3.133\,\lambda_B(t)$, or the relative hazard ratio $\lambda_A(t)/\lambda_B(t)$ is estimated by 3.133. If the hazard rate for each behavior type is constant with respect to time, then $\hat{\lambda}_A = 0.00013$ and $\hat{\lambda}_B = 0.000044$ (expression

11.20), giving another estimate of the relative hazard ratio of $\hat{\lambda}_A/\hat{\lambda}_B = 3.036$. These two estimates of the relative hazard ratio (3.133 and 3.036) are usually similar when the hazard functions do not depend on time (exponential survival). It is remarkable, however, that the hazard ratio c can be estimated without assuming that the hazard functions are constant. Estimates of c are possible when the hazard functions are proportional and, furthermore, it is not necessary to specify the form of $\lambda(t)$.

A formal evaluation of an estimate hazard ratio $\hat{c}$ is usually in terms of the $\log(\hat{c})$ to produce a test statistic with an approximate normal distribution (similar to using $\log(\hat{OR})$ instead of $\hat{OR}$; expression 10.7). The same computer algorithm used to estimate the constant c provides an estimate of $\log(c)$ and its variance. For the behavior type data, the estimated variance of $\log(\hat{c})$ is $S_{\log(\hat{c})}^2 = 0.667$. To test whether the estimated value $\hat{c} = 3.133$ is likely to have arisen by chance when the compared proportional hazards functions are identical (hazard ratio $= c = 1$ or $\log(c) = 0$), the test statistic

$$z = \frac{\log(\hat{c}) - \log(c)}{S_{\log(\hat{c})}} \tag{11.39}$$

has an approximate standard normal distribution when $c = 1$ (no difference in hazard rates) and, specifically, $z = 1.142/0.817 = 1.398$, yielding a p-value of 0.162. An approximate α-level confidence interval for c is

$$lower\ bound = \hat{c}e^{-z_{1-\alpha/2}S_{\log(\hat{c})}} \quad and \quad upper\ bound = \hat{c}e^{+z_{1-\alpha/2}S_{\log(\hat{c})}} . \tag{11.40}$$

Based on the estimate $\hat{c} = 3.133$, the 95% confidence interval is (0.632, 15.529). The estimation and testing of c involve a special case of the proportional hazards model that is the topic of the next section.

B O X **11.1**   **THE RELATIONSHIP BETWEEN A HAZARD FUNCTION AND A SURVIVAL PROBABILITY: EXPONENTIAL CASE**

For two constant hazard functions,

$$\lambda_1 = \lambda_0 c , \quad then \quad \lambda_1 t = \lambda_0 tc$$

gives

$$e^{-\lambda_1 t} = e^{-\lambda_0 t c} , \quad then \quad e^{-\lambda_1 t} = \left[e^{-\lambda_0 t}\right]^c$$

and

$$S_1(t) = \left[S_0(t)\right]^c .$$

Proportional Hazard Analysis

The proportional hazards model, sometimes called the Cox model after D. R. Cox, who originated the approach, allows multivariable comparisons of hazards functions between two groups or individuals. Parallel to linear and logistic regression analyses, the proportional hazards model also makes it possible to investigate the impact of each independent variable on differences between two hazard functions, adjusted for the confounding influence of the other independent variables. For example, the hazard function associated with individuals with type A behavior can be compared to a hazard function associated with type B individuals while assessing the separate influences of such variables as smoking exposure and cholesterol levels. The principal tool used to make these comparisons, as before, is a set of estimated coefficients.

A general expression for a proportional hazards model is

$$\lambda_j(t \mid x_{1j}, x_{2j}, x_{3j}, \cdots, x_{kj}) = \lambda_0(t)\, c = \lambda_0(t) e^{\sum_{i=1}^{k} b_i x_{ij}} \tag{11.41}$$

where $\lambda_j(t)$ is a hazard function and the k values represented by x_{ij} are independent variables or covariates (i^{th} variable associated with the jth hazard function). The model specifies that a specific hazard function for a group or an individual is equal to a baseline hazard function $\lambda_0(t)$ multiplied by a factor that incorporates the influence of the k independent variables. The proportional hazards model is semiparametric. It is nonparametric in the sense that the functions $\lambda_j(t)$ and $\lambda_0(t)$ do not need to be specified to evaluate the covariates associated with differences in survival times, but it is parametric in the sense that the parameters b_i reflect the role of these variables.

When two hazard functions are proportional the corresponding survival functions have a specific relationship, or, as mentioned,

$$S_j(t) = \left[S_0(t) \right]^c = \left[S_0(t) \right]^{e^{\sum_{i=1}^{k} b_i x_{ij}}} \tag{11.42}$$

where $c = \Sigma_i b_i x_{ij}$. This relationship between survival functions assures two important properties. If the hazards functions are proportional, the survival functions do not cross. Also the difference between log-log transformations of two survival functions associated with two proportional hazard functions is constant, or

$$\log(-\log[S_1(t)]) - \log(-\log[S_2(t)]) = constant \tag{11.43}$$

A "log-log" transformation is useful when the question of whether two hazard functions are proportional is addressed by graphic techniques [6] because proportional hazards will produce transformed survival functions that form parallel lines.

Proportional hazards models behave like most multivariable regression models. The coefficient b_i quantifies the change in the ratio of two hazard functions from a 1-unit increase in value of the ith variable while the other $k-1$ variables are held constant. Specifically, when two proportional hazard functions are compared, then

$$\lambda_1(t) = \lambda_2(t) e^{\Sigma b_i (x_{i1} - x_{i2})} . \tag{11.44}$$

If all covariates are equal $x_{i1} = x_{i2}$, except for variable 3 and x_{31} differs from x_{32} by one unit, then

$$\lambda_1(t) = \lambda_2(t)\, e^{\hat{b}_3},$$

or the estimated relative hazard ratio is

$$\frac{\lambda_1(t)}{\lambda_2(t)} = \hat{c}_3 = e^{\hat{b}_3}.$$

Furthermore, when two hazard functions are proportional (expression 11.44) the constant of proportionality c factors into a series of relative hazard ratios, each reflecting the separate role of an independent variable. In symbols, the relative hazard is

$$\frac{\lambda_1(t)}{\lambda_2(t)} = \hat{c} = \hat{c}_1\hat{c}_2\hat{c}_3 \cdots \hat{c}_k \quad \text{where} \quad \hat{c}_i = e^{\hat{b}_i(x_{i1}-x_{i2})}.$$

(11.45)

Both properties are analogous to properties described in the context of additive logistic regression model (expressions 10.36 and 10.37).

Although it is not obvious from the description of the proportional hazards model, the estimation of the coefficients b_i is unaffected by censored observations [3]. The estimation process uses the information on survival time from the censored data but in an unbiased fashion. Like the estimation of the product-limit survival curve and the estimation of the mean survival time, censored data are used where they contribute information and are eliminated from consideration when they are no longer relevant.

Typically survival data contain observations that failed from a variety of causes. For example, in a study of factors influencing lung cancer survival, patients may die from other diseases. Competing causes of failure complicate certain types of analytic approaches. However, when a proportional hazards model is used, observations failing from competing causes are treated as censored and create no special problems.

Comparing survival times between two groups with proportional hazard functions (one binary independent variable indicating group membership) is a special case of the general model and takes the same form as expression 11.39 where $c = e^{bx}$ where x is either 0 or 1, indicating group membership. Estimates for this simplest proportional hazards model from the coronary heart disease data (Table 11.17) produce the values given in Table 11.18 (complete STATA output is included at the end of the chapter). The estimated coefficients measure the relative influence of type A ($x = 1$) and type B ($x = 0$) behavior on the risk of a coronary event (i.e., $\lambda_A(t) = 3.133\lambda_B(t)$ where $\hat{c} = e^{\hat{b}} = e^{1.142} = 3.133$).

The comparison of two groups with a Mantel-Haenszel chi-square statistic generally produces similar results (expression 10.34) to a proportional hazards model with a single binary independent variable. The two approaches to evaluating survival data are related and for large sample sizes produce essentially the same p-values. For the heart disease data (Table 11.17), a chi-square measure of association between

T A B L E **11.18**

Analysis of the impact of behavior type on "survival" using a proportional hazards model

Variable	Coefficient	Estimate	Standard Error	p-Value	Hazard Ratio
A/B	b	1.142	0.817	0.162	3.133

$$-2\log(L) = 46.225$$

behavior type and time to a coronary event produces $X_{MH}^2 = 2.174$ (p-value = 0.140), which is not very different from $z^2 = (1.142/0.817)^2 = (1.398)^2 = 1.954$ (p-value = 0.162; Table 11.18). The Mantel-Haenszel procedure employs a series of 2-by-2 tables to estimate a measure of association at each time of failure. These stratum-specific measures are combined to summarize the association between group membership and survival, unaffected by censored data. The estimation of the parameters of a proportional hazards model also uses estimates at each time of failure and combines information from each stratum to calculate the overall constant of proportionality, $\hat{c}$. The estimate of c is unaffected by the form of the hazard functions as long as the functions are proportional over the range being considered. Furthermore, as with the Mantel-Haenszel chi-square analysis, the estimates are not biased by incomplete data, which is a principal feature of a proportional hazards model approach. The proportional hazards model can be applied to investigate the influence of a number of independent variables, and the transformed likelihood value $-2\log(L)$ serves, as before, as a relative measure of goodness-of-fit. Parallel to the logistic regression model, these likelihood values are employed to evaluate differences between nested multivariable models.

A Bivariate Example

A bivariate form of the proportional hazards model further illustrates an analysis of the time to a coronary event comparing type A and type B individuals (data in Table 11.17) while accounting for differing smoking patterns. The model requires the hazard functions associated with behavior types to be proportional and, specifically,

$$\lambda_j(t) = \lambda_0(t)e^{(b_1 x_{1j} + b_2 x_{2j})} \tag{11.46}$$

where $x_{1j} = 0$ for type B individuals and $x_{1j} = 1$ for type A individuals while $x_{2j} = 0$ for nonsmokers and $x_{2j} = 1$ for smokers. The estimates of the coefficients and the relative hazard ratios associated with variables x_{1j} and x_{2j} are in Table 11.19. The estimated proportionality constant is

$$\hat{c} = e^{1.191x_1 + 1.151x_2} = (3.288)^{x_1}(3.160)^{x_2},$$

which reflects the relative influence of the four combinations of behavior type and smoking exposure variables (Table 11.20) on the ratio of hazard functions. Both

T A B L E **11.19**

Analysis of behavior type and smoking using a proportional hazards model

Variable	Coefficient	Estimate	Standard Error	p-Value	Hazard Ratio
A/B	b_1	1.190	0.818	0.155	3.288
Smoking	b_2	1.151	1.081	0.295	3.160

$$-2\log(L) = 44.765$$

behavior type and smoking have about the same influence on "survival" time when these influences are measured by binary variables ($\hat{c}_1 \approx \hat{c}_2$). The two-variable proportional hazards model shows explicitly that $\hat{c}$ factors into two pieces $\hat{c}_1$ and $\hat{c}_2$ ($\hat{c} = \hat{c}_1\hat{c}_2$). From the example, $\hat{c}_1 = 3.288$ and $\hat{c}_2 = 3.160$ and, therefore, for a type A individual who smokes the hazard ratio is $\hat{c} = \hat{c}_1\hat{c}_2 = (3.160)(3.288) = 10.390$ relative to a nonsmoking, type B individual.

Example

A proportional hazards model can include any number of independent variables to more clearly evaluate differences in survival time as reflected by comparing hazard functions. Using four independent variables from Table 11.17 produces the coefficients and relative hazard ratios given in Table 11.21 from the proportional hazards model

$$\lambda_j(t) = \lambda_0(t)e^{b_1x_{1j}+b_2x_{2j}+b_3x_{3j}+b_4x_{4j}} . \tag{11.47}$$

The four variables are behavior type (x_{1j}, coded 0 or 1), body weight (x_{2j}, recorded in pounds), systolic blood pressure (x_{3j}, recorded in mm), and cholesterol level (x_{4j}, recorded in mg/100ml) where j indicates one of the $n = 35$ observations in Table 11.17.

The estimated coefficients allow the investigation and description of the impact of the four independent variables on "survival" time (an example is in Box 11.2). The coefficients directly estimate multiplicative changes in the relative hazard ratio

T A B L E **11.20**

Summary of the relative hazards associated with behavior type and smoking

	A/B	Smoker/ Nonsmoker	$\hat{c}$ = Relative Hazard
B and nonsmoker	0	0	1.000
B and smoker	0	1	3.160
A and nonsmoker	1	0	3.288
A and smoker	1	1	10.390

T A B L E **11.21**

Analysis of behavior type, body weight, blood pressure, and cholesterol level using a proportional hazards model

Variable	Coefficient	Estimate	Standard Error	p-Value	Hazard Ratio
A/B	b_1	1.183	0.907	0.202	3.263
Weight	b_2	−0.0174	0.019	0.373	0.983
Blood pressure	b_3	0.00350	0.032	0.913	1.004
Cholesterol	b_4	0.00830	0.005	0.111	1.008

$$-2\log(L) = 43.322$$

for a 1-unit increase in each independent variable. The estimated changes reflect the influence of each variable adjusted for the presence of the other $k - 1$ variables in the model. Like the previous multivariable models, the magnitude of these changes is influenced by the measurement units; therefore, making direct comparisons among the coefficients is useless when variables are measured in different units. A commensurate measure is achieved by dividing each coefficient by its standard error $(\hat{b}_i/S_{\hat{b}})$. From the analysis in Table 11.21, these standardized coefficients are 1.30 (A/B), 0.90 (weight), 0.11 (blood pressure), and 1.64 (cholesterol). Comparisons show that cholesterol level and behavior type have the greatest influence on "survival" time whereas weight and, particularly, systolic blood pressure have relatively less impact.

B O X **11.2** **RELATIVE HAZARD FUNCTION FOR TWO INDIVIDUALS**

Two individuals with specific values for behavior type, body weight, blood pressure, and cholesterol level the coefficients from Table 11.21 give

	Behavior	Weight	Blood Pressure	Cholesterol
Person 1 (x_{i1})	B	180	160	300
Person 2 (x_{i2})	A	160	180	330
$\hat{b}_i(x_{i2} - x_{i1})$	1.183	0.349	0.070	0.249

The estimated relative hazard is

$$\frac{\lambda_2(t)}{\lambda_1(t)} = e^{\Sigma\hat{b}_i(x_{i2}-x_{i1})}$$

$$= (e^{1.183})(e^{0.349})(e^{0.070})(e^{0.249}) = (3.263)(1.418)(1.072)(1.283) = 6.365 .$$

The risk for person 2 is more than six times that of person 1, measured by the estimated ratio of hazard functions, as long as the hazard functions describing coronary experience are proportional.

T A B L E **11.22**

Analysis of behavior type, blood pressure, and cholesterol using a proportional hazards model—body weight deleted from the model

Variable	Coefficient	Estimate	Standard Error	p-Value	Hazard Ratio
A/B	b_1	0.914	0.850	0.290	2.494
Blood pressure	b_3	–0.010	0.030	0.735	0.990
Cholesterol	b_4	0.007	0.005	0.146	1.007

$$-2\log(L) = 44.132$$

The formal assessment of the influence of the independent variables follows the same pattern as the logistic regression analysis. A likelihood statistic is calculated with the variables of interest deleted from the model and compared to the likelihood measure generated with the same variables retained in the model. For example, if body weight is deleted from the proportional hazards model, the likelihood value from the reduced model (Table 11.22) compared to that from the full model (Table 11.21) reflects the influence of body weight on the "survival" time associated with a coronary event. The comparison of likelihood measures shows almost no impact from deleting the variable body weight from the model, $Y^2 = -2\log(L_0) - [-2\log(L_1)] = 44.132 - 43.32 = 0.810$. As before, the deviance Y^2 has an approximate chi-square distribution with degrees of freedom equal to the difference in the number of parameters necessary to specify each of the two models. Deleting body weight from the model yields a p-value of 0.368 (one degree of freedom).

A Method for Variable Selection

An "optimum" model can be developed by a stepwise process using a likelihood criterion. The criterion for the jth step is

$$criterion = Y_j^2 = [-2\log(L_{j-1})] - [-2\log(L_j)] . \tag{11.48}$$

To construct the "optimum" model, the single variable that produces the largest transformed likelihood value becomes the first variable chosen to build a multivariable model. Then, the $k - 1$ remaining variables are each entered into the model one at a time and the likelihood values from these bivariate models are calculated. The variable that produces the largest value of the criterion is then added to the model as long as Y_2^2 is greater than 3.841. The value 3.841 is the 95th percentile of a chi-square distribution with one degree of freedom, which is a traditional choice used to decide whether a variable adds sufficient information to the analysis to be retained in the model. Of course, other levels for admission can be chosen. If a second variable is added to the model, the process is repeated for each of the remaining $k - 2$ variables. The variable producing the largest value of Y_3^2 is then retained, forming a three-variable model, as long as Y_3^2 exceeds 3.841. The process continues until no further "significant" variables are found, which means the values of Y_j^2 are less than 3.841 for all remaining variables. The model is then considered to contain the "optimum" variables for the prediction of differences in survival time. This "stepwise forward" procedure is just one example of a number of stepwise procedures used to select optimum multivariable models. Similar stepwise analyses also apply to the linear and logistic regression models and are often part of computer statistical analysis systems ([8] or [9]). Of course, "optimum" means optimal in a limited sense. Other criteria produce other models that are considered "optimum" by other standards. No single standard exists and a variety of procedures produce efficient but different models to represent relationships within the data.

Many issues discussed in the context of linear or logistic regression also arise in applying a proportional hazards model. The process is subject to "shrinkage" where the model derived from one sample of data performs less well in another sample. Correlations among the independent variables adversely affect the precision of the estimated coefficients (collinearity). Interaction terms and other nonlinear terms can be included in the proportional hazards model to evaluate possibly nonadditive effects from the independent variables. Goodness-of-fit is a critical issue in drawing inferences from any multivariable model. However, the unique feature of the proportional hazards model is the ability to assess differences in hazard functions without specifying the form of the functions, producing an assessment of differences in survival time unbiased by incomplete data.

Survival Analysis

The following STATA code produces most of the analyses referred to in Chapter 11, particularly the proportional hazards analyses. The data come from the $n = 35$ individuals with extremely high cholesterol levels given in Table 11.17.

```
chds.survival

. infile weight sysbp chol cigs ab chd time using chol.dat

. summarize

Variable |     Obs        Mean     Std. Dev.        Min        Max
---------+-------------------------------------------------------
  weight |      35    170.0857     23.55516        120        225
   sysbp |      35    129.7143     14.28403        104        154
    chol |      35    369.2857     51.32284        343        645
    cigs |      35    17.14286     13.07702          0         40
      ab |      35    .5142857     .5070926          0          1
     chd |      35    .2285714      .426043          0          1
    time |      35    2580.629     616.0796        773       3141

. tabulate chd ab
```

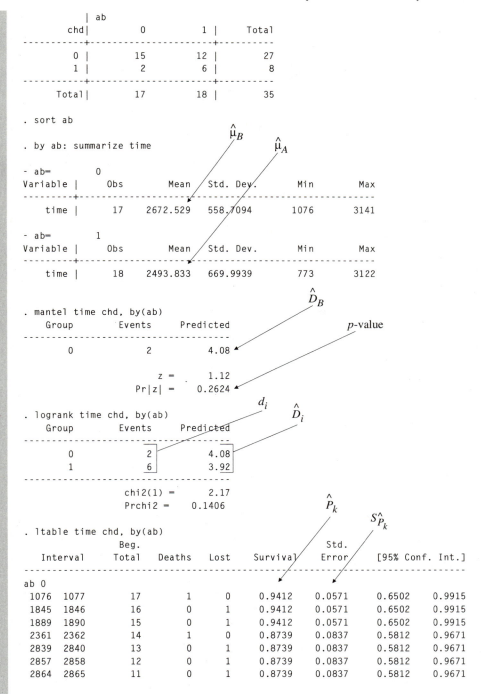

```
             | ab
         chd|         0         1 |     Total
    ---------+----------------------+----------
          0 |        15        12 |        27
          1 |         2         6 |         8
    ---------+----------------------+----------
      Total |        17        18 |        35

. sort ab

. by ab: summarize time
```

$\hat{\mu}_B$ $\hat{\mu}_A$

```
- ab=        0
Variable |     Obs        Mean    Std. Dev.        Min         Max
---------+------------------------------------------------------
    time |      17    2672.529    558.7094        1076        3141

- ab=        1
Variable |     Obs        Mean    Std. Dev.        Min         Max
---------+------------------------------------------------------
    time |      18    2493.833    669.9939         773        3122
```

$\hat{D}_B$

p-value

```
. mantel time chd, by(ab)
    Group       Events     Predicted
-----------------------------------------
        0          2          4.08

                   z =        1.12
               Pr|z| =      0.2624
```

d_i $\hat{D}_i$

```
. logrank time chd, by(ab)
    Group       Events     Predicted
-----------------------------------------
        0          2          4.08
        1          6          3.92
-----------------------------------------
               chi2(1) =      2.17
               Prchi2 =      0.1406
```

$\hat{P}_k$ $S\hat{P}_k$

```
. ltable time chd, by(ab)
                     Beg.                                  Std.
      Interval       Total  Deaths   Lost   Survival      Error    [95% Conf. Int.]
---------------------------------------------------------------------------------
ab 0
 1076  1077            17      1       0     0.9412      0.0571    0.6502    0.9915
 1845  1846            16      0       1     0.9412      0.0571    0.6502    0.9915
 1889  1890            15      0       1     0.9412      0.0571    0.6502    0.9915
 2361  2362            14      1       0     0.8739      0.0837    0.5812    0.9671
 2839  2840            13      0       1     0.8739      0.0837    0.5812    0.9671
 2857  2858            12      0       1     0.8739      0.0837    0.5812    0.9671
 2864  2865            11      0       1     0.8739      0.0837    0.5812    0.9671
```

2875	2876	10	0	1	0.8739	0.0837	0.5812	0.9671
2890	2891	9	0	1	0.8739	0.0837	0.5812	0.9671
2925	2926	8	0	1	0.8739	0.0837	0.5812	0.9671
2926	2927	7	0	1	0.8739	0.0837	0.5812	0.9671
2927	2928	6	0	1	0.8739	0.0837	0.5812	0.9671
2930	2931	5	0	1	0.8739	0.0837	0.5812	0.9671
2987	2988	4	0	1	0.8739	0.0837	0.5812	0.9671
3048	3049	3	0	1	0.8739	0.0837	0.5812	0.9671
3053	3054	2	0	1	0.8739	0.0837	0.5812	0.9671
3141	3142	1	0	1	0.8739	0.0837	0.5812	0.9671

ab 1

773	774	18	0	1	1.0000	0.0000	.	.
1257	1258	17	1	0	0.9412	0.0571	0.6502	0.9915
1825	1826	16	1	0	0.8824	0.0781	0.6060	0.9692
2042	2043	15	0	1	0.8824	0.0781	0.6060	0.9692
2059	2060	14	1	0	0.8193	0.0946	0.5377	0.9380
2296	2297	13	1	0	0.7563	0.1063	0.4730	0.9011
2424	2425	12	0	1	0.7563	0.1063	0.4730	0.9011
2613	2614	11	1	0	0.6875	0.1168	0.4022	0.8572
2840	2841	10	0	1	0.6875	0.1168	0.4022	0.8572
2862	2863	9	0	1	0.6875	0.1168	0.4022	0.8572
2881	2882	8	0	1	0.6875	0.1168	0.4022	0.8572
2940	2941	7	0	1	0.6875	0.1168	0.4022	0.8572
2943	2944	6	0	1	0.6875	0.1168	0.4022	0.8572
2959	2960	5	0	1	0.6875	0.1168	0.4022	0.8572
3000	3001	4	0	1	0.6875	0.1168	0.4022	0.8572
3009	3010	3	0	1	0.6875	0.1168	0.4022	0.8572
3044	3045	2	0	1	0.6875	0.1168	0.4022	0.8572
3122	3123	1	1	0	0.0000	.	.	.

```
. cox time ab, dead(chd)
```

$\log(L)$ $\log(\hat{c})$ $S_{\log(\hat{c})}$

```
Cox regression                              Number of obs =      35
                                            chi2(1)       =    2.27
                                            Prob  chi2    =  0.1322
Log Likelihood = -23.112598                 Pseudo R2     =  0.0467
------------------------------------------------------------------
   time |
    chd |     Coef.   Std. Err.      t     P|t|    [95% Conf. Interval]
--------+---------------------------------------------------------
     ab |  1.141906   .8169539    1.398   0.171    -.5183441   2.802156
------------------------------------------------------------------
```

```
. cox time ab, dead(chd) hr nolog
```

$\hat{c}$ $S_{\hat{c}}$

```
Cox regression                              Number of obs =      35
                                            chi2(1)       =    2.27
                                            Prob  chi2    =  0.1322
Log Likelihood = -23.112598                 Pseudo R2     =  0.0467
------------------------------------------------------------------
   time |
    chd | Haz. Ratio  Std. Err.      t     P|t|    [95% Conf. Interval]
--------+---------------------------------------------------------
     ab |  3.132733   2.559299    1.398   0.171     .5955058   16.48014
------------------------------------------------------------------
```

```
. generate smk=cigs
(24 chages made)

. recode smk 0=0 1/max=1

. cox time ab smk, dead(chd)
```

Cox regression   Number of obs =   35
                                                  chi2(2)   =   3.73
                                                  Prob chi2   = 0.1552
Log Likelihood = -22.382672   Pseudo R2   = 0.0768
----
    time |
     chd | Coef.   Std. Err.   t   P|t| [95% Conf. Interval]
---------+--
      ab | 1.190297   .8176501   1.456   0.155   -.4732242   2.853819
     smk | 1.150665   1.081088   1.064   0.295   -1.048825   3.350155

Labels pointing to table: $\log(L)$ → Log Likelihood; $\hat{b}_i$ → Coef.; $S\hat{b}_i$ → Std. Err.; *p*-values → P|t|

```
. cox time ab smk, dead(chd) hr
```

hazard ratios = $e^{\hat{b}_i}$

Cox regression   Number of obs =   35
                                                  chi2(2)   =   3.73
                                                  Prob chi2   = 0.1552
Log Likelihood = -22.382672   Pseudo R2   = 0.0768
----
    time |
     chd | Haz. Ratio Std. Err.   t   P|t| [95% Conf. Interval]
---------+--
      ab | 3.288059   2.688482   1.456   0.155   .6229904   17.35393
     smk | 3.160294   3.416555   1.064   0.295   .3503493   28.50714

```
. cox time ab sysbp chol, dead(chd)
```

Cox regression   Number of obs =   35
                                                  chi2(3)   =   4.36
                                                  Prob chi2   = 0.2251
Log Likelihood = -22.065904   Pseudo R2   = 0.0899
----
    time |
     chd | Coef.   Std. Err.   t   P|t| [95% Conf. Interval]
---------+--
      ab | .9140557   .8499092   1.075   0.290   -.8171527   2.645264
   sysbp | -.0101772   .0298561   -0.341   0.735   -.0709921   .0506377
    chol | .0073174   .0049112   1.490   0.146   -.0026865   .0173212

```
. cox time ab sysbp chol, dead(chd) hr nolog
```

Cox regression   Number of obs =   35
                                                  chi2(3)   =   4.36
                                                  Prob chi2   = 0.2251
Log Likelihood = -22.065904   Pseudo R2   = 0.0899
----
    time |
     chd | Haz. Ratio Std. Err.   t   P|t| [95% Conf. Interval]
---------+--

```
       ab |   2.494419   2.120029     1.075   0.290    .4416875   14.08716
    sysbp |   .9898744   .0295538    -0.341   0.735    .9314692   1.051942
     chol |   1.007344   .0049473     1.490   0.146    .9973172   1.017472
----------------------------------------------------------------------------

. cox time ab weight sysbp chol, dead(chd)

Cox regression                               Number of obs =       35
                                             chi2(4)       =     5.17
                                             Prob  chi2    = 0.2703
Log Likelihood = -21.660809                  Pseudo R2     = 0.1066
----------------------------------------------------------------------------
   time |
    chd |      Coef.   Std. Err.       t    P|t|    [95% Conf. Interval]
--------+-------------------------------------------------------------------
     ab |   1.182778   .9067051     1.304   0.202   -.6664588   3.032016
 weight |  -.0174563   .0193213    -0.903   0.373   -.0568623   .0219497
  sysbp |   .0034974   .0317206     0.110   0.913   -.0611972   .068192
   chol |   .0083005   .0050613     1.640   0.111    -.002022   .0186231
----------------------------------------------------------------------------

. cox time ab weight sysbp chol, dead(chd) hr nolog

Cox regression                               Number of obs =       35
                                             chi2(4)       =     5.17
                                             Prob  chi2    = 0.2703
Log Likelihood = -21.660809                  Pseudo R2     = 0.1066
----------------------------------------------------------------------------
   time |
    chd | Haz. Ratio  Std. Err.       t    P|t|    [95% Conf. Interval]
--------+-------------------------------------------------------------------
     ab |   3.263429   2.958968     1.304   0.202    .5135239   20.73899
 weight |   .9826952   .0189869    -0.903   0.373    .9447241   1.022192
  sysbp |   1.003504   .0318317     0.110   0.913    .9406377   1.070571
   chol |   1.008335   .0051035     1.640   0.111     .99798    1.018798
----------------------------------------------------------------------------

. exit,clear
```

Notation

Exponential model:

$$S(t) = P(surviving\ beyond\ t) = e^{-\lambda t}$$

$$average\ rate = \frac{S(t) - S(t + \delta)}{\int\limits_{t}^{t+\delta} S(u)du} = \lambda$$

Estimated mean (exponential model):

$$\hat{\mu} = \frac{\sum\limits_{i=1}^{d} t_i + \sum\limits_{i=1}^{n-d} (t_i^+ + \hat{\mu})}{n} \quad \text{and} \quad \hat{\mu} = \frac{\sum\limits_{i=1}^{n} t_i}{d}$$

$$S_{\hat{\mu}}^2 = \frac{\hat{\mu}^2}{d}$$

Estimated rate (exponential model):

$$average\ rate = \hat{\lambda} = \frac{d}{\Sigma t_i}$$

$$S_{\hat{\lambda}}^2 = \frac{\hat{\lambda}^2}{d}$$

Median (exponential model):

$$S(t_m) = e^{-\lambda t_m} = 0.5$$

and the estimated median is

$$\hat{t}_m = \frac{-\log(0.50)}{\hat{\lambda}}$$

Survival probability (nonparametric):

$$\hat{P}_k = P(surviving\ beyond\ k\ intervals) = \hat{p}_1\,\hat{p}_2 \cdots \hat{p}_k = \prod_{i=1}^{k} \hat{p}_i$$

$$S_{\hat{P}_k}^2 = \hat{P}_k^2 \sum_{i=1}^{k} \frac{\hat{q}_i}{n_i\,\hat{p}_i}$$

Median (nonparametric):

$$\tilde{t}_m = t_{j+1} - \frac{(t_{j+1} - t_j)\,(0.5 - \hat{P}_{j+1})}{\hat{P}_j - \hat{P}_{j+1}}$$

Estimated mean (nonparametric):

$$\hat{\mu} = \sum_{i=1}^{d} \hat{P}_{i-1}(t_i - t_{i-1}) \quad \text{with}\ \hat{P}_0 = 1.0\ \text{and}\ t_0 = 0$$

$$S_{\hat{\mu}}^2 = \sum_{i=1}^{d} \frac{\hat{q}_i}{n_i\,\hat{p}_i} A_i^2 \quad \text{where}\ A_i = \hat{\mu} - \sum_{j=1}^{i} \hat{P}_{j-1}(t_j - t_{j-1})$$

Weibull distribution:

$$S(t) = e^{-(\lambda t)^a}$$

Mantel-Haenszel chi-square statistic:

$$X_{MH}^2 = \frac{\left[\sum\limits_{i=1}^{d} a_i - \sum\limits_{i=1}^{d} \hat{A}_i \right]^2}{\sum\limits_{i=1}^{d} variance(a_i)} \quad \text{or} \quad X_{MH_c}^2 = \frac{\left[\left| \sum\limits_{i=1}^{d} a_i - \sum\limits_{i=1}^{d} \hat{A}_i \right| - 0.5 \right]^2}{\sum\limits_{i=1}^{d} variance(a_i)}$$

Hazard rate:

$$hazard\ rate = \lambda(t) = -\frac{dS(t)/dt}{S(t)}$$

Estimated hazard rate in the middle of the interval (t_c):

$$\hat{\lambda}(t_c) \approx \frac{d_t}{\delta(l_{t+\delta} + \frac{1}{2} d_t)}$$

General additive proportional hazards model:

$$\lambda_j(t \mid x_{1j}, x_{2j}, x_{3j}, \cdots, x_{kj}) = \lambda_0(t)\ c = \lambda_0(t)e^{\sum\limits_{i=1}^{k} b_i x_{ij}}$$

Relationship of survival functions for proportional hazard rates:

$$S_j(t) = \left[S_0(t) \right]^c = \left[S_0(t) \right]^{e^{\sum b_i x_{ij}}}$$

Problems

1   Using the AIDS data in Table 11.23, estimate the "survival" probabilities for individuals less than 30 years old and for those individuals greater than or equal to 30 years old assuming that the "survival" times are exponentially distributed. Then repeat the calculation making no assumption about the distribution of "survival" times. Plot the survival functions for these two groups. Also plot the log-log transformation of these survival functions. Compare the survival experience of these two age groups (< 30 and ≥ 30) with a Mantel-Haenszel chi-square statistic to assess the differences in "time to AIDS."

2   Using the AIDS data in Table 11.23, calculate the mean time to the occurrence of AIDS for individuals < 30 years old and for individuals ≥ 30 years old based on the assumption that the "time to AIDS" has an exponential distribution. Also construct the associated approximate 95% confidence interval from the estimated mean values. Repeat the calculation of the mean time to AIDS without a parametric assumption. Calculate the approximate 95% confidence interval based on this second estimate.

3   Use a proportional hazards model to assess the role of age where age is not a binary variable but is entered into the model as reported (years). Contrast the results with another proportional hazards model analysis using age as a categorical variable: age < 30 and age ≥ 30.

T A B L E **11.23**

Data on time between diagnosis of HIV and AIDS for 51 individuals with their age at diagnosis recorded

Obs.	Time	Censored (C)	Age	Obs.	Time	Censored (C)	Age
1	1	1	34	27	37	1	32
2	17	1	42	28	41	1	50
3	37	0	47	29	37	1	44
4	81	1	26	30	33	1	34
5	2	1	40	31	95	0	29
6	92	0	32	32	91	0	30
7	12	1	42	33	65	1	39
8	89	0	33	34	94	0	27
9	59	1	31	35	95	0	30
10	93	0	33	36	92	0	29
11	24	0	29	37	94	0	39
12	15	1	28	38	67	1	39
13	90	0	35	39	67	1	32
14	9	1	50	40	32	0	35
15	54	1	29	41	51	1	46
16	53	1	27	42	97	0	37
17	72	1	37	43	86	0	36
18	66	0	33	44	34	0	33
19	94	0	31	45	96	0	31
20	20	1	29	46	18	1	33
21	60	1	38	47	97	0	27
22	84	1	26	48	92	0	30
23	91	0	30	49	15	1	26
24	64	1	30	50	93	0	37
25	94	0	36	51	81	0	29
26	91	0	36				

Note: Table 11.23 describes the time (months) between diagnosis of HIV infection and the symptoms of AIDS for 51 individuals along with their age at diagnosis. The variable *censored* indicates those individuals diagnosed with AIDS during the study period ($C = 1$) and those who had not developed AIDS before the end of the study period ($C = 0$). "Survival" means the time from diagnosis as HIV-positive to the symptoms of AIDS.

4   Show that the estimate of the variance of the estimated median $\hat{t}_m = -\log(0.5)/\hat{\lambda}$ is *variance*$(\hat{t}_m) = 0.480\hat{\mu}^2/d$ where the n survival times have an exponential distribution and d failures are observed.

5   Consider two hypothetical groups each with 10 survival times:

Observation	1	2	3	4	5	6	7	8	9	10
Group 1	5	7^+	19^+	22	34^+	35	62^+	101	109	140
Group 2	9	18^+	54	68^+	67	68	99^+	108	144	166

Construct the $k = 13$ individual 2-by-2 tables to contrast the survival experience of these two groups. Conduct a Mantel-Haenszel chi-square analysis of these data to evaluate the survival experience between groups 1 and 2. Calculate the expected number of deaths in groups 1 and 2 when group status is unrelated to survival. Use these expected values to conduct a chi-square test again contrasting the survival experience of the two groups. Repeat the same analysis using a proportional hazards model.

References

1 Johnson, R. C., and Johnson, N. L. *Survival Models and Data Analysis*. New York: John Wiley, 1980.

2 Kalbfleisch, D. J., and Prentice, R. L. *The Statistical Analysis of Failure Time Data*. New York: John Wiley, 1980.

3 Cox, D. R., and Oakes, D. *Analysis of Survival Data*. London: Chapman and Hall, 1984.

4 Fleiss, J. L. *Statistical Methods for Rates and Proportions*. New York: John Wiley, 1981.

5 Breslow, N. E., and Day, N. E. *Statistical Methods in Cancer Research*, volume 1. New York: Oxford University Press, 1980.

6 Miller, R. G. *Survival Analysis*. New York: John Wiley, 1981.

7 Roseman, R. H., Brand, R. J., and Jenkins, C. C. Coronary Heart Disease in the Western Collaborative Group Study. *Journal of the American Medical Association* 223 (1975): 872–877.

8 Afifi, A. A., and Clark, V. *Computer-Aided Multivariate Analysis*, 2nd ed., New York: Chapman and Hall, 1990.

9 Neter, J., Wassermann, W. and Kutner, M. H. *Applied Linear Statistical Models*. Homewood, Ill.: Irwin, 1990.

12

Poisson Regression Analysis

Background

A basic element of a linear regression analysis is a continuous dependent variable with a normal distribution. Both the character of the variable (continuous) and its distribution (normal) determine the analytic strategy. Similarly, a logistic regression analysis employs a binary dependent variable with a binomial distribution to make inferences from a series of independent variables. Between analyses based on continuous data and analyses based on binary data lies the Poisson regression approach. For a Poisson regression, the dependent variable consists of counts and the Poisson distribution provides the basis for inferences. Often counts are expressed as rates; in either case, a Poisson regression equation relates a count (or a rate) to a series of independent variables providing a structure for statistical analysis. As might be expected, similarities exist between Poisson regression and other regression techniques. In fact all regression analyses are special cases of a general linear model approach. However, for most applications a high degree of generality is not necessary and a description of Poisson regression as a separate entity is useful.

Poisson regression techniques are designed to be particularly effective when data are collected in a specific pattern. Suppose a group of individuals is exposed to an occupational or environmental agent suspected of causing a disease. A study to verify whether such an agent is a hazard often begins by defining a group of exposed and unexposed individuals, sometimes called a cohort. This cohort is observed over time and events such as disease or death are recorded as well as other relevant variables, such as age, smoking habits, and years employed. When the study period ends, observations are classified into a series of categories to begin to study the influence of the recorded variables on the outcome. Typically tables are formed and category-specific rates are calculated for each cell of the table. These tables are a summary themselves, but the relationships among the variables are usually sufficiently complex that further analysis is necessary. A Poisson regression analysis is one way that a table of rates can be analyzed.

Person-Years

Calendar Time

An average rate, to repeat, is the number of events during a specified time interval divided by the total person-time accumulated during the interval (expression 11.5). The total person-time is a sum of the accumulated time for those individuals experiencing the event as well as those who could have but did not experience the event. The term "person-time" is used here but other measures of exposure are certainly employed to calculate and study rates. For example, exposure could be in person-miles when car or airplane fatalities are the subject of interest. In the context of a Poisson regression analysis, time is accumulated into a series of categories. For example, a study could last over a period of several years where individuals enter the study during this period and are observed until the event of interest occurs, they are lost from the study, or the study ends. All three types of observations add time to the accumulation of the total time observed. The observations are then classified into a series of risk categories such as the first year, second year, and third year after diagnosis of a disease. A few observations from such a study are given in Table 12.1 and displayed in Figure 12.1.

The table shows, for example, that the first person entered the study in February of 1980 and continued to be observed until September of 1982, contributing 31 months of observation. If the data are tabulated into one-year intervals for the purpose of analysis, then this person contributed 10 months of observation to the first year of the study, 12 months to the total for the second year, and 9 months to the third year. Similarly, all study participants contribute the amount of time observed after diagnosis. The three categories, therefore, contain the partial contributions of different subjects. For the 10 observations in Table 12.1, the total person-months are 34 for the first year, 61 person-months for the second year, and 43 person-months for the

T A B L E **12.1**

Follow-up "data"—calendar time

	Start—Month(year)	End—Month(year)	Months
1	Feb (1980)	Sept (1982)	31
2	April (1980)	Aug (1981)	16
3	Sept (1980)	July (1982)	22
4	March (1980)	Aug (1980)	5
5	March (1982)	Oct (1982)	7
6	March (1981)	May (1981)	2
7	April (1982)	Dec (1982)	8
8	March (1981)	Oct (1982)	19
9	Oct (1980)	June (1981)	8
10	June (1980)	Feb (1982)	20

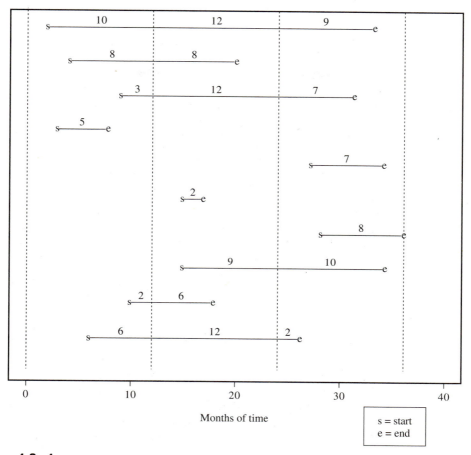

FIGURE **12.1**
Display of follow-up times—person-years calculation

third year after diagnosis. These values would be the denominator of rates calculated for each of these three years.

In studies of the health risks among industrial workers, the years employed are sometimes used as a substitute measure of exposure. Categories based on person-years employed would not produce rates that accurately reflect risk. Individuals who died early in a study contribute relatively few person-years to the low-exposure groups, and the high-exposure group similarly contains a greater number of person-years. The time of death determines whether an individual contributes person-years to the low- or high-exposure group, which tends to increase the low-exposure rates and decrease the high-exposure rates. However, a scheme where the person-years for each classification is a sum of the times experienced by all individuals at risk within calendar-time based categories avoids this problem.

T A B L E   **12.2**
Follow-up "data"—calendar-time/age

	Start—Year	End—Year	Start—Age	End—Age	Years
1	1941	1949	43	51	8
2	1946	1950	52	56	4
3	1942	1955	55	68	13
4	1942	1947	63	68	5
5	1950	1961	40	51	11
6	1955	1970	52	67	15
7	1959	1962	61	64	3
8	1962	1966	54	58	4
9	1965	1970	46	51	5
10	1965	1970	40	45	5

Calendar Time and Age

The situation for calculating accumulated exposure taking age into account is not much different from considering time alone. Each subject is followed over time and simultaneously ages. This property of cohort data is displayed in Figure 12.2, where a 45-degree line depicts the time associated with each study participant. Ten observations of calendar-time and age are given in Table 12.2 and displayed in Figure 12.2.

The first person admitted to the study entered in 1941 at age 43 and continued in the study until 1949, when the person was 51 years old. If a table is constructed with 10-year intervals for age and study period as shown in Table 12.3, this person contributed a total of 8 person-years of observation—7 years to the cell with less than 10 years of observation for ages 40–50 and 1 year to the cell with 10–20 years of observation for ages 50–60. The total of 73 person-years associated with the 10 observed individuals is similarly distributed into each age/time cell. Table 12.3, constructed from the "data" in Table 12.2, illustrates the accumulation of the person-years into nine time/age categories. These person-years would be the denominators of the three

T A B L E   **12.3**
The person-years for the illustrative data classified by time/age

Time/Age	0–10 Years	10–20 Years	20–30 Years	Total Person-Years
40–50 years	7	10	9	26
50–60 years	10	5	9	24
60–70 years	8	6	9	23
Total person-years	25	21	27	73

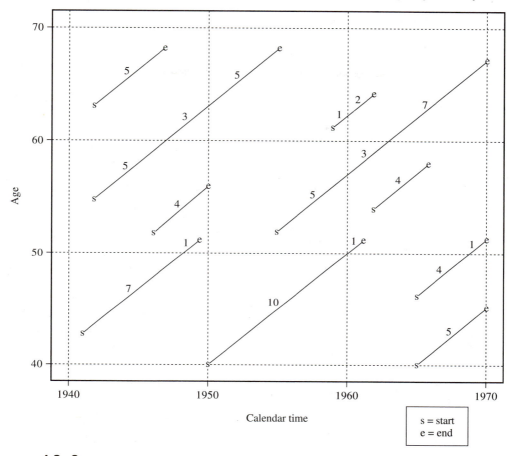

F I G U R E **12.2**
Display of follow-up times—time/age calculation

age-specific rates calculated for each of the three calendar-time groups. The job of computing the person-years from a data set is not always as easy as the two synthetic examples presented. Two computer programs that calculate person-time for cohort studies have been developed ([1] and [2]).

Age Adjustment of Rates

A primary reason to calculate rates is to make comparisons. Clearly comparing the absolute number of events among groups is not satisfactory because any comparison depends almost entirely on the sizes of the groups that generated the observations. A rate, like a percentage or a probability, is a basis for direct comparisons. However, comparisons of summary disease or mortality rates among sources of data must be done with care. If two populations differ with regard to age distribution, observed differences between rates will be, at least partially, due to the influences of age (the role of age is described rigorously in [3]). For example, populations with large numbers of older individuals, who have the highest mortality rates, naturally have increased overall mortality rates regardless of other factors.

Table 12.4 shows age-specific lung cancer data for three groups: the entire state of California, females living in Los Angeles County, and females living in the San Diego area. The estimated average mortality rates constructed from the population totals disregarding age, called the crude mortality rates, are:

California = 45.18 deaths per 100,000 person-years

Los Angeles = 39.96 deaths per 100,000 person-years

San Diego = 47.60 deaths per 100,000 person-years

A crude mortality rate is the total number of deaths divided by the total person-years lived by all observed individuals. To explore whether factors such as air quality might affect the lung cancer rates of these two areas (Los Angeles and San Diego) the following question needs to be answered: If age is taken into account, will the difference in crude mortality rates between Los Angeles and San Diego change or, stated a bit differently, is the age structure of these two areas similar enough so that observed differences in mortality reflect factors other than age? Answers to this question are achieved by calculating a direct age-adjusted rate or an indirect age-adjusted rate.

Direct Age-Adjustment

A comparison of two rates is free of the influence of age when the age distributions in each group are identical; otherwise the influence of age can interfere with the interpretation of any observed differences in crude disease or mortality rates. The direct method of age-adjustment uses the age distribution from a standard population to calculate a "crude" mortality rate for each comparison group. Because the age distribution is the same for the calculated rates, observed differences are due to factors other than age. If r_{ij} represents an age-specific rate (i^{th} group and the jth age stratum) and P_j represents the number of individuals in the jth age stratum of a standard population, then

$$\text{direct age-adjusted rate} = r_i^* = \frac{\text{estimated deaths}}{P} = \frac{\sum_{j=1}^{m} r_{ij} P_j}{P} \qquad (12.1)$$

where P is the total number of individuals that make-up the standard population (i.e., $P = \Sigma P_j$) and m is the number of age strata. The estimated number of "deaths" in each age stratum comes from applying the rates in the jth age category of the ith comparison group to the count of individuals in the corresponding age category of the stan-

T A B L E **12.4**

Lung cancer data for females from Los Angeles and San Diego (California, 1989)

	California			Los Angeles			San Diego		
Age	Deaths	Person-Years	Rates	Deaths	Person-Years	Rates	Deaths	Person-Years	Rates
0–4	1	1,149,564	0.1	1	359,582	0.3	0	98,351	0.0
5–9	0	1,165,077	0.0	0	396,940	0.0	0	91,812	0.0
10–14	0	953,776	0.0	0	316,812	0.0	0	75,467	0.0
15–19	2	958,438	0.2	1	274,500	0.4	0	87,941	0.0
20–24	2	1,063,345	0.2	1	289,078	0.3	0	103,078	0.0
25–29	9	1,196,919	0.8	2	338,045	0.6	0	104,628	0.0
30–34	19	1,300,402	1.5	4	389,651	1.0	2	119,014	1.7
35–39	54	1,217,896	4.4	24	361,840	6.6	2	106,308	1.9
40–44	118	1,045,801	11.3	36	304,426	11.8	10	86,233	11.6
45–49	252	810,260	31.1	78	240,319	32.5	17	65,569	25.9
50–54	437	665,612	65.7	114	198,114	57.5	31	55,424	55.9
55–59	637	633,646	100.5	170	190,769	89.1	45	53,671	83.8
60–64	937	650,686	144.0	228	198,371	114.9	93	55,659	167.1
65–69	1278	600,455	212.8	333	186,181	178.9	124	51,423	241.1
70–74	1188	474,609	250.3	282	145,909	193.3	114	41,751	273.0
75–79	911	376,781	241.8	261	113,466	230.0	76	33,693	225.6
80–84	492	255,412	192.6	137	76,887	178.2	48	23,103	207.8
85+	319	213,603	149.3	106	68,021	155.8	43	17,758	242.1
Total	6656	14,732,282	45.18	1778	4,448,911	39.96	605	12,708,83	47.60

Note: rates per 100,000 person-years.

dard. The sum of all such "deaths" produces the number of "deaths" that would be expected if the age distribution in the ith group were identical to the standard population. In this way all direct adjusted rates have the same age distribution. Using the California population as a standard, the direct adjusted mortality rates (r_i^*) from Table 12.1 are:

Los Angeles = $r_1^* = 39.64$ deaths/100,000; San Diego = $r_2^* = 46.57$ deaths/100,000.

For the California data, the differences in female lung cancer mortality persist after adjustment for differences in the age distributions from the two areas. In fact, the adjusted rates hardly differ from the crude mortality rates showing only negligible effects from differences in age distributions of these two areas.

Indirect Age-Adjustment

An alternative approach to comparing the mortality experience among several groups "free" from influences of differences in age distribution is based on rates from a standard population. The number of deaths expected in each group is calculated by applying the rates from a standard population to the number of individuals in each age category of the comparison group. That is, the expected number in the ith group is

$$expected\ number\ of\ deaths = e_i = \sum_{j=1}^{m} p_{ij} R_j \tag{12.2}$$

where R_j is the rate from the jth age category of the standard population and p_{ij} is the number of individuals in the jth age category from the ith group. The ratio of the number of observed deaths to the number of expected deaths is called the standardized mortality ratio (SMR). Formally, an SMR for the ith group is

$$SMR_i = \frac{total\ observed\ number\ of\ deaths}{total\ expected\ number\ of\ deaths} = \frac{\sum_{j=1}^{m} d_{ij}}{e_i} = \frac{D_i}{e_i} \tag{12.3}$$

where $D_i = \Sigma d_{ij}$ is the total number of deaths in the ith group. The expected number of deaths (e_i) for the lung cancer data are

$$Los\ Angeles = e_1 = 2027.8\ and\ San\ Diego = e_2 = 574.8\ ,$$

based on the rates from the entire state of California (standard). The SMRs are, then

$$Los\ Angeles:\ SMR_1 = \frac{D_1}{e_1} = \frac{1778}{2027.8} = 0.877\ and$$

$$San\ Diego:\ SMR_2 = \frac{D_2}{e_2} = \frac{605}{574.8} = 1.052\ .$$

To produce an indirect age-adjusted rate, these SMRs are applied to the crude rate of the standard to yield two "rates" where the influence of the differing age distributions is not an issue. Again from the California data, Los Angeles = 45.18(0.877) = 39.76 and San Diego = 45.18(1.052) = 47.44. Indirect age-adjustment shows almost identical "rates" to those calculated from the direct adjustment.

Implicit in the calculation and interpretation of a SMR is the assumption that the age-specific death rates are a constant multiple of the corresponding standard age-specific rates. For example, the SMR for Los Angeles County suggests that each age-specific rate is 12.3% (SMR = 0.877) lower than each corresponding age-specific lung cancer rate in the state of California. If the age-specific rate ratios are not constant, the SMR represents an "average" of a series of heterogeneous quantities and does not provide a meaningful single summary of the underlying relationship between a comparison group and the standard population. Indirect age-adjustment has been criticized [4] for a number of failings. However, when indirect adjustment is applied to data where the ratios of the age-specific rates in the comparison groups to the age-specific rates in the standard population are constant, or approximately constant, the problems with indirect age-adjustment disappear. The homogeneity of a series of age-specific mortality ratios is a fundamental issue in calculating and interpreting a SMR.

As will be seen, a Poisson regression model can also be used to analyze age-specific rates from a series of populations and produce measures that parallel SMRs and indirect rates. In addition, the Poisson regression approach provides a strategy to investigate the homogeneity of mortality ratios between groups that allows a statistical evaluation of the basic issue underlying an SMR summary.

To statistically assess an SMR, as always, an estimate of its variability is necessary. Because the expected number of deaths e_i is usually based on a large fixed population, the quantity is treated as a constant. The number of deaths D_i is a variable subject to sampling variation. The variable D_i takes on the values 0, 1, 2, $\cdots$, which can often be accurately described by a Poisson distribution (see appendix). Because the mean and the variance of a Poisson distribution are equal, the variance is then estimated by the mean. The variance of D_i is, therefore, estimated by D_i (i.e., *variance*$(D_i) = D_i$). Once the distribution of D_i is established as being at least approximately Poisson, a variance for the logarithm of the estimated SMR is estimated by $1/D_i$ (i.e., *variance*$[\log(SMR)] = 1/D_i$ [5]). The logarithm of the SMR can then treated as a variable with an approximately normal distribution and an approximate 95% confidence interval constructed. That is, an approximate confidence interval for the logarithm of the SMR is

$$lower\ bound = \log(SMR) - 1.96\frac{1}{\sqrt{D_i}} \quad and \quad upper\ bound = \log(SMR) + 1.96\frac{1}{\sqrt{D_i}}.$$

An approximate 95% confidence for the SMR is then the antilogarithm of the upper and lower bounds giving (e^{lower}, e^{upper}) as an approximate confidence interval for an SMR. For example, the SMR for Los Angeles county is 0.877 and the approximate 95% confidence interval based on $\log(SMR) = \log(0.877) = -0.131$ is $-0.131 \pm 1.96\sqrt{1/1778} = (-0.146, -0.116)$. The approximate 95% confidence interval based on the estimated $SMR = 0.877$ is then $(e^{-0.178}, e^{-0.085}) = (0.837, 0.919)$. Constructing a confidence interval from the logarithm of an estimate is a common statistical pattern because a logarithmic transformation makes skewed distributions resulting from ratio estimators more symmetric. These more symmetric distributions are more accurately approximated by a normal distribution, yielding improved accuracy of the confidence interval, as noted earlier (Chapters 8, 9, and 10). For the SMR, taking logarithms also prevents the occurrence of negative values that are not valid estimates of a SMR. When the number of deaths is small, the Poisson distribution can be skewed to such an extent that this normal distribution based confidence interval is not very accurate. Other more elaborate but more accurate approximate confidence intervals are available [5].

Two types of standard populations are chosen for age-adjustment: external and internal. An internal standard consists of combining the groups to be compared to

T A B L E **12.5**

United States standard million population by age (1970)

Age	Population
0–4	84,416
5–9	98,204
10–14	102,304
15–19	93,845
20–24	80,561
25–29	66,320
30–34	56,249
35–39	54,656
40–44	58,958
45–49	59,622
50–54	54,643
55–59	49,077
60–64	42,403
65–69	34,406
70–74	26,789
75–79	18,871
80–84	11,241
85+	7435
Total	1,000,000

form one standard group or selecting a specific group among those sampled to serve as a standard. This internal "standard" group is then used as a source of age-specific counts (direct standardization) or of rates (indirect standardization). External standards are conventionally chosen from large populations published for use in the adjustment process. An example of such a population is given in Table 12.5, where the 1970 population of the United States combining counts of both males and females provides the basis for a "standard million."

No real substitute exits for examining age-specific rates to detect differences among sets of mortality data. However, several reasons exist to summarize the mortality experience of a group. In some studies only small numbers of individuals are available, yielding even smaller numbers of deaths. For example, in a study of mortality in an industrial setting, workers involved in particularly hazardous occupations are almost always rare and deaths among these workers are even rarer. An indirect age-adjustment can sometimes produce a reliable summary value since it depends only on the total number of observed deaths and not the small numbers of deaths in the age-specific categories. Similarly, age categories may contain such small numbers of individuals that rates calculated for specific age groups are too unstable for effective use. An indirect adjusted rate can provide a more stable summary based on all available data. Also, an obvious feature of age-adjusted summaries is the convenience and ease of comparison that is provided by a single number.

Poisson Regression Model

A regression model consists of a function of the dependent variable modeled as a linear combination of a number of independent variables. For example, a linear model requires the dependent variable to be represented as a weighted sum of independent variables. A logistic model consists of the logarithm of the odds also expressed as a linear combination of the independent variables. The Poisson model is not much different. The logarithm of a rate is modeled as a weighted sum of the independent variables (risk variables). When the logarithm of a rate is a linear combination of set of independent variables, the rate is a product of influences associated with each independent variable, sometimes called a product-model. Before developing a general notation and presenting a detailed discussion, it is probably useful to look at two simple illustrations where the logarithm of a rate is a sum of a series of independent variables.

Preliminary Examples

Consider a rate that takes on only two values in each of five strata, one value in group 1 and another in group 2 (Table 12.6). Using two parameters, the stratum-specific rates are described by the model

$$rate = AC_j \quad \text{or} \quad log\text{-}rate = \log(rate) = a + c_j \qquad (12.4)$$

where $a = \log(A)$ is a constant term and $c_j = \log(C_j)$ is a binary variable. An observed rate is denoted as r_{ij} where i denotes the strata and j the group.

T A B L E **12.6**

Hypothetical data with a constant rate in each of two group

Stratum	Group 1			Group 2		
	Person-Years	Deaths	Rates (r_{i1})	Person-Years	Deaths	Rates (r_{i2})
1	1000	20	0.020	100	4	0.040
2	2000	40	0.020	200	8	0.040
3	3000	60	0.020	300	12	0.040
4	4000	80	0.020	400	16	0.040
5	5000	100	0.020	500	20	0.040
Total	15,000	300	0.020	1500	60	0.040

T A B L E **12.7**

The parameters that generated the hypothetical data—constant rate

Coefficient	Parameters
a	$\log(0.02) = -3.912$
c_1	$\log(1.0) = 0.0$
c_2	$\log(2.0) = 0.693$

The "data" in Table 12.6 are generated using the parameters given in Table 12.7. That is,

$$\log(r_{i1}) = -3.912 \quad \text{and} \quad \log(r_{i2}) = -3.912 + 0.693 = -3.219$$

and, therefore

$$r_{i1} = e^{-3.912} = 0.02 \quad \text{and} \quad r_{i2} = e^{-3.219} = 0.040 \,.$$

The SMR calculated directly from Table 12.6 (group 1 = standard) is

$$SMR = \frac{D_2}{e_2} = \frac{4 + 8 + \cdots + 20}{0.02(100) + 0.02(200) + \cdots + 0.02(500)} = \frac{60}{30} = 2.0 \,.$$

The SMR is also a function of the model parameters, where

$$r_{i1} = e^{a + c_1} \quad \text{and} \quad r_{i2} = e^{a + c_2}$$

and the rate ratio is

$$rate\ ratio = SMR = \frac{r_{i2}}{r_{i1}} = e^{c_2 - c_1} = e^{0.693} = 2.0 \,.$$

This three-parameter model is perhaps the simplest illustration of the most fundamental property of a Poisson regression model. The underlying rates are a multiplicative function of the parameters making the logarithm of the rates a linear function of a set of related parameters.

A slightly more sophisticated model can be constructed where differing influences from each of five strata are incorporated. In symbols,

$$rate = AB_iC_j \quad \text{or} \quad log\text{-}rate = a + b_i + c_j \,. \tag{12.5}$$

The parameters $b_i = \log(B_i)$ allow the log-rates to differ among the five strata and the parameter c_j is a binary variable that reflects membership in one group or the other. The "data" in Table 12.9, generated from this model and parameters in Table 12.8,

T A B L E **12.8**

The parameters that generated the hypothetical data—constant rate ratio

Coefficient	Estimates
a	$\log(0.02) = -3.912$
b_1	$\log(1.00) = 0.0$
b_2	$\log(1.50) = 0.405$
b_3	$\log(2.25) = 0.811$
b_4	$\log(4.50) = 1.504$
b_5	$\log(9.00) = 2.197$
c_1	$\log(1.00) = 0.0$
c_2	$\log(2.00) = 0.693$

necessarily have a constant stratum-specific rate ratio. For this artificial data, the parameters perfectly reproduce the rates. For example,

$$r_{21} = e^{a + b_2 + c_1} = e^{-3.912 + 0.405} = 0.030 \quad \text{and}$$

$$r_{22} = e^{a + b_2 + c_2} = e^{-3.912 + 0.405 + 0.693} = 0.060 .$$

Again under the conditions of this model, the rate ratio and the SMR are equal (group 1 = standard). That is,

$$SMR = \frac{D_2}{e_2} = \frac{4 + 12 + \cdots + 180}{0.020(100) + 0.030(200) + \cdots + 0.180(500)} = \frac{295}{147.5} = 2.0 .$$

T A B L E **12.9**

Hypothetical data with a constant stratum-specific rate ratio

	Group 1			Group 2		
Stratum	Person-Years	Deaths	Rates	Person-Years	Deaths	Rates
1	1000	20	0.020	100	4	0.040
2	2000	60	0.030	200	12	0.060
3	3000	135	0.045	300	27	0.090
4	4000	360	0.090	400	72	0.180
5	5000	900	0.180	500	180	0.360
Total	15,000	1475	0.098	1500	295	0.197

The rate ratio from the model is

$$rate\ ratio = SMR = \frac{r_{i2}}{r_{i1}} = e^{c_2 - c_1} = e^{0.693} = 2.0\ .$$

Poisson Regression

The Poisson model is appropriate for situations where the dependent variable is a count (e.g., number of cases of disease or death) within a series of subdivisions of the sampled data, as mentioned. For example, in a study of a sample of individuals observed for a period of time, mortality rates are recorded for a series of k strata (such as age categories) and also classified by a series of m exposure categories (such as exposure to radiation). The notation for such a sample is displayed in Table 12.10.

Each cell of the table contains a count of events (y_{ij}) and the person-time accumulated (p_{ij}) for those individuals who experienced the event and those who could have experienced the event. These two quantities yield an estimate of the cell-specific rate for an exposure variable (B_i) and a stratum variable (C_j) given by $r_{ij} = y_{ij}/p_{ij}$.

Poisson Model

The additive Poisson model for the data described in Table 12.10 has three major components.

T A B L E **12.10**

Notation for a Poisson regression with m exposure categories (B) and k strata (C)

Risk/Strata	C_1	C_2	C_3	$\cdots$	C_k
B_1					
Events	y_{11}	y_{12}	y_{13}	$\cdots$	y_{1k}
Person-years	p_{11}	p_{12}	p_{13}	$\cdots$	p_{1k}
B_2					
Events	y_{21}	y_{22}	y_{23}	$\cdots$	y_{2k}
Person-years	p_{21}	p_{22}	p_{23}	$\cdots$	p_{2k},
B_3					
Events	y_{31}	y_{32}	y_{33}	$\cdots$	y_{3k}
Person-years	p_{31}	p_{32}	p_{33}	$\cdots$	p_{3k}
.	.	.	.		.
.	.	.	.		.
.	.	.	.		.
B_m					
Events	y_{m1}	y_{m2}	y_{m3}	$\cdots$	y_{mk}
Person-years	p_{m1}	p_{m2}	p_{m3}	$\cdots$	p_{mk}

1   The dependent count variable has a Poisson distribution (see chapter appendix) and the probability a specific number of events y_{ij} occurs is

$$probability\ of\ y_{ij}\ events = \frac{e^{-\lambda_{ij}}\lambda_{ij}^{y_{ij}}}{y_{ij}!}.$$

(12.6)

2   The expected number of events in each cell of the table, represented by parameter λ_{ij}, is a function of the exposure/stratum categories, or

$$\lambda_{ij} = p_{ij}e^{a+b_i+c_j}.$$

(12.7)

The values b_i and c_j are parameters that incorporate into the model the influences from the categorical independent variables. The parameter a represents a constant value related to the overall frequency of events.

3   In addition,

$$log\text{-}rate = \log\left(\frac{\lambda_{ij}}{p_{ij}}\right) = a + b_i + c_j,$$

(12.8)

producing the expected log-rate in the ith, j^{th} cell, showing that the logarithm of a rate is a linear function of the parameters.

A somewhat more subtle property of the Poisson model is the requirement that the exposure/stratum subdivisions are sufficiently small units so the probability of occurrence of an event is at least roughly constant for all observations within each cell.

The two basic components of a Poisson regression model are the probability distribution (Poisson) and the relationship among the parameters (linear). These two components make up the necessary "input" to estimate the model parameters (e.g., $\hat{a}, \hat{b_i}, \hat{c_j}$) from a data set. The estimation process almost always requires a statistical computer program and produces a succinct parametric representation of each cell-specific rate.

Example

The application of the Poisson model is illustrated by comparing Hodgkin's disease mortality rates observed in males with the mortality rates observed in females for residents of California in 1989 (Table 12.11).

The Poisson model used to compare the male ($j = 1$) and the female ($j = 2$) mortality rates among the 12 age strata ($i = 1, 2, \cdots, 12$) is

$$log\text{-}rate = a + b_i + c_j$$

(12.9)

where again b_i produces different levels of log-rates for each age category and c_j is a binary variable differentiating male and female groups. Note that the model requires

T A B L E **12.11**

Hodgkins disease mortality data from California for males and females, 1989

Age	Males			Females			r_{i2}/r_{i1}
	Person-Years	Deaths	Rates (r_{i1})	Person-Years	Deaths	Rates (r_{i2})	
30–34	1,299,868	55	4.23	1,300,402	37	2.85	0.67
35–39	1,240,595	49	3.95	1,217,896	29	2.38	0.60
40–44	1,045,453	38	3.63	1,045,801	23	2.20	0.61
45–49	795,776	26	3.26	810,260	12	1.48	0.45
50–54	645,991	19	2.94	665,612	7	1.05	0.36
55–59	599,729	17	2.83	633,646	12	1.89	0.67
60–64	568,109	22	3.87	650,686	9	1.38	0.36
65–69	506,475	21	4.15	600,455	19	3.16	0.76
70–74	368,751	18	4.88	474,609	13	2.74	0.56
75–79	252,581	11	4.36	376,781	14	3.72	0.85
80–84	140,053	10	7.14	255,412	5	1.96	0.27
85+	81,850	4	4.87	313,603	3	1.40	0.29
Total	7,545,231	290	3.84	8,345,163	183	2.19	0.58

Note: Rates per 100,000 person-years.

the difference in the log-rates to be the same in each age category (i.e., [*log-rate (female rate)*] − [*log-rate (male rate)*] = $c_2 - c_1$). Of course, the observed age-specific rate ratio will not be identical (Table 12.11, last column). The formal comparison of the model (expression 12.9) to the data indicates whether the variation among these observed ratios is likely under the hypothesis of constant ratios of age-specific rates.

The logarithm of a rate involves two components: the count of events and the person-years of exposure, log(*rate*) = log(*count*) − log(*person-years*)). Therefore, the regression model (expression 12.9) can be viewed as log(*count*) − log(*person-years*) = $a + b_i + c_j$ or

$$log(count) = a + b_i + c_j + log(person\text{-}years)$$

The "coefficient" associated with the log(*person-years*) term is always 1. Such a term is called an offset and is part of the computer implementation (see examples near the end of the chapter).

The Poisson model (expression 12.9) applied to the Hodgkin's disease mortality data produces the estimates of the model parameters given in Table 12.12. An explanation of the estimation process is beyond the scope of this text but is found in more theoretical sources [6].

To establish a Poisson model for a two-way classification of data, $1 + (m - 1) + (k - 1) = m + k - 1$ parameter estimates are required (m = rows and k = columns). For

T A B L E   **12.12**

Results from a Poisson regression applied to the Hodgkin's disease mortality data (males and females) stratified by age

Coefficient	Estimates	Std. Error	Z-Statistic
a	−10.008	0.110	—
b_2	−0.112	0.154	−0.725
b_3	−0.193	0.165	−1.169
b_4	−0.400	0.193	−2.074
b_5	−0.575	0.222	−2.590
b_6	−0.401	0.213	−1.884
b_7	−0.311	0.208	−1.500
b_8	0.045	0.189	0.235
b_9	0.073	0.208	0.351
b_{10}	0.171	0.226	0.758
b_{11}	0.152	0.279	0.547
b_{12}	−0.271	0.392	−0.691
c_2	−0.561	0.095	−5.905

$$-2\log(L) = 8.558$$

rather technical reasons, one parameter describing the row variable and one parameter describing the column variable are set to zero and the other $m + k - 1$ parameters are estimated from the data. Similar to contingency table analysis, one parameter for each set of variables used to describe a table of rates is completely determined by the other values of the parameters (not an independent contributor) and is, therefore, conventionally set to zero. For the Hodgkin's disease data, $12 + 2 - 1 = 13$ parameters are estimated and b_1 and c_1 are set to zero.

These estimated coefficients produce estimated values that conform perfectly to the structure required by the model. The estimated mortality rate for the age group 65–69 for females is, for example,

$$\log(\hat{r}_{82}) = -10.0081 + 0.045 - 0.561 = -10.524 \quad \text{and} \quad \hat{r}_{82} = e^{-10.524} = 0.0000269$$

or 2.67 deaths per 100,000 person-years. Therefore, the expected number of deaths based on the model is $\hat{r}_{82}p_{82} = 0.0000269(600455) = 16.145$ (females) where the observed value is 19 (Table 12.11). Similarly, the model based estimated rate is $\hat{r}_{81} = 0.0000471$ for males and the estimated rate ratio is $0.0000269/0.0000471 = 0.571$ (female/male), which can also be directly calculated from the model parameters. The model dictates a constant rate ratio for all age strata and the estimate of that constant ratio is $e^{\hat{c}_2 - \hat{c}_1} = e^{-0.561} = 0.571$, because $c_1 = 0.0$.

A valid approach to calculating a SMR requires the same assumption explicit in the additive Poisson regression model (expression 12.9)—constant mortality ratios between all age strata. The SMR from the Hodgkin's disease mortality data (using the male sample as a standard) is

$$SMR = \frac{D_{females}}{e_{females}} = \frac{183}{323.497} = 0.566 \ . \tag{12.10}$$

The female rates are estimated to be about 57% smaller than those for males, providing a summary of the difference between the female and male Hodgkin's disease age-specific mortality rates. The rate ratios from the Poisson model and the SMR calculated directly from the data will not in general be identical but when the observed rate ratios are similar among all age strata, the two approaches give similar estimates.

Each estimated coefficient from the Poisson model has an associated standard error derived from the estimation process [6]. This measure of the impact of random variation on an estimated coefficient can be used to construct approximate confidence intervals. For example, the approximate 95% confidence interval for the logarithm of the rate ratio is

$$\hat{c}_2 \pm 1.96\sqrt{variance(\hat{c}_2)} = -0.561 \pm 1.96(0.0948) = (-0.747, -0.375)$$

and the approximate confidence interval for the rate ratio itself, estimated by $e^{\hat{c}_2} = e^{-0.561} = 0.571$, follows as $(e^{-0.747}, e^{-0.375}) = (0.474, 0.687)$.

Another use of the estimated standard error involves computing a z-statistic. An estimated coefficient divided by its estimated standard error produces a unitless quantity which serves as a standardized value or yields a statistical test. The test statistic

$$z = \frac{estimated\ coefficient}{estimated\ standard\ error} \tag{12.11}$$

has an approximate standard normal distribution when the estimated coefficient differs from zero by chance alone and is one way to evaluate its magnitude. For example, the Hodgkin's disease mortality data yield the estimated coefficient $\hat{c}_2 = -0.561$ measuring the difference in female/male mortality with an estimated standard error of 0.0948 (from the computer estimation program), giving a z-statistic $= -0.561/0.0948 = -5.915$. The value z has an approximate normal distribution when the rate ratio of female to male rates is 1.0 (i.e., $c_2 = 0$). A value of -5.915 is unlikely to arise from strictly random variation, leading to the inference that the rate ratio of female to male Hodgkin's disease rates is not 1.0.

Assuming that the rate ratio between female and male Hodgkin's disease mortality rates is constant, then it is natural to assess the relationship of age. If the estimated coefficient $\hat{c}_2$ substantially changes when age is deleted from the model, then clearly age effects the ratio and should be retained in the regression equation to yield an unbiased estimate of c_2. The change in an estimate caused by excluding a variable from the analytic model is confounder bias (introduced in Chapter 3). The Poisson model with the age parameters deleted (*log-rate = a + c_j*) applied to the Hodgkin's disease mortality data yields an estimate of the rate ratio of $e^{-0.549} = 0.577$ ($\hat{c}_2' = -0.549$), which is not very different from the model including age where $e^{-0.561} = 0.571$ ($\hat{c}_2 = -0.561$). The confounding influence of any variable can be assessed by comparing the same estimated parameters computed from two models, one containing the potentially confounding variable and one excluding the same variable. The observed difference in the estimated coefficients (e.g., $\hat{c}_2$ versus $\hat{c}_2'$) measures the degree of confounder bias, as noted in Chapter 3. The assumption that the rate ratios are constant can also be evaluated and a discussion of the fit of a Poisson model is presented in the next section.

Poisson Regression Model: *m* Strata and *k* Categories

A typical data collection pattern consists of recording the level of exposure for each individual, then classifying these individuals into a series of exposure categories, and finally stratifying the data to control for a confounding variable. To investigate the effects on the risk of cancer from exposure to radiation among workers at the Oak Ridge National Laboratory, data on individual exposure to external penetrating radiation were collected along with a number of other variables such as age (see [7] for a complete analysis and description). The data consist of white males who were employed during the years 1943 to 1972 classified by exposure status and stratified by age. Fifty-six observed counts (y_{ij}) and rates (r_{ij}) for the exposure/age data make up Table 12.13. A two-way classification is generated to address a series of questions about the independent variables, such as:

What is the impact of each independent variable on the observed outcome event?

What is the role of the stratum variable?

What are the relative influences of the independent variables in determining the rate associated with the outcome event?

Are two variables sufficient to describe the collected data?

How do the rate ratios behave within each stratum?

These questions and others can be answered by a Poisson regression analysis. When the data are sparsely distributed among the cells of a table, a Poisson model is especially helpful in describing the relationships under investigation and assessing the impact of random variation.

An additive Poisson model describing rates classified into a two-way table with *m* rows and *k* columns is

$$rate = AB_iC_j \quad \text{or} \quad log\text{-}rate = \log(rate) = a + b_i + c_j . \tag{12.12}$$

T A B L E **12.13**

Deaths, person-years, and cancer mortality rates for workers at Oak Ridge National Laboratory (1943–1972) classified by exposure and age

mSv/age	< 45	45–49	50–54	55–59	60–64	65–69	> 69
mSv = 0							
Deaths	0	1	4	3	3	1	3
Person-years	29901	6251	5251	4126	2778	1607	1358
Rate/100,000	0.0	16.00	76.18	72.71	107.99	62.23	220.91
mSv = 0–19							
Deaths	1	5	4	10	11	16	11
Person-years	71382	16705	13752	10439	7131	4133	3814
Rate/100,000	1.40	29.93	20.90	95.79	154.26	387.13	288.41
mSv = 20–39							
Deaths	0	2	1	2	0	2	3
Person-years	6523	2423	2281	1918	1322	723	538
Rate/100,000	0.0	82.54	43.84	104.38	0.0	276.63	557.62
mSv = 40–59							
Deaths	0	0	1	1	0	2	3
Person-years	2341	972	958	816	578	375	303
Rate/100,000	0.0	0.0	104.38	122.55	0.0	533.33	990.10
mSv = 60–79							
Deaths	0	0	0	0	0	1	0
Person-years	1363	478	476	387	225	164	150
Rate/100,000	0.0	0.0	0.0	0.0	0.0	609.76	0.0
mSv = 80–99							
Deaths	0	0	0	1	0	0	0
Person-years	779	296	282	251	193	125	69
Rate/100,000	0.0	0.0	0.0	398.41	0.0	0.0	0.0
mSv = 100–119							
Deaths	0	0	0	0	1	0	1
Person-years	520	188	217	184	109	60	23
Rate/100,000	0.0	0.0	0.0	0.0	917.43	0.0	4347.83
mSv ≥ 120							
Deaths	0	0	1	3	1	2	2
Person-years	2104	1027	1029	827	555	297	153
Rate/100,000	0.0	0.0	97.18	362.76	180.18	673.40	1307.19

Note: mSv = milliseiverts (.1 rem), a measure of radiation exposure.

T A B L E **12.14**

Results from a Poisson regression applied to the Oak Ridge National Laboratory
cancer mortality data classified by exposure and by age

Coefficient	Estimates	Std. Error	Z-Statistic
a	−11.992	1.021	—
b_2	0.387	0.290	1.335
b_3	0.426	0.408	1.044
b_4	0.814	0.457	1.778
b_5	−0.355	1.029	−0.345
b_6	0.080	1.025	0.078
b_7	1.365	0.750	1.819
b_8	1.249	0.422	2.957
c_2	3.432	1.056	3.249
c_3	3.892	1.040	3.741
c_4	4.732	1.021	4.637
c_5	4.895	1.027	4.763
c_6	5.849	1.017	5.754
c_7	5.991	1.017	5.890

$$-2\log(L) = 31.819$$

There are m values of the parameters b_i (rows) and k values of the parameter c_j (columns). The relationship between the parameters b_i and c_j and log-rates can be expressed as two sets of $m-1$ and $k-1$ design variables (described in Chapters 5 and 10), which is often necessary for computer implementation. Construction of design variables is usually an automatic part of computer regression programs. Under the Poisson model, the estimates of the parameters produce estimated numbers of deaths and estimated rates providing a statistical tool to explore the relationship of the row and column variables to the outcome count or rate. The estimated parameters for the radiation exposure data are given in Table 12.14.

The fit of the Poisson model to the data can be directly evaluated by generating mk expected values based on the model and its estimated parameters. For the data in a two-way table, the estimated numbers of events in each cell is

$$estimated \ number \ of \ events = \hat{y}_{ij} = p_{ij}e^{\hat{a}+\hat{b}_i+\hat{c}_j} = p_{ij}\,e^{\hat{a}}\,e^{\hat{b}_i}\,e^{\hat{c}_j} = p_{ij}\,\hat{A}\,\hat{B}_i\,\hat{C}_j. \quad (12.13)$$

Using the Oak Ridge data, for example, $\hat{y}_{34} = (1918)e^{-11.992+0.426+4.732} = (1918)e^{-6.833} = 2.067$ deaths. In addition, the estimated rates are $\hat{r}_{ij} = (\hat{y}_{ij}/p_{ij}) \times 100{,}000$ (e.g., $\hat{r}_{34} = (2.067/1918) \times 100{,}000 = 107.800$ deaths/100,000). All 56 estimated counts $(\hat{y}_{ij})$ and rates $(\hat{r}_{ij})$ from the exposure/age data make up Table 12.15.

T A B L E   **12.15**

Estimated numbers of deaths, person-years, and estimated cancer mortality rates for workers at Oak Ridge National Laboratory (1943–1972) classified by exposure and age

mSv/age	< 45	45–49	50–54	55–59	60–64	65–69	> 69
mSv = 0							
Deaths	0.19	1.20	1.59	2.90	2.30	3.45	3.36
Person-years	29901	6251	5251	4126	2778	1607	1358
Rate/100,000	0.62	19.18	30.36	70.38	82.76	214.95	247.75
mSv = 0–19							
Deaths	0.65	4.72	6.15	10.81	8.69	13.08	13.91
Person-years	71382	16705	13752	10439	7131	4133	3814
Rate/100,000	0.91	28.23	44.39	103.59	121.82	316.41	364.69
mSv = 20–39							
Deaths	0.06	0.71	1.06	2.07	1.68	2.38	2.04
Person-years	6523	2423	2281	1918	1322	723	538
Rate/100,000	0.95	29.37	46.51	107.80	126.77	329.25	379.49
mSv = 40–59							
Deaths	0.03	0.42	0.66	1.30	1.08	1.82	1.69
Person-years	2341	972	958	816	578	375	303
Rate/100,000	1.40	43.28	68.53	156.84	186.79	485.15	559.18
mSv = 60–79							
Deaths	0.01	0.06	0.10	0.19	0.13	0.25	0.26
Person-years	1363	478	476	387	225	164	150
Rate/100,000	0.43	13.44	21.27	49.31	57.98	150.60	173.58
mSv = 80–99							
Deaths	0.01	0.06	0.09	0.19	0.17	0.29	0.19
Person-years	779	296	282	251	193	125	69
Rate/100,000	0.67	20.77	32.89	76.24	89.65	232.86	268.39
mSv = 100–119							
Deaths	0.01	0.14	0.26	0.51	0.35	0.50	0.22
Person-years	520	188	217	184	109	60	23
Rate/100,000	2.43	75.08	118.88	275.56	324.05	841.64	970.07
mSv ≥ 120							
Deaths	0.05	0.69	1.09	2.03	1.60	2.23	1.32
Person-years	2104	1027	1029	827	555	297	153
Rate/100,000	2.16	66.86	105.87	245.38	288.56	749.48	863.85

Goodness-of-Fit

The differences between the estimated values (Table 12.15) and observed values (Table 12.13) can be evaluated by a Pearson chi-square statistic or, as usual,

$$X^2 = \sum_{all\ cells} \frac{(y_{ij} - \hat{y}_{ij})^2}{\hat{y}_{ij}}$$

(12.14)

has an approximate chi-square distribution with $km - s$ degrees of freedom when the additive Poisson model is an accurate representation of the data (expression 12.12). The value s is the number of independent parameters necessary to specify the model. For the Oak Ridge workers, $X^2 = 30.82$ with $56 - 14 = 42$ degrees of freedom. The probability of observing a more extreme chi-square value is 0.905 when the estimated and observed values differ by chance alone, implying that the observed values are accurately represented by the Poisson model.

Sometimes the Pearson chi-square statistic can be misleading large when one or a few of the estimated numbers of events are small (i.e., $\hat{y}_{ij}$ is small; discussed in Chapter 8). These small values can lead to inflated values of X^2 giving the impression of a poor correspondence between observed and estimated values when, in fact, only a few observations are poorly represented by the model.

Inference

Perhaps the most basic issue in a Poisson regression analysis is the assessment of the role of each categorical variable. The estimated coefficients represent the influence of a series of exposure or stratum categories in a Poisson regression, but it is difficult to evaluate the total effect of an independent variable from these coefficients. The z-statistics formed by dividing an estimated coefficient by its standard error (i.e., $\hat{b}_i/S_{\hat{b}_i}$ or $\hat{c}_j/S_{\hat{c}_j}$ usually part of computer analysis programs) reflect the importance of each category. However, a series of evaluations (z-tests) of each individual coefficient is not the best way to assess the total influence of the exposure or stratum variable. A likelihood statistic provides a single assessment of a variable regardless of the number of categories used to describe a variable.

Likelihood statistics measure the goodness-of-fit of a Poisson model in the same way as described for the logistic and proportional hazards models. As before, the fundamental analytic tool is the difference in transformed likelihood statistics (deviances).

Recall, −2 times the log-likelihood value has an approximate chi-square distributions (Chapter 10) when the data deviate from the model only because of random variation. The Poisson model applied to the Oak Ridge data yields a transformed log-likelihood value of $-2log(L) = 31.819$ with 42 degrees of freedom. The probability that a greater likelihood value would arise strictly from random variation is then 0.873 (p-value), implying that no reason exists to reject the proposed Poisson model (expression 12.12) as a representation of the data. A likelihood goodness-of-fit approach usually produces results similar to the Pearson chi-square comparison of the observed and estimated number of events (31.819 versus 30.820 for the Oak

Ridge data), particularly for large data sets with few sparse cells. Both approaches measure the agreement between the data and the model generated values and have the same degrees of freedom.

To illustrate the comparison of two nested models, consider again the data from the workers at the Oak Ridge National Laboratory. If the variable reflecting cumulative radiation exposure is ignored, the Poisson model becomes

$$rate = AC_j \quad \text{or} \quad log\text{-}rate = a + c_j \tag{12.15}$$

That is, seven exposure parameters are set to zero (i.e., $b_2 = b_3 = \cdots = b_8 = 0$) and the model is based entirely on parameters representing the age categories of the workers (c_j-coefficients). Estimated parameters for this restricted model are not the main focus but rather the increase in the log-likelihood value, which indicates the impact of the deleted variable. The log-likelihood value for the model including both age and exposure variables is $-2\log(L_1) = 31.819$ with 42 degrees of freedom. Excluding the exposure variable increases the log-likelihood to $-2\log(L_0) = 42.84$ with $56 - 7 = 49$ degrees of freedom. The contrast of these two models produces a deviance of $Y^2 = 42.84 - 31.819 = 11.02$, which has an approximate chi-square distribution with $49 - 42 = 7$ degrees of freedom (the number of parameters set to zero) when radiation exposure has no systematic influence on the risk of cancer mortality. The probability of observing a greater value of Y^2 when the exposure variable is unrelated to risk is 0.138, providing weak evidence, at best, of an association between radiation exposure and cancer.

The radiation exposure data also demonstrate the known association between age and cancer risk. The model with age excluded is

$$rate = AB_i \quad \text{or} \quad log\text{-}rate = a + b_i \tag{12.16}$$

and produces a log-likelihood value of $-2\log(L_0) = 250.007$ with $56 - 8 = 48$ degree of freedom. The deviance is $Y^2 = 250.007 - 31.819 = 218.188$, which has an approximate chi-square distribution with $48 - 42 = 6$ degrees of freedom when age is unrelated to the risk of cancer. A chi-square value of $Y^2 = 218.19$ is extremely unlikely to occur because of chance variation (p-value < 0.001).

A goodness-of-fit procedure must be applied to the entire table that produced the data. Occasionally the marginal frequencies are calculated for a particular variable and the goodness-of-fit procedure applied to these values. Data that fit a model requiring the logarithms of the rates to be an additive function of the parameters will produce marginal frequencies that also fit an additive model. However, the converse is not true. The fact that the marginal frequencies of a table fit an additive Poisson model is not good evidence that the data follow an additive Poisson model in general. Specifically, a value of a chi-square statistic could indicate a close fit when applied to the marginal frequencies of a table while the chi-square statistic calculated from the whole table shows a substantial lack of fit.

The estimated rate ratios calculated directly from an additive Poisson model (expression 12.12) are one way to describe effectively the separate effects of the exposure and stratum variables on the outcome. An additive Poisson model implies

T A B L E **12.16**

Rate ratios associated with exposure categories for the Oak Ridge data

mSv	0	0–19	20–39	40–59	60–79	80–99	100–119	≥ 120
$\hat{r}_{ij}/\hat{r}_{0j}$	1.00	1.47	1.53	2.26	0.70	1.08	3.92	3.49
Unadjusted ratio	1.00	1.56	2.17	3.77	1.05	1.71	5.25	5.13

that each category has a separate multiplicative influence on the rate. The separate multiplicative influences are easily expressed as a rate ratio when a specific category is chosen as a reference group. For example, the nonexposed Oak Ridge workers make a natural reference group. Using these baseline workers, the model estimated rate ratios for the other seven exposure categories are given in Table 12.16. Table 12.16 also contains the observed unadjusted rate ratios (age ignored) as a contrast to the age-adjusted values estimated from the Poisson model. The unadjusted rate ratios are calculated from the marginal totals (Table 12.13) formed by adding the deaths and the person-years across the seven age categories.

The additive (multiplicative) Poisson model guarantees that the influences of the row variable is unaffect by the influences of the column variable and vice versa. Similar to the logistic model, two variables with additive influences are often said to produce independent effects. When effects are additive, the rate ratios are a convenient way of expressing the impact of a specific variable. That is,

$$estimated\ rate\ ratio = \frac{\hat{r}_{ij}}{\hat{r}_{0j}} = \frac{e^{\hat{a}+\hat{b}_i+\hat{c}_j}}{e^{\hat{a}+\hat{b}_0+\hat{c}_j}} = e^{\hat{b}_i-\hat{b}_0} \tag{12.17}$$

where $\hat{b}_0$ is the coefficient associated with the reference group. The estimated rate ratio quantifies the influence of the row variable (e.g., radiation exposure) adjusted for the influence of the column variable (e.g., age). The estimated rate ratios for the radiation exposure data show an inconsistent pattern of risk with increasing radiations exposure but, not surprisingly, the high rates of exposure have associated high risks of cancer (rate ratios of 3.92 and 3.49). Without a model, patterns of influence are often not easily discerned from the data itself. Rate ratios calculated directly from the data are typically subject to rather extreme sampling variation due to the small numbers of observed events. More important, these rate ratios are not adjusted and, therefore, do not account for the influence of the other independent variables, a primary goal of a multivariable analysis.

Geometry

The two-variable additive Poisson model dictates that the differences between the logarithm of the stratum-specific rates are constant at all levels of the other variable. Requiring constant differences between logarithms is just another way of requiring

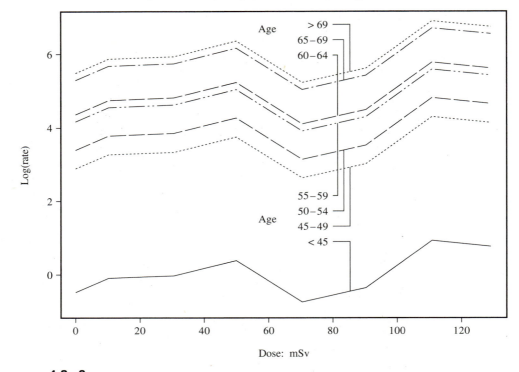

Dose by age for the Oak Ridge workers—values generated from the Poisson model

separate multiplicative effects. For example, expression 12.17 shows that the influence of radiation exposure estimated from the Poisson model is unaffected by age (depends only on the parameters b_i). Figure 12.3 is a plot of the logarithms of the estimated rates from the radiation exposure data for each of the seven age groups. Geometrically, the additive Poisson model guarantees that the estimated log-rates when plotted for constant levels of the other independent variable form parallel lines.

If the additive Poisson model does not describe the data, then influences of one variable can not be summarized without specific reference to other variable, which means an interaction exists. The relationship between one independent variable and the number of events (dependent variable) depends on the level of another independent variable. Lack of additivity means that each rate ratio must be considered for each category of the other variable. Geometrically, the logarithm of the rates do not form parallel lines. It is the additivity of the parameters that allows a separate summary description of each variable, such as the description of the effects of radiation exposure (Table 12.16) without reference to age.

Table 12.17 summarizes four possible Poisson models applied to the two-way classification of the Oak Ridge Laboratory data.

T A B L E **12.17**
Summary of the four Poisson regression models applied to the Oak Ridge Laboratory data

Model	$\log(r_{ij})$	Number of Parameters	$-2\log(L)$	Degrees of Freedom
No age/exposure influence	a	1	269.42	55
No exposure influence	$a + c_j$	7	42.84	49
No age influence	$a + b_i$	8	250.01	48
Age/exposure influence	$a + b_i + c_j$	14	31.82	42

Poisson Regression Model: Multidimensional Table

The Poisson model discussed in the context of an m-by-k table generalizes to any size table. A part of a data set from a prospective study of heart disease [8] illustrates the Poisson analysis of a multidimensional table of rates. The study observed 3154 men ages 40 to 50 for an average of about eight years and recorded the incidence of cases of coronary heart disease (CHD). The risk variables used here are reported smoking exposure (0, 1–20, 21–30, and more than 30 cigarettes per day), blood pressure (<140 and ≥ 140), and behavior type (type A and type B). The data are given in Table 12.18.

The rates of CHD show increased smoking generally associated with increased rates of heart disease. High blood pressure is also associated with higher rates of disease. Similarly, high rates of CHD are associated with type A behavior. On the other hand, a number of questions are not easily answered without further analysis. For example:

Do these three variables adequately reflect the rate of heart disease?

What is the pattern of rates from increased smoking?

What is the relative influence of each of the three risk factors on the rate of CHD?

What is the estimated rate of CHD from various combinations of these risk variables?

What is the impact of random variation on the answers to these questions?

Modern computer technology makes it possible to use practically any model to describe the relationship between dependent and independent variables. A good place to start is a simple linear relationship. Before considering behavior type and blood pressure, the question of the relationship of smoking exposure to the rate of CHD is addressed with a simple linear model. For this approach, a quantitative measure of smoking exposure is entered directly into the model (F_i); no cigarettes smoked per day = $F_0 = 0$; 1–20 cigarettes smoked per day = $F_1 = 10$; 21–30 cigarettes smoked per day = $F_2 = 20$; and more than 30 cigarettes smoked per day = $F_3 = 30$.

T A B L E **12.18**

Cases of CHD, person-years, and rates for individuals classified by smoking exposure, behavior type, and blood pressure

	Type A and Blood Pressure ≥ 140			
Cigs/Day	0	1 to 20	21 to 30	30+
CHD	29	21	7	12
Person-years	1251.9	640.0	374.5	338.2
Rate/1000	23.17	32.81	18.69	35.48

	Type A and Blood Pressure < 140			
Cigs/Day	0	1 to 20	21 to 30	30+
CHD	41	24	27	17
Person-years	4451.1	2243.5	1153.6	925.0
Rate/1000	9.21	10.70	23.41	18.38

	Type B and Blood Pressure ≥ 140			
Cigs/Day	0	1 to 20	21 to 30	30+
CHD	8	9	3	7
Person-years	1366.8	497.0	238.1	146.3
Rate/1000	5.85	18.11	12.60	47.84

	Type B and Blood Pressure < 140			
Cigs/Day	0	1 to 20	21 to 30	30+
CHD	20	16	13	3
Person-years	5268.2	2542.0	1140.7	614.6
Rate/1000	3.80	6.29	11.40	4.88

A simple linear Poisson model is

$$rate = AB^{F_i} \quad \text{or} \quad log\text{-}rate = a + bF_i \tag{12.18}$$

where $\log(B) = b$ represents the change in the logarithm of the rate per unit increase in the variable F_i. Notice that F_i does not represent a specific category but rather a quantitative independent variable.

The estimated parameters of this model using the coronary heart disease data (Table 12.18) are $\hat{a} = -4.799$ and $\hat{b} = 0.0318$. The rate of CHD increases by a factor

T A B L E **12.19**

Summary of the analysis of smoking risk and CHD

Cigs/Day	0	10	20	30
Cases	98	70	50	39
Person-years	12337.9	5922.5	2906.9	2024.5
Rates/1000	7.94	11.82	17.20	19.27
Estimated cases	101.61	67.00	45.18	43.21
Estimated rates	8.24	11.31	15.54	21.35

of $e^{0.0318} = 1.032$ for every cigarette smoked. That is, the rate of CHD in smokers of one pack per day (20 cigarettes) is estimated to be $(1.032)^{20} = 1.88$ times higher than the rate for nonsmokers based on a simple linear model. A test for trend or "dose-response" consists of evaluating the magnitude of the estimate $\hat{b}$ (i.e., $z = \hat{b}/S_{\hat{b}}$).

More important, the log-likelihood value $-2log(L) = 1.184$ with two degrees of freedom indicates that a straight line relationship between the amount smoked and the logarithm of the rate of CHD is a good summary (Table 12.19). A less good fit would occur by chance with probability 0.553 (a chi-square statistic comparing observed and estimated numbers of cases of CHD yields essentially the same result). A Poisson model describing a linear trend gives the same results for any set of equally spaced independent variables. For the smoking/CHD model, $F_i = 0, 1, 2,$ and 3 gives the same fit between observed and estimated numbers of CHD cases (same log-likelihood value) as the values $F_i = 0, 10, 20,$ and 30. This invariance property was previously noted in connection with the analysis of a 2-by-k table (Chapter 8).

A four-parameter additive Poisson model is a basis for answering questions concerning the three risk variables and the rate of CHD where

$$log\text{-}rate = a + bF_i + cC + dD .$$
(12.19)

This model is no more than the linear relationship between smoking and CHD (expression 12.18) extended to include possible influences from behavior type and blood pressure. Type A behavior is coded $C = 1$ and type B individuals are coded $C = 0$, whereas low blood pressure <140mm is coded $D = 0$ and high blood pressure $\geq$ 140 mm is coded $D = 1$. These variables and Poisson regression estimation applied to the CHD data in Table 12.18 give the four estimated model parameters (expression 12.19) in Table 12.20.

Again, the smoking data are not a categorical variable but the number of cigarettes smoked per day (actually coded values 0, 10, 20, and 30) directly applied to estimating the model parameters. Using the reported number of cigarettes smoked provides a single coefficient that efficiently represents the risk from smoking but requires the relationship to be constrained in a specific way, namely linear. The estimated coefficients (Table 12.20) produce the number of estimated cases of CHD as

T A B L E **12.20**

Results from a Poisson regression applied to the CHD data

Coefficient	Estimates	Std Error	Z-Statistic
a	−5.420	0.130	—
b	0.027	0.006	4.875
c	0.753	0.136	5.531
d	0.753	0.129	5.834

$$-2\log(L) = 21.241$$

well as the estimated rates given in Table 12.21. The estimated values labeled "deaths" and "rates" (Table 12.21) are calculated from the Poisson model

$$estimated\ log\text{-}rate = \log(\hat{r}_{ij}) = -5.420 + 0.027F_i + 0.753C + 0.753D\ .$$

The fit of this model to the data can be a evaluated by a Pearson chi-square statistic or the log-likelihood value.

T A B L E **12.21**

Further results from a Poisson regression applied to the prospectively collected CHD data

Person-Years	Smoking	Blood Pressure	Behavior	Deaths	"Deaths"	Rate	"Rate"
5268.2	0	0	0	20	23.32	3.80	4.43
2542.0	10	0	0	16	14.79	6.29	5.82
1140.7	20	0	0	13	8.72	11.40	7.65
614.6	30	0	0	3	6.18	4.88	10.05
4451.1	0	0	1	41	41.82	9.21	9.39
2243.5	10	0	1	24	27.71	10.70	12.35
1153.6	20	0	1	27	18.73	23.41	16.23
925.0	30	0	1	17	19.74	18.38	21.34
1366.8	0	1	0	8	12.85	5.85	9.40
497.0	10	1	0	9	6.14	18.11	12.36
238.1	20	1	0	3	3.87	12.60	16.25
146.3	30	1	0	7	3.12	47.84	21.35
1251.9	0	1	1	29	24.98	23.17	19.96
640.0	10	1	1	21	16.79	32.81	26.23
374.5	20	1	1	7	12.91	18.69	34.48
338.2	30	1	1	12	15.33	35.48	45.32

Note: "Deaths" = estimated number of deaths and "rates" = estimated rates per 1000 person-years.

The direct comparison of the observed numbers of deaths and the estimated numbers of deaths produces a chi-square statistic of $X^2 = 22.168$ with $16 - 4 = 12$ degrees of freedom. A more extreme result would occur by chance alone with probability of 0.036. Similarly, the transformed likelihood for the CHD model (expression 12.19) is $-2\log(L) = 21.241$ and the probability associated with this chi-square statistic (degrees of freedom also $= 12$) is 0.047. Both approaches show that a simple four-parameter additive model is not an extremely good representation of the data.

A lack of correspondence between observed and estimated values can be generally poor or the fit can be adequate or good with the exception of a few poorly estimated values. A direct comparison of each estimated and observed value usually reveals which of these two possibilities cause a large chi-square statistic. This type of detailed comparison also suggests ways a model can be modified to better fit the data. Comparing the observed to the estimated values (columns labeled deaths to "deaths"; Table 12.21) shows generally good agreement with the exception of one category (heavy smokers, with high blood pressure who are type B). The number of observed deaths is 7 and the number estimated from the model is 3.12 or, in terms of rates, 47.84 per 1000 person-years (observed) versus 21.35 per 1000 person-years (estimated). Because one cell is responsible for much of the lack of fit, inferences based on the four-parameter model give a simple and general picture of the relationships of the three risk factors to coronary heart disease.

To assess the separate influences of each independent variable, a sequence of Poisson regression analyses are performed with the variables of interest removed from the model and the resulting log-likelihood compared to the log-likelihood value from the model retaining the variable. That is, two measures of fit are calculated for the model, one with all variables included ($-2\log(L_1)$) and the other with one variable excluded ($-2\log(L_0)$). To illustrate, the deviances for smoking, blood pressure, and behavior type are given in Table 12.22 where the log-likelihood value for the Poisson model containing all three variables is $-2\log(L_1) = 21.241$ with 12 degrees of freedom. All three variables have important influences on the rate of CHD.

The deviance not only serves to test the importance of a specific variable but, as noted, allows a comparison of the relative influences from different variables. The deviance is directly proportional to the unique impact of each variable. For example, the blood pressure (*deviance* $= 31.316$) and behavior type (*deviance* $= 32.870$) mea-

T A B L E **12.22**

Summary of the models to assess the risk variables from the CHD data

Variable Removed	$-2\log(L_0)$	Degrees of Freedom	Deviance	$df_1 - df_0$	Significance
Smoking	43.601	13	22.360	1	< 0.001
Behavior type	54.111	13	32.870	1	< 0.001
Blood pressure	52.557	13	31.316	1	< 0.001

Note: Significance = probability of a more extreme value by chance alone.

sured by binary variables have almost identical impacts on the rate of coronary disease and smoking has relatively less influence (*deviance* = 22.360). When a single variable is assessed, the deviance is approximately equal to the square of the z-statistic (i.e., $Y^2 \approx z^2$) when z is a standardized coefficient (see Table 12.20). For example, blood pressure has a deviance of 31.316 and $z = 5.834$ where $(5.834)^2 = 34.036$. Therefore, as mentioned, a z-statistic also serves as a standardized value allowing direct comparisons of the influence of the independent variables.

The Joint Influence of the Independent Variables

The parameters estimated from a Poisson model can be combined to give a summary of the joint influences from a series of variables. The estimated rate ratios, calculated relative to an arbitrary reference group, can be combined to reflect the influence of combinations of the independent variables. For the CHD data, a natural reference group is the nonsmokers who are type B with low blood pressure. The estimated rate ratios associated with each variable are:

Smoking = $e^{0.027} = 1.028$ for each cigarette smoked

Behavior type = $e^{0.753} = 2.124$, type A versus type B

Blood pressure = $e^{0.753} = 2.122$, high versus low

Each rate ratio measures the influence of a specific variable while accounting for the influence of the other variables in the additive Poisson model. Combining the rate ratios calculated from the coefficients given in Table 12.20 produces estimates of the joint influences of the risk variables, or

$$estimated\ rate\ ratio = e^{0.027F_i + 0.753C + 0.753D}$$
$$= (e^{0.027})^{F_i} (e^{0.753})^C (e^{0.753})^D = (1.028)^{F_i} (2.124)^C (2.122)^D .$$

This expression for the estimated rate ratios yields the values in Table 12.23. For example, the rate of CHD among type A individuals with high blood pressure who are heavy smokers is estimated to be about 10 times the rate of nonsmokers who are type B with low blood pressure (the reference group).

Typically a Poisson regression is applied to data consisting of small rates (*rates* < 0.01). Small rates are almost always the case in the study of disease or mortality.

T A B L E **12.23**

Estimated rate ratios for CHD by smoking exposure, behavior type, and blood pressure

Blood Pressure	Behavior Type	$F = 0$	$F = 10$	$F = 20$	$F = 30$
<140	B	1.00	1.31	1.73	2.27
≥140	B	2.12	2.79	3.67	4.82
<140	A	2.12	2.79	3.67	4.82
≥140	A	4.51	5.93	7.79	10.24

When a rate is small, the logarithm of a rate is approximately equal to the logarithm of the corresponding probability (i.e., $rate \approx q$), which is approximately equal to the logarithm of the odds $(1 - q \approx 1)$, or

$$\log(rate) \approx \log(q) \approx \log(q/[1 - q]) \tag{12.20}$$

when q is small. Therefore, the dependent variables for both Poisson and logistic regression models are approximately equal when the event being studied is rare. In addition, a Poisson distribution is not very different from a binomial distribution when the number of observations is large and the probability of an event is small (see chapter appendix). For example, if the probability of any specific event is 0.01 and $n = 100$, then the probability of, say, zero events is $P(Y = 0) = 0.366$ based on a binomial distribution and $P(Y = 0) = 0.368$ based on the Poisson distribution. Both these facts imply that a logistic regression differs little from a Poisson regression for small rates.

Poisson Regression

The implementation of a Poisson regression is illustrated with the hypothetical data given in Table 12.8. Remember, these artificial data are generated from a product-model, so the observations perfectly fit the values from the "estimated" parameters.

example.poisson

```
. infile pop deaths group age using pdata.dat
(10 observations read)

. list
          pop      deaths       group        age
  1.     1000          20           0          1
  2.     2000          60           0          2
  3.     3000         135           0          3
  4.     4000         360           0          4
  5.     5000         900           0          5
  6.      100           4           1          1
  7.      200          12           1          2
  8.      300          27           1          3
  9.      400          72           1          4
 10.      500         180           1          5

. tab age, gen(a)
      age|       Freq.      Percent        Cum.
----------+-----------------------------------
        1 |           2        20.00       20.00
        2 |           2        20.00       40.00
        3 |           2        20.00       60.00
        4 |           2        20.00       80.00
        5 |           2        20.00      100.00
----------+-----------------------------------
    Total |          10       100.00
```

```
. poisson deaths a2-a5 group, exposure(pop)

Poisson regression, normalized by pop          Number of obs    =        10
Goodness-of-fit chi2(4)       =    -0.000        Model chi2(5)   =   829.682
Prob  chi2                    =     1.0000       Prob  chi2      =    0.0000
Log Likelihood               =    -29.889       Pseudo R2       =    0.9328
```

$\hat{b}_i$

```
------------------------------------------------------------------------------
   deaths |     Coef.   Std. Err.       t     P|t|     [95% Conf. Interval]
----------+-------------------------------------------------------------------
       a2 |   .4054651   .2357022     1.720    0.086    -.0568189    .8677491
       a3 |   .8109303   .2187224     3.708    0.000     .3819489    1.239912
       a4 |  1.504077    .2097176     7.172    0.000    1.092757    1.915398
       a5 |  2.197225    .2063797    10.647    0.000    1.792451    2.601998
    group |   .6931472   .0637793    10.868    0.000     .5680566    .8182378
    _cons |  -3.912023   .2044007   -19.139    0.000    -4.312915   -3.511131
------------------------------------------------------------------------------
```

a

```
. predict d1

. gen temp1=pop*exp(d1)

. list deaths temp1
```

r_{ij} $\hat{r}_{ij}$

```
          deaths      temp1
 1.           20         20
 2.           60         60
 3.          135        135
 4.          360        360
 5.          900        900
 6.            4          4
 7.           12         12
 8.           27         27
 9.           72         72
10.          180        180

. exit,clear
```

Applied Example

Poisson Regression Analysis

United States mortality rates are analyzed with STATA to compare the pattern of mortality for 1940 to the pattern observed in 1960 with the goal of assessing the increased risk of death from smoking-related diseases (Table 12.24).

The specialized term "offset" is used in the STAT Poisson regression command (i.e. "Poisson deaths group s2–s11, offset(lpop)"). Most Poisson regression programs require an "offset" variable be defined as part of the analysis. The term "offset" has a general definition. In the context of Poisson regression an offset value is included in the model as an additive effect that is not an independent variable. That is, an offset by the person-years provides a weighting for each number of deaths (dependent variable) but is not a risk or stratum variable and, therefore, not a statistical quantity to be analyzed. For the Poisson model to include person-years as a weight for the num-

T A B L E   **12.24**

Mortality data from the United States for 1940 and 1960

	1940			1960		
Age	Person-Years	Deaths	Rates	Person-Years	Deaths	Rates
<1	1,784,033	141	7.90	906,897	45	4.96
1–4	7,065,148	926	13.11	3,794,573	201	5.30
5–14	15,658,730	1253	8.00	10,003,544	320	3.20
15–24	10,482,916	1080	10.30	10,629,526	670	6.30
25–34	9,939,972	1869	18.80	9,465,330	1126	11.90
35–44	10,563,872	4891	46.30	8,249,558	3160	38.31
45–54	9,114,202	14,956	164.10	7,294,330	9723	133.30
55–64	6,850,263	30,888	450.90	5,022,499	17,935	357.09
65–74	4,702,482	41,725	887.30	2,920,220	22,179	759.50
75–84	1,874,619	26,501	1413.67	1,019,504	13,461	1320.35
85+	330,915	5928	1791.40	142,532	2238	1570.17
Total	78,367,152	130,158	166.09	59,448,513	71,058	119.53

Note: Rates per 100,000 person-years

ber of deaths, it is necessary to use the logarithm of the person-years rather than the actual person-years measured. Also note the use of design variables "s2–s11" to reflect the 11 age categories in the analysis.

```
mort.poisson

. infile group strata deaths pop using mort.dat
(22 observations read)

. tab strata, generate(s)
    strata|     Freq.     Percent        Cum.
------------+-----------------------------------
         1 |         2        9.09        9.09
         2 |         2        9.09       18.18
         3 |         2        9.09       27.27
         4 |         2        9.09       36.36
         5 |         2        9.09       45.45
         6 |         2        9.09       54.55
         7 |         2        9.09       63.64
         8 |         2        9.09       72.73
         9 |         2        9.09       81.82
        10 |         2        9.09       90.91
        11 |         2        9.09      100.00
------------+-----------------------------------
     Total |        22      100.00
```

```
. generate lpop=log(pop)

. poisson deaths group  s2-s11, offset(lpop)

Poisson regression, normalized by exp(lpop)      Number of obs   =      22
Goodness-of-fit chi2(10)    =    517.250          Model chi2(11)  =487070.00
Prob  chi2                  =      0.0000          Prob  chi2      =  0.0000
Log Likelihood              =   -365.875          Pseudo R2       =  0.9985

------------------------------------------------------------------------------
  deaths |    Coef.    Std. Err.       t      P|t|      [95% Conf. Interval]
---------+--------------------------------------------------------------------
   group | -.1822272   .0046791    -38.945   0.000    -.1913981    -.1730563
      s2 |  .4085851   .0791433      5.163   0.000     .2534662     .563704
      s3 | -.1107801   .0775379     -1.429   0.153    -.2627525     .0411923
      s4 |  .2114682   .0771255      2.742   0.006     .0603042     .3626322
      s5 |  .8302612   .0755691     10.987   0.000     .6821476     .9783749
      s6 |  1.841193   .0741671     24.825   0.000     1.695827     1.986559
      s7 |  3.099207   .0736009     42.108   0.000     2.954951     3.243463
      s8 |  4.101147   .0734641     55.825   0.000     3.957159     4.245134
      s9 |  4.806312   .0734305     65.454   0.000      4.66239     4.950234
     s10 |  5.299837   .073494      72.112   0.000     5.155791     5.443884
     s11 |  5.513262   .0741541     74.349   0.000     5.367922     5.658602
   _cons | -9.52187    .0733368   -129.838   0.000    -9.665608    -9.378132
------------------------------------------------------------------------------

. predict temp

. generate pdeaths=pop*exp(temp)

. list group strata pop deaths pdeaths

         group    strata        pop    deaths    pdeaths
  1.         0         1    1784033       141   130.6494
  2.         0         2    7065148       926   778.5237
  3.         0         3    1.57e+07      1253   1026.478
  4.         0         4    1.05e+07      1080   948.4751
  5.         0         5    9939972      1869   1669.812
  6.         0         6    1.06e+07      4891   4876.941
  7.         0         7    9114202     14956   14804.43
  8.         0         8    6850263     30888   30305.19
  9.         0         9    4702482     41725   42110.12
 10.         0        10    1874619     26501    27498.4
 11.         0        11     330915      5928   6008.972
 12.         1         1     906897        45   55.35059
 13.         1         2    3794573       201   348.4763
 14.         1         3    1.00e+07       320   546.5209
 15.         1         4    1.06e+07       670   801.5258
 16.         1         5    9465330      1126   1325.189
 17.         1         6    8249558      3160   3174.058
 18.         1         7    7294330      9723   9874.565
 19.         1         8    5022499     17935    18517.8
 20.         1         9    2920220     22179   21793.89
 21.         1        10    1019504     13461   12463.59
 22.         1        11     142532      2238   2157.028

. exit,clear
```

Notation

Direct age-adjustment:

$$\text{direct age-adjustment rate} = r_i^* = \frac{\text{estimated deaths}}{P} = \frac{\sum_{j=1}^{m} r_{ij} P_j}{P}$$

Indirect age-adjustment:

$$\text{expected number of deaths} = e_i = \sum_{j=1}^{m} p_{ij} R_j$$

Standardized mortality ratio:

$$SMR_i = \frac{\text{total observed number of deaths}}{\text{total expected number of deaths}} = \frac{\sum_{j=1}^{m} d_{ij}}{e_i} = \frac{D_i}{e_i}$$

Poisson distribution:

$$\text{probability of } y_{ij} \text{ events} = \frac{e^{-\lambda_{ij}} \lambda_{ij}^{y_{ij}}}{y_{ij}!}$$

Poisson model (example):

$$\text{log-rate} = \log(\text{rate}) = \log\left(\frac{\lambda_{ij}}{p_{ij}}\right) = a + b_i + c_j$$

Estimated Poisson model (example)

$$\text{estimated number of events} = \hat{y}_{ij} = p_{ij} e^{\hat{a}+\hat{b}_i+\hat{c}_j} = p_{ij} e^{\hat{a}} e^{\hat{b}_i} e^{\hat{c}_j} = p_{ij} \hat{A} \hat{B}_i \hat{C}_j$$

Rate ratio from Poisson model (example):

$$\text{rate ratio} = \frac{\hat{r}_{ij}}{\hat{r}_{0j}} = \frac{e^{\hat{a}+\hat{b}_i+\hat{c}_j}}{e^{\hat{a}+\hat{b}_0+\hat{c}_j}} = e^{\hat{b}_i - \hat{b}_0}$$

Appendix: Poisson Probability Distribution

The Poisson probability distribution is, of course, basic to the application of Poisson regression methods. It is probably worthwhile to review some of the properties of this distribution. Complete descriptions are found in most introductory statistics texts (e.g., [9] and [10]).

The Poisson distribution can be developed from a number of points of view. One perspective starts with the binomial distribution and develops the relationship between this distribution and the Poisson distribution.

A binomially distributed variable is a sum of a series of variables with two outcomes. That is,

1 A binary variable X_i is defined that takes on either 1 or 0 with probabilities p and $1 - p$, respectively.

2 Each variable X_i is statistically independent.

3 The probability p is the same for all values X_i.

The binomial variable (represented by X) is the sum of n values of the random variable X_i or

$$X = X_1 + X_2 + \cdots + X_n .$$

More simply, X is the count of the number of times X_i equals 1. The binary character of X_i makes the binomial distribution applicable to the description of a wide range of data with two outcome possibilities: alive or dead, male or female, case or control and, in general, event A or event not A.

The probabilities associated with a specific value of binomial variable X, denoted as k, are given by the expression

$$probability\ of\ k\ events = P(X = k) = \binom{n}{k} p^k (1 - p)^{n-k} \quad where\ k = 0, 1, 2, \cdots, n .$$

The term

$$\binom{n}{k}$$

is the number of different ways k values of 1 can occur among n values of 0 and 1. As usual the term

$$\binom{n}{k} = n! / k!(n - k)! .$$

The term $p^k(1 - p)^{n-k}$ is the probability of a specific configuration of k values of 1 and $n - k$ values of 0. The product of the two terms is the probability that k values of X_i will equal 1 and $n - k$ values will equal 0 and, therefore, the sum X equals k. A direct result of these binomial probabilities is that the expected value of X sampled from a binomial probability distribution is

$$expected\ binomial\ value = \sum_{k=0}^{n} P(X = k)\ k = np$$

and the variance associated with the binomial variable X is

$$variance\ of\ X = \sum_{k=0}^{n} P(X = k)\ (k - np)^2 = np(1 - p) .$$

To illustrate, consider the case for $n = 10$ and $p = 0.4$, shown in Table A.1. The expected value is $np = 10(0.4) = 4$ and the variance is $np(1 - p) = 10(0.4)\ (0.6) = 2.4$.

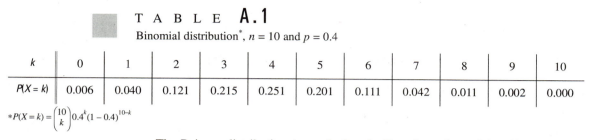

T A B L E **A.1**

Binomial distribution*, $n = 10$ and $p = 0.4$

k	0	1	2	3	4	5	6	7	8	9	10
P(X = k)	0.006	0.040	0.121	0.215	0.251	0.201	0.111	0.042	0.011	0.002	0.000

$*P(X = k) = \binom{10}{k} 0.4^k (1 - 0.4)^{10-k}$

The Poisson distribution (named after the French mathematician Simeon Denis Poisson) can be justified from the following considerations:

Starting with the binomial distribution

$$P(X = k) = \binom{n}{k} p^k (1 - p)^{n-k} = \frac{n(n-1)(n-2)\cdots(n-k+1)}{k!} p^k (1 - p)^{n-k}$$

and rearranging a bit gives

$$P(X = k) = \frac{1\left(1 - \frac{1}{n}\right)\left(1 - \frac{2}{n}\right)\cdots\left(1 - \frac{k-1}{n}\right)}{k!} (np)^k \left(1 - \frac{np}{n}\right)^{n-k}.$$

If n is large and p is small, using

$$\left(1 - \frac{np}{n}\right)^{n-k} \approx e^{-np},$$

then

$$\text{probability of } k \text{ events} = P(X = k) = \frac{(np)^k e^{-np}}{k!} = \frac{\lambda^k e^{-\lambda}}{k!}.$$

The Greek letter $\lambda = np$ is a traditional notation associated with a Poisson distribution. This expression describes the Poisson probabilities viewed as a limiting case of the binomial distribution. That is, the binomial distribution becomes indistinguishable from the Poisson distribution when n (the number of X_i-values) is large and $P(X_i = 1) = p$ is small. The Poisson distribution can be derived from other considerations and, like the binomial, plays a important role in describing categorical data.

The expectation and variance are also derived from the probabilities $P(X = k)$. The expected value of a variable X sampled from a Poisson distributed is

$$\text{expected Poisson value} = \sum_{k=0}^{\infty} P(X = k)\, k = \lambda$$

with variance of X given by

$$\text{variance of } X = \sum_{k=0}^{\infty} P(X = k)\, (k - \lambda)^2 = \lambda.$$

T A B L E **A.2**

Poisson distribution[*], $\lambda = 4$

k	0	1	2	3	4	5	6	7	8	9	10+
$P(X = k)$	0.018	0.073	0.147	0.195	0.195	0.156	0.104	0.060	0.030	0.013	0.009

$*P(X = k) = \dfrac{e^{-4} 4^k}{k!}$

To illustrate the Poisson distribution, the distribution where $\lambda = 4$ is given in Table A.2. Because the parameter λ is the expected number of observations (population mean) associated with a specific Poisson distribution, an estimate of λ is the sample mean $\bar{x}$ ($\hat{\lambda} = \bar{x}$). In addition, because the mean and the variance of a Poisson distribution are equal, the sample mean is also an estimate of the variance.

Situations arise where the variable observed is binary, the number of occurrences n is large and p is small, implying the the Poisson distribution can be used as a description of a phenomenon with two outcomes. For example, the probability of k deaths from leukemia occurring in a specific geographic area (n = all persons at risk for leukemia (a large number) and p = the probability that any one person dies from leukemia (a small probability)) has been modeled by a Poisson distribution [11]. For the application of a Poisson distribution knowledge of p and n is unnecessary; only the parameter λ must be known or estimated to generate the Poisson probabilities. A partial list of phenomena that have been observed to "fit" a Poisson distribution is bacterial counts, deaths by horse kicks, numbers of radioactive decay particles, arrival of patients at doctor's waiting rooms, typographical errors, presidential appointments to the supreme court, numbers of individuals over 100 years old, occurrences of suicide, and telephone calls arriving at a switch board.

Problems

1   Use a two-variable additive Poisson regression model to describe the data in Table 12.25. Calculate the rate ratios for the four classification of smokers, using the never-smoker group for baseline rates from the observed data. Make the same calculation using the values generated by the Poisson model. Construct a plot of the estimated rates.

2   Compute the chi-square statistic comparing the observed and expected (based on the model) number of deaths for the lung cancer data. Contrast this calculated chi-square value with the log-likelihood chi-square.

3   Measure the impact of the smoking categories using likelihood statistics. Similarly, measure the impact of the age categories using likelihood statistics. Which has more influence on the risk of lung cancer?

4   If d_x is the number of deaths in interval x to $x + 1$ and l_x are the number of individuals who began the interval, and if

$$q_x = \frac{d_x}{l_x}$$

is the estimated probability of death in the interval and

$$r_x = \frac{d_x}{l_x - \frac{1}{2}d_x}$$

is the estimated rate of mortality in interval x, then show the probability of death and the mortality rate are approximately equal (i.e., $q_x \approx r_x$).

5 Demonstrate that the binomial and Poisson distributions are similar when p is small and n is large. That is, for $p = .01$ and $n = 50$, calculate for these two distributions the probabilities $P(X = k)$ for $k = 0, 1, 2, \cdots, 5$ where

$$\text{probability of } k \text{ events} = P(X = k) = \binom{n}{k} p^k (1 - p)^{n-k}$$

T A B L E **12.25**

Rate of lung cancer among men classified by age and smoking category

Smoking/Age	< 45	45–54	55–64	65–74	> 74
Never smoker					
Deaths	11	20	31	40	93
Person-years	114616	101015	78405	49216	58269
Rate/10,000	0.96	1.98	3.95	8.13	15.96
Past smoker					
Deaths	6	13	19	17	50
Person-years	58259	64836	48952	30296	31854
Rate/10,000	1.03	2.01	3.88	5.61	15.70
≤ 20 cigarettes per day					
Deaths	4	2	4	8	20
Person-years	19482	11641	8295	5031	6401
Rate/10,000	2.05	1.72	4.82	15.90	31.24
> 20 cigarettes per day					
Deaths	5	6	11	16	31
Person-years	12947	7450	6823	4024	5032
Rate/10,000	3.86	8.05	16.12	39.76	61.61

Note: Table 12.25 describes the rate per 10,000 person-years of lung cancer among California men for five age classes and four smoking classes (never smoked, past smokers, less than or equal to a pack per day, and more than a pack per day). "Past" smoker means a person who has not smoked for at least five years.

and

$$\text{probability of k events} = P(X = k) = \frac{(np)^k e^{-np}}{k!} = \frac{\lambda^k e^{-\lambda}}{k!} \, .$$

References

1 Frome, L. E., and Checkoway, H. Use of Poisson Regression Models in Estimating Rates and Ratios. *American Journal of Epidemiology* 121 (1985): 309–323.

2 Pearce, N., and Checkoway, H. A Simple Computer Program for Generating Person-Time Data in Cohort Studies Involving Time-Related Factors. *American Journal of Epidemiology* 125 (1987): 1085–1091.

3 Fleiss, J. L. *Statistical Methods for Rates and Proportions.* New York: John Wiley, 1981.

4 Gaffey, W. R. A Critique of the Standardized Mortality Ratio. *Journal of Occupational Medicine* 18 (1976): 157–160.

5 Breslow, N. E., and Day, N. E. *Statistical Methods in Cancer Research*, volume 2. New York: Oxford University Press, 1980.

6 Kendall, K. G., and Stuart, A. *The Advanced Theory of Statistics*, volume 2. New York: Hafner Publishing Co., 1967.

7 Wing S., Shy, C. M., Wood, J.L., Wolf, S., Cragle, E.L., and Frome, E.F. Mortality among Workers at Oak Ridge National Laboratory. *Journal of the American Medical Association* 23 (1991): 1397–1402.

8 Roseman, R. H., Brand, R. J., and Jenkins, C. C. Coronary Heart Disease in the Western Collaborative Group Study. *Journal of the American Medical Association* 223 (1975): 872–877.

9 Chiang, C. L. *Introduction to Stochastic Processes in Biostatistics.* New York: John Wiley, 1968.

10 Johnson, N. L., and Kotz, S. *Discrete Distributions.* Boston: Houghton Mifflin, 1969.

11 Knox, G. Epidemiology of Childhood Leukemia in Northumberland and Duran. *British Journal of Medicine* 18 (1964): 17–24.

Index